Oil, Gas and Mining Law in Africa

AF716834

Oil, Gas and Mining Law in Africa

Thierry Lauriol
Docteur d'Etat, Paris XI
Paris Bar, attorney at law
Co-chairman of the Diploma in International Economic Law in Africa at Paris II — Panthéon Assas
Chairman of the Energy Law Group

Émilie Raynaud
Paris Bar, Attorney at law
Lecturer at Paris II — Panthéon Assas

First published 2018

Juta and Company Ltd, First floor, Sunclare Building,
21 Dreyer Street, Claremont, 7708
©2018 Juta & Company (Pty) Ltd

ISBN: 9781485127079

This book is copyright under the Berne Convention. In terms of the copyright Act 98 of 1978, no part of this book may be reproduced or transmitted in any form or by any means, electronic or mechanical, including photocopying, recording or by any information storage and retrieval system,without permission from the publisher.

The author and the publisher believe on the strength of due diligence exercised that this work does not contain any material that is the subject of copyright held by another person. In the alternative, they believe that any protected pre-existing material that may be comprised in it has been used with the appropriate authority or has been used in circumstances that may make such use permissible by law.

Although every care is taken to ensure the accuracy of this publication, supplements, updates and replacement material, the authors, editors, publishers and printers do not accept responsibility for any act, omission, loss, damage or the consequences thereof occasioned by a reliance by any person upon the contents hereof.

Production Coordinator: Deidre du Preez
Editor: Precillar Moyo
Proofreader: Kallie Pauw
Indexer: Lexinfo
Typesetter: Brian Hopking
Cover designer: Drag and Drop

Acknowledgements

We would like to thank the mining and oil administrations from all the African countries with whom we have been in contact, as well as the universities, practitioners and lawyers who have provided us with their assistance in gathering the texts and information.

We would especially like to thank Philip Botha *(South Africa)*, Lynda Belaid *(Algeria)*, Nuno Galvão Teles *(Angola and Mozambique)*, Raoul Placide Houngbedji *(Benin)*, John Carr-Hartley *(Botswana)*, Mamadou Savadogo *(Burkina Faso)*, Gilbert Nyatanyi *(Burundi)*, Daniel Keuffi *(Cameroon)*, Jacques Mbosso *(Central African Republic)*, Célestin M'Foutou *(Republic of the Congo)*, Désiré Kolongele Eberande *(Democratic Republic of the Congo)*, Rosine Kouame Kete *(Côte d'Ivoire)*, Alain Martinet *(Djibouti)*, Bassam Moussa *(Egypt)*, Berhane Gila Michael *(Eritrea)*, Mesfin Tafesse *(Ethiopia)*, Albert Bikalou *(Gabon)*, Amie Bensouda *(Gambia)*, Albert Mumma *(Kenya)*, Webber Newdigate *(Lesotho)*, Mohamed Tumi *(Libya)*, Raphaël Jakoba *(Madagascar)*, Krishna Savjani *(Malawi)*, Mamadou Diarra *(Mali)*, Thierry Koenig *(Mauritius)*, Yarba Ould Ahmed Saleh *(Mauritania)*, Beate Loch *(Namibia)*, Bernard Olivier Kouaovi *(Niger)*, Sola Adepetun *(Nigeria)*, Désiré Kamanzi *(Rwanda)*, Carlos Gomes *(Sao Tome e Principe)*, Guedel Ndiaye *(Senegal)*, Glenna Thompson *(Sierra Leone)*, Robert Cloete *(Swaziland)*, Abdoulaye Adam Bahar *(Chad)*, Ahlin Gabriel Komlan *(Togo)*, Adly Bellagha *(Tunisia)*, Charles Mkokweza *(Zambia)* and Alex Masterson *(Zimbabwe)*.

ACKNOWLEDGEMENTS

We would like to thank the mining and oil administrations from all the African countries with whom we have been in contact, as well as the universities, practitioners and lawyers who have provided us with their assistance in gathering the texts and information.

We would especially like to thank Philip Botha *(South Africa)*, Lynda Belaid *(Algeria)*, Nuno Galvão Teles *(Angola and Mozambique)*, Raoul Placide Houngbedji *(Benin)*, John Carr-Hartley *(Botswana)*, Mamadou Savadogo *(Burkina Faso)*, Gilbert Nyatanyi *(Burundi)*, Daniel Keuffi *(Cameroon)*, Jacques Mbosso *(Central African Republic)*, Célestin M'Foutou *(Republic of the Congo)*, Désiré Kolongele Eberande *(Democratic Republic of the Congo)*, Rosine Kouame Kete *(Côte d'Ivoire)*, Alain Martinet *(Djibouti)*, Bassam Moussa *(Egypt)*, Berhane Gila Michael *(Eritrea)*, Mesfin Tafesse *(Ethiopia)*, Albert Bikalou *(Gabon)*, Amie Bensouda *(Gambia)*, Albert Mumma *(Kenya)*, Webber Newdigate *(Lesotho)*, Mohamed Tumi *(Libya)*, Raphaël Jakoba *(Madagascar)*, Krishna Savjani *(Malawi)*, Mamadou Diarra *(Mali)*, Thierry Koenig *(Mauritius)*, Yarba Ould Ahmed Saleh *(Mauritania)*, Beate Loch *(Namibia)*, Bernard Olivier Kouaovi *(Niger)*, Sola Adepetun *(Nigeria)*, Désiré Kamanzi *(Rwanda)*, Carlos Gomes *(Sao Tome e Principe)*, Guedel Ndiaye *(Senegal)*, Glenna Thompson *(Sierra Leone)*, Robert Cloete *(Swaziland)*, Abdoulaye Adam Bahar *(Chad)*, Ahlin Gabriel Komlan *(Togo)*, Adly Bellagha *(Tunisia)*, Charles Mkokweza *(Zambia)* and Alex Masterson *(Zimbabwe)*.

Foreword

Whether it involves an opencast or underground mine, an offshore oil platform or a derrick drilling through the earth, they all serve the same goal: to develop the natural resources sought after by industry worldwide. If demand slumps in the foundries in Asia, mining excavators will immediately become less rapacious across the African continent. Globalisation in the sense of economies becoming inextricably interlinked with one another is nowhere more evident than in the oil and mining industry. This can also be seen in the interconnection of legal systems as, more often than not, the production of a state's natural resources is funded by foreign capital.

From a legal point of view, the entire scope of a mining operation covers everything from the existence of sovereignty, establishing the right to grant (*jus concendentis*), to the protection of mining rights (*jus concessum*). To illustrate this point, the legal framework is seen as being made up of different layers, with each layer governing a separate area yet remaining interconnected. Thus, the establishment of sovereignty comes under international law; the granting of mining rights falls under national law and national law in compliance with international law protects the exercise of mining rights. Naturally, these three items are most often considered separately. In some respects, however, they form a single legal reality. Discussions on the compensation due to an operator in return for losses resulting from the infringement of its mining rights, presupposes that the State was entitled to grant mining rights under international law and that the investor validly acquired these rights under national law.

International law: establishing the right to grant access to natural resources

This is the primary layer, establishing the right to grant (an extension of the State's exclusive right to assign) a licence for the exploration and production of resources in the area over which it has sovereignty. The State's right to grant a licence for the exploitation of natural resources draws its validity from an international standard. Therefore, international public law governs the existence and extension of a state's sovereignty over oil and mining resources. The State can only validly grant a licence to develop a national resource if its claim to sovereignty in that country is recognised under international law. On that basis, all mining projects are rooted in the international legal system.

Under international law, territorial rights are acquired according to certain methods, such as discovery, actual occupation or transfer. Any state that is granted such rights shall benefit from the right to validly grant licences to exploit natural resources within the area concerned. In Africa, more than anywhere, the assessment of territorial sovereignty is subject to the 'principle of the intangibility of frontiers inherited from colonization'. In the case of the border dispute between Mali and Burkina Faso, the International Court of Justice, when confronted with the implementation of this principle—the application of which was, in this case, requested by both parties—wished '[. . .] to emphasize its general scope in view of its exceptional importance for the African continent' (*Frontier Dispute, Burkina Faso/Republic of Mali*, Judgement of 22 December 1986, ICJ Reports 1986, at 565, no. 20). The distribution of oil and mining resources between African states has thus been inherited from colonisation. This decision was supposed to be placatory having been created to 'avoid the independence and the stability of the new states being put in danger by fratricidal struggles stemming from border disputes following the withdrawal of the administrating power' (ibid). As a result, today, the largest portions of the continent are under firmly established sovereignties. Nevertheless, the issue of the legal status of these territories is far from being resolved everywhere. There remain numerous sources of uncertainty and, let there be no doubt, others will flare up every now and then. One cannot help but notice that, more often than not, these disputes affect areas rich in oil and mining resources.

National law: allocating mining rights

Once its existence is established and its scope defined according to the rules of international law, sovereignty over oil and mining resources is implemented according to the national laws of the state granting the rights to exploit natural resources. On this point, the International Court of Justice has underlined that 'every state possesses a fundamental right to choose and implement its own political, economic and social systems' (Case concerning military and paramilitary activities in and against Nicaragua, *Nicaragua v United States of America*, Judgement of 27 June 1986, ICJ Reports at 131, no 258). In line with this fundamental right, each state is free to establish its mining system and the manner in which it intends to grant foreign operators access to its resources.

In this phase, mining rights are allocated to an operator based on a mining licence and/or a contract that can take one of many forms. Given that the allocation system is subject to national law, each state establishes the rules governing concession rights based on its own organisation. The state's national laws therefore determine the regime for the allocation of resources. The main model on the African continent remains the model where the State controls the resources and governs the access to those resources. With regard to this last point, various procedures are possible. The most common procedure grants the State or one of its subdivisions the power to allocate mining rights according to certain suitability criteria. Regardless of the allocation procedure followed, however, it is clear that these mining rights can cover a number of different areas, ranging from prospecting to production.

It is no secret that capital on the one hand, and oil and mining resources on the other, are unfairly distributed from a geographical perspective. Consequently, mineral-rich soils are generally found in countries that are lacking in capital, whereas countries that are rich in capital are lacking or have insufficient resources. More often than not, a foreign investor will fund mining and oil projects and will establish itself in the host state and carry out its operations based on a mining licence. The national legal system of the host state therefore governs the entry of foreign investors, in addition to the mining rights granted, which include a wide spectrum of rights ranging from prospecting to production.

National law in compliance with international law: the protection of mining rights

If the mining licence is extended by a contract entered into with a foreign company, it is immediately protected by international law. From that time on, sovereignty is exercised according to national law but in compliance with international law. As highlighted by the International Court of Justice, 'when a State admits into its territory foreign investments or foreign nationals, whether natural or juristic persons, it is bound to extend to them the protection of the law and assumes obligations concerning the treatment to be afforded them' (Case concerning the *Barcelona Traction, Belgium v Spain*, Judgment of 5 February 1970, ICJ Reports, 1970, at 32, no 33). Without defining its substance, the Court specified further on in its judgement that this treatment was '[. . .] guaranteed by general international law, in the absence of a treaty applicable to the particular case' (op cit, at 46, no 87). Thus, regardless of any special protection of a contractual (mainly stabilisation and intangibility clauses) or conventional (bilateral investment treaties) nature that may be obtained by the investor, its assets shall be placed under a legal protection scheme in accordance with customary international law.

An operation via which an economic agent commits a significant amount of capital with a view to the exploration and production of mineral resources on the territory of a state of which it is not a national has all the characteristics of an international investment model. The investment of the capital is fixed: it is used to create significant infrastructure. It is carried out over the long term, from the exploration phase to the depletion of the deposit, usually over a period of 10 to 30 years. Lastly, the holder of the capital is bound to the success or failure of the project.

In this kind of operation, the risks are increased by the fact that the holder of the capital is not a national of the state on whose territory the investment is made. The relationship with the host state is all the more delicate due to the fact that the foreign capital makes up for a deficit in national capital in what is a sensitive sector. Moreover, foreign investment has a direct bearing on the sovereignty of the host state as it is established on the latter's territory and concerns the resources contained therein. Investment in the mining industry therefore inevitably raises questions relating to sovereignty.

Drilling through the various legislative layers

This book by Thierry Lauriol and Émilie Raynaud is like a borehole that sinks through these various legislative layers. Its scope is considerable, encompassing the applicable legal systems across the entire African continent. However, what makes this publication all the more remarkable is that, based on impressive documentation, it manages to establish a clear structure that cuts across the various national legal systems.

Above all, this book tackles the key issue of national law in African countries. The result is an enlightening overview that begins by outlining 'the fundamental legal aspects of production' (part 1), focusing on the relevant instruments in the area: mining licences and agreements that contractually define the specific terms and conditions. The book then highlights 'the legal functioning of operations' (part 2), focusing on the substantial number of standards in the area designed to ease the preoccupations of the State (technological issues and development), as well as those of the operator (operation procedures and risk management).

In spite of the wide variety of standards in this field, the authors succeed in delineating the major principles that are set out in national law. The book provides a good level of understanding and manages to avoid being either too general or too technical, alternating between overviews and more in-depth analyses of specific issues. It allows the reader to cover the key areas of the subject and to examine the most unusual aspects.

Consequently, even though the prospected area is only supposed to be subject to national rules, the authors highlight applicable bodies of international standards, such as the West African Economic Monetary Union mining code or the Economic Community of West African States mining directive. These instruments, which are very rare and—to our knowledge—only exist in Africa, are made up of international laws that govern all mining operations. The authors also demonstrate that, within the same sector of national law, several generations of standards can be found to coexist. Indeed, by means of stabilisation clauses, previous standards sometimes remain in force within a fixed contractual area. Thus, within the same national legal system, several generations of standards can coexist depending on the stabilisation commitments undertaken by the state in question.

The authors proceed by delving into the tricky issue of protecting mining rights, where international law governs the impact of national rules. At what exact point does this occur? This question has led to much discussion on the issue of State contracts, an area in which French law has made significant progress, despite the fact that a clear answer has yet to be found. On this point, as elsewhere, the authors argue for a practical response and, above all, for the 'de-politicisation' rather than the stabilisation of mining contracts.

Upon reading this book, it is clear that its success owes much to the respective qualities of its authors. With 30 years of practice within the Jeantet firm and his experience as head of the mining and energy department, Thierry Lauriol is extremely well versed in this area. He is aware of the issues at stake, having worked to resolve them across most of Africa, and here he imparts his valuable knowledge to us readers. He also deserves credit for surrounding himself with the right people and for embarking on this project with his young and very talented colleague, Émilie Raynaud, who has thus carved out a name for herself among the select group of specialists in this field.

This publication seeks to restore African law to its rightful place, at the forefront of legislative production, a position warranted by the innovative nature of the instruments that are being created there. It will soon become an important reference work on our bookshelves, but will no doubt be found more often on the desks of all those interested in oil and mining law. It is an absolute must for anyone who wishes to tackle the key issues in this field.

Yves Nouvel
Professor at Panthéon-Assas University

PREFACE

It was inevitable that I would one day write this book. At a very early age, my father introduced me to the energy and mines sectors. He was among those who participated in the discovery and development of the Saharan oil sector or what was commonly referred to at the time as '*les pétroles*'. The *Compagnie Française des Pétroles, Hassi Messaoud, Hassi R'Mel, Maison Verte* all formed part of my childhood. This early introduction to the sector helped me better understand its issues and importance. My father's thoughts on the development of energy and, more generally, on natural resources in national and world economies often come back to me and remain relevant to this day.

Unsurprisingly, early in my career I decided to specialise and dedicate my professional life to this sector. I completed my doctoral thesis in natural resources contracts. I was lucky enough during my career to gain knowledge and experience in not only the negotiation and implementation of business transactions but also the conduct of commercial dispute resolution. Most notably in Africa, where these issues are, dare I say, even more interesting.

It is a rich and contrasting area. It is the source of political, economic and military conflicts. However, it has also greatly reduced, and sometimes even eliminated, the divide between private and public law. This book is as much a summary as it is a passion.

Thierry Lauriol
PhD
Lawyer at the Paris Bar

List of abbreviations

AFDB	African Development Bank
AFDI	*Annuaire français de droit international*
AJIL	American Journal of International Law
Ann Fac Dt Liège	*Annales de la faculté de droit, d'économie et de sciences sociales de Liège*
AOGD	Arab Oil & Gas Directory
APPA	African Petroleum Producers' Association
ASA	*Association suisse d'arbitrage*
ATAF	African tax administration forum
BFCE	*Banque française du commerce extérieur* (French Foreign Trade Bank)
BRICS	Brazil, Russia, India, China and South Africa
BYIL	British Yearbook of International Law
CCHAIL	Collected Courses of the Hague Academy of International Law
CEMAC	*Communauté économique et monétaire de l'Afrique centrale* (Central African Economic and Monetary Community)
CIMA	*Conférence interafricaine des marchés d'assurances* (Inter-African Conference on Insurance Markets)
COMESA	Common Market for Eastern and Southern Africa
COTUNACE	*Compagnie tunisienne pour l'assurance du commerce extérieur* (Tunisian Foreign Trade Insurance Company)
CSR	Corporate Social Responsibility
CTRTCM	*Comité technique de revue des titres et conventions miniers* (Technical Committee for the Review of Mining Rights and Conventions)
D	Dalloz
Denver JILP	Denver Journal of International Law and Policy
DPA	Diamond Producers Association
DPCI	*Droit et pratique du commerce international*
ECOWAS	Economic Community of West African States
EITI	Extractive Industries Transparency Initiative
Exp Eco Pet Ind	Exploration and Economics of the Petroleum Industry
FOB	Free on Board
Grotian Soc.Pap	Grotian society papers
GYIL	German Yearbook of International Law
IBLJ	International Business Law Journal
IBRD	International Bank for Reconstruction and Development
ICC	International Chamber of Commerce
ICSID	International Centre for Settlement of Investment Disputes
ICSID Rev	ICSID Review
IDA	International Development Agency

IFC	International Finance Corporation
IJIL	Indian Journal of International Law
ILM	International and Legal Materials
ILR	International Law Reports
Int. Lawyer	International Lawyer
Int'l & Comp LQ	International and Comparative Law Quarterly
JDA	Joint Development Area
JDI	*Journal de droit international (Clunet)*
JWTL	Journal of World Trade Law
Law & Pol Int Bus	Law and Policy in International Business
LGDJ	*Librairie générale de droit et de jurisprudence*
LLJ	Liberian Law Journal
MER	Maximum Efficient Rate
MIGA	Multilateral Investment Guarantee Agency
MPR	Maximum Production Rate
NRF	*Nouvelle revue française*
NWULR	Northwestern University Law Review
NYUL Rev	New York University Law Review
OAPEC	Organisation of Arab Petroleum Exporting Countries
OAU	Organisation of African Unity
OECD	Organisation for Economic Co-operation and Development
OGEL	Oil, Gas & Energy Law
OHADA	*Organisation pour l'harmonisation en Afrique du droit des affaires* (Organisation for the Harmonisation of Business Law in Africa)
OPEC	Organisation of Petroleum Exporting Countries
OPIC	Overseas Private Investment Corporation
Oz	Ounce
Proc ASIL	Proceedings of the American Society of International Law
R Alg SJPE	*Revue algérienne des sciences juridiques, politiques et économiques*
RBDI	*Revue belge de droit international*
RDIDC	*Revue de droit international et de droit comparé*
Rev arb	*Revue de l'arbitrage*
Rev crit DIP	*Revue critique de droit international privé*
Rev Egypt DI	*Revue égyptienne de droit international*
Rev PIC	*Revue juridique et politique : indépendance et coopération*
RGDIP	*Revue générale de droit international public*
RTD Civ	*Revue trimestrielle de droit civil*
RTD Com	*Revue trimestrielle de droit commercial*
SADC	Southern African Development Community
SPV	Special Purpose Vehicle
SYSCOA	*Système comptable ouest africain* (West African Accounting System)
TDM	*Transnational Dispute Management*
toe	Tonne of Oil Equivalent

WAEMU	West African Economic and Monetary Union
Vand J Transnat'l L	Vanderbilt Journal of Transnational Law
YB Com Arb	Yearbook of Commercial Arbitration

Table of Contents

PART 2: LEGAL BASIS OF THE EXPLOITATION OF NATURAL RESOURCES

Chapter 8
The concept of mining rights

Chapter 9
The granting of mining rights

Chapter 10
Agreements with the host state

Chapter 11
The organisation of exploitation between operators

Chapter 15
Financing

Chapter 16
Host state participation

Part 3: OPERATIONALASPECTS OF EXPLOITATION

Chapter 17
Disciplinary rules

Chapter 18
Surveillance rules

Chapter 19
Contractual arrangements to manage risks specific to exploitation agreements

Part 3: OPERATIONAL ASPECTS OF EXPLOITATION

Chapter 17
Disciplinary rules

Chapter 18
Surveillance rules

Chapter 19
Contractual arrangements to manage risks specific to exploitation agreements

Chapter 20
Rules not specific to exploitation agreements

Chapter 21
Instruments to implement the transfer of technology

Chapter 22
Variable effectiveness of technology transfer

Chapter 23
Stimulating the local economy

Chapter 24
Enriching national assets of the host state

TABLE OF CASES

Table of Statutes

Year	No	Title	Page
SOUTH AFRICA			
1986	56	Diamonds Act	6.28
1996	29	Mine Health and Safety Act	24.6
2002	28	Mineral and Petroleum Resources Development Act	
		Preamble	3.10
		s 3(1)	5.13
		s 5(4)	8.73, 19.134
		s 11(1)	8.65
		s 12	9.15
		s 17	24.7
		s 23	24.7
		s 23(1)	9.3
		s 27(7)	24.14
		s 38	19.136
		s 38(1)	19.148
		s 39	19.134
		s 41	19.137
		s 43	24.7
		s 75	24.7
		s 80	24.7
		s 84	24.7
		s 89	15.51
2005	37	Precious Metals Act	6.28
ALGERIA			
1996		Constitution of the Republic of Algeria	
		art 17	4.27
2002	02-65	Executive Decree Defining the Conditions and Procedures for Allocating Mining Rights	
		art 2	9.29, 9.30
		art 4	9.30
		art 7	9.33
2006	06-440	Executive Decree Establishing the Procedure, the Conditions for Application and the Methodology for Calculating the Tax on Exceptional Profits (TPE)	
		art 2	14.45
2013	13-01	Law Amending and Supplementing the Law on Hydrocarbons	
		art 20	8.14
		art 22	8.13, 18.6
		art 24	12.3
		art 25	13.1
		art 26	14.35
		art 29	17.16
		art 32	16.9
		art 43(2)	15.52
		art 50	12.8
		art 52	13.28
		art 54	11.30
		art 54(2)	11.31
		art 54(5)	11.32
		art 55	23.72

Year No	Title	Page
	art 37(2)	11.31
	art 37(3)	11.32
	art 41	23.48
	art 44	21.19, 21.20
	art 45	15.56, 15.57, 19.144, 19.145
	art 46	16.2, 16.7, 16.8, 16.9
	art 48	16.7
	art 49	8.71
	art 50	8.74, 8.76, 8.77, 8.81, 8.85
	art 51	8.83
	art 54	22.10, 22.46, 23.32, 23.35
	art 55	23.3, 23.12, 23.25, 23.27, 23.57
	art 56	24.7, 24.8
	art 57	24.9
	art 59	19.139
	art 61	19.148
	art 64	18.27
	art 83	14.69
	art 104	1.28
	art 105	23.72
	art 106	23.62, 23.64, 23.65, 23.66, 23.69, 23.75
	art 107	8.65, 11.52, 11.53
	art 109	20.14
	art 110	20.41
	art 111	19.16, 19.17, 19.23
	art 112	20.48
	art 123	18.44
1987 001/PRG/87	Ordinance	
	art 9	6.4
GUINEA-BISSAU		
1996	Constitution of the Republic of Guinea-Bissau	
	art 9	4.27
	Mining Code	
	art 126	8.68
	art 127	8.68
KENYA		
2016	Mining Act	6.30
	art 67(1)	8.15
2007	Occupational Safety and Health Act	24.6
	Petroleum Act	
	art 9(1)	23.4
	art 10(2)	19.149
	art 11	22.13
LESOTHO		
2005	Mines and Minerals Act	
	art 3	5.13
	art 5(1)	9.20, 9.21
	art 11(1)	22.10
	art 20(2)	9.29
	art 24(1), (2)	8.16
	art 25(1)	8.10

Year No	Title	Page
SOUTH SUDAN		
2012	Mining Act	19.140
	art 7(2)	9.16, 9.20, 9.21
	art 45(1)	8.21
	art 45(2)	8.30
	art 50(1)	8.29
	art 52(1)	8.26
	art 55	9.3
	art 61(1)	8.43
	art 62	9.3
	art 67(1)	8.43
	art 88	23.12
	art 88(2)	23.24
	art 88(7)	23.32
	art 88(8)	23.58
	art 121(2)	19.134
	art 122	19.136
	art 128	22.46
	art 128(2)	22.48
	art 140	8.77
	art 141	8.77
	art 145	24.14
	art 146	24.14
	art 159	24.10
2012	Petroleum Act	19.140
	art 17(4)	8.16
	art 50	24.9
	art 50(3)	24.9
	art 52	24.9
	art 59	19.134
	art 60	19.134, 19.148
	art 61	19.148
	art 63	23.3, 23.4, 23.7, 23.32
	art 65	23.3, 23.12, 23.77
	art 65(2)	23.27
	art 65(3)	23.24
	art 65(4)	23.13
	art 65(5)	22.10
	art 84	15.56
	art 93	19.145
1972	Petroleum Code (Petroleum Resources Act)	
	art 34(4)	7.18
SUDAN		
2007	Mineral Resources and Mining Development Act	3.14
	art 4	5.19
	art 9	9.3
	art 11	9.3
	art 17	19.142
	art 18	19.144
SWAZILAND		
2005	Constitution of the Kingdom of Swaziland	

Year	No	Title	Page
		art 7	19.8
		art 7(4)	19.5
		art 9	23.62
		art 10	7.9
		art 11(2)	22.48
		art 11(4)	23.12
		art 11(5)	23.4, 23.7, 23.32, 27.33
		art 16(1), (7)	22.46
		Mining Directive	7.8–7.11
		art 22	7.9
ORGANISATION FOR THE HARMONISATION OF BUSINESS LAW IN AFRICA (OHADA)			
1999		Uniform Act on Arbitration	20.68
		art 15	20.69
2014		Uniform Act on Commercial Companies and Economic Interest Groups	
		art 120	9.17
1998		Uniform Act on Simplified Recovery Procedures and Measures of Execution	
		art 30	20.83
		Uniform Act on the Organisation and Harmonisation of Company Accounting	14.71
2010		Uniform Act on the Organisation of Security Rights	8.57
		art 28	15.54
		Uniform Act relating to Accounting and Financial Information	14.71
WEST AFRICAN ECONOMIC AND MONETARY UNION (WAEMU)			
2001	07/2001/CM/ UEMOA	Directive Establishing the Harmonised Regime for Advance Payments of Profit Tax within WAEMU	14.6
2008	01/2008/CM/ UEMOA	Directive Harmonizing the Modalities for the Determination of the Taxable Result for Legal Entities within WAEMU	14.6
1998	02/98/CM/ UEMOA	Directive on the Harmonisation of the Laws of the Member states on the Value Added Tax	14.6
2008	08/2008/CM/ UEMOA	Directive on the Harmonisation of the Rates of Taxes Levied on the Profits of Legal Entities in WAEMU Member States	14.6
2013	05/2013/CM/ UEMOA	Regulation Amending the West African Accounting System (SYSCOA)	14.71
2003	18/2003/CM/ UEMOA	Regulation Enacting the Community Mining Code	7.4–7.7
		art 2	7.3, 7.4
		art 4	7.10
		art 6(7)	19.137
		art 11	24.9
		art 12	7.6, 16.3, 16.4, 16.5
		art 13–17	7.6
		art 14	17.30, 23.4, 23.7, 23.32
		art 15	23.62, 23.75
		art 16	23.12
		art 17	19.8, 19.15
		art 22	14.18, 14.19
		art 25–33	7.6
		art 25	14.73
		art 26	14.76
		art 28	14.76
		art 41	7.6

Part 1

INTRODUCTION

1

GENERAL BACKGROUND

1.1 The most ancient civilisations were marked by the discovery and use of mineral and petroleum resources. For millennia, stories and legends have recounted the use of petroleum and mineral resources[1] and the Iron and Bronze ages illustrate their importance.

1.2 The use of mineral and petroleum resources has evolved; they have become strategic resources over time, necessary for the world's economy. Africa, rich in mineral and petroleum resources, currently appears to be a highly sought-after continent. The exploitation of its resources has become a major component of African economies, which poses a challenge for their processing.

1 EXTRACTIVE NATURAL RESOURCES, STRATEGIC RESOURCES

1.3 The race for natural resources in Africa has been a major issue for centuries. It increased simultaneously with the rise in the strategic importance of certain resources such as petroleum.[2] Population growth and the emergence of new economic powers suggest that the role of natural resources will continue to increase.

1.4 The worldwide consumption of primary energy has been steadily increasing for decades, from 5,4 to 13 billion toe (Tonne of Oil Equivalent)

[1] The Mesopotamians used petroleum as a pharmaceutical and cosmetic product. In 6,000 BC petroleum was used for the caulking of ships. It is believed that asphalt was used by the Egyptian civilisation in the mummification process, while bitumen was used by King Nebuchadnezzar to build the hanging gardens of Babylon. Biblical texts also mention bitumen (Genesis, chapter 6: 'Make yourself an ark of gopher wood. Make rooms in the ark and cover it inside and out with pitch'). Throughout history, petroleum has been used as fuel: The Greeks' Persian fire, the Romans' oleum incendiarium, the Byzantines' Greek fire, up until the invention of the kerosene lamp. The use of burning oil subsequently gave way to electricity, while the use of petroleum by the automotive industry contributed to the development of its mass consumption. When it comes to mineral resources, ferrous and non-ferrous metals as well as precious metals marked the most ancient civilisations. The Incas extensively used certain precious metals such as gold and silver. According to Inca mythology, gold was the sweat of the sun, whereas silver was believed to be the tears of the moon. Other metals such as lead or tin were also traded by the Phoenicians and Carthaginians. Similarly, in Greek Antiquity, due to the richness of their tin mines, the British Isles were given the name of Cassiterides ('tin islands', from the Greek Kassiteros meaning tin). Natural resources such as gold marked the first West African civilisations, notably the former Kingdom of Mali. Historical writings mention King Mansa Mussa entering the city of Cairo in 1324, preceded by 500 slaves, each carrying 1,7 kilograms of gold. He is believed to have caused the first financial crisis following a drop in the price of this raw material.

[2] In 1917, Georges Clémenceau stated: 'For nations and populations nowadays, one drop of petroleum is worth one drop of blood'.

between 1972 and 2012, ie approximately a two per cent annual increase.[3] Petroleum plays a predominant role in the global economy. In 2015, global petroleum consumption soared by 1,9 million barrels per day.[4] In 2014, global petroleum demand reached 93,06 million barrels per day.[5] Petroleum is currently and undoubtedly for many years to come, the planet's number one source of energy. For developed countries, it represents a strategic raw material on which the effective running of their industries depends.

1.5 Natural resources represent a major concern for producing states as they constitute the main source of income and, therefore, of their economic development. Natural resources give them significant economic and political power on the international stage.[6]

1.6 The non-renewable nature of mineral and petroleum resources shed doubt on their future. Non-renewable resources are bound to experience a production peak, after which they will become scarcer, precipitating the need for other sources. However, despite the growing development of new sources of energy, the use of non-renewable energy is not about to end. Petroleum should ultimately retain its importance, which may be further amplified by the development of non-conventional hydrocarbons.[7]

The dwindling supply of non-renewable natural resources and the constant rise in energy demand will undoubtedly reinforce the interest in mineral and petroleum resources.

1.7 Although essential, natural resources are not immune to a drop in their consumption, which remains largely driven by the global economy.[8] The drop in petroleum prices illustrated this in 2014. Within the space of six months, the price of a barrel of Brent crude oil[9] collapsed from $112 in July 2014 to $49 in January 2015, having dropped as low as $46. This decrease continued throughout 2015, until the price per barrel reached $37 in December 2015.

[3] French Commission on Sustainable Development *Chiffres clés de l'énergie, Édition 2014* (February 2015).

[4] BP *BP Statistical Review of World Energy* (June 2016). It should be noted that one barrel is exactly 158,9873 litres.

[5] According to the estimates of the International Energy Agency (for information on this topic, see Izak Odendaal 'Outlook on oil' *Mining Prospectus* Issue 20, February 2015 at 25).

[6] For example, Nigeria (Africa's largest petroleum producer) became Africa's number one economic power in 2014.

[7] World reserves remain substantial: in 2015, worldwide proven hydrocarbon reserves amounted to 1 697,6 billion barrels, enough to satisfy the world production at the current rate for another 50,7 years (BP *BP Statistical Review of World Energy* op cit).

[8] The slowing down of the Chinese economy is one of the factors which led to the drop in the price of raw materials in 2014. China uses nearly one quarter of the world's energy and half of the worldwide production of metals (for information on this topic, see Terence Creamer 'Risk Factor' *Mining Weekly* January 23–29 2015 at 26). As far as petroleum is concerned, the drop in the oil price was also due to the exploitation of non-conventional hydrocarbons (and in particular shale gas) by countries seeking to obtain their energy independence such as the United States of America.

[9] Brent crude oil refers to crude petroleum from a production mixture of 19 oil fields in the North Sea. Despite a limited production, the price of the Brent barrel has been used as an international benchmark.

After a significant decrease in January 2016, due to the lifting of sanctions against Iran,[10] the prospect of an agreement to freeze production among the Organisation of Petroleum Exporting Countries (OPEC) had resulted in an increase in petroleum prices. Ultimately, on 28 September 2016, OPEC members reached an agreement to limit their production.

1.8 The petroleum sector is not the only one to suffer from economic fluctuations. In 2014 in the mining sector, the prices of iron dropped sharply on international markets, from $140 per tonne in early 2014 to $68 per tonne in December 2014, ie a drop of more than 40% over one year. Goldman Sachs' pessimistic report of September 2014, entitled *The End of the Iron Age*, illustrates the skepticism of the sector. According to the US merchant bank, the prices of iron have begun a decline that should continue for the next 15 years.[11] The prices of other raw materials also dropped in 2014. For example, the price of copper dropped by 15% in 2014.[12] Gold followed suit, with prices dropping to $1,049/oz in December 2015 before recovering in 2016. The price of precious metals remained low for a while, in particular due to their loss of appeal as a safe investment for institutional investors.[13] The price of raw materials remained low throughout 2015.

1.9 Overall, economists are not very positive about price evolution over the next 10 years; the immediate effect of this is the review of exploitation projects that had been developed and selected.

Estimated evolution of the price of certain raw materials in the medium and long term[14]

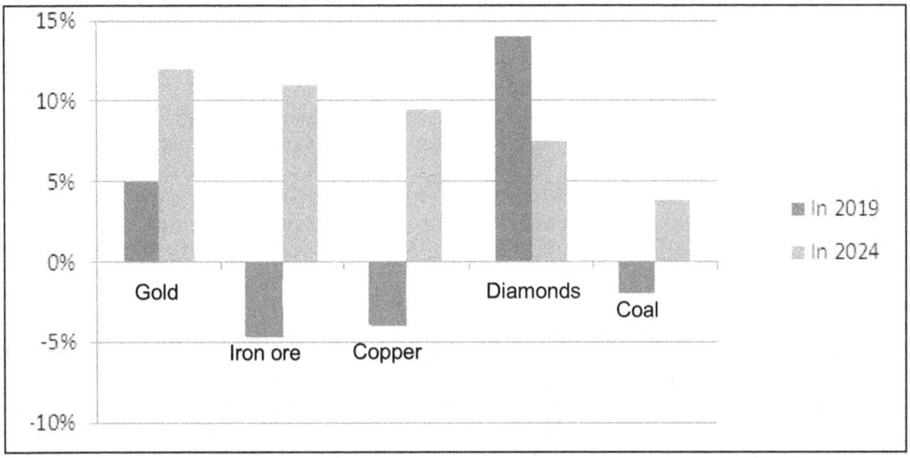

[10] Antoine Izambar 'Chute du prix du pétrole : pourquoi l' OPEP peut exploser ?' *Challenges* 20 January 2016 [online].

[11] For information on this topic, see Jasmine Ng 'Goldman Calls End to Iron Age After "Dramatic" Drop in Ore Price' *BloombergBusiness* 10 September 2014 [online].

[12] For information on this topic, see *African business* 'Gold, Diamonds Hold Steady but Copper Suffers' *African Business* February 2015 at 39.

[13] See Ina Opperman 'Commodities forecast 2015' *Mining* December 2014 at 58 et seq.

[14] Source: Opperman op cit 63.

1.10 These variations have a significant impact on private operators and producing states. An elevated price helps the recovery of natural resources for which production costs are high, while a low price is likely to put into question projects with uncertain economic viability. In this context, the drop in petroleum prices has affected the profitability of many projects.[15] The decline in iron prices, because of slowing demand and a sharp increase in production for full-capacity projects, has led to the suspension of numerous mining projects on the African continent.[16] As market fluctuations can alter the economic balance of a project, declining prices can also affect the revenue of producing states. The volatility of international raw material prices has significant consequences for natural resources exporting countries, as it results in unstable exchange rates and inflationary pressure, as well as drastic income variations. The African continent in its entirety accounts for approximately 10% of the world's petroleum production, but only 4% of this production is consumed on this continent; therefore, the drop in prices results in a drastic reduction in revenue for African states.[17] In light of the decrease in petroleum prices, certain African states such as Algeria[18] or Gabon[19] have had to envisage new budget approaches.[20] This situation led other countries to devalue their currency, such as Nigeria whose national currency, the Naira, depreciated by around nine per cent in 2014.[21] It is in the interest of these states, largely dependent on the petroleum industry,[22] to opt for the diversification of their economy that a lot of them have understood yet few have implemented.

---

[15] For example, in early 2015, British group Tullow Oil announced a reduction in their exploitation budget from $1 billion to $200 million. For information on this topic, see *Jeune Afrique* 'Tullow Oil réduit à nouveau ses dépenses d'exploration' *Jeune Afrique* 15 January 2015 [online].

[16] For information on this issue, see the suspension of the Askaf mining project in Mauritania by the Sphere Minerals Company, a subsidiary of the Glencore group (*Jeune Afrique* 'Mauritanie : Glencore repousse le démarrage du projet minier d'Askaf' *Jeune Afrique* 28 October 2014 [online]) or Rio Tinto's hesitation as part of the exploitation of the Simandou 'mega project' in Guinea (for information on this issue, see Christophe Le Bec 'En Guinée, le doute plane à nouveau sur le mont Simandou' *Jeune Afrique* 20 October 2014 [online]).

[17] See Jim Lennon 'Africa—coping with the new realities' *Metal Bulletin Magazine* African Mining (Special Issue) February 2015 at 62.

[18] For information on this topic, see Chloé Rondeleux 'Face à la chute des cours du pétrole, l'Algérie change de ton' *Jeune Afrique* 17 December 2014 [online].

[19] For information on this topic, see *Jeune Afrique* 'Chute des cours du pétrole: le Gabon envisage de réajuster son budget 2015' *Jeune Afrique* 9 January 2015 [online].

[20] Nigeria and Algeria respectively set barrel prices at $119 and $121 to avoid a deficit budget. The budget of Angola, Gabon and the Republic of Congo is based on a barrel price of $98, $97 and $95 to achieve balance (for information on this topic, see Christophe Le Bec 'Chute du cours du pétrole: qui perd, qui gagne en Afrique?' *Jeune Afrique* 18 November 2014 [online]).

[21] For information on this topic, see Stéphane Ballong 'Quand le cours du pétrole a entraîné le naira dans sa chute' *Jeune Afrique* 1 December 2014 [online].

[22] In Nigeria, the petroleum sector accounts for eight per cent of the State's revenue, and more than ninety per cent of its exports (for information on this issue, see *African Mining* 'Nigeria's black gold' *African Mining* January–February 2015 at 30).

Even though natural resources are still important as sources of revenues in several African countries, their role in driving growth is slowly diminishing. For instance, in Nigeria, oils accounts for only about 10% of the Gross Domestic Product, down from 25,6% in 2000.[23]

1.11 Finally, in the face of this uncertain economic context marked by the decreasing price of raw materials, certain listed companies have had to deal with legal proceedings based on the Securities Act and instituted by their shareholders complaining that they had not been informed of the difficulties experienced by the extractive company.[24]

1.12 However, the cyclical difficulties faced by the extractive industries are not new.[25] Despite these fluctuations, regarding the exploitation of natural resources, Africa remains a continent with largely untapped potential, attempting to enhance its appeal, with the legal exploitation framework playing a key role.

2 AFRICA, A COVETED CONTINENT

1.13 Numerous investors are looking for a supply of mineral and petroleum resources for their own national market and the African continent is an attractive prospect.[26] Some factors make this continent a key destination for investors, including the diversity of the resources contained in its subsoil in significant quantities, as well as its geographical location that facilitates the supply of natural resources for all consumer countries.[27]

1.14 Africa holds a considerable share of global reserves. Indeed, the African continent has an abundance of natural resources, based on current geological knowledge; its subsoil contains more than 7,6% of the world's petroleum reserves,[28] 7,5% of the world's natural gas reserves,[29] 40% of the

---

[23] African Development Bank, Organisation for Economic Co-operation and Development, United Nations Development Programme, Economic Commission for Africa *African Economic Outlook 2017: Entrepreneurship and Industrialisation* at 30 [online]. Id.

[24] For information on this issue, see *Africa Mining Intelligence* 'Les class actions menacent les compagnies minières!' *Africa Mining Intelligence* 7 April 2015 at 1.

[25] Among numerous examples, in 1998 the Petroleum Code of Senegal mentioned the difficulties already faced by the petroleum industry for the past 10 years:

> '*Durant ces dix dernières années, l'environnement pétrolier international a été caractérisé par une réduction importante des budgets d'exploration des compagnies pétrolières.*'
>
> [Free translation into English: 'Over the past 10 years, the international petroleum environment has been characterised by a significant reduction in the petroleum companies' exploration budgets'
>
> (Preamble to Law No. 98-05 of 8 January 1998 Enacting the Mining Code)].

[26] Economic Commission for Africa *Mineral resources and development in Africa—Report from the International study group on African mining regimes* Economic Commission for Africa and African Union 2011 at 47 [online].

[27] See MAGIC 'Africa: A future Centre of Gravity for Global Mining' *Magic* 2015 at 6. For a detailed analysis of the African continent's attractiveness, see Ernst & Young *EY's attractiveness survey Africa 2015 Making choices* (2015).

[28] OPEC *Annual Statistical Bulletin 2014* (2016) [online].

[29] Id.

world's gold reserves and 80-90% of the world's known chromium and platinum reserves.[30]

Hydrocarbon resources of the African continent[31]

	2000		2010		
Hydrocarbon resources	Share of the African continent in global production	Amount of African production in 2000 ($ million)	Share of the African continent in global production in 2010	Amount of African production in 2010 ($ million)	Growth between 2000 and 2010
Petroleum	10%	216.001	11%	284.875	32%
Gas	5%	39.036	7%	68.423	75%

Mineral resources of the African continent[32]

	2000		2010		
Mineral resources	Share of the African continent in global production in 2000	Amount of African production in 2000 ($ million)	Share of the African continent in global production in 2010	Amount of African production in 2010 ($ million)	Growth between 2000 and 2010
Platinum	55%	10.588	74%	14.191	34%
Cobalt	43%	490	62%	1.775	262%
Diamonds	45%	4.265	54%	4.967	16%
Coal	6%	21.266	4%	23.759	12%
Chromite	51%	1.578	42%	2.442	55%
Manganese	32%	493	30%	3.131	535%
Phosphates	28%	4.607	26%	5.662	23%
Gold	24%	25.568	19%	19.947	–22%
Uranium	17%	111	19%	1.013	813%
Copper	3%	2.871	8%	7.806	172%
Nickel	5%	1.225	5%	1.535	25%
Iron ore	5%	4.637	4%	6.404	38%

1.15 Among African countries, a distinction exists between producing and exporting countries. In the petroleum sector for example, some African countries do not export the hydrocarbon resources they produce, such as Benin, Côte d'Ivoire, Ghana, Tanzania or the Democratic Republic of the Congo.[33] Other states such as Algeria, Angola, Cameroon, the Republic of

---

[30] See *Mining Review Africa* 'Taking Safety more seriously in Africa' *Mining Review Africa* January 2015 at 72.

[31] African Development Bank, Organisation for Economic Co-operation and Development, United Nations Development Programme, Economic Commission for Africa *African Economic Outlook 2013: Structural Transformation and Natural Resources* at 40.

[32] Id.

[33] Nadine Bret-Rouzaut and Jean-Pierre Favennec *Recherche et production du pétrole et du gaz : réserves, coûts, contrats*, 2nd ed (Technip Pub 2011) 189.

Congo, Gabon, Equatorial Guinea, Libya, Mauritania, Nigeria, Sudan or Tunisia, export the hydrocarbons produced on their national territory.[34]

1.16 To date, the African continent's potential remains largely under-explored although Africa is home to some of the planet's largest unexploited mineral deposits.[35] The constant discovery of new deposits attests of the scale of Africa's potential, demonstrating that the exploitation of the continent's natural resources is not living up to its potential. For instance, estimates show that the unexploited mineral resources of the Democratic Republic of the Congo amount to $24 billion.[36] In light of the global average, Africa's limited ratio in this domain seems to confirm that numerous deposits are still to be discovered.[37]

1.17 Faced with the exhaustion of conventional hydrocarbon reserves, the energy future of the African continent will also depend on the exploitation of new deposits. Certain African states can lay claim to holding large shale gas reserves.[38] In Algeria for example, estimates indicate that the volume of shale gas reserves is about 50% higher than that of conventional gas reserves. Predictions indicate that shale gas reserves in Libya and South Africa are greater than the conventional gas reserves of the continent's major producers.[39] Some African states already embarked on the exploitation of shale gas,[40] despite the environmental issues and concerns raised by the exploitation of this resource. While, for some, mineral resources account for nearly half of the wealth produced,[41] known or potential natural resources constitute significant development potential, reinforcing the importance of attractive and modern regulatory and contractual legal frameworks.

1.18 Mineral and petroleum resources constitute, for economies in general and African economies specifically, substantial resources for funding change in their societies. The idea is to transform the exploitation of natural resources into a development driver, through the training and employment of

---

[34] Id.

[35] See *Metal Bulletin Magazine* 'International View' *Metal Bulletin Magazine* (African Mining Special Issue) February 2015 at 26; see 'Africa: A future Centre of Gravity for Global Mining' loc cit.

[36] See MAGIC 'Regional Overview: East and Central Africa' *Magic* 2015 at 82.

[37] Id.

[38] Such as Algeria, South Africa, Libya, Morocco, Tunisia or Mauritania (for information on this issue, see African Development Bank 'Shale gas and its implications for Africa' (2013).

[39] Id.

[40] This applies to Algeria who, in May 2014, announced its intention to start exploring non-conventional hydrocarbons on its territory (see Amir Rabah, 'L'Algérie autorise l'exploitation du gaz de schist' *Jeune Afrique* 23 May 2014 [online]), incidentally resulting in numerous objections on the grounds of environmental sustainability, as attested by the protests which broke out in early 2015 after the first test wells were drilled in Southern Algeria (see *Jeune Afrique*, 'Pour le président algérien, l'exploitation du gaz de schiste n'est pas encore à l'ordre du jour' *Jeune Afrique* 28 January 2015[online]). Morocco is also counting on the exploitation of shale gas, conducting tests on the Timahdit deposit (See *Agence ecofin*, 'Maroc: San Leon effectue des tests positifs sur l'huile de schiste de Timahdit, selon le procédé Enefit' *Agence Ecofin* 26 January 2015 [online]).

[41] This is true for countries such as Angola, Nigeria, Mauritania or the Republic of Congo.

local personnel, the local supply of goods and services or the construction of infrastructures; these topics are now part of the exploitation's legal framework.

3 CHALLENGES RELATING TO EXTRACTIVE ACTIVITIES ON THE AFRICAN CONTINENT

1.19 To maximise the profitability of mining and petroleum projects, host states and private operators pay increasing attention to taking some issues into consideration.

1.20 These issues have remained largely constant over the past decades. Since the Lagos plan adopted as part of the Organisation of African Unity (OAU) in 1980, the main problems associated with extractive activities on the African continent are still topical, ie:

> '[the] lack of information on natural resource endowment of large and unexplored areas [. . .]; lack of adequate capacity (capital, skills and technology) for the development of these resources; considerable dependence on transnational corporations [. . .]'.[42]

1.21 The states' desire to address these issues resulted in the adoption of mining and petroleum policies to deal with difficulties at local level and maximise the potential benefits derived from the exploitation of natural resources.

a Problems associated with the exploitation of mineral and petroleum resources in Africa

1.22 Extractive industry stakeholders, particularly public authorities, are faced with numerous challenges when it comes to the exploitation of natural resources; these issues include the environmental and social impact of mining and petroleum projects, the management of artisanal and small-scale mining, the access to infrastructures and geological data, good governance, etc.

1.23 Extractive industries are often criticised for their impact on the environment of the host state. These projects effectively modify the landscape a great deal; this is particularly true of opencast mines that generate significant soil movements and sometimes require deforestation. Numerous environmental impacts can emerge from the extractive projects such as soil erosion and degradation or water contamination generated using chemicals. Nevertheless, the environmental impact of mining and petroleum projects varies substantially from one site to the next, depending on the nature of the natural resources extracted, the site's geology, the technology used, the extraction techniques or the scale of the activities[43] . African states have become aware of the need to consider environmental issues and have

---

[42] For an overview of the challenges, see also the preamble to ECOWAS Directive C/DIR 3/05/09 of 27 May 2009 on the Harmonisation of the Guiding Principles and Policies in the Mining Sector.

[43] Economic Commission for Africa, 'Mineral resources and development in Africa—Report from the International study group on African mining regimes' at 50.

developed legal means to preserve their environment. There are now very few African mining and petroleum laws which fail to address the issue of environmental protection; in this respect, African legislation actually ranks among the world's most modern. The obligation imposed upon operators to rehabilitate mining and petroleum sites illustrates the attention paid by African states to environmental issues. In addition to national laws, regional instruments are also designed to take into consideration environmental issues on the scale of the continent.[44]

1.24 As extractive activities often affect local populations; petroleum projects must take the social impact of mining into account. State and private operators must consider elements such as the acquisition of land, the diversion of rivers, the displacement of local populations or the employment of expatriates.[45] To offset the negative social impact of certain projects, the association of local populations helps enhance the legitimacy of extractive projects at local level.[46] Moreover, the artisanal mining industry involves significant social problems such as child labour.[47]

1.25 Public authorities are also faced with numerous challenges in terms of artisanal and small-scale mining. In this domain, the stakes are raised by the weight of this subsector in national economies; in Ghana, for example, the gold panning revenue streams injected into the national economy exceeds those derived from industrial exploitation.[48] This subsector, which forms a significant part of the African economy, is faced with many difficulties such as the absence of appropriate technological resources, the lack of funds, legislation which is inadequate and fails to take into account the specific characteristics of this subsector, or the operators' market access difficulties.[49] The consideration of the artisanal mining activity therefore poses a genuine legal challenge. In addition, the artisanal or informal mining activity, when poorly managed, can have negative effects such as environmental damage, the depletion of reserves due to rudimentary extraction techniques, hazard-

[44] In this respect, see the Libreville Declaration of 29 August 2008 on Health and Environment in Africa.

[45] For information of the social impact of mining projects, see in particular Economic Commission for Africa, 'Mineral resources and development in Africa—Report from the International study group on African mining regimes' at 53–54.

[46] Economic Commission for Africa *Mineral resources and development in Africa—Report from the International study group on African mining regimes* at 59.

[47] According to a Human Rights Watch report, more than 20 000 children work illegally in Mali's artisanal gold mines (for information on this issue, see *African Mining* 'Mali: gold and good people' at 38).

[48] Interministerial Committee for International Cooperation and Development (CICID) *Ressources minérales et développement en Afrique. Document d'orientation stratégique* (May 2005) at 9.

[49] Economic Commission for Africa *Mineral resources and development in Africa—Report from the International study group on African mining regimes* 73 et seq.

ous exploitation or the development of an underground economy undermining the national economy.[50]

1.26 Besides these difficulties, the African continent's lack of infrastructure also represents a challenge that the extractive industries must overcome, especially in Sub-Saharan Africa, where the infrastructures are either inadequate or non-existent. Yet, the existence of modern, quality infrastructures is essential for growth of the extractive sector.[51] The main difficulty results from the financing of these infrastructures. The lack of infrastructures currently represents a sizeable challenge, particularly for landlocked countries, where the states cannot maximise the benefits derived from the exploitation of their natural resources. Many African countries have no access to the infrastructures needed for the processing of natural resources on their national territory, forcing them to export their natural resources unprocessed, thus foregoing the potential capital gain resulting from the export of raw materials, ie processed products intended for consumption. This lack of infrastructures deprives African states from many benefits; the promotion of natural resources on the national territory would allow them to supply the local market without having to import raw materials, to create local jobs and to increase the host state's tax revenue by realising capital gains.[52]

1.27 African states only rarely own the geological and land information needed to optimise the granting of mining rights.[53] The mining policies implemented by the states often mention this difficulty.[54] Certain states are currently trying to tackle these difficulties, in particular by establishing funds aimed at financing geological research.[55]

---

[50] Interministerial Committee for International Cooperation and Development (CICID) *Ressources minérales et développement en Afrique. Document d'orientation stratégique* at 9.

[51] As reiterated by the Government of Malawi as part of its Mining Policy '[s]upport infrastructure is key to the growth of the mineral sector. Government is aware that the sector is unlikely to develop unless improvements are made to the existing infrastructure such as transport networks, telecommunication, water supply and electricity'. (Mines and Minerals Policy of Malawi, Ministry of Mining, March 2013).

[52] Tax revenue derived from the export of raw materials obviously exceeds that generated by the export of unprocessed natural resources.

[53] Interministerial Committee for International Cooperation and Development (CICID) *Ressources minérales et développement en Afrique. Document d'orientation stratégique* at 10.

[54] See, in particular, the Mining Policy implemented by the Government of Uganda: '[a]n important basis for investment in the mineral sector is the availability of basic and reliable data. To date there is a lack of adequate, modern, comprehensible and detailed geological, geochemical and remote sensing information covering the entire country' (Mineral Policy of Uganda at 13). See also Madagascar's Mining Policy which reminds the 'insufficient basic geological data in Madagascar'.

[55] In Burkina Faso, see Article 29 of Law No. 036-2015/CNT of 16 June 2015 Enacting the Mining Code of Burkina Faso, which stipulates:

> '[l]*e Fonds de financement de la recherche géologique et minière et de soutient à la formation sur les sciences de la terre est alimenté par 15% des redevances proportionnelles, des taxes superficiaires, des droits fixes et des frais de demande d'agrément d'achat et de vente d'or collectés.*'

1.28 In addition, public authorities must overcome a multitude of challenges in terms of governance[56] . African governments are increasingly taking transparency-related issues into consideration as part of their energy policies.[57] The involvement of African states in the Extractive Industries Transparency Initiative initiative ('EITI') attests to the public authorities' desire to establish a framework conducive to investment in the mining and petroleum sectors.[58] To date, 18 African states comply with EITI require-

[Free translation into English: '[t]he Fund to finance geological and mining exploration and support training in earth sciences is subsidised by 15% of proportional charges, land taxes, fixed duties and collected gold purchase and sale agreement application fees'].

[56] For example, the granting of mining rights by the government of Kenya prompted a response from the International Monetary Fund, which pointed to the lack of transparency in this respect (for information on this topic, see MAGIC 'Regional Overview: East Africa' *Magic* 2014 at 18).

[57] The Guinean Government essentially created the Mining Rights and Agreements Technical Review Committee to enhance transparency in the management of the mining sector in the Republic of Guinea. For more information of African states' consideration of issues of good governance, see The National Petroleum Policy of Liberia (Section III at 7 et seq).

[58] The Lancaster House conference, organised by the British Government in London in 2003, resulted in the adoption of a Declaration of principles intended to increase transparency relating to payments and revenues in the extractive sector. These twelve principles are known as EITI Principles and form the cornerstone of the Initiative:

'We share a belief that the prudent use of natural resource wealth should be an important driver engine for sustainable economic growth that contributes to sustainable development and poverty reduction, but if not managed properly, can create negative economic and social impacts.

We affirm that management of natural resource wealth for the benefit a country's citizens is in the domain of sovereign governments to be exercised in the interest of their national development.

We recognise that the benefits of resource extraction are manifested in the form of revenue streams running over multiple years and can be highly price dependent.

We recognise that a public understanding of government revenues and expenditure over time could help public debate and inform choice of appropriate and realistic options for sustainable development.

We underline the importance of transparency by governments and companies in the extractive industries and the need to enhance public financial management and accountability.

We recognise that achievement of greater transparency must be set in the context of respect for contracts and laws.

We recognise the enhanced environment for domestic and foreign direct investment that financial transparency may bring.

We believe in the principle and practice of accountability of the government to all citizens for the stewardship of revenue streams and public expenditure.

We are committed to encouraging high standards of transparency and accountability in public life, government operations and in business.

We believe that a broadly consistent and workable approach to the disclosure of payments and revenues is required, which is simple to undertake and to use.

We believe that payments' disclosure in a given country should involve all extractive industry companies operating in that country.

In seeking solutions, we believe that all stakeholders have important and relevant contributions to make—including governments and their agencies, extractive industry companies, service companies, multilateral organisations, financial organisations, investors and non-governmental organisations'.

ments.[59] Incidentally, certain states even insert provisions on the application of EITI standards to the mining and petroleum sectors into their corresponding legislation.[60] Moreover, listed international companies (who have numerous obligations in this respect) also subject states to this approach.

1.29 Progress in good governance also requires tackling corruption issues,[61] a factor which is taken into consideration by an increasing number of African states.[62] Corruption, which has a devastating effect on the stability

[59] The following African countries comply with all EITI requirements: Burkina-Faso, Cameroon, Côte d'Ivoire, Ghana, Guinea, Liberia, Mali, Mauritania, Mozambique, Niger, Nigeria, the Democratic Republic of Congo, the Republic of Congo, Sierra Leone, Tanzania, Chad, Togo and Zambia (data collected on https://eiti.org/fr/countries). It should be noted that, while the following African countries implement these principles, they have not yet been deemed compliant with EITI requirements: Ethiopia, Madagascar, Sao Tome and Principe, Senegal and the Seychelles.

[60] See for instance Law No. 2014-138 of 24 March 2014 Enacting the Mining Code of the Republic of Côte d'Ivoire, in the Chapter on commitment to the principles of good governance, stipulates that the application of the Equator Principles and the EITI principles (Articles 117 et seq of the Mining Code). Law No. L/2014/N° 034/AN Enacting the Petroleum Code in the Republic of Guinea also specifies the application of EITI standards in Article 104 '*[o]n the application of EITI—Guinea*'. Finally, in Burkina Faso, Article 6 paragraph 3 of Law No. 036-2015/CNT of 16 June 2015 Enacting the Mining Code of Burkina Faso stipulates that '*[l]'État réaffirme son adhésion à tout initiative de bonne gouvernance dans le secteur minier notamment, le Processus de Kimberley et l'Initiative pour la transparence dans les industries extractives (ITIE)*' [Free translation into English: '[t]he State reaffirms its commitment to any good governance initiative in the mining sector, specifically, the Kimberley Process and the Extractive Industries Transparency Initiative (EITI)'].

[61] '*[O]n entend par corruption le fait de solliciter, d'offrir, de donner ou d'accepter directement ou indirectement une commission illicite ou un autre avantage indu qui affecte l'exercice normal d'une fonction ou le comportement requis du bénéficiaire de la commission illicite, ou de l'avantage indu, ou de la promesse d'un tel avantage indu*' [Free translation into English: '[C]orruption means requesting, offering, giving or accepting, directly or indirectly, a bribe or any other undue advantage or prospect thereof, which distorts the proper performance of any duty or behaviour required of the recipient of the bribe, or the undue advantage or prospect thereof]' (Article 2 of the Civil Law Convention On Corruption of the Council of Europe). Corruption can be active or passive. The notion of active corruption refers to the offering or giving to a public official, directly or indirectly, any undue advantage, for the official himself or herself or for another person, so that the official acts or refrains from acting in the performance of his or her official duties. Passive corruption refers to the solicitation or acceptance by a public official, directly or indirectly, of an undue advantage, for the official himself or herself or for another person or entity.

[62] As evidenced by a recent and highly publicised incident which, as part of the exploitation of the Simandou mining project, led the Guinean Government to terminate mining agreements and brought many proceedings before the national and arbitral tribunals. On 30 April 2010, Brazilian mining group Vale announced that they acquired the right to exploit the Simandou mines, in the South-East Guinea, by taking over 51% of VBG's share capital, a subsidiary owned by Benny Steinmetz Group Resources (BSGR). Simandou is one of the world's largest unexploited iron ore deposits. Located in West Africa, Simandou gives access to billions of tonnes of high-grade iron ore. The Simandou deposit, a mine with an exploitation cycle of more than one generation, could potentially make Guinea one of the world's major iron ore exporters. On 9 April 2014, further to the report issued by the Mining Rights and Agreements Technical Review Committee (CTRTCM), the Government of Guinea indicated that the mining rights and the mining agreement held by the VBG

of states and the effective development of their economy, remains a recurring problem on the African continent. This phenomenon hampers economic growth, notably by discouraging foreign investment and causing distortions in terms of resource allocation. This scourge affects states just as much as private investors: while private stakeholders experience an increase in costs,[63] the states see their revenue plummet.[64] According to the African Union's estimates, corruption amounts to 25% of the African states' gross domestic product, ie $148 billion.[65] The extractive industries sector is particularly affected and accounts for 19% of corruption cases.[66] African governments are increasingly committed to tackling corruption; this commitment notably consists in developing regional and subregional instruments in an effort to stem this phenomenon. Against this background, it is worth mentioning the African Union Convention on Preventing and Combating Corruption, dated 11 July 2003. This instrument (signed by 48 states and ratified by 37) represents a major innovation in combating corruption. The Convention allows the public to access information and facilitates the participation of civil society in the fight against corruption. It also defines obligations in terms of criminal sanctions and facilitates judicial cooperation between states. The scope of this instrument, however, remains limited. The degree of flexibility given to the states in the application of the Convention often leads to the disparate and inefficient application of its provisions. Despite this, the convention provides control mechanisms as part of the African Union, which effectively results in the setting up of an Advisory Committee on corruption, the purpose of which is to monitor the implementation of the convention.[67]

1.30 Political instability remains a major obstacle to investment in certain regions of the African continent. The northern Mali region has been unstable since 2012.[68] Other examples include recent events in Burkina Faso in 2014[69] and Burundi in 2015. While the Arab Spring undermined the stability in some North African countries,[70] certain central African countries still suffer from

---

Company had been acquired under fraudulent conditions and terminated the said mining agreement. Arbitration proceedings were instituted by Vale against BSGR, under the aegis of the London Court of International Arbitration. On Wednesday 30 April 2014, the Mining Group Rio Tinto announced that they filed a complaint against Vale and BSGR in the USA. The complaint relates to Rio Tinto's loss of half of the Simandou mining concession and 'outlines the illegal activity' which led to Rio Tinto's loss of two blocks North of the concession which they had been granted two years earlier.

[63] According to the estimates of the World Economic Forum, corruption increases the cost of business by 10% on average (OECD *CleanGovBiz: Integrity in practice* OECD 2014).

[64] See OECD *CleanGovBiz: Integrity in practice* at 2-3.

[65] See OECD *CleanGovBiz: Integrity in practice* at 3.

[66] OECD *Foreign Bribery Report: An Analysis of the Crime of Bribery of Foreign Public Officials* (OECD 2014) at 8.

[67] Article 22 of the African Union Convention on Preventing and Combating Corruption.

[68] See *African Mining* 'Mali: gold and good people' at 33.

[69] On 30 October 2014, President Blaise Compaore left office after 27 years in power.

[70] Specifically, Libya and Egypt.

instability.[71] Even though the continent is generally rooted in democracy, it is not unusual for elections to be postponed.

1.31 Besides, health crises did not spare the African continent. In this regard 2014 was a difficult year for Africa. The Ebola virus, which affected West Africa, had a considerable impact on the mining industry in some countries such as Guinea, Sierra Leone and Liberia.[72] This health hazard led the governments to take action to restrict the transportation of extracted resources from one state to another, thereby affecting operators in areas hit by the Ebola virus. The propagation of the Ebola virus resulted in a significant rise in production costs in the mining sector, notably linked to procurement delays and the unavailability of labour, as people were sometimes quarantined.[73]

1.32 Investors and, likewise, African states who wish to maximise returns from extractive activities within their national territory, should not underestimate the above-mentioned elements. In particular, State policies enacted by governments in the mining and petroleum sectors should reflect states' awareness of these issues.

b State policies in the petroleum and mining sectors

1.33 To optimise the management of their natural resources and address the issues arising from their exploitation, some African states have implemented specific policies.[74]

The primary objective of these instruments is to enhance the development of extractive industries which best contribute to economic and social development.[75]

These state policies are designed to take into consideration the problems faced by producing countries; whether environmental issues, as already

---

[71] See MAGIC 'Regional Overview: East and Central Africa' at 82.

[72] In August 2014, the ArcelorMittal group suspended its activities as part of the Yekepa project in Liberia, claiming force majeure (see *Metal Bulletin Magazine*, 'West African iron ore reassessed' *Metal Bulletin Magazine* (African Mining Special Issue) February 2015 at 43).

[73] For information on this issue, see Meredith Campion and Tim Scales 'New dawn for Africa in 2015' *Australia's Playdirt* December 2014—January 2015 at 65.

[74] The following texts are worth mentioning: Mines and Minerals Policy of Malawi, Mineral Policy of Tanzania, National Petroleum Policy of Liberia, National Oil and Gas Policy for Uganda, Mineral Policy of Uganda, Decree No. 98-394 of 28 May 1998 Defining the Sectoral Mining Policy in Madagascar.

[75] For example, the Mining Policy of Uganda specifies that:

> '[t]he Government policy objectives for the mineral sector are: (i) To stimulate investment in the mineral sector by promoting private participation; (ii) To ensure that mineral wealth supports national economic and social development; (iii) To regularise and improve artisanal and small-scale mining; (iv) To minimise and mitigate the adverse social and environmental impacts of mineral exploitation, (v) To remove restrictive practices on women's participation in the mineral sector and protect children against mining hazards; (vi) To develop and strengthen local capacity for mineral development; and (vii) To add value to mineral ores and increase mineral trade'.

mentioned,[76] the development of infrastructures[77] or the supervision of artisanal mining activities.[78]

1.34 Regional cooperation is also taken into consideration by mining and petroleum policies to provide for the development of these sectors. The states' involvement in regional organisations is essential for maximising the exploitation of natural resources. For example, the mining policy of Malawi highlights the lack of cooperation between states when it comes to sharing geological and technological data and advocates increased cooperation

---

[76] From an environmental perspective, Section V of the Mines and Minerals Policy of Malawi of March 2013 stipulates:

> '[t]he National Environmental Policy (1996) provides an overall structure against which relevant sectoral environmental policies can be reviewed to ensure that the policies are consistent with the principles of sustainable development, while the Environment Management Act (1996) and the Mines and Minerals Act (1981) contain sections that specifically deal with the protection and management of the environment. Section 24 of the Environment Management Act outlines the requirements for projects for which an Environmental Impact Assessment (EIA) is needed. Mining is not a benign activity and it is impossible to extract minerals from the earth and process them without impacting, in various degrees, the air, land and water, as well as plant and animal life. The closure of a mine can also seriously impact the local community. In these contexts, it must be recognised that mining must be committed to minimising the disruption of the environment during exploration and exploitation and to maximising the effectiveness of land rehabilitation and closure measures at the end of the mine's life'.

On this topic, see also The Mineral Policy of Tanzania of 2009:

> '[m]ining activities can adversely affect safety and occupational health of mineworkers and result in environmental land degradation, pollution, and social disruption to local communities around mine sites. However, improved management systems for health, safety and environment have dramatically increased safety and environmental performance in mining operations over the last decade. To ensure sustainable mining activities, there is a need for the Government to strengthen the monitoring and regulation of the mining industry to reduce or eliminate the adverse effects on health and safety, environment and social issues. It is also important for the Government to make an effort to increase environmental awareness and promote environmentally-friendly practices in the mining industry, particularly among small-scale miners. In addition, coordination with other institutions mandated to administer these issues needs to be improved for effective monitoring and enforcement'.

[77] Decree No. 98-394 of 28 May 1998 Defining the Sectoral Mining Policy in Madagascar stipulates the following objective relative to this topic:

> '*[p]romouvoir le développement de l'infrastructure régionale pour faciliter autant que possible l'accès aux zones minières et la circulation des produits*'.
>
> [Free translation into English: '[p]romote the development of the regional infrastructure to facilitate access to mining areas and movement of products as much as possible'].

[78] On this issue, Decree No. 98-394 of 28 May 1998 Defining the Sectoral Mining Policy in Madagascar stipulates:

> '*[l]'État mettra en place des conditions et les services d'appui en vue d'aider les orpailleurs et les artisans miniers à s'intégrer dans le circuit formel issu de la réforme du secteur. Particulièrement, le Comptoir de l'Or sera transformé en Agence de l'Or qui sera un organe de régulation et de promotion de la filière Or*'.
>
> [Free translation into English: '[t]he State shall create conditions and support services in order to help gold panners and mining artisans to integrate into the formal economy born of the reform of the sector. More specifically, the Comptoir de l'Or shall be transformed into Agence de l'Or, which shall be the Gold industry regulation and promotion body'].

within organisations such as the Southern African Development Community (SADC) or the Common Market for Eastern and Southern Africa (COMESA).[79]

1.35 Africa is characterised by the increased influence of regional organisations in the extractive sector. In West Africa, the Economic Community of West African States (ECOWAS) or the West African Economic and Monetary Union (WAEMU) now have a key role to play in the adoption of mining policies.[80] In Central Africa, the Central African Economic and Monetary Community (CEMAC) adopt a similar approach.

[79] Mines and Minerals Policy of Malawi, Ministry of Mining, March 2013.

[80] For information on the role of regional organisations in the establishment of mining standards, see below, para 7.3 et seq.

2

The stakeholders

2.1 Projects relating to the exploitation of natural resources involve two stakeholders with complementary and not, as is often described, conflicting interests: the host state, conscious of the potential for development of any foreign investment, and the national or foreign private operator who has the technology and the capital required for the exploitation of natural resources.

1 STATE STAKEHOLDERS

2.2 The State entity may assume a number of roles in mining and petroleum operations: directly or indirectly via a local State operator.

a The States

2.3 A distinction exists between natural resource producing states, in particular the one whose natural resources remain unsubstantiated in terms of commercial quantities. This unknown element is key and largely influences the policy and regulations of the states, who are generally more inclined to grant rights to private investors when they bear the risk of non-discovery. Conversely, private investors' room for negotiation is seriously restricted when the operations are performed on the territory of already producing states.

2.4 We should also mention the grouping of producing states, essentially in the hydrocarbon sector, as part of cartels or associations such as the Organisation of the Petroleum Exporting Countries (OPEC),[1] the African Petroleum Producers' Association (APPA),[2] and the Organisation of Arab Petroleum Exporting Countries (OAPEC)[3] or, in the mining sector, the newly created African Diamond Producers Association (ADPA).[4] Their role and power however tend to decline.[5] It is very difficult to find a small number of

[1] An organisation created in 1960 and currently comprising 15 states: Algeria, Angola, Ecuador, Gabon, Iran, Indonesia, Iraq, Kuwait, Libya, Nigeria, Qatar, Saudi Arabia, the United Arab Emirates, Venezuela and, since 2017, Equatorial Guinea.

[2] Intergovernmental organisation created in 1987 in Lagos, Nigeria, whose members include Algeria, Angola, Benin, Cameroon, the Republic of Congo, Côte d'Ivoire, Egypt and Gabon. The purpose of this organisation is to promote cooperation, collaboration and the sharing of knowledge and skills among African petroleum-producing countries.

[3] Organisation involving the United Arab Emirates, Bahrain, Algeria, Saudi Arabia, Syria, Iraq, Qatar, Kuwait, Libya and Egypt.

[4] Association created in May 2015 by seven major global diamond companies, ie Alrosa, De Beers, Rio Tinto, Dominion Diamond Corporation, Lucara Diamond Corporation, Petra Diamonds Ltd. and Gem Diamonds Ltd.

[5] As attested by the disappearance of certain organisations such as the intergovernmental Council of Copper Exporting Countries (CIPEC) in 1988 or the International Bauxite Association (IBA) in 1994.

producing states with a concentration of a large part of global production who share cultural or ideological affinities.

2.5 As owners of the natural resources located on their territories, states are key players.[6] Therefore, the State, usually represented by its government, has the authority to intervene directly or indirectly in mining and petroleum projects.

2.6 The State can intervene directly, via its government, in mining agreements or petroleum contracts. It can also act as a licensing public authority and grant mining rights. The State may also collaborate as an active industrial partner involved in financing and decision-making. The host state's indirect intervention may consist in the signing of related agreements with the operator, in particular establishment agreements; through these agreements, the host state provides the investor with political and legal guarantees to ensure the smooth running of the operations described in the main contract. The host state will as well be involved as sovereign legislator on the territory where the extractive activities occur, by enacting fiscal, social, safety or labour regulations.

2.7 Legal provisions[7] or the mining and petroleum policies laid down by governments,[8] sometimes reiterate the different missions of the State.

2.8 The states may notably implement these different attributes through the governmental authorities competent in the mining and petroleum field.

Mining and petroleum laws designate the governmental authorities in charge of monitoring natural resource exploitation operations. It should be pointed out that mining and petroleum activities are, more often than not, within the remit of different ministries.[9]

---

[6] About the states' ownership of natural resources see para 5.1 et seq below.

[7] For example, Article 4 of Law No. 2010-033 of 20 July 2010 Enacting the Crude Hydrocarbons Code, amended by Law No. 2011-044 of 25 November 2011, mentions the two roles of the Mauritanian State as part of petroleum operations:

'[l]'État peut entreprendre toutes les opérations pétrolières soit directement, soit par l'intermédiaire de l'entreprise nationale. L'État peut également autoriser une ou plusieurs personnes morales établies en Mauritanie, ou y disposant d'une succursale, à entreprendre des opérations pétrolières en vertu d'une autorisation de reconnaissance, d'un contrat d'exploration-production ou d'une autorisation de transport, conformément aux dispositions de la présente loi.'

[Free translation into English: '[the] State may undertake all petroleum operations either directly or by the intermediary of the national company. The State may also authorise one or more legal entities established in Mauritania, or having a subsidiary there, to undertake petroleum operations on the basis of a prospecting authorization, an exploration-production contract or a transportation authorisation, pursuant to the provisions of this law'].

[8] The Mining Policy laid down by the Tanzanian Government clearly defines the Government's role in mining operations: '[t]he role of the Government is to formulate policy, establish guidelines and regulate the mineral sector; promote and facilitate private investments; and provide essential services needed by the mines. The Government may participate strategically in mining activities such as exploration, mineral exploitation and value addition'. (The Mining Policy of Tanzania, 2009).

[9] For example, in the Democratic Republic of Congo, a Ministry of Mines co-existed with a Ministry of Hydrocarbons. Similarly, in Algeria, the mining sector was recently withdrawn

b The local State operator

2.9 It is common practice for the host state to get involved in petroleum or mining projects via a local State operator.[10] This local State operator is an emanation of the host State—a specialised state-owned or public-private company—which is involved in a project undertaken on the territory under the sovereignty of the State that created the said company. The local State operator may act as a partner, or as sole operator, with or without the support of private foreign or national companies.

2.10 Historically, the purpose of these domestic companies was to create or develop mining and petroleum industries in the state of origin, or to return these industries to the industrialising countries that had gained political independence. On a global level, the first state operators emerged in 1922 in the hydrocarbons sector.[11] This trend was subsequently extended to the mining sector so that it is now extremely rare to find African states without a company specialising in the exploitation of one or more natural resources. In some countries, these companies have become very powerful, so much so that they constitute genuine states within the State, subdivided into multiple companies, with Sonatrach in Algeria being emblematic of this trend.[12]

---

from the jurisdiction of the Ministry of Energy and transferred to the Ministry of Industry. The Ministry of Energy is currently exclusively in charge of the energy sector. Furthermore, quarrying activities may not be attributable to the Ministry of Mines, as is the case in Morocco, where quarrying falls within the remit of the Ministry of Equipment, Transport and Logistics.

[10] Many legal provisions provide for the host state's involvement via the local State operator, including: Law No. 2007-01 of 31 January 2007 Enacting the Petroleum Code in Niger (Article 8:

'*[l]'État peut mener pour son propre compte soit directement, soit par l'intermédiaire d'un organisme public, seul ou en association avec des tiers nationaux ou étrangers, des opérations pétrolières conformément aux dispositions de la présente loi et des textes pris pour son application.*'

[Free translation into English: '[the] State may engage, on its behalf, directly or via a public organisation, alone or in association with national or foreign third parties, in petroleum operations in accordance with the provisions of this law and texts pursuant to its implementation']

Law No. 007-2002 of 11 July 2002 Enacting the Mining Code in the Democratic Republic of Congo (Article 8 paragraph 3:

'*[l]'État peut également, à travers des personnes morales publiques ou des organismes spéciaux créés à cet effet, soit seul soit en association avec des tiers, se livrer à une activité régie par le présent Code. Dans ce cas, les personnes morales publiques et les organismes spécialisés de l'État à vocation minière sont soumis aux dispositions du présent Code au même titre que les personnes privées*'.

[Free translation into English: '[the] State can also, through public corporations or special bodies created for this purpose, alone or in association with third parties, engage in an activity regulated by this Code, in which case the public corporations and State bodies specialising in the mining sector shall be subject to the provisions of this Code, in the same capacity as private bodies'].

[11] Creation of Yacimientos Petrolíferos Fiscales ('YPF') in Argentina.

[12] The Sonatrach group includes 154 subsidiaries and affiliates, both nationally and internationally. In 2013, the Oil and Gas Group was ranked first in Africa and twelfth worldwide, all businesses combined, with an export turnover of more than $63 billion.

These public entities are particularly well established in the oil & gas sector and are destined to be of growing importance in the mining sector. In mining activities, the local State operator has an essential part to play and it will usually be in charged to take and manage the State's participation in the share capital of the operation company. However, this mission may prove to be a difficult task considering the lack of funding and the poor management that some of these public companies may have to face. The situation that the Ivorian national company, Sodemi, is currently facing illustrates the challenges that state operators must overcome.[13]

2.11 On a legal point of view, these companies raise a substantive question. Is the local State operator legally separate from the state from which it emanates? A positive response was given in the *Jordan Investment Ltd v Sojutzneftexport*[14] arbitration. However, certain arbitral tribunals have been more inclined to accept the theory that the state's emanation should be treated as the said State. Thus, in the ICC award of 25 April 1985 in a case between a French company and two bodies in charge of nuclear affairs in Iran, the arbitral tribunal found that 'with regard to third parties, the public company—*a fortiori* the public institution—and the State only represent a single legal entity'.[15] It is indisputable that the local State operator only exists because of the host state's desire, but each case should be analysed to determine its level of connection and independence.

2.12 In reality, the difficulty is to differentiate between what falls within the state's prerogatives and obligations (notably the objective of fulfilling local market needs) and actions falling under the operation of a private sector company.

An overlap may exist between the local State operator and the host state when the exploitation contract is performed on the state's territory; otherwise, both entities are clearly distinct although the local State operator can argue that their operations abroad are linked to the performance of a state service designed to guarantee supply security for their state of origin. Herein lies the ambiguity of such a distinction.

The separation is not clear-cut. There is only a presumption of independence between a State operator acting beyond the borders of the state who created it and the said state; how else would it be possible to accept, in this very specific domain, a state's political interference on the territory of another state? However, judges or arbitrators can only provide a definitive

[13] In 2017, the Sodemi had to sell its participation in various operation companies due to the lack of funding.

[14] Award made on 19 June 1958 by the USSR's Foreign Trade Arbitration Commission in the case between *Jordan Investment Ltd and Sojutzneftexport*, Award rendered Foreign Trade Arbitration Commission 1960 *Rev arb* 92. On this issue, see also the Award rendered by the ICC case No. 3493 (1983), *Spp (Middle East) and Southern Pacific Properties Ltd v the Arab Republic of Egypt and Egyptian General Company for Tourism and Hotels* 22 ILM 752 (1983).

[15] See Philippe Kahn, 'State sovereignty and dispute settlement. Legal regime of the State contract' 1985 *Rev arb* 654.

response factually, notably by examining the legislation applicable to the local State operator and its practical application.[16]

2.13 Private stakeholders can also get involved in mining and petroleum projects alongside these state stakeholders.

2 PRIVATE STAKEHOLDERS

2.14 The states with natural resources who do not intend to bear the risks associated with extractive activities or do not have appropriate technical and financial capabilities, most frequently resort to private operators for the exploitation of the natural resources located on their territory.

2.15 These local or foreign private operators have a nationality, which will largely influence their legal relationship with the host state, as certain national laws impose the creation of a local company in order to exploit natural resources on the territory of the host state, in particular once the exploration stage is over. Besides, the nationality of the private operator is not without impact, in particular in terms of dispute resolution and arbitration.[17]

Two cases must be considered, depending on whether the private operator is a private company, or a company established using public funds.

a Privately-owned operator

2.16 The landscape of mining and petroleum industries is marked by the existence of large transnational companies and local businesses with a more restricted area of intervention. The former are huge, broadly diversified industrial empires whereas the latter's ambitions are more limited.

2.17 The main consequence of this distinction is the adoption of different (even sometimes contradictory) conventional strategies, in particular due to the significant disparity in terms of technical and, more importantly, financial resources.

i *Transnational companies*

2.18 As natural resources exploitation operations require considerable investments, mineral and petroleum deposits have long been exploited for the sole benefit of major foreign corporations. To this day, despite the increasing involvement of local investors[18] and certain states, transnational companies continue to play a predominant part in the exploitation of natural resources on the African continent.

---

[16] The issue of the independence of emanations vis-à-vis the State is of particular importance as part of the CIRDI arbitration, with regard to the interpretation of Article 25(1) of the 1965 Washington Convention on the Settlement of Investment Disputes between States and Nationals of Other States (on this issue, see Sébastien Manciaux *Investissements étrangers et arbitrage entre États et ressortissants d'autres États* (Litec 2004) 108.

[17] It is worth mentioning for example the ICSID arbitration and Article 25(1) of the Washington Convention, which confers jurisdiction upon the International Centre for Settlement of Investment Disputes over legal disputes between a contracting state (the host state of the investment) and a national of another contracting state.

[18] On the issue of local investors' participation in mining and petroleum projects, see below para 2.21 et seq.

These transnational companies constitute huge, broadly diversified industrial empires the initial purpose of which may or may not have been the exploitation of one or several natural resources. They are involved from the exploration phase to the processing of the natural resource into raw material, sometimes even until the marketing phase.

These companies also enjoy significant growth. In the mining sector for example, the net cumulated income of the world's top three mining groups (Rio Tinto, BHP Billiton and Anglo American) soared from US$4,3 billion in 2002 to US$29,9 billion in 2006.[19] In spite of a slightly unfavourable global climate, the income of these 'majors' has generally continued to increase.[20]

2.19 The power of these international corporations is far from negligible; some of them post higher profits than those generated by entire states. For example, Rio Tinto's 2007 income was equivalent to twice the GNP of Guinea.[21] The weight of these 'majors' may, in some cases, have a significant impact on the negotiations of the investment's contractual framework.

2.20 Finally, mention should be made of the increase in South-South investment agreements in the natural resources sector, resulting in extensive African investments from international groups from the BRICS.[22] This trend notably manifests itself by the emergence of Japan and China as a source of funding for mining operations on the African continent. The China-Africa Development Fund, launched in 2007, attests to the financing possibilities of African projects.[23]

ii *Local operators*

2.21 Alongside international companies, with sometimes almost unlimited resources, local companies have appeared. These companies' ambitions are more limited than those of the 'majors' and they often specialise geographically in the exploration of a single natural resource.

These 'juniors' initially focused their operations on their national territory before investing in other African countries. It should however be pointed out that the exploitation of natural resources by local companies beyond their own borders is a trend that is far more apparent in the petroleum sector, while local mining companies tend to focus their activity on their national territory.

2.22 These juniors, who remain largely dependent on external capital to finance their projects, were hit hard by the difficulties experienced by

---

19 Interministerial Committee for International Cooperation and Development (CICID) op cit 8.

20 From 2011 to 2014, Rio Tinto's net earnings rose from $5,826 million to $9,305 million.

21 Interministerial Committee for International Cooperation and Development (CICID) *Ressources minérales et développement en Afrique Document d'orientation stratégique* at 10.

22 Group of countries including Brazil, Russia, India, China and South Africa. As was the case for Brazilian group Vale or Chinese group China Shenhua.

23 Economic Commission for Africa 'Mineral resources and development in Africa—report from the International study group on African mining regimes' at 35.

extractive industries, and in particular the drop in the prices of raw materials and the difficulty in finding external financing.[24]

2.23 Besides, artisanal or small-scale mining is common practice on the African continent, as this subsector is an integral part of Africa's mining economy.

b Publicly-owned private operator

2.24 The dominance of global production by state-owned companies has varied over time depending on the resources concerned. In the mining sector, private companies accounted for 40–60% of the global production until the end of the Soviet era. According to estimates, this figure has dropped by nearly 25%. However, the role of publicly-owned private operators is not insignificant.[25]

2.25 A private operator wholly or partly publicly-owned, is a company operating outside its territory of origin and acting as a private enterprise.

In addition to companies acting as private operators, service companies highly specialised in a specific technique of the operating cycle have developed. This market fragmentation shows that the major companies traditionally involved in extractive activities no longer have a monopoly on technology.

In fact, we are witnessing the transition from a 'Fiscal State' to a 'Trading State' model. The role of traditional private companies tends to take a back seat to the emergence of actions by the states, specialist state companies and service companies.

2.26 In contractual terms, the problem is to establish a balance between the role of publicly-owned companies and private corporations. There are two possible scenarios depending on whether the exploitation contract only involves private operators or involves both state and private operators.

A bipolarisation is therefore frequently observed with, on the one hand, a party representing local public interests to whom we shall refer as the 'local State operator' and, on the other, a representative of the private interests of local or foreign investors, to whom we shall refer as the 'private operator'. The host state and the local legislature can sometimes be part of this bipartite relationship.

---

[24] See Dominic PIPER 'African exploration down and more support is needed' *Australia's Playdirt* December 2014—January 2015 at 92.

[25] Economic Commission for Africa 'Mineral resources and development in Africa—Report from the International study group on African mining regimes' at 36.

3

NATURAL RESOURCES EXPLOITATION ACTIVITIES

3.1 The exploitation of national resources is subject to a great diversity of operations, from prospecting to production, within a specific overall context particularly on the risks associated with the exploitation of natural resources.

1 SUBJECT OF THESE ACTIVITIES: THE NATURAL RESOURCES

3.2 The term 'natural resources' embraces a large diversity of resources that can be divided in two categories: renewable and non-renewable natural resources. The legal rules applicable to natural resources reflect this distinction.

This study is limited to non-renewable natural resources such as mineral and petroleum resources that are legally subject to some specific standards.

3.3 The term 'mine' refers to the area where deposits of mineral or fossil substances are exploited, opencast or via shafts and galleries. Mineral substances are minerals or natural associated minerals of the soil and subsoil, in and under water, likely to be used in economic activities due to their chemical composition or their remarkable physical properties.

3.4 A distinction should be made between mines and quarries. The term quarry refers to the open-cast or underground exploitation site of quarry products. Materials from quarries are referred to as 'extracted materials' or 'aggregates' and are generally divided into three categories: unconsolidated materials, solid rock and recycled materials. The unconsolidated material category includes alluvial aggregates, marine aggregates and sand in general. Solid rock includes limestone and volcanic aggregates. Finally, so-called 'recyclable materials' include other materials such as carbonaceous shale, slag, clinker and demolition materials.[1] In some cases, quarries can be subject to specific provisions of the mining regulations and can result in the granting of special mining rights. Quarry products, whose exploration is governed by specific regulations, are generally defined by the applicable mining code.[2]

[1] Jean-Pierre Boivin and Claude Lavaire *Carrières et Granulats* (Le Moniteur 2008) 21.

[2] For illustration purposes, the Mining Code of the Republic of Guinea (law L/2013/053/CNT of 8 April 2013 Amending Law No. L/2011/006/CNT of 9 September 2011 Enacting the Mining Code) provides a definition of quarries:

> *'[c]arrière: gîte de matériaux de construction, de matériaux pour l'industrie céramique, de matériaux d'amendement, de sel gemme, et d'autres substances analogues, à l'exception des phosphates, nitrates, sels alcalins, et autres sels associés dans les mêmes Gisements. Les tourbières sont également classées parmi les carrières.'*

et des substances de carrières *'Substance de carrières: tourbe, matériau de construction, matériau pour l'industrie céramique, matériau d'amendement, sel gemme, ou autre substance analogue, à l'exception des phosphates, nitrates, sels alcalins, et autres sels associés dans les mêmes Gisements.'*

3.5 Hydrocarbons include naturally occurring liquid or gaseous hydrocarbons, also known as crude oil or natural gas depending on the case, as well as all related products and substances extracted in combination with these hydrocarbons.

3.6 Petroleum and mining natural resources should therefore be distinguished from the raw materials from which they are derived.

a Distinction between natural resources and raw materials

3.7 An initial distinction should be made between natural resources and raw materials. Business and legal practices have taken into account this dual nature, implicitly or not, to draw a clear line between extractive activities and those of the processing industry.

3.8 A natural resource refers to the extracted resource without undergoing any processing (refining of mineral substances and hydrocarbons); prior to this stage, the natural resource is most of the time immediately unfit for consumption.

3.9 However, raw material refers to a marketable product, directly usable in industry or for final consumption; it is derived from the natural resource. This distinction involves significant legal consequences, particularly in terms of property and sovereignty; the notion of sovereignty of the states over their natural resources only applies to natural resources, not raw materials, ie processed materials.

b Petroleum and mining resources, non-renewable resources

3.10 Natural resources fall into two major categories depending on whether they are renewable; each follows its own rules and is governed by different principles. Renewable natural resources are resources which are not destroyed when used or which are capable of naturally renewing themselves. These resources are therefore inexhaustible in the very long term, as they directly result from regular or constant natural phenomena. The principal sources of renewable energy are hydraulic, wind, biomass, solar, geothermal or marine energy. Renewable natural resources, excluded from the scope of our analyses, raise specific legal issues that considerably differ from non-renewable natural resources (resources with limited stocks on earth which

---

[Free translation into English: '[q]uarry: deposit of construction materials, materials for the ceramics industry, material for soil improvement, rock salt, and other similar substances, with the exception of phosphates, nitrates, alkaline salts, and other associated salts in the same deposits. Peat lands are also classified as quarries'].

Similarly, in Burkina Faso, Article 5 of Law No. 036-2015/CNT of 16 June 2015 Enacting the Mining Code of Burkina Faso defines 'quarry' as:

'*[c]arrière: l'endroit d'où sont extraits des substances minérales telles que les minéraux industriels, les roches industrielles, les pierres ornementales et les matériaux de construction.*'

[Free translation into English: '[Q]uarry: site from where mineral substances such as industrial minerals, industrial stone, ornamental stone and construction materials are extracted.']

take a very long time to replenish). These include mining resources and hydrocarbons.[3]

3.11 When it comes to non-renewable natural resources, a distinction must be made between 'young' and 'old' resources. Some of these resources have only been exploited recently whereas the exploitation of others has been happening since Ancient Times. Conventional wisdom relating to their exploitation is therefore variable, which will affect the negotiations and, more importantly, the drafting of the terms of exploitation agreements. Within the hydrocarbon sector, natural gas is a young resource, unlike petroleum, which has been used since time immemorial; therefore, in conventional terms, the latter has been used as a model and often a reference for the former while moving away from this model. Besides, certain sources of energy are currently posing new challenges. This is particularly true of non-conventional hydrocarbons,[4] which are difficult to exploit using traditional extraction methods. For the mining sector, rare earth materials are also considered as young resources,[5] compared with other resources such as iron, copper or precious metals.

2 PURPOSE OF THESE ACTIVITIES: EXPLOITING NATURAL RESOURCES

3.12 A distinction must be made between exploitation activities and marketing, which is one of its consequences. Exploiting a natural resource means developing it, ie highlighting it and making it available; these operations require prospecting, exploration, development and ultimately production or extraction.

3.13 These notions are subject to relatively detailed legal definitions that may, when relevant, be complemented by conventional definitions. Numerous African mining and petroleum statutes define upstream activities which are prospecting, exploration, development and production operations. Some equate production with exploitation, which sometimes creates confusion between the different phases that precede production.

3.14 Exploration operations refer to all exploration and assessment activities, including all geological, geophysical and geochemical activities (the

[3] As a reminder of the non-renewable nature of mineral and petroleum resources, see the Preamble of South Africa's Mineral and Petroleum Resources Development Act 28 of 2002: '*RECOGNISING that minerals and petroleum are non-renewable natural resources*'. See also article 6 of Algerian Law No. 14-05 of 24 February 2014 Amending the Mining Law:

'*[l]e patrimoine minéral est une richesse naturelle épuisable et non renouvelable.*'

[Free translation into English: '[T]he mineral heritage is an exhaustible and non-renewable wealth.']

[4] The term non-conventional hydrocarbons—as opposed to so-called conventional hydrocarbons- refers to hydrocarbons difficult to extract and having non-standard characteristics. These non-conventional hydrocarbons, trapped in compact rocks, require the use of specific extraction techniques (such as hydraulic fracturing, which consists of causing micro cracks in the rock by injecting water at very high pressure to release trapped hydrocarbons).

[5] See Grégory Gautier (interviewed by Alain Faujas) 'Les terres rares sont le pétrole du xxie siècle' *Le Monde* 16 June 2014 [online].

objective of which is to discover natural resources) as well as technical and economic feasibility studies carried out to assess the potential commercial nature of these resources.[6]

3.15 Developing a deposit consists of constructing all production infrastructures used to extract the natural resources from the deposit. Development operations therefore refer to all activities relating to the development of a deposit, from the date this deposit is deemed commercially exploitable and the decision to develop has been made via the initiation of drilling, construction and installation works, as well as corresponding studies carried out to successfully extract, process, transport, store and remove hydrocarbons.

3.16 The transition to development phase for a mining or petroleum project is determined based on a feasibility study. The purpose of the study is to confirm the existence of a resource, assess the importance and quality of existing reserves as well as clarify the production conditions and techniques; the feasibility study is generally submitted by operators to the competent authorities to request the exploitation of the deposit.

---

[6] For a definition of exploration, see Law No. 96-669 of 29 August 1996 Enacting the Petroleum Code in Côte d'Ivoire (Article 1):

'*"recherche", également dénommée "exploration", les activités de reconnaissance détaillée ainsi que les forages de recherche destinés à découvrir des gisements d'hydrocarbures économiquement exploitables, y compris les activités d'évaluation et de délimitation d'une découverte d'hydrocarbures présumée commercial.*'

[Free translation into English: '"research" also referred to as "exploration", detailed prospecting activities as well as exploration drilling to discover economically exploitable hydrocarbon deposits, including activities designed to assess and delineate a presumably commercial discovery of hydrocarbons'].

See also Law No. 99-003 of 18 February 1999 Enacting the Hydrocarbons Code of Togo:

'*"exploration", tout ensemble de travaux superficiels ou de profondeur (les activités de reconnaissance détaillée, les forages de recherche, les activités d'évaluation et de délimitation d'une découverte d'hydrocarbures), exécutés en vue d'établir la continuité d'indices découverts en surface par la prospection pour en étudier leur exploitabilité et leur utilisation industrielle dans le but de conclure à l'existence de gisements d'hydrocarbures commercialement exploitable.*'

[Free translation into English: '"exploration", any superficial or in-depth work (detailed prospecting activities, exploration drilling, activities designed to assess and delineate a discovery of hydrocarbons), carried out with a view to establishing the continuity of clues found on the surface via exploration to study their exploitability and industrial use to determine the existence of commercially exploitable hydrocarbon deposits.']

In the sphere of mining, for a definition of exploration operations see Sudan's 2007 Mineral Resources and Mining Development Act: ' "[t]he Exploration" means the extensive enquiry and expansion in testing the ground—and its bottom—by all the geological, geophysical and geochemical techniques, remote sensing, all the other necessary surveys, drilling of all types and techniques and all the activities used in the ground and its bottom extensively; and which lead to knowing the extent of the spread of the mineral materials and the other mining raw materials, determining their places, the degrees and forms of their concentration, their storage, hosting environment at the search area or licensing. It also includes all that is necessary in the realm of studies, detailed laboratory tests, to assess them to determine the feasibility of their economic utilisation'].

3.17 For a deposit, production operations refer to all extraction, injection, processing, transport, storage, removal operations which must be carried out to start production on this deposit from the date commercial production begins.[7]

3.18 Marketing a natural resource, ie placing it on the market while pursuing immediate profit, implies its prior transformation into raw material. Exploitation generates no immediate profit, as profit only appears when the extracted products are marketed.[8]

3.19 Transport operations, when they only concern natural resources, may fall within exploitation but due to their specific nature, shall be excluded from our study.

---

[7] For a definition of exploitation operations, see Law No. 99-003 of 18 February 1999 Enacting the Hydrocarbons Code in Togo:

> '*"exploitation", les acticités destinées à extraire les hydrocarbures à des fins commerciales, notamment les activités de développement, de production et d'abandon des gisements d'hydrocarbures.*'
>
> [Free translation into English: '"exploitation", activities designed to extract hydrocarbons for commercial purposes, specifically, hydrocarbon deposit development, production and abandonment activities'].

See also Article 2 of Angola's Petroleum Law No. 10/04 of 12 November 2004:

> '*"[p]rodução, o conjunto de actividades que visam a extracção de petróleo, nomeadamente o funcionamento, assistência, manutenção e reparação de poços completados, bem como do equipamento, condutas, sistemas, instalações e estaleiros concluídos durante o desenvolvimento, incluindo todas as actividades relacionadas com a planificação, programação, controlo, medição, ensaios e escoamento, recolha, tratamento, armazenagem e expedição de petróleo, a partir dos reservatórios subterrâneos de petróleo, para os locais designados de exportação ou de levantamento e ainda as operações de abandono das instalações e dos jazigos petrolíferos e actividades conexas.*'
>
> [Free translation into English: '"[p]roduction" ie all activities the purpose of which is to extract petroleum, specifically, the operation, assistance, maintenance, repair of wells created, as well as the equipment, conduits, systems, facilities and work sites completed during the development phase, including all activities linked to the planning, programming, control, measuring, testing and flow, collection, processing, storage and shipment of petroleum, from underground petroleum reservoirs to dedicated export or removal sites, or petroleum deposit and facilities abandonment operations and related activities.']

[8] The notion of exploitation was already described in similar terms in 1930 by Rudolf Isay, according to whom:

> '*[p]ar exploitation des mines on ne comprend pas seulement l'extraction des minéraux, mais encore dans certaines limites la transformation de ceux-ci. A l'exploitation des mines est toujours rattaché le traitement des minerais, c'est-à-dire la séparation du minéral utile du minerai brut, et la séparation de plusieurs minéraux coexistants.*'
>
> [Free translation into English: '[m]ine exploitation does not only involve the extraction of minerals but also, to a certain extent, the processing of these minerals. Mining is always related to processing, ie the separation of usable ore and raw ore, and the separation of several co-existing minerals.']

(Rudolf Isay *Le droit minier des principaux États civilisés* (Éditions Marcel Giard 1930) at 17–18).

3 SPECIFIC PARAMETERS RELATING TO THE EXPLOITATION OF MINERAL AND PETROLEUM RESOURCES

3.20 The exploitation of mining and petroleum resources are long-term activities.[9] As such, they pose significant risks, require considerable investment as well as cutting-edge technologies and play a strategic role in national and international economies.[10]

The risk factor associated with mining and petroleum operations increases due to the accumulation of risks inherent in any investment activity, added to those specific to the exploitation of natural resources. The magnitude of the risks is assessed using highly sophisticated calculation methods.

Henceforth, a distinction should be made between the different risks to which mining and petroleum operators are exposed: natural risks, financial risks and political risks.

In addition, the current context, shaped by the instability of international commodity prices, reinforces the risks faced by private operators within the context of their extractive projects. This risk, which did not exist before the 1980s, when petroleum prices were not determined by the market, is of particular importance in the current economic context.

a Natural or geological risk

3.21 Significant natural or geological risks affect petroleum and mining operations.[11] The principal unknown for the operators remains the natural limitation of reserves, for which estimates are often uncertain, especially in the hydrocarbons sector.[12] The discovery of marketable amounts of natural resources, regardless of the preliminary studies carried out, is an unknown factor over which the operators have no control.

---

[9] The average lifetime of a petroleum deposit is 15 to 20 years, while that of a gas deposit is 20 to 30 years. The average lifetime of a gold deposit is 10 to 15 years. These are only averages, as the lifetime of some deposits can be much longer.

[10] See above para 1.3 et seq for information on the strategic and economic role of mineral and petroleum resources.

[11] The legislation refers to the risks faced by the operators: see article 31 of Law No. 96-669 of 29 August 1996 Enacting the Petroleum Code in Côte d'Ivoire:

> '*[l]'autorisation d'exploitation d'hydrocarbures confère à son titulaire le droit exclusif d'effectuer, à ses risques et dépens, dans les limites du périmètre qui en est l'objet, toutes les opérations d'exploitation d'un gisement commercialement exploitable, et le cas échéant, de recherche, ainsi que de disposer de tout ou partie de la production des hydrocarbures, conformément aux stipulations du contrat pétrolier qui lie le titulaire à l'État.*'
>
> [Free translation into English: '[the] hydrocarbon exploitation authorisation gives its holder the exclusive right to perform, at its own risk and expense, within the scope defined, all exploitation operations on a commercially exploitable deposit and, when relevant, exploration operations, as well as to use all or part of the production of hydrocarbons, in accordance with the stipulations of the petroleum agreement between the holder and the State'].

[12] This lack of reliability is all the more apparent in the exploitation of non-conventional hydrocarbons.

3.22 The risks are such that operators may only know the actual quantity contained in the deposit after the closure of the project.[13] Therefore, at best, proven reserves can only be confirmed within a margin of 10%.[14] The recovery rate, ie recoverable qualities in relation to existing quantities, varies from one deposit to another.

3.23 Once the presence of natural resources has been established, the operators are still faced with many natural risks. It is difficult for the operators to identify in advance the quality of the natural resources contained in the deposit. Some deposits may contain hydrocarbons that cannot be recovered, in particular because of their viscosity, or heavy oils for which refining costs will be significantly higher.[15] The same applies to mineral resources; the quality of which can vary dramatically from one deposit to another or even within the same deposit.

3.24 In addition, as resources are getting scarcer, the operators are faced with new challenges. Generally speaking, the deposits yet to be discovered tend to be smaller and located in less accessible areas, making exploitation operations more complex. In the hydrocarbons sector, the deep offshore, liquefied natural gas, non-conventional gas, extra-heavy oil and bituminous sand involve new challenges that tend to increase the technical complexity of the projects and the related risks.

3.25 However, the natural risk of non-discovery tends to decrease, which is partly due to technological advances. For hydrocarbons, the old techniques often made it impossible to maximise the performance of the deposits, as the different natural resources extraction phases[16] did not allow the extraction of all the hydrocarbons contained in a deposit. With technical progress, the

---

[13] Bret-Rouzaut and Favennec op cit 100.

[14] Bret-Rouzaut and Favennec op cit 102.

[15] As such, the states offer no guarantee as to the quantity and quality of the resources contained in their subsoil when mining rights are issued. See Article 23 paragraph 4 of Law No. 2006-18 of 17 October 2006 Enacting the Petroleum Code of the Republic of Benin:

'*[l]orsque le Gouvernement accorde un permis de recherche d'hydrocarbures ou permis H, il le fait aux risques du titulaire et ne garantit pas l'existence ou la qualité des hydrocarbures, pas plus que leur quantité.*'

[Free translation into English: 'when the government grants a hydrocarbon exploration permit, or H permit, it does so at the holder's own risk and does not guarantee the existence, quality or quantity of hydrocarbons.'].

[16] Three different phases should be distinguished:

(i) primary recovery. The petroleum spontaneously reaches the surface. Recovery is possible when the well is a 'blowing' well. This recovery helps bring 5% to 40% of the resources contained in the deposit to the surface;

(ii) secondary recovery (first part of enhanced recovery). This stage consists of increasing pressure in the well thanks to water injections and the re-injection of the gases obtained during the initial recovery. This deposit exploitation phase helps extract 25% to 35% of the petroleum from the well;

(iii) tertiary recovery (second part of enhanced recovery). This phase consists of reducing the viscosity of the remaining petroleum so that a greater amount can be brought to the surface (in particular via gas injection—this time into the liquid part of the deposit—or fracturing).

operators are now able to put back in operation former oil fields.[17] Furthermore, technical advances have allowed and will allow for the discovery of new deposits, notably on the African continent.

Finally, the 'gas risk'[18] is now alleviated by the development in technological capabilities, for example by liquefying the gas extracted from the deposits.

b Financial risk

3.26 A distinction should be made between the notions of risk and financing. Financial participation in a project sometimes does not involve the sharing of risks. An agreement may assign the exploration risk and corresponding expenses to a single party, while stipulating that, in the event of a discovery in commercial quantities, all parties will share the exploration expenses retroactively.

Private operators who wish to invest in mining and petroleum projects face a significant financial risk, in light of the considerable amounts invested and the difficulties relating to the funding of extractive operations.

3.27 Substantial investments are the common denominator of extractive industries. For example, annual investments in the oil and gas industry amount to around $400 billion on average. Besides, on a global scale, 8% of the revenue generated by petroleum sales are reinvested[19] . Considerable amounts are also invested in the mining sector: in 2008, global exploration expenses for commercial purposes amounted to $13,8 billion compared with just $2,6 billion in 2000.[20] The current trend in extractive industries, which consists of significantly reducing exploration expenses, should however be pointed out; in 2013, exploration expenses in the mining sector dropped by 30%, ie $10 billion dollars.[21]

3.28 The breakdown of investment by exploitation phase varies depending on the sector considered (hydrocarbons or solid minerals); as the former requires much higher average exploration budgets than the latter.

[17] The amount of reserves resulting from the improvement in recovery techniques is currently estimated at 300 billion barrels (Bret-Rouzaut and Favennec op cit 97).

[18] The notion of 'gas risk' refers to the difficulties previously encountered by operators, in particular in terms of storage, when operations revealed the existence of gases in the deposit.

[19] Bret-Rouzaut and Favennec op cit 126.

[20] 'Mineral resources and development in Africa' report from the International study group on African mining regimes (African Union and Economic Commission for Africa) 33.

[21] See Bloomberg 'Imminent Boom?' *Mining Weekly* 24–30 January 2014 at 18.

Technical costs evolution in the upstream petroleum sector between 2002 and 2008, in dollars per barrel[22]

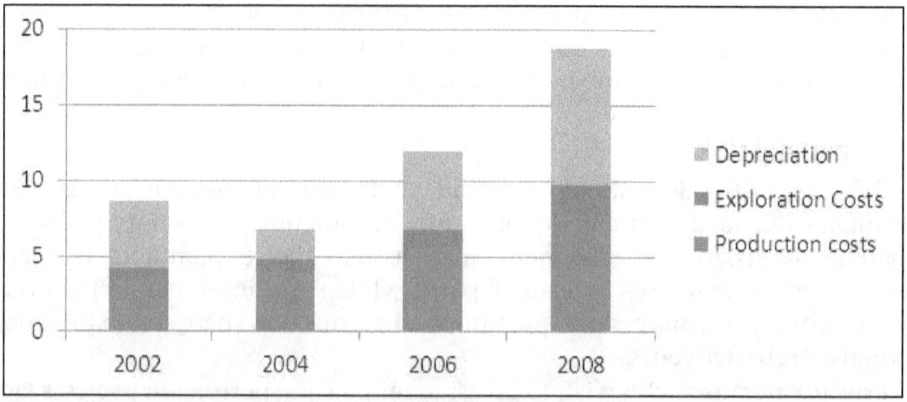

3.29 The requirements for substantial investments result in specific financing methods. More often, the need for financial resources is problematic because natural resources are often located in countries with limited financial means. This situation results in the internationalisation of contractual relations. Yet, in the context of these projects, the banks often only want to get involved once natural resources have been discovered in commercial quantities; up until then, the operators are forced to self-finance.

3.30 The management of this financial risk by mining and petroleum companies may result in the provision of mining rights as a guarantee, the granting of guarantees on future production or the pledging of the equipment needed for the exploitation of the natural resources. As part of mining and petroleum projects, guarantees are not given on immovable properties (non-extracted natural resources) but on movable properties (the resources extracted), which significantly affects the funding of the project. In addition, although it is possible to provide a guarantee over mining rights, this operation poses difficulties. The third-party beneficiary of the guarantee must find, should they decide to activate this guarantee, an operator with the technical capacity required to mine the deposit.

3.31 Institutional credits are another way to address financing requirements. There are two possible types of credit: refundable credits and non-refundable credits. Concerning non-refundable credits, cooperation agreements should be mentioned, where the investor's state of origin provides the said investor with cooperation credits to promote investment in certain countries.

3.32 However, operators are faced with increasing challenges when it comes to securing the funds required to successfully complete mining and petroleum projects on the African continent. The current context of falling

[22] Bret-Rouzaut and Favennec op cit 166.

commodity prices makes it increasingly harder to finance this type of project.[23]

c Political risk

3.33 As the amounts invested in mining and petroleum projects are most often considerable, their amortisation is a long-term process. As a result, the stability of the investment framework and the evaluation of the political risk are key determinants in the decision to invest.

3.34 The operators who decide to invest in the natural resources sector should expect the legislative and regulatory framework applicable on the day of their investment to evolve.[24] Moreover, the strategic nature of natural resources is likely to promote the evolution of the legal framework applicable to the petroleum and mining sectors, in particular in the event of a change of government. The example of Mali is a clear example of this evolution. Since its independence in 1960, Mali's mining legislation has been amended six times, in 1970,[25] 1981,[26] 1991,[27] 1999,[28] 2000[29] and 2012,[30] and this trend is not yet over as the latest code is already being revised.

3.35 In addition to the risk associated with changes of legislation, private operators also face the risk of nationalisation or expropriation of their investment.

3.36 Besides, it is common for countries in possession of natural resources to embark on renegotiations or reviews of mining agreements which can, in some cases, result in their unilateral termination.[31]

3.37 The trend involving the renegotiation of mining agreements has significantly evolved on the African continent over the past few years, notably in the Republic of Guinea (Conakry) or in the Democratic Republic of the Congo.

On 11 June 2007, the government of the Democratic Republic of the Congo initiated a process that would result in the modification of the mining agreements signed with private operators. This revision process, which was completed in 2009, led to the termination of numerous mining agreements and disputes brought before arbitral tribunals. For example, following the

[23] For information on this matter, see Ilan S 'Private equity's role in financing mine projects to be addressed' *Mining Weekly* 23–29 January 2015 at 60.

[24] See para 6.30 below on the evolution of the legislative and regulatory framework applicable to mining and petroleum projects.

[25] Ordinance No. 34/CMLN of 3 September 1970 Enacting the Mining Code in the Republic of Mali.

[26] Law No. 81-80/AN-RM of 13 July 1981 Amending the Mining Code in the Republic of Mali.

[27] Ordinance No. 91-065/P-CTSP of 19 September 1991 Enacting the Mining Code in the Republic of Mali.

[28] Ordinance No. 99-032/P-RM of 19 August 1999 Enacting the Mining Code in the Republic of Mali.

[29] Ordinance No. 00-013/P-RM of 10 February 2000 Amending the Mining Code in the Republic of Mali.

[30] Law No. 2012-015 of 27 February 2012 Enacting the Mining Code in the Republic of Mali.

[31] For information on this issue, see Thierry Lauriol and Emilie Raynaud 'La rupture unilatérale des contrats d'exploitation' in *Le contentieux extractif* (ICC 2015) 140.

withdrawal of its mining rights, Canadian mining company *International Quantum Resources Limited* instituted proceedings against the Democratic Republic of the Congo before the International Chamber of Commerce and the International Centre for Settlement of Investment Disputes.[32]

More recently, in Guinea Conakry, the renegotiation process initiated by the government affected mining companies. On 20 January 2012, the Government of the Republic of Guinea, led by President Alpha Condé, announced the systematic review of existing mining agreements, under the aegis of the Mining Rights and Agreements Technical Review Committee (CTRTCM).[33] This process, which resulted in the cancellation of several mining rights,[34] ended in April 2016.[35]

The reviewing of the agreements signed with private operators also led Mali's President, Ibrahim Boubacar Keita, to ask for mining agreements, rights and permits in the mining and petroleum sectors to be inventoried in 2013, pointing out that any agreements deemed contrary to the interests of Mali would be renegotiated with the partners concerned. In the mining sector, this revision has already given rise to the cancellation of many permits; 14 exploration permits and 10 quarry exploitation authorisations were withdrawn and a decision was made to put an end to 106 applications to obtain exploration permits.[36] The petroleum sector has also been affected, as attested by the cancellation of 10 concession agreements[37] and production sharing agreements.[38]

---

[32] *International Quantum Resources Limited, Frontier SPRL and Compagnie Minière de Sakania SPRL v Democratic Republic of Congo* (ICSID Case No. ARB/10/21).

[33] For further details on the CTRTCM's remit, see Ministerial Order No. 2745/CAB.MIN.MINES/01/2007 Establishing the Ministerial Commission in Charge of Reviewing Mining Agreements as well as the CTRTCM website.

[34] Several companies saw their mining rights cancelled, including VBG (a joint venture between BSG Resources and Vale) and IMD (International Mining Development). For information on this topic, see 'Guinée: Le Comité de revue clos en mars', *Africa Mining Intelligence* (1 March 2016) 2.

[35] This process ended with the adoption of decree No. D/2016/112/PRG/SGG of 19 April 2016.

[36] For information on this matter, see *Jeune Afrique* 'Le Mali fait le ménage dans son secteur minier' *Jeune Afrique* 1 August 2014 [online].

[37] The Government cancelled three concession agreements allocated to the Petro Plus Angola Ltd Company, relating to block 1A of the Taoudeni Basin (approved by decree No. 2011-560/P-RM of 1 September 2011), the Petro Plus Angola Ltd Company relating to block 1B of the Taoudeni Basin (approved by Decree No. 2011-549/P-RM of 1 September 2011) and the New Catalyst Capital Investments Company relating to block 4 of the Taoudeni Basin (approved by Decree No. 2013-695/P-RM of 2 September 2013).

[38] The Government cancelled seven production sharing agreements entered into with the Simba Energy Inc Company relating to Block 3 of the Taoudeni Basin (approved by Decree No. 2011-721/P-RM of 2 November 2011), the Moh Oil Company relating to Block 5 of the Taoudeni Basin (approved by Decree No. 08-507/P-RM of 10 September 2008), the Corvus Resources Managements Ltd Company, relating to Block 6 of the Taoudeni Basin (approved by Decree No. 2013-623/P-RM of 25 July 2013), the Mali Petroleum SA Company relating to Block 8 of the Taoudeni Basin (approved by Decree No. 05-318/P-RM of 12 July 2007), the Mali Petroleum SA Company relating to Block 10 of the Gao Graben (approved by Decree No. 05-318/P-RM of 12 July 2007), the Oranto Petroleum Ltd Company relating to Block 12

The events, which took place in Burkina Faso in October 2014 and precipitated the departure of President Blaise Compaoré, could also lead to the review of mining rights. In November 2014, the Minister of Mines and Energy, Colonel Boubacar Ba, announced his intention to reinforce transparency on the mining rights granted.[39] Following the elections, the National Assembly, by Resolution no. 019-2016/AN of 12 April 2016, set up a parliamentary investigation committee on mining rights and the corporate social responsibility of mining companies.

3.38 Encouraging transparency, ensuring compliance with the legal framework and protecting national interests are the main arguments put forward by licensing authorities as part of these renegotiation programmes,[40] which often result in the unilateral termination of an increasingly significant number of exploitation agreements by the host state, essentially in the mining sector.

3.39 These renegotiation processes involve many difficulties. Despite their desire to preserve their economic interests, the producing states cannot continually renegotiate exploitation agreements without creating genuine insecurity, at the risk of inhibiting investments in the extractive sector.

3.40 To address these political risks, private operators are not without means of protection. Besides the protections provided for in investment protection treaties, contractual instruments such as stabilisation or intangibility clauses inserted in agreements help private operators limit the political risks to which they are exposed.[41]

---

of the Nara trench (approved by Decree No. 08-270/P-RM of 12 May 2008) and the Afex Global Limited Company relating to Block 13 of the Nara trench (approved by Decree No. 07-154/P-RM of 10 May 2007).

[39] Recent events, marked by the arrest of the former Minister of Mines in April 2015, seem to confirm the new Government's resolve in this domain. For information on this topic, see *Africa Mining Intelligence* 'Burkina Faso: Kaboré arrêté pour plus de transparence?' *Africa Mining Intelligence* 21 April 2015 at 2.

[40] For information on this issue, see the terms of reference for the review of mining agreements in the Republic of Guinea.

[41] See below para 19.4 et seq on the contractual provisions offsetting political instability.

4

THE PERMANENT SOVEREIGNTY OF STATES

4.1 The second half of the 20th century was marked by the disruption in relations between foreign investors and host states, driven by the latter who intended to reaffirm their interests in managing their natural resources. The strategic importance of national resources often led the states to resort to nationalisation to reassert control over their resources and attempt to draw greater profits from them.

4.2 It is in the oil sector and over the South American continent that the movement towards affirmation by the states of their sovereignty over natural resources originated. From 1938 on, the Mexican government carried out numerous nationalisations of foreign oil companies. The wave of nationalisation then spread to Venezuela who from 1948 and driven by its Minister of Development, Mr Pérez Alphonso, required equal sharing of profits between the State and foreign oil companies. This requirement was to extend, until the King of Libya was overthrown in 1969 and the arrival of Colonel Gaddafi led to the abandonment of the principle of equal sharing of profits to the benefit of the new Libyan State, which granted itself 55% of the income generated from oil operations.[1]

4.3 The desire of states to ensure full and complete control over the management of their natural resources and maximise their returns led to the appearance of the first national oil companies, such as Aramco in Saudi Arabia, Sonatrach in Algeria or Pemex in Mexico.

4.4 This movement towards affirmation by the states of sovereignty over their natural resources is evidenced by the creation in 1960 of the Organisation of the Petroleum Exporting Countries (OPEC),[2] in response to the influence of international oil companies and the Seven Sisters[3] cartel in particular.

4.5 Parallel to this trend, the process of decolonisation largely provided for the claim of sovereignty of states over their natural resources. As it happens, apart from the political independence, which they had just acquired, newly independent territories wanted more than anything to obtain their economic independence. From that time on, the principle of permanent sovereignty of states over the natural resources appeared to be a means for developing states to achieve their new aspirations.

---

[1] Frédéric Teulon *La nouvelle économie mondiale* (Presses Universitaires de France 1993) 157.

[2] Venezuela, Saudi Arabia, Iraq, Iran and Kuwait were the first member states of OPEC in 1960.

[3] The term 'Seven Sisters' designated the seven multinational oil companies which dominated the global petroleum industry from the mid-1940s to the mid-1970s.

4.6 The United Nations were not able to ignore the waves of nationalisations, particularly inspired by the decolonisation movement. This is how United Nations, through several resolutions, established the principle of permanent sovereignty of states over their natural resources. This principle, having been taken up well beyond the framework of the United Nations, emerged as a principle of customary international law. This international principle has been transposed into domestic law by numerous African states who raised the principle of sovereignty over their natural resources at the constitutional level.

1 THE PRINCIPLE OF PERMANENT SOVEREIGNTY OVER NATURAL RESOURCES ESTABLISHED BY THE INTERNATIONAL COMMUNITY

4.7 The affirmation of the principle of permanent sovereignty of states over their natural resources was the fruit of ten years maturation process within the United Nations. This maturing period culminated in 1962 with Resolution 1803 (XVII) and the assertion of the principle of permanent sovereignty of the states over their natural resources. This principle was then imposed as a principle of customary international law and has been asserted well beyond the United Nations framework.

a The emergence of the principle of permanent sovereignty over natural resources

4.8 It was as part of the work dedicated to the drafting of international treaties relating to human rights that the United Nations first studied states' rights to permanent sovereignty over natural resources.[4]

Thus in 1952, at the eighth session of the Commission on Human Rights, Chile presented a draft resolution asserting the principle of permanent sovereignty of states over their natural resources. In the terms of this draft resolution, it is asserted that the right of peoples to self-determination also includes a right to permanent sovereignty over their wealth and their natural resources. The rights which other states may claim cannot under any circumstances justify that a people is deprived of its own means of subsistence.[5]

It was in 1952 that the General Assembly of the United Nations first recognised the principle of sovereignty of states over natural resources. In its Resolution 523 (VI) of 12 January 1952, the Assembly ruled as follows:

> 'under-developed countries have the right to determine freely the use of their natural resources [. . .] that they must use [. . .] in order to be in a better position to further the

[4] This work was conducted pursuant to Resolution 421 (V) of the General Assembly of 4 December 1950—*Draft International Covenant on Human Rights and measures of implementation: future work of the Commission on Human Rights.*

[5] See George Fischer 'La souveraineté sur les ressources naturelles' 1962 *AFDI* 516.

realization of their plans of economic development in accordance with their national interests, and to further the expansion of the world economy'.[6]

Through this resolution, the Assembly thus affirms the right of states to freely use the natural resources available to them.

4.9 In the same year and in a second resolution—Resolution 626 (VII) dated 21 December 1952—the General Assembly stated 'that the right of peoples freely to use and exploit their natural wealth and resources is inherent in their sovereignty and is in accordance with the Purposes and Principles of the Charter of the United Nations'.[7] The General Assembly also recommended that all member states should abstain from any action, direct or indirect, intended to prevent a state from exercising its sovereignty over its natural resources.

This resolution, which was considered a declaration of principle, does not however at this stage establish the sovereignty of states over their natural resources as a rule of international law.[8] This notion was then merely a doctrine with content that was still poorly defined. However, this doctrine does have a distinctive feature, in that it is not the result of the wish of one or more countries but of the international community through the United Nations.[9]

4.10 Following these Resolutions, the United Nations movement towards establishing the principle of sovereignty of states over their natural resources constantly strengthened and on 12 December 1958 led to the creation of a United Nations Commission on Permanent Sovereignty over Natural Resources. This Commission's task was to carry out a study of the principle of permanent sovereignty of states over their natural resources and present observations aiming to reinforce this right. Once this work was complete, on 3 August 1961, the Commission's report of the Economic and Social Council was sent to the General Assembly of the United Nations.

4.11 This long maturing period led, one year later, to Resolution 1803 (XVII) which establishes the right of permanent sovereignty over natural resources.

b The recognition of the principle of permanent sovereignty over natural resources within the United Nations

4.12 It was through Resolution 1803 (XVII) of 14 December 1962,[10] adopted by 87 votes in favour and two against with 12 abstentions, that the General Assembly of the United Nations confirmed the principle of permanent sovereignty of states over the natural resources.

---

[6] General Assembly Resolution 523 (VI) of 12 January 1952—Integrated economic development as—*Integrated economic development and Agreements commercial agreements.*

[7] General Assembly Resolution 1803 (XVII) of 14 December 1962—*Right to exploit freely natural wealth and resources.*

[8] See on this issue, Fischer op cit 516.

[9] Id.

[10] General Assembly Resolution 1803 (XVII) of 14 December 1962—*Permanent sovereignty over natural resources.*

The principle of sovereignty is asserted here as an inalienable and permanent right likely to promote the economic independence of states:

> 'the recognition of the inalienable right of all States freely to dispose of their natural wealth and resources in accordance with their national interests, and on respect for the economic independence of States'.

The General Assembly of the United Nations recognised the general competence of any state to manage its national resources as it sees fit.

4.13 The Resolution also expressed that the right to permanent sovereignty over natural resources must be exercised in the interest of the population of the state in question.[11]

Furthermore, the Resolution specified that the profits derived from the use of natural resources must be shared among the foreign investors and the host state, without impairing the sovereignty of the said states.

It should be noted that the Resolution grants wide discretion to the states regarding the prospecting of resources, making them available and importing foreign capital. Indeed, the principle of sovereignty over natural resources means that access to investments are left to the entire discretion of the state.

4.14 The acknowledged sovereignty enables the State to carry out nationalisations, compulsory purchases or requisitions, on condition that these measures are justified on the grounds of public utility and give rise to adequate compensation. Thus, the exercising by the state of its sovereignty over natural resources remains subject to the observance of international law, and particularly to the principle of prompt, adequate and effective compensation.

4.15 Even though the sovereignty of states over natural resources has been asserted unequivocally in Resolution 1803 (XVII), there could be some questions surrounding the scope of this principle. The resolutions of the General Assembly of the United Nations constitute declarations of principle and are not legally binding. However, some resolutions of the General Assembly of the United Nations can be considered to be 'legislative'. Resolution 1803 (XVII) is one of these rare resolutions.[12] Indeed, the right of sovereignty over natural resources has been affirmed as a customary international principle,[13] the nature of customary rule of permanent sovereignty over natural resources having been subsequently confirmed through large use of this notion on an international scale.

---

[11] The General Assembly stated: 'The right of peoples and nations to permanent sovereignty over their natural wealth and resources must be exercised in the interest of their national development and of the well-being of the people of the State concerned'. Thus, the obligation to make use of the natural resources in accordance with the national interest is incumbent upon the said state.

[12] Joe Verhoeven *Droit International Public* (Larcier 2000) 362.

[13] Patrick Daillier, Mathias Forteau & Alain Pellet *Droit International Public* 8th ed (LGDJ 2009) 260; Joe Verhoeven *Droit International Public* at 362; Jean Combacau & Serge Sur *Droit International Public* 8th ed (Montchrestien 2008) 61.

c Consolidation of the principle of sovereignty over natural resources and the emergence of a principle of customary international law

4.16 The principle of sovereignty, which has been asserted as being one of the main corollaries of the right of peoples to self-determination, has been confirmed by numerous international instruments.

4.17 Since 1966, it has been through the International Covenant on Economic, Social and Cultural Rights[14] that the principle of sovereignty of states over their natural resources has been asserted. This text provides that 'all peoples may, [. . .], freely dispose of their natural wealth and resources'[15] and reaffirms 'the inherent right of all peoples to enjoy and utilize fully and freely their natural wealth and resources'.[16] Two years later, this principle went beyond the United Nations framework and was affirmed within OPEC. The organisation reaffirmed the inalienable right of states to exercise permanent sovereignty over their natural resources with a view to pursuing their development.[17] The Stockholm Declaration of 1972[18] reiterates the principle of sovereignty over natural resources, following the United Nations Conference on the Environment, and the Rio Declaration on the Environment and Development.[19] In 1974, the principle of sovereignty over natural resources was once more asserted within the framework of the United Nations through the Charter of Economic Rights and Duties of States of 12 December 1974.[20] The Vienna Convention on Succession of States in respect of Treaties of 1978 also establishes this principle in its article 13, which provides for the following: '[n]othing in the present Convention shall affect the principles of international law affirming the permanent sovereignty of every people and every State over its natural wealth and resources.'

At the European level, the principle of sovereignty was established as part of the Energy Charter Treaty, which recognises national sovereignty and the

[14] Adopted and opened for signature, ratification and membership by the General Assembly by Resolution 2200 A (XXI) of 16 December 1966—*International Covenant on Economic, Social and Cultural Rights.*

[15] Article 1, paragraph 2 of the International Covenant on Economic, Social and Cultural Rights.

[16] Article 25 of the International Covenant on Economic, Social and Cultural Rights.

[17] OPEC Resolution XVI.90 of 24 June 1968—*Declaratory Statement of Petroleum Policy in Member Countries.*

[18] Principle 10 of the Stockholm Declaration: 'For the developing countries, stability of prices and adequate earnings for primary commodities and raw materials are essential to environmental management, since economic factors as well as ecological processes must be taken into account'.

[19] Principle 2 of the Rio Declaration on Environment and Development: 'States have, in accordance with the Charter of the United Nations and the principles of international law, the sovereign right to exploit their own resources pursuant to their own environmental and developmental policies, and the responsibility to ensure that activities within their jurisdiction or control do not cause damage to the environment of other States or of areas beyond the limits of national jurisdiction.'

[20] General Assembly Resolution 3281 (XXIX) of 12 December 1974—*Charter of Economic Rights and Duties of States.*

sovereign rights of contracting states over energy resources. Similarly, within the framework of the European Union, directive 94/22/EC of the Parliament and the Council of 30 May 1994 on the conditions for granting and exercising permissions to prospect, use and extract hydrocarbons provides that '[m]ember States have sovereignty and sovereign rights over hydrocarbon resources on their territories'.

4.18 With regard to the African continent, it is the African Charter of Human and Peoples Rights which in 1981 established the principle of sovereignty of states over their natural resources by affirming: '[a]ll peoples shall freely dispose of their wealth and natural resources. This right shall be exercised in the exclusive interest of the people. In no case shall a people be deprived of it.'[21]

4.19 This provision was interpreted by the African Commission on Human and Peoples Rights in the case *Social and Economic Rights Action Centre (SERAC) and the Centre for Economic and Social Rights (CSR) v Nigeria*.[22] The applicants argued that the Nigerian Government was directly involved in the exploitation of the oil through a state company, which was a major shareholder in a *consortium*. According to them, the exploitation of oil resources would be detrimental to the local population. In its response, the Commission was able to specify the meaning of article 21 of the Charter by stating that:

> '[T]he aftermath of colonial exploitation has left Africa's precious resources and people still vulnerable to foreign misappropriation. The drafters of the African Charter obviously wanted to remind African governments of the continent's painful legacy and restore cooperative economic development to its traditional place at the heart of African Society'.[23]

In its decision, the Commission asserts its position that governments have the duty to protect their citizens, which assumes positive action of the latter.[24] The commission considered that the Nigerian government failed in its obligation to protect people against the disturbance to the enjoyment of their rights; the State facilitated the destruction of the country in violation of article 21 of the Charter.[25] While the Charter does not have any horizontal effect, the Commission imposes positive actions on the states, particularly in order to guarantee the right of peoples to dispose of their wealth and natural resources.

---

[21] Article 21, paragraph 1 of the African Charter on Human and Peoples Rights of 1981.

[22] Communication no. 155-96 of 13 October 2001: *Social and Economic Rights Action Centre (SERAC) and Centre for Economic and Social Rights (CESR) v Nigeria*.

[23] *Social and Economic Rights Action Centre (SERAC) and Centre for Economic and Social Rights (CESR) v Nigeria* no 56.

[24] *Social and Economic Rights Action Centre (SERAC) and Centre for Economic and Social Rights (CESR) v Nigeria* no 57.

[25] *Social and Economic Rights Action Centre (SERAC) and Centre for Economic and Social Rights (CESR) v Nigeria* no 58.

It should be highlighted that the sovereignty of the state over its natural resources extends over all of its territory and particularly its continental shelf[26].

4.20 These various legal instruments demonstrate the extent to which the international community has ceaselessly asserted the existence of a principle of permanent sovereignty of states over their natural resources. Therefore, the broad acceptance of this principle has led to the birth of a customary international law principle that is based on both an objective element—general practice—and a subjective element—*opinio juris.*[27]

4.21 However, states often intend to emphasise the customary scope of the principle of permanent sovereignty over natural resources in order to exempt themselves from their commitments with regard to foreign investors.[28]

4.22 It appears that the question of the permanent ownership of states over natural resources has in reality only been raised in order to hide the true motivations of receiving states which were to increase their financial revenue without taking on the exploitation risks.

4.23 Nevertheless, the principle of sovereignty of states over their natural resources has extended to national rights, which, sometimes, have gone so far as to make this principle a right with constitutional value.

2 THE CONSTITUTIONAL ENSHRINEMENT OF THE PRINCIPLE OF PERMANENT SOVEREIGNTY OF STATES OVER NATURAL RESOURCES

4.24 As permanent sovereignty over natural resources is perceived as a vector of economic independence for producing states, the latter have ceaselessly reaffirmed the principle of permanent sovereignty over their natural resources by asserting its constitutional value.

4.25 This trend is especially sensitive on the African continent, where numerous countries have made the choice to insert provisions in their constitution relating to sovereignty and entire ownership of the state over the natural resources present on their territory.

4.26 However, some differences must be mentioned regarding constitutional references to the principle of sovereignty of states over their natural resources. While some African constitutions unequivocally assert the principle of sovereignty over natural resources, others are more qualified and simply provide the principle of ownership of natural resources.

---

[26] See to that effect, Article 77(1) of the United Nations Convention on the Law of the Sea, concluded at Montego Bay on 10 December 1982, which provides as follows: 'The coastal State exercises over the continental shelf sovereign rights for the purpose of exploring it and exploiting its natural resources'. Also see Article 81 of the same Convention, which provides that 'The coastal State shall have the exclusive right to authorise and regulate drilling on the continental shelf for all purposes'.

[27] Daillier, Forteau & Pellet op cit 260; Verhoeven op cit 362.

[28] Combacau and Sur op cit 445.

a African constitutions asserting the principle of sovereignty of states over their natural resources

4.27 Over the entire African continent, nine constitutions have been identified as explicitly enshrining the principle of the sovereignty of states over their natural resources.

Three out of these nine countries are Portuguese-speaking countries. The constitution of the Republic of Angola asserts the principle of the sovereignty of the Angolan State over its natural resources by stating that the state exercises its sovereignty in terms of the conservation, exploitation and use of natural resources, biological and non-biological, over the contiguous zone, the exclusive economic zone and the continental shelf, in accordance with the law and international law.[29] The Cape Verdean Constitution, for its part, provides that the State of Cape Verde holds sovereign rights in terms of conservation, exploitation and promotion of the natural resources, living or otherwise, in its contiguous zone, in the exclusive economic zone and on the continental shelf, and has jurisdiction over these regions by virtue of domestic law and the rules of international law.[30] The same applies for Guinea-Bissau, the constitution of which provides that the Republic exercises a right of sovereignty over all natural resources that exist on its territory.[31]

Furthermore, four French-speaking African countries grant a constitutional value to the principle of sovereignty over natural resources; among these is the Republic of the Congo, the constitution of which provides as follows in its preamble:

> '[w]e, the Congolese people, [. . .] do solemnly reaffirm our permanent right to inalienable sovereignty over all our wealth and our natural resources is a fundamental element of our development'.[32]

---

[29] Article 3, paragraph 3 of the Constitution of the Republic of Angola of 2010:

'*[o] Estado exerce jurisdição e direitos de soberania em matéria de conservação, exploração e aproveitamento dos recursos naturais, biológicos e não biológicos, na zona contígua, na zona económica exclusiva e na plataforma continental, nos termos da lei e do direito internacional.*'

[Free translation into English: 'The State exercises its jurisdiction and its rights of sovereignty in terms of conservation, exploration and exploitation of natural resources, biological or nonbiological, in its contiguous zone, in its exclusive economic zone and on its continental shelf, observing domestic and international law'].

[30] Article 6 (2) of the Constitution of the Republic of Cape Verde of 2010:

'*[n]a sua zona contígua, na zona económica exclusiva e plataforma continental, definidas na lei, o Estado de Cabo Verde possui direitos de soberania em matéria de conservação, exploração e aproveitamento dos recursos naturais, vivos ou não vivos, e exerce jurisdição nos termos do direito interno e das normas do Direito Internacional.*'

[Free translation into English: 'In its contiguous zone, in its exclusive economic zone and on its continental shelf, as defined by law, the State of Cape Verde possesses sovereign rights in terms of conservation, exploration and exploitation of natural resources, living or otherwise, and shall exercise its jurisdiction, observing domestic and international law'].

[31] Article 9 of the Constitution of the Republic of Guinea-Bissau of 1996.

[32] Constitution of the Democratic Republic of the Congo of 2002.

The constitution of the Democratic Republic of the Congo provides that 'the State exercises permanent sovereignty particularly on the soil, subsoil, waters and forests, over the Congolese air, river, lake side and maritime spaces as well as over the Congolese territorial sea and on the continental shelf'.[33] The Chad Constitution provides, for its part, that 'the State exercises its full and permanent sovereignty over all the national wealth and natural resources for the well-being of the entire national community'.[34] In its constitution of 25 November 2010, Niger states that 'the State exercises its sovereignty over the natural resources and the subsoil'[35] in accordance with the principle according to which the natural resources and the subsoil are the property of the people of Niger.[36]

Tunisia is the sole North African country granting a constitutional value to the principle of the sovereignty of the state over natural resources. The Tunisian Constitution provides that 'the natural resources belong to the Tunisian people. The state exercises its sovereignty over these resources on behalf of the people'.[37] Finally, the Constitution of Equatorial Guinea states that on its territory, the State exercises its sovereignty in full and can exclusively explore and exploit all the mineral and hydrocarbon resources and wealth.

b African Constitutions restricted to asserting the ownership of the State over natural resources

4.28 Some constitutions do not go as far as enshrining the principle of sovereignty of states over their natural resources and limit themselves to using the term ownership to designate the allocations of the State over its natural resources. Fifteen African countries recognise the full and entire ownership of the State over the natural resources present on its national territory.

In this regard, the constitution of Algeria extends the notion of public ownership to the subsoil, mines and quarries, natural energy resources, mineral, natural and living wealth of the various zones of the national maritime area, waters and forests.[38] Similarly, the constitution of Mozambique

---

33 Article 9 of the Constitution of the Democratic Republic of Congo of 2006.

34 Article 57 of the Constitution of the Republic of Chad of 1996.

35 Article 149 of the Constitution of the Republic of Niger of 2010.

36 Article 148 of the Constitution of the Republic of Niger of 2010.

37 Article 13 of the Constitution of the Tunisian Republic of 2014.

38 Article 17 of the Constitution of the Republic of Algeria of 1996:

> *'[l]a propriété publique est un bien de la collectivité nationale. Elle comprend le sous-sol, les mines et les carrières, les sources naturelles d'énergie, les richesses minérales, naturelles et vivantes des différentes zones du domaine maritime national, les eaux et les forêts.'*
>
> [Free translation into English: 'Public property is an asset of national collectivity. It includes the subsoil, mines and quarries, natural energy resources, mineral, natural and living wealth of the various fare of the national maritime area, waters and forests'].

states that the natural resources are the property of the State of Mozambique and provides that the public domain include mineral and energy resources.[39]

4.29 Other constitutions, such as the constitution of Burkina Faso,[40] the Arab Republic of Egypt[41] and that of Ethiopia[42] provide that natural resources belong to their people.

4.30 Some texts provide for state ownership over natural resources, particularly the constitutions of Eritrea,[43] Ghana,[44] Namibia,[45] Tanzania,[46] and Somalia.[47]

---

[39] Article 98 of the Constitution of the Republic of Mozambique of 2004:

'[o]s recursos naturais situados no solo e no subsolo, nas águas interiores, no mar territorial, na plataforma continental e na zona económica exclusiva s ão propriedade do Estado. 2. Constituem domínio público do Estado: [. . .] f) o potencial energético; [. . .] h) as jazidas minerais.'

[Free translation into English: 'The mineral resources located in the soil, subsoil, internal waters, territorial sea, continental shelf, and exclusive economic zone are the property of the State. 2. The public domain of the State constitutes [. . .] f) the energy potential [. . .]; h) mining deposits'].

[40] Article 14 of the Constitution of Burkina Faso of 1991:

'[l]es richesses et les ressources naturelles appartiennent au peuple. Elles sont utilisées pour l'amélioration de ses conditions de vie.'

[Free translation into English: 'The wealth and natural resources belong to the people. They are used to improve their living conditions'].

[41] Article 32 of the Constitution of the Arab Republic of Egypt of 2014:

'The State's natural resources belong to the people. The State shall preserve and effectively exploit them, may not deplete them, and shall observe the rights of future generations to them. The State shall make the best use of renewable energy sources, motivate investment therein, and encourage relevant scientific research. The State shall encourage the manufacture of materials and increase their added value as per economic feasibility.'

[42] Article 40(3) of the Constitution of the Federal Democratic Republic of Ethiopia of 1994:

'[t]he right to ownership of rural and urban land, as well of all natural resources, is exclusively vested in the State and in the peoples of Ethiopia. Land is a common property of the Nations, Nationalities and Peoples of Ethiopia and shall not be subject to sale or to other means of exchange.'

[43] Article 23(2) of the Constitution of Eritrea of 1997:

'[a]ll land and all natural resources below and above the surface of the territory of Eritrea belongs to the State. The interests citizens shall have in land shall be determined by law.'

[44] Article 257(6) of the Constitution of the Republic of Ghana of 1992:

'[e]very mineral in its natural state in, under or upon any land in Ghana, rivers, streams, water course throughout Ghana, the exclusive economic zone and any area covered by the territorial sea or continental shelf is the property of the Republic of Ghana and shall be vested in the President on behalf of, and in trust for the people of Ghana.'

[45] Article 100 of the Constitution of the Republic of Namibia of 1990 (amended):

'[l]and, water and natural resources below and above the surface of the land and in the continental shelf and within the territorial waters and the exclusive economic zone of Namibia shall belong to the State if they are not otherwise lawfully owned.'

[46] Article 27(1) of the Constitution of the Republic of Tanzania, adopted in 1977:

'[e]very person has the duty to protect the natural resources of the United Republic, the property of the state authority, all property collectively owned by the people, and also to respect another person's property.'

[47] Article 44 of the Constitution of the Federal Republic of Somalia of 2012:

The constitutions of the Republic of Nigeria[48] and of Uganda[49] provide for the ownership of the Government to the detriment of the State itself. Similarly, Swaziland also allocates ownership of natural resources to a local sovereign, in this case the king.[50]

The Constitution of Liberia, for its part, establishes the ownership of the state over natural resources in a roundabout way by affirming that private ownership cannot extend to natural resources, which are the property of the Republic.[51]

4.31 In respect of this analysis, it appears that African constitutions are far from being unanimous when it comes to granting a constitutional value to the principle of sovereignty of states over natural resources. However, numerous constitutions refer explicitly to the principle of state ownership over their natural resources.

4.32 Another trend should be mentioned: this tendency to insert provisions in African constitutions relating to the management of natural resources by the state. Numerous constitutions echo the principle asserted by Resolution 1803 (XVII) of 14 December 1962, according to which the right of

---

'*[t]he allocation of the natural resources of the Federal Republic of Somalia shall be negotiated by, and agreed upon, by the Federal Government and the Federal Member States in accordance with the Constitution.*'

48 Article 45(3) of the Constitution of the Republic of Nigeria of 1999 (amended):

'[n]otwithstanding the foregoing provisions of this section, the entire property in and control of all minerals, mineral oils and natural gas in under or upon any land in Nigeria or in, under or upon the territorial waters and the Exclusive Economic Zone of Nigeria shall vest in the Government of the Federation and shall be managed in such manner as may be prescribed by the National Assembly.'

49 Article 244 of the Constitution of the Republic of Uganda of 1995 (amended):

'[s]ubject to article 23 of this Constitution, the entire property in, and the control of, all minerals and petroleum in, on or under, any land or waters in Uganda are vested in the Government on behalf of the Republic of Uganda.'

50 Article 213 of the Constitution of the Kingdom of Swaziland of 2005:

'[a]ll minerals and mineral oils in, under or upon any land in Swaziland shall, after the commencement of this Constitution, continue to vest in iNgwenyama [title of the male ruler, ie king] in trust for the Swazi Nation as vested on the 12th April 1973.'

51 Article 22*(b)* of the Constitution of the Republic of Liberia of 1986:

'[p]rivate property rights, however, shall not extend to any mineral resources on or beneath any land or to any lands under the seas and waterways of the Republic. All mineral resources in and under the seas and other waterways shall belong to the Republic and be used by and for the entire Republic.'

See also Article 339 of the Constitution of Zambia of 1991:

'[t]he management and development of Zambia's natural resources shall—*(f)* not bestow private ownership of any natural resource or authorise its use in perpetuity.'

permanent sovereignty of states over their wealth and natural resources should be exercised in the interest of national development and of the well-being of the people.[52]

[52] The following constitutions may be mentioned as examples: The Constitution of Burkina Faso of 1991 (Article 14):

'*[l]es richesses et les ressources naturelles appartiennent au peuple. Elles sont utilisées pour l'amélioration de ses conditions de vie.*'

[Free translation into English: 'The wealth and natural resources belong to the people. They are used to improve their living conditions.']

The Egyptian Constitution, adopted on 15 January 2014 (Article 32):

'[t]he State's natural resources belong to the people. The State shall preserve and effectively exploit them, may not deplete them, and shall observe the rights of future generations to them. The State shall make the best use of renewable energy sources, motivate investment therein, and encourage relevant scientific research'

The Constitution of the Republic of Niger of 2010 (Article 149):

'*[l]'État exerce sa souveraineté sur les ressources naturelles et du sous-sol. L'exploitation et la gestion des ressources naturelles et du sous sol doit se faire dans la transparence et prendre en compte la protection de l'environnement, du patrimoine culturel ainsi que la préservation des intérêts des générations présentes et futures.*'

[Free translation into English: 'the State exercises its sovereignty over natural resources and the subsoil. The use and management of natural resources and the subsoil must be done transparently and take into account protection of the environment, and cultural heritage as well as the preservation of the interests of present and future generations.']

The Constitution of the Kingdom of Swaziland of 2005 (Article 210):

'(1) [s]ubject to the provisions of the Constitution or any other law, land, minerals and water are national resources. (2) In the interests of the present and future generations, the State shall protect and make rational use of its land, mineral and water resources as well as its fauna and flora, and shall take appropriate measures to conserve and improve the environment'.

The Constitution of the Republic of Chad of 1996 (article 57):

'*[l]'État exerce sa souveraineté entière et permanente sur toutes les richesses et ressources naturelles nationales pour le bien-être de toute la communauté nationale.*'

[Free translation into English: '[t]he State exercises its full and permanent sovereignty over all the national wealth and natural resources for the well-being of the entire national community.']

5

Ownership of Natural Resources

5.1 Beyond the question of the permanent sovereignty of states over their natural resources, an essential issue remains regarding the determination of the ownership over natural resources.

Indeed, the same natural resource, whether it is extracted or not, constitutes two different properties, the ownership of which may or may not be transferable. The natural resources contained in the subsoil follow a separate legal regime from extracted resources. A distinction must, therefore, be made regarding the nature of these properties, namely non-extracted natural resources and extracted resources.

5.2 A natural resource that is still contained in the subsoil is either unknown or unrecognised in terms of its quantity and its quality. It cannot be processed or marketed. Its legal situation is comparable to that of a property immovable by use, the owner being the state on the territory on which it is located but would have the particular feature of being physically attached to the material support establishing the sovereignty of the latter. It is therefore impossible to avoid it being under State control; it is struck by temporary inalienability.

5.3 However, an extracted natural resource is detached from this material support; it becomes movable property which can be sold as any other movable property. In the same way that a state is free to dispose of its own cultural heritage or its industrial or agricultural products, it must also be able to do so with regard to the natural resources extracted from its subsoil. A national human potential to create wealth is not different from a natural potential. The sale of a lorry produced in a public-sector factory or of local agricultural produce should not be considered different to the sale of the national production of natural resources.

5.4 This distinction is taken up by African oil and mining legislation, as for example the Madagascan Petroleum Code, which provides that deposits of hydrocarbons are items of immovable property while the materials extracted from these deposits are movable property.[1]

The differentiation between immovable property and movable property has significant consequences regarding the financing of mining and oil activities. Guarantees will not be taken on immovable property (non-extracted natural resources) but on movable property (extracted natural resources).

5.5 The property right which is granted in mining rights can only be effective based on the dual condition that there is proof of the existence of a natural resource and of its extraction. Mining rights do not confer a perpetual

[1] See Articles 7 and 8 of Law No. 96-018 of 4 September 1996 Enacting the Petroleum Code.

right of ownership over a natural resource that is not yet extracted, but only an ownership right over the part of a deposit which is extracted. The holder of the mining rights may therefore extract the natural resources during the period of validity of the previously mentioned rights. Mining rights are no more and no less than administrative permission entitling the holder to access the natural resources contained in the subsoil.

A specific personal right must not be seen here, nor even the division of an identical thing between two owners, but only a change in nature of that thing.

5.6 However, a distinction must be made regarding the system that applies to quarries. Thus, some legislation provides that deposits subject to the quarry scheme follow the conditions for ownership of the land.[2] Any natural person or legal entity may prospect them, explore them and exploit them provided that they are the owner of the land where these deposits are located or even if the said person has received permission from the owner through an excavating rights contract.[3]

5.7 Therefore, with regard to the ownership of natural resources, various systems should be singled out:

- The occupation system (this system has practically disappeared and is not found in Africa);[4]
- The sovereign system;[5]
- The state land system;[6]

---

[2] See, in particular, to that effect, Article 4 of Law No. 2012-0 15 of 27 February 2012 Enacting the Mining Code of Mali: '*[l]es substances minérales soumises au régime des mines dans le territoire de la République du Mali appartiennent à l'État. Toutefois, les titulaires des titres miniers d'exploitation acquièrent la propriété Ces substances minérales qu'ils extraient. Les droits à ces substances constituent une propriété distincte de celle de la surface. Le régime des mines se différencie de celui des carrières.*' [Free translation into English: 'Mineral substances subject to the mining regime in the territory of the Republic of Mali belong to the State. However, the holders of mining operating rights shall acquire ownership of these mineral substances which they extract. The rights to the substances constitute a separate ownership from that of the surface. The mine scheme is different from the quarry scheme.'].

[3] A contract by which the owner of the land decides not to exploit the deposit which is on their land and grants the right to exploit it to a third party.

[4] The mine belongs to the first occupant of the land or to the first applicant for the occupancy title. This system means that everyone has the right to undertake exploration works sometimes after various formalities have been completed; if these works result in the discovery of a deposit, the rights to start production is granted, subject to proof of discovery. The host state does not have the right to choose an operator for the production phase other than the inventor themselves, and this contrary to the sovereign system. This system was, for example, sometimes applied during the colonisation of the American continent.

[5] It is the State which allocates the mining rights to companies of its choice which are bound to observe the conditions laid down by the law without discrimination; this is the case of France.

[6] Exploitation of the mines is subject to agreements made between the host state and the company it chooses. Contractual aspects prevail over statutory aspects. This system is dominant in the Near East. The development of deposits may be carried out either by whoever exercises mining rights, or by a third party especially designated by it.

- The accession system;[7]
- The system derived from Muslim law.

It is specified that several systems can be superimposed in the same state. In the same way that several systems can have the authority to apply, these may also evolve[8] in order to take account of local realities.

1 SOVEREIGN SYSTEMS

5.8 A distinction should be made between the standard sovereign system on the one hand and the sovereign-inspired system on the other hand.

a The standard sovereign system: *'res nullius'*

5.9 The mining resources are considered ownerless property. The State does not proclaim itself the owner but assumes the right to establish rules for allocating it and using it through statutes.

5.10 The local state grants mining rights at its own discretion which are subject to the legislation enacted by the State. The regulatory aspect prevails and does not truly lead to the signature of the contract between the local state and the private operator, but rather more to the adhesion by the latter to general principles established by the state.

5.11 However, the mining titles allocated by the state cannot be an actual right of ownership; they are all limited in time or limited to one phase of exploitation. This breakdown leads to the creation of prospecting, exploration, development, production and transport rights, all separate from each other.

b Sovereign-inspired systems

5.12 In addition to the standard sovereign system, the state can assert, to its benefit, a right of ownership over the natural resources contained in its subsoil; the latter are then no longer ownerless property.[9] It can thus fix the terms and conditions for exercising mining rights by drawing up a specific standard contract.

5.13 Several pieces of African mining and oil legislation apply the sovereign-inspired system, enshrining the principle of the ownership of the people or the nation over natural resources.

---

[7] This system specifically implies that the ownership of the land carries with it ownership of what is on the surface and what is below; this is particularly the case of the United States in respect of so-called 'private' land.

[8] The issue of the system of ownership of natural resources is currently being debated as part of the exploitation of shale gas. Therefore, according to the *Institut Français du Pétrole*, one of the reasons for the development of shale gas operations in the United States came from the fact that the ownership rights of individuals extend to the subsoil and gives entitlement to substantial fees causing landowners to agree to the exploitation of shale gas resources. See, to that effect, Max Falque 'Gaz de Schiste: le problème de fond est la propriété des droits sur le sous-sol' *Institut Turgot* [on line].

[9] It is the case of French Mining Law.

The people's ownership over natural resources is particularly asserted in South Africa, where legislation provides that mining and oil resources are the joint heritage of all South African citizens, and that the State is the custodian of this.[10]

Some national laws allocate ownership of natural resources to the nation. Thus, Lesotho's mining code also provides that the nation has ownership over natural resources.[11] Similarly, in French-speaking Africa, the petroleum code which applies in the Republic of Benin states: '[d]eposits of liquid and gaseous hydrocarbons are separate from ownership of the land. They belong to the Nation and constitute a special public domain, the management of which is governed by this law'.[12]

5.14 Finally, some national legislation can assert the existence of the joint ownership of the state and the people over natural resources. This is the case for Ethiopia, whose mining code provides that the mining resources existing in their natural condition on Ethiopian territory or in its subsoil are the property of the government and all Ethiopians.[13]

5.15 In order to carry out mining or petroleum operations, it is therefore necessary to obtain a right of occupation of the land either by purchasing it, or by agreement with the owner or the authorities. This is divided into a joint mining right and into measures which apply in addition to various natural resources. It concerns all natural resources and it makes a distinction between quarries and mines.

The allocation of mining rights is set up in principle. Generally, exploration works can be carried out with the agreement of the landowner. In the event that commercial quantities are discovered, the bearer of an exploration permit will also be entitled to a so-called production licence. In both of these cases, a special contract is signed with the state called 'technical specification'; this is mainly geared around the protection of the environment and health and safety. Other states, as part of investment codes, have used 'establishment agreements'. When the operating project displays a major interest for the host state, the latter grants various guarantees and advantages in exchange for specific undertakings by the operators. Establishing agreements are found essentially in French-speaking sub-Saharan Africa.

[10] See section 3(1) of the Mineral and Petroleum Resources Development Act 28 of 2002: '[m]ineral and petroleum resources are the common heritage of all the people of South Africa and the State is the custodian thereof for the benefit of all South Africans.'

[11] See Article 3 of the Mines and Minerals Act, 2005: '[s]ubject to Section 1407 of the Constitution of Lesotho and without prejudice of any rights conferred by this Act, all rights of ownership in minerals are vested in the Basotho Nation.'

[12] See Article 3 of Law No. 2006-18 of 17 October 2016 On the Petroleum Code in the Republic of Benin.

[13] See Article 5(1) of Proclamation No. 678/2010 Promoting Sustainable Development of Mineral Resources: '[m]ineral resources existing in their natural condition on, in, and under the territory of Ethiopia are the property of the Government and all the peoples of Ethiopia'.

2 STATE LAND SYSTEM

5.16 The State, that is to say the local government or the local sovereign, is deemed to be the owner of the natural resources of its subsoil, which constitute an integral part of its domain.

The difference which exists with the sovereign system must be clearly defined. In the latter, it is the people who are the owners of the natural resources of the subsoil; the government only has the power to enforce the regulations enacted by the representatives of the people meeting in an assembly. On the other hand, in the state land system, the people do not own the natural resources contained in the subsoil.

The state land system is without doubt the system which is most widely applied on the African continent.[14]

5.17 In French-speaking Africa, the ownership of the State over natural resources contained in the subsoil is thus asserted in Burkina Faso where the mining code reiterates that: 'The natural deposits of mineral substances contained in the soil and subsoil of Burkina Faso are, ipso jure, the property of the State.'[15]

---

[14] See to that effect Evaristus Oshionebo 'Fiscal regimes for natural resource extraction: implications for Africa's development' in FN Botchway (ed) *Natural Resource Investment and Africa's Development* (Edward Elgar Publishing 2011) 200.

[15] See article 5 of Law No. 031-2003 Enacting the Mining Code. For other examples where the state land system is adhered to in French-speaking Africa, see, in particular, in Burundi, Article 7 of Law No. 1-21 of 15 October 2013 Enacting the Mining Code:

> '*[l]es gisements des substances minérales ou fossiles soumis au régime légal des Carrières ou des Mines constituent la propriété exclusive de l'État. Cette propriété relève de la souveraineté permanente de l'État sur ses ressources naturelles et, est inaliénable, imprescriptible et distincte de la propriété du sol.*'
>
> [Free translation into English: '[d]eposits of mineral substances or fossil subject to the quarries or mines legal system are the exclusive property of the State. This ownership comes under the permanent sovereignty of the State over its natural resources and is inalienable, not subject to the statute of limitations and separate from ownership of the land.']

In the Côte d'Ivoire, Article 3 of Law No. 2014-138 of 24 March 2014 Enacting the Mining Code:

> '*[t]outes les substances minérales, toutes les eaux minérales et tous les gites géothermiques contenus dans le sol et le sous-sol, les eaux territoriales, la zone économique exclusive et sur le plateau continental ainsi que son extension au-delà de deux cents miles marins jusqu'aux limites conventionnelles internationalement reconnues de la Côte d'Ivoire, sont propriétés de l'État de Côte d'Ivoire.*'
>
> [Free translation into English: '[a]ll mineral substances, all mineral waters and all geothermal deposits contained in the soil and subsoil, the territorial watercourses, the exclusive economic zone and over the continental shelf as well as its extension beyond 200 nautical miles up to the conventional boundaries internationally recognised of the Côte d'Ivoire, are the property of the State of the Côte d'Ivoire']

In Djibouti, Article 3 of Law No. 117-AN-05 of 16 October 2005 Enacting the Petroleum Code:

> '*[t]ous les gisements ou accumulations naturelles d'hydrocarbures existant dans le sous-sol de la République de Djibouti sont la propriété de l'État.*'
>
> [Free translation into English: '[a]ll deposits or natural accumulations of hydrocarbons existing in the subsoil of the Republic of Djibouti are the property of the State'.]

5.18 In Portuguese-speaking Africa, state ownership over natural resources is also applied in Angola, where the mining law states that the mining resources present over all of its territory are the property of the state and are part of its private domain.[16]

5.19 Examples of application of the State land system are also numerous in English-speaking Africa. Thus, Nigeria,[17] Sudan,[18] Tanzania[19] and Ghana[20] should be mentioned as countries enshrining the principle of the property of the State over natural resources.

5.20 There might be a certain confusion between the wish of the government in place when signing the contract and that of the local State represented by all the interests of the national community. It is thus possible to observe a tendency to mix the applicable ownership regimes; exploitation will thus be governed by contractual and regulatory provisions.

Or in Guinea, Article 4 of Law No. L/2014/no.0034/N of 23 December 2014 Enacting the Petroleum Code:

'*[l]es gisements ou accumulations naturelles d'Hydrocarbures existant dans le sous sol du Territoire National sont la propriété de l'État et ne sont susceptibles d'aucune forme d'appropriation privée*.'

[Free translation into English: '[t]he deposits or natural accumulations of hydrocarbons lying in the subsoil of the National Territory are the property of the State and are not subject to any kind of private appropriation'].

16 Article 42 of the Mining Law No. 31/11 of 23 September 2011:

'*[o]s recursos minerais existentes no solo, no subsolo, no mar territorial, na plataforma continental, na zona económica exclusiva e nas demais áreas do domínio territorial ou marítimo sob jurisdiç ão da República de Angola, s ão propriedade originária do Estado e fazem parte do seu domínio privado*.'

[Free translation into English: '[t]he mineral resources lying in the soil, the subsoil, in the territorial sea, the continental shelf, the exclusive economic zone and in the other zones of the land or maritime area under the jurisdiction of the Republic of Angola are the property of the State and are part of its private domain'].

17 See Article 1(1) of the Mines and Minerals Act, 2007: '[t]he entire property in and control of all Mineral Resources in, under or upon any land in Nigeria, its contiguous continental shelf and all rivers, streams and water courses throughout Nigeria, any area covered by its territorial waters or constituency and the Exclusive Economic Zone is and shall be vested in the Government of the Federation for and on behalf of the people of Nigeria.', and Article 1(1) of the Petroleum Act: '[t]he entire ownership and control of all petroleum in, under or upon any lands to which this section applies shall be vested in the State.'

18 See Article 4 of the Mineral Resources and Mining Act, 2007: '[a]ll the mining materials on the surface or underneath the earth; in the regional waters or in its continental shelf, shall be the property of the State, as represented by the National Government, and the latter shall have the absolute right to search and explore all mining materials and dispose of them.'

19 See Article 5 of the Mining Act, 2010: '[s]ubject to the provisions of this Act the entire property and control over minerals on, in or under the land to which this Act applies is vested in the United Republic.'

20 See Article 1 of the Minerals and Mining Act, 2006: '[e]very mineral in its natural state in, under or upon land in Ghana, rivers, streams, water-courses throughout the country, the exclusive economic zone and an area covered by the territorial sea or continental shelf is the property of the Republic and is vested in the President in trust for the people of Ghana.'

3 THE ACCESSION SYSTEM

5.21 The accession system provides that the land owner is also the owner of the subsoil and therefore the natural resources it contains; their exploitation is carried out contractually through the signing of private-law contracts. Established by Roman law,[21] the accession system finds its justification in the principle of the precedence of property right over the law.

5.22 The separation of the right of use from ownership of the land is operated by a division in space of the ownership of the land into two or more independent rights. In other terms, the principle is that the ownership extends up to the centre of the earth and that it can be broken up horizontally into two or more independent parts. This separation can take place through the contractual transfer of the subsoil, or by alienation or expropriation of the surface, or through alienation of the soil subject to the natural resources concerned; acquisitive prescription has the same effect as an alienation.

5.23 The accession system has a major disadvantage in that the surfaces are divided up, making operations diverse and therefore costly, as well as limiting intervention by the local state, which has led it to re-establish regulations safeguarding the national public interest.[22]

5.24 This system was applied in Great Britain up to 1934, except for gold and silver minerals deemed to belong to the sovereign;[23] it stipulated that the right of use could in principle only be detached from ownership of the land by contract. There was neither allocation of the right of use nor investigations against the wish of the land owner.

5.25 This system constituted a very specific framework; its decline in terms of territorial application has been able to be observed over numerous decades and is not preferentially applied in English-speaking African countries[24] which favour the application of other systems, particularly the sovereign-inspired system and the State land system.

4 MUSLIM LAW

5.26 Petroleum history has been strongly marked by the countries of the Middle East (Iran, Iraq, Saudi Arabia, Kuwait, and the Emirates of the Persian Gulf) and North Africa (Libya, Algeria and Egypt). In this regard, it is helpful to

[21] See Peter Fischer 'Historic Aspects of International Concession Agreement' *Grotian Soc Pap* (1972) 222.

[22] This is the context which has given rise to 'unitisation' or 'proration' provisions (see John Vafai 'Production Control in the Petroleum Industry: A Critical Analysis'1971 *Santa Clara Lawyer* 189 at 195 et seq)

[23] Great Britain has adopted an accession right system long ago, up to the Petroleum Act of 1934. This position had a great influence in the countries of the Commonwealth. See Kenneth S Carlston 'International Role of Concession Agreements, (1957) 52 *NWULR* 618. Spain, under King Juan I, also in an ephemeral manner had the accession system and applied it to Mexico by granting one third of the mining rights to the landowner, see John Perry Bullington 'The Land and Petroleum Laws of Mexico' (1928) 22 *AJIL* 50.

[24] Some scholars assert to the contrary that the accession scheme is designed to apply in the English-speaking Africa. See for this purpose Serge Bandoki *Le droit minier et pétrolier en Afrique* (Edilivre 2008) 12 et seq.

present the main legal rules derived from Muslim law which are applied in these states.

5.27 Muslim law does not have a general theory of contracts comparable to that of the laws of Western countries[25] ; nevertheless, the material sources of Muslim law all agree on single solutions in terms of honouring commitments undertaken and the use of the wealth of the subsoil; they are all based on the principle of *pacta sunt servanda.*[26]

5.28 The sources of Muslim law[27] are: the *Koran*[28] and the *Sunna*[29] but also the *ijma*[30] and *qiyâs.*[31] So-called spontaneous sources can also be added, which are customs *(orf)* and case law *(amal)*.[32] Mention should also be made of the *hadiths* (or oral communications of the Prophet Mohammed which constitute a compendium listing all the traditions relative to the acts and words of the Prophet and his companions)[33] and the *ijtihad* made up of works of doctrine and the codification efforts relating thereto.

5.29 Muslim law makes a very clear distinction between buried mineral wealth and that which is exposed.

a Exposed mineral wealth (*'dharir'* or *'zahir'*)

5.30 This is a very concrete assumption; hydrocarbons, in particular, have been known in most of these countries for many centuries in the form of seepages on the surface. In this case, everyone can freely dispose of it for the satisfaction of their needs; on the other hand, commercial use opens up the right to claim fees by the landowner.

This rule in practice returns to allocating ownership thereof to the Muslim community and management thereof to the *imam*, that is, to the sovereign.

[25] 'The Islamic system of law did not develop a general theory of contracts comparable to that of the West. Nevertheless, it did proclaim the duty of honouring agreements, in conformity with injunctions of the Koran and the Traditions.' George Lenczowski *Oil and State in the Middle East* (Cornell University Press 1960) 94 and 95.

[26] See quotations from Muslim legal experts in Simon G Siksek *The Legal Framework for Oil Concessions in the Arab World* (Middle East Research and Publishing Center 1960) 80 et seq; James N D Anderson 'The Moslem Ruler and Contractual Obligations' (1958) 33 *NYUL Rev* 917.

[27] It should be emphasised that sources are likely to vary depending on different branches of Islam.

[28] The *Koran* is the word of Allah brought to the Earth by the Prophet Mohammed. In Islam, legal order is based on the Koran (see Louis Millot and François-Paul Blanc *Introduction à l'étude du droit musulman* (Dalloz 2001) 83 et seq and 102).

[29] The Sunna is, for Sunnis, the unchanging law of God. This traces the path or route to follow for the believer (see Millot & Blanc op cit 87 et seq).

[30] *Ijma* is the universal consent of the scholars, meaning the assent of the community. A rule of *ijma* is considered to be established when the community has approved a practice or a determined precept and has never deviated from it (see Millot & Blanc op cit 105 et seq).

[31] Qiyas is a reasoning by analogy used in order to determine the rules of law which the Koran or the Sunna has not decreed (see Millot & Blanc op cit 115).

[32] See Millot & Blanc op cit 171.

[33] Hadiths and the Koran forms the law or *sharia*. See Millot & Blanc op cit 79.

However, *Shafi'i, Hanbalite* and *Malekite*[34] rites do not acknowledge that the owner of a surface has property rights over the mineral resources which are exposed there.

b Buried mineral wealth (*'bâtin'*)

5.31 Several interpretations are possible with regard to buried mineral wealth.

5.32 Firstly, it is possible to apply the accession system, which practically speaking, given the low number of private owners, comes back to acknowledging a right of ownership to the State. In the *hanifi* rite, a distinction is made depending on whether the land, belonging to a private owner, is closed or not; in the first case, and in agreement with the *hanbali* rite, the land is qualified as *mulk*. All of the products extracted from the subsoil belong to the owner of the land. In the second case, one fifth of the revenue belongs to the sovereign representing the Muslim community. If the line is discovered on land belonging to the public domain, it also comes under the public domain and it cannot be put into operation by a private individual without the written agreement of the sovereign. The *Shafi'i* rite is almost identical. However, in the presence of so-called *mulk* land, the owner may pay a religious tax, or *zakât*, the amount of which generally varies from one fifth to one fourteenth of production depending on the difficulty of the works. A doctrinal divergence in the *Shafi'i* rite does not allow it to be determined if, in the case of a mine located in the public domain, the mining rights allocated by the sovereign confer a simple right of use or a more complete right of ownership.

5.33 Secondly, it is possible to conduct the exposed mineral wealth system by way of assimilation, and this as provided for by the *Maliki* rite.

5.34 Finally, it is possible to consider that buried mineral wealth constitutes public property, the use of which would be managed by the State. This is how in the *Maliki* rite that this wealth is deemed to be the property of the Muslim community represented by the sovereign. Only the latter can, exceptionally and exclusively, grant the exploration and production rights to a third party for a limited period, called a general concession.

The point of agreement between all these rites concerns the possibility for the owner of the mineral wealth to confer the exploitation thereof to third parties, that is, the use and development.

---

[34] Muslim jurists gather around four large ideological movements, each originally having an *imam* at their head:

(i) the Hanafi rite, represented by the imam Hanifa (660–766AD) is widespread in Eastern Africa, Turkey and Arabia as well as the Muslim countries of pre-Asia and India;
(ii) the Maliki rite, represented by the imam Malek (708–795 AD), which monopolises North Africa and Central Africa;
(iii) the Shafi'i Chafeï rite, represented by the imam Shafi'i (766–819AD), which only influences certain areas of Egyptian law and the countries of the Persian Arabian Gulf;
(iv) the Hanbali rite, represented by the imam Hanbal (780–855AD), which currently does not have any influence in Koran positive law, except in certain branches of Saudi law.

It is undeniable that there is no transfer of ownership on the deposit but only on the part of it which is taken away.

5.35 In Arab countries, two names designate these exploitation permits allocated by the sovereign: *ijarat* and *iqta'a*.[35] These two concepts are very close to each other, the first being nevertheless much less developed than the second.

5.36 However, positions (which come particularly from doctrine)[36] assert the existence of gaps in Muslim law to govern the exploitation of natural resources should be mentioned. It is with regard to the fact that Muslim jurists have promoted a development of mining and oil legislation as part of a wide codification movement inspired mainly both by international practice and comparative law.

Summary Table

Nature of the system	*Who is the owner?*	*Dominant source*
Standard sovereign	No owner	Regulatory
Sovereign-inspired	Nation	Regulatory
State land	State	Contractual
Accession	Landowner	Contractual
Muslim	Land owner or the custodian of community property	Administrative operating contract

[35] According to Muslim traditions, which form the bases of the law applicable in the countries of the Middle East and North Africa, subsoil hydrocarbons can be developed by a third party by virtue of an exploitation contract granted for a determined time called an 'Iqta', a name deriving from the verb Qataa, which means 'cut'. It is established in Muslim law that, under any circumstances, the operator authorised by the State does not become the owner of the deposits, and its rights are limited to the appropriation of the products coming out of the well. (Ahmed Sadek El-Kosheri 'Le régime juridique créé par les accords de participation dans le domaine pétrolier' (1975) 147 *RCADI* 342 note 27). See also to that effect, Shavarsh Toriguian *Legal Aspects of Oil Concessions in the Middle East* (Hamaskaine Press 1972) 100.

[36] See, in particular, the opinion of Toriguian *Legal Aspects of Oil Concessions in the Middle East* at 99.

6

THE LEGAL AND CONTRACTUAL FRAMEWORK FOR THE EXPLOITATION OF NATURAL RESOURCES

6.1 Mining and petroleum activities are regulated exemption texts. The specific characteristics of projects to exploit natural resources assume the application of rules derogating from common law, both from a legal and a contractual point of view.

1 SECTORS GOVERNED BY EXEMPTION TEXTS

6.2 The strategic importance of the mining and petroleum sectors has led states which hold natural resources to subject mining and oil activities to exemption laws. Thus, it is frequent that exploration and production operations are not subject to common law but are regulated by derogatory texts, in this case mining and petroleum codes.

Besides, the specific characteristics of each mining and oil project sometimes require that the applicable regulation to be supplemented by supporting contractual documentation signed between the host state and the private operator.

a Sectors outside the scope of investment codes

6.3 States which hold natural resources and those which do not, have, since their independence, adopted laws relating to investments. Armed with their sovereignty, the states have, through investment codes, regulated investment operations over the national territory.

6.4 The strategic importance of the mining and oil sectors as well as their weight in the national economy have led host states to exclude investments made in these strategic areas of investment laws. States enacted specific pieces of legislation and have often excluded extractives activities from the scope of investment laws.[1]

[1] As an illustration, in Cameroon, Article 2 paragraph 3 of Law No. 2013/004 of 18 April 2013 Setting the Incentives for Private Investment in the Republic of Cameroon provides that:

> '*[l]es dispositions de la présente loi ne s'appliquent pas aux investissements dans les secteurs régis par des textes particulières, notamment le secteur pétrolier amont, le secteur minier et le secteur gazier, ainsi qu'au régime général des contrats de partenariat.*'
>
> [Free translation into English: '[t]he provisions of this Law do not apply to investments if the sector is governed by special texts, particularly the upstream oil sector, the mining sector and the gas sector, as well as to the overall partnership contract system'].

See also, in Burkina Faso, Article 4 of Law No. 62/95/ADP of 14 December 1995 Enacting the Investment Code as amended by Law No. 15-1997 17 April 1997, the Finance Law for 2009 and Law No. 07-2010 of 29 January 2010; in the Central African Republic, Article 5 of Law No. 01-010 of 16 July 2001 Enacting the Investment Charter; in the Republic of Congo,

6.5 However, some have chosen to apply concomitantly special texts regulating mining and petroleum activities and laws on investments.[2]

6.6 Indeed, some African countries expressly provide for the application of investment codes to operations carried out in the mining and oil sectors,[3] or grant a private operator the right to seek the benefit of provisions of the investment code which it deems to be more favourable.[4]

Article 5 of Decree No. 12/046 of 1 November 2012 Implementing the Investment Code; in the Democratic Republic of Congo, Article 3 of Law No. 004 of 21 February 2002 Enacting the Investment Code; in Côte d'Ivoire, Article 4 of Ordinance No. 2012-487 of 7 June 2012 Enacting the Investment Code; in Ethiopia, Article 3 of Investment Proclamation No. 280/2002; in Ghana, Article 17 of the Ghana Investment Act; in Guinea, Article 9 of Ordinance No. 001/PRG/87 of 3 January 1987 amended by Law No. L/95/029/CTRN of 30 June 1995; in Libya, Article 27 of law No. 9/2010; in Madagascar, Article 2 of Law No. 2007-036 of 14 January 2008 On Investments in Madagascar; in Mauritania, Article 2 of Law No. 2002-03 Enacting the Investment Code; in Mozambique, Article 3 of Law No. 3/93; and in São Tomé and Principe, Article 1 paragraph 3 of Law No. 7/2008 of 27 August 2008 Enacting the Investment Code.

[2] As an illustration, see Article 4 of Law No. 15/1998 Enacting the Investment Charter of Gabon:

> '*[d]es textes particuliers complètent les dispositions de la Charte pour préciser les conditions techniques, fiscales et financières de l'investissement et de l'exploitation dans certains secteurs spécifiques, notamment ceux relatifs à l'exploitation et à la transformation des ressources naturelles.*'
>
> [Free translation into English: 'Special texts supplement the provisions of the Charter to specify the technical, fiscal and financial conditions of the investment and of operations in certain specific sectors, particularly those relating to the exploitation and processing of natural resources.']

See also Article 3 of the Investment Promotion Act 2004 of Sierra Leone: '[i]n addition to this Act, there are other enactments providing specially for investment in business enterprises concerned with tourism, fisheries, mines and minerals, banks, non-bank financial institutions and other business activities, including such matters as the incentives for any such investments.'

[3] In particular, in Senegal, Article 2 of Law No. 2004-06 of 6 February 2004 Enacting the Investment Code:

> '*[l]e présent Code s'applique à toutes les entreprises qui exercent leurs activités dans l'un des Secteurs suivants :* [. . .] *extraction ou transformation de substances minerals.*'
>
> [Free translation into English: '[t]his Code applies to all companies who carry out their activities in one of the following Sectors: [. . .] the extraction or processing of mineral substances'];

In Chad, Article 3 of Law No. 006/PR/2008 Enacting the Investment Charter of the Republic of Chad:

> '*[l]a présente Charte s'applique aux investissements productifs réalisés principalement dans les domaines d'activités suivantes :* [. . .] *les activités d'exploration et de recherches géologiques, minières et pétrolières ;* [. . .] *les activités d'exploitation des substances minérales, de leur transformation et activités connexes.*'
>
> [Free translation into English: '[t]his Charter applies to productive investments made principally in the following sectors: [. . .] geological, mining and oil exploration and survey activities; [. . .] activities for the use of mineral substances, their processing and related activities'].

[4] See, to that effect, in Benin, Article 23, paragraph 5 of Law No. 2006-17 of 17 October 2006 Enacting the Mining and Taxation Code in the Republic of Benin, which provides that:

> '*[e]n se référant au code des investissements, le demandeur peut solliciter le bénéfice des dispositions dudit code qu'il juge plus favourable.*'

6.7 In any event, the exclusion of mining and oil projects from the scope of investment codes may be open to criticism. Mining and oil operations must, in effect, be apprehended in their entirety; mining and oil projects are cross-disciplinary projects which are not simply limited to the extraction of natural resources. Investment codes should be able to be applied to agreements which are within the orbit of mining and oil projects and which supplement the oil or mining agreement. Such an approach should be chosen in order to promote not only extractive projects in Africa but also their incorporation within the general local social and economic framework.

b The rules applicable to the oil and mining sectors supplemented by supporting contractual documents

6.8 Apart from the regulations enacted by the host state and applicable to mining and petroleum operations, contractual provisions also govern the latter. Mining and oil agreements are designed to take into consideration the particular features of each project.

The negotiation of an adequate contractual framework may be used to reconcile the interests of the host state, the local populations, the land owner, the operators as well as defenders of the environment. In order to do so, it is necessary during negotiations to place the agreement in its political, economic, social and technical contexts; economic and industrial realities should be sought in the legal formulation of clauses. Finally, it is recommended to envisage an estimate of the lifetime of the agreement according to the probable evolution of the local but also the international environment. Before any negotiation, a great deal of research is carried out from a geological, technical, political, legal, fiscal and economic point of view. The principle is to determine the probable profitability of a project by reviewing the two uncertainties represented by expenses and income while measuring the sensitivity of certain variables.

6.9 A same project will usually imply the intervention of a multitude of companies and the proliferation of contractual agreements. We are therefore witnessing an increase in the quantity of agreements connected to the exploitation of natural resources. Yet a dividing line should be established between principal agreement (ie the exploitation agreement signed between the State and the private operator) and those which are only ancillary (the latter constituting a means of carrying out the mining or oil project).

6.10 With regard to so-called ancillary agreements, a private operator responsible for the exploitation of a deposit should as part of the mining and petroleum operations sign private agreements with other operators; this includes for example drilling contracts, contracts for the construction of the mine, etc. Because of their importance in the performance of the project, these co-contracting and subcontracting agreements have been apprehended

---

[Free translation into English: '[b]y referring to the investment code, the applicant can seek the benefit of provisions of the said code which it deems to be more favourable.']

by legislative and regulatory texts.[5] Financing necessities may also give rise to the signing of financial agreements between the private operator and funders, in order to find the considerable financial resources needed for mining and oil operations.[6]

6.11 Apart from agreements signed with private operators, a secondary contractual fabric will be woven between the host state and the private operator. This contractual whole in particular includes investment agreements concluded between the State and the private operator.

6.12 Parallel to these so-called ancillary agreements, the State will conclude a main agreement with the operator (petroleum or mining agreements), the aim of which is to provide for the exploitation of the deposit. An analysis of these agreements reveals the existence of a great diversity, even though this type of agreement does possess common characteristics.

i *Specific characteristics of exploitation agreements*

6.13 Agreements relating to the exploitation of natural resources contain numerous specific features.

6.14 Agreements relating to the exploitation of natural resources contain significant derogations compared to the general law applicable in the host state. Mining and oil legislation is drawn up within a special general framework which is frequently the source of a unique system[7] .

6.15 These agreements are subject of a very long preparation and meticulous editing work; they are therefore designed to be very complete and are thus very complex.

Furthermore, these agreements remain very sensitive to political, economic, social and even cultural developments, both on a global and local scale. Their purpose is unique, and because of this, they develop within a permanent climate of doubt, suspicion and instability.

6.16 The extreme legal interweaving of contractual and regulatory elements must be highlighted.[8] Even though agreements explicitly refer to regulatory or legal provisions, some laws give a more or less important place to negotiations.

6.17 Finally, contracts relating to the exploitation of natural resources are subject to standardisation through the adoption of model agreements. Generally approved through a decision of the local legislative power[9] or

---

[5] Therefore, some provisions are designed to be applied to subcontractors, particularly in fiscal and customs matters.

[6] See below para 15.25 et seq on the techniques for financing mining and oil projects.

[7] See above para 6.4 et seq on the derogatory nature of the system applicable to mining and oil operations.

[8] This interweaving can be found in the terminology used. For instance, the word 'concession' sometimes means mining rights, therefore is of regulatory origin, and sometimes means a type of agreement, therefore is of contractual origin. We will indicate the first by the word 'concession' and the second by the words 'concession agreement'.

[9] See in Guiena Article 21 of Law No. L/2014/no. 034/AN of 23 December 2014 Enacting the Petroleum Code:

adopted through regulations,[10] they establish the overall conditions for exploitation, allowing some specific elements to be negotiated, such as financial conditions, even though no generality can be made on this point.

6.18 In some cases, these agreements can go as far as supplementing applicable texts, or even derogating from them. Such a derogation is frequently subject to the authorisation of the legislative power. By way of an example, in Cameroon, article 16(2) of law no. 001 of 16 April 2001 Enacting the Mining Code, as amended by law no. 2010/011 of 29 July 2010, provides that if the provisions of the agreement supplement those of the law, even without derogating from them, the said agreement shall be subject to a law authorising the government to conclude it. In contrast, other regulations explicitly exclude the possibility for an operator to be discharged of any legal obligation whatsoever. For instance, article 79 of the Mines and Mineral Act of 2011 of Swaziland provides that any contractual provision contrary to the mining law will be declared null and void.[11] The same article also provides that the provisions of a mining agreement cannot absolve the operator from its legal obligations.[12] It is vital to ensure that the agreements in question do not contradict the texts governing the matter.

'[a]près émission d'un avis favorable de la Cour Suprême portant sur sa conformité avec le droit en vigueur, y compris le présente Code, le Contrat Pétrolier est transmis à l'Assemblée Nationale pour ratification. La ratification fait l'objet d'un acte promulgué par décret du Président de la République et publié au Journal Officiel de la République.'

Free translation into English: '[a]fter the issuance of a favourable opinion by the Supreme Court concerning its compliance with the law in force, including this Code, the Petroleum Agreement is sent to the National Assembly for ratification. Ratification is the subject of an act enacted through a decree of the President of the Republic and published in the Official Journal of the Republic.']

10 Therefore, in Algeria, Article 11, paragraph 3 of Law No. 05-07 of 28 April 2005 On Hydrocarbons provides that:

'[l]e ministre chargé des hydrocarbures introduit les demandes d'approbation des contrats de recherche et/ou d'exploitation qui sont approuvés par décret pris en conseil des ministres.'

[Free translation into English: '[t]he Minister responsible for hydrocarbons makes requests for approval of exploration and/or exploitation agreements approved by decree passed in Council of Ministers.']

See also, in Mauritania, Article 19 of Law No. 2010-033 of 20 July 2010 Enacting the Hydrocarbon Code amended by Law No. 2011-044 of 25 November 2011:

'[l]e Contrat d'exploration-production est signé par le Ministre au nom de l'État et par le Contractant. Le Contrat d'exploration-production ainsi que ses avenants sont approuvés par décret pris en Conseil des Ministres.'

[Free translation into English: '[t]he exploration-production contract is signed by the Minister on behalf of the State and by the Contractor. The contract as well as any rider shall be approved by decree taken by the Council of Ministers.']

11 See in Swaziland, section 79(1) of the Mines and Minerals Act, 2011: '[a] term or condition of a minerals agreement that is inconsistent with this Act shall, to the extent of the inconsistency, have no legal effect.'

12 See in Swaziland, section 79(2) of the Mines and Minerals Act, 2011: '[n]othing contained in a mineral agreement shall be construed as absolving a party to the agreement from any requirement prescribed by law.'

It is therefore necessary to ensure the legality of the agreement particularly in regard to the hierarchy of standards and derogation possibilities provided for in the applicable regulation.

ii *Diversity of exploitation agreements*

6.19 It is in the oil sector that agreements relating to the exploitation of natural resources first saw the light of day. In the 1950s up to 1960, there were very few mixed agreements.[13] Indeed, the exploitation of a natural resource was either entirely private, which in fact covered practically all situations, or was entirely State-controlled. One exception to this general trend appeared in 1954 in Iran as part of the petroleum agreement signed with a consortium made up of the main international companies. Since then, the contractual landscape supporting the exploitation of natural resources has become strongly diversified. It has switched with the concession agreement from a more or less general unity to wide multiplicity. *De facto*, there are as many contractual categories as there are exploitation agreements. Work to categorise these forms in view of their typology is therefore made very difficult and is certainly risky. Each state constitutes a particular case. Furthermore, the exploration, development and production conditions vary from one state to another and even from one area to another. Each exploitation agreement requires a specific new approach; even in cases where there are standard agreements or framework agreements,[14] the agreements actually signed may differ from each other. Currently, all the major types of agreements used are not, in substance, very different from each other.

The private operator can be placed in three different scenarios depending on whether it holds mining rights, it is a shareholder in the operating company but without holding mining rights or it is simply a service provider. These three assumptions will be considered quickly in the following sections.[15]

[13] Which means signed between local State operators and private operators.

[14] A framework agreement is always drawn up exclusively by the negotiators themselves; on the contrary, a standard agreement is generally drafted by a third party with respect to these negotiators or by only one of them, in particular the host state. A framework agreement generally delimits the overall negotiation boundaries for each type of clause. A standard agreement is a more elaborate form of agreement in which the possibilities of amendments during negotiations are limited. Sometimes, the framework agreement term applies to a very specific situation during negotiations and it designates a text drawn up by the negotiators which defines the main rules to which their future contractual relations will be subject. In this case, the framework agreement should not be confused with an agreement protocol which, as it has a more modest function, only prepares for the sole exploitation agreement which is now negotiated. In this regard, see Joanna Schmidt '*Négociation et conclusion de contrats*' 1982 *D* 229.

[15] For a complete analysis of agreements relating to the exploitation of natural resources, see below para 10.3 et seq.

(1) The private operator holds mining rights

6.20 From an historical point of view, this is the oldest situation. Cases where the private operator is the sole holder of mining rights must be distinguished from those where co-ownership has been set up.

6.21 Historically, the concession agreement illustrates sole ownership of mining rights. The host state waives its right of ownership against the payment of fees and income from the imposition of taxes. The duration of this agreement is very long. The private operator makes all the important decisions and provides all of the capital and technical knowledge needed. This assertion shall be qualified as it suffers numerous exceptions.

6.22 Besides these concession agreements, the idea of a shared ownership of mining rights with the local State operator has appeared; these are contracts of association. Their duration is usually limited compared to concession contracts and—as the extractive industry becomes a State monopoly—they include new provisions relating to the local development of social and economic facilities as well as the training of local personnel.

(2) The private operator is a shareholder in the operating company

6.23 In this context, only the local State operator holds the mining rights; the latter can itself market its share of production, make important management decisions, exercise a right of intervention, etc. The host state will receive a part of the profits from exploitation.

Under this heading, we find several types of contract: joint-venture agreements, co-ownership agreements, participation agreements, service agreements and some production sharing contracts.

(3) The private operator is a service provider

6.24 This is the most varied category and the one which lends itself least to being typified. Very few of these agreements are service agreements within the meaning which is commonly attributed to them in international trade. The common thread of all these agreements is that the not-yet extracted natural resource does not belong to the private operator, even though the latter can become the owner of a more or less significant parts of production as a consequence of provisions included in the exploitation agreement. However, and as opposed to the previous category, the private operator is not in principle partnered with the local State operator within the share capital of the operating company.

We can particularly mention as being in this category:

- technical assistance agreements;
- service agreements;
- work contracts;
- some production sharing agreements;
- exploration service contracts;[16]

[16] The private operator acting as a contractor provides a full exploration programme up to the discovery of a natural resource in commercial quantities. This programme includes

- full service contracts;[17]
- management contracts;[18]
- industrial cooperation agreements: commercial agreements for support in terms of personnel and technical assistance, risk contracts, operating agreements.

6.25 On the fringes of these agreements, the existence of special related agreements or agreements designed specifically for the exploitation of natural resources must also be indicated, such as farm-in and farm-out agreements, consortium agreements, etc.

2 THE LEGAL AND REGULATORY FRAMEWORK GOVERNING MINING AND OIL OPERATIONS

6.26 Originally, national laws relating to non-renewable natural resources displayed similarities and joint solutions. European mining legislation has progressively been imposed in practically all states, either because of territorial hegemony policies, or following the qualification established by European lawyers on the subject of local laws as being incomplete, non-existent, inoperative or even non-compliant with the 'general principles of civilised nations'. This development towards a direction outside their territory of origin of European mining legislation finds its source in the industrial revolution which, having increased national needs for natural resources, forced European mining and oil companies to expand abroad. Since the end of the Second World War, the growing economic importance of natural resources and the gaining of political independence of numerous countries have resulted in the emergence of new national mining and petroleum laws which are fundamentally different from each other, even though there are common principles from one state to another.

6.27 The wealth of the subsoil largely influences national laws. Therefore, trends can be highlighted which allow a distinction to be made between producing countries, the potential of which has over time been largely established, and states which could be qualified as being small producers with resources that are still poorly estimated. With regard to the differences between these two categories, the participation of the State as part of mining and oil projects can be mentioned by way of illustration. Therefore, while the laws of producing states most often enshrined free participation of the state within mining or oil projects, a contributive participation of the state is generally envisaged by the mining and petroleum codes of states where resources are low or not proven. Similarly, while the participation of the State

provisions governing the training of local personnel and the transfer of suitable technology to the local State operator.

[17] With respect to exploration service contracts, a programme regarding the development of the discovery is also included; the contractor shall continue to supervise operations until full reimbursement of its costs.

[18] The host state entrusts the private operator with the financial, commercial and administrative management of a company that it has created for exploitation.

is only on option in the legislation of states with proven reserves, it is generally automatic in the laws of producing states with proven potential.

6.28 The issues of the legal framework applicable to extractive activities can be made more complex because of the simultaneous application of various texts at national level. In order to understand the matter in its entirety, reference should be made (apart from to mining and petroleum codes) to a multitude of national legislation designed to be applied indirectly to mining and oil operations. These are particularly laws relating to the protection of the environment, land laws, tax laws or legislation that applies in terms of employment or health and safety. Furthermore, through the application of stabilisation clauses, some states simultaneously apply several mining regulations. The simultaneous application of various texts also results in provisions which allow operators the possibility of opting to apply the old or the new legislation.[19] The multiplicity of standards applicable to mining and petroleum operations can in some cases be multiplied by the application of texts governing specific fields, such as the exploitation of precious materials or that of quarries.[20]

---

[19] See, in particular, Article 4 of Law No. 2003-30 of 28 April 2003 Enacting the Tunisian Mining Code:

> '*[a] la date d'entrée en vigueur du Code Minier, les titulaires de permis de recherche en cours de validité et/ou de concessions d'exploitation instituées ont la faculté d'opter relativement à ces permis et concessions pour l'application des dispositions du Code Minier. L'exercice de l'option prévue à l'alinéa premier du présent article doit faire l'objet d'une notification signée par le titulaire du permis et/ou de la concession d'exploitation ou par son représentant légal. Chaque titre minier doit faire l'objet d'une notification séparée qui doit être adressée sous pli recommandé avec accusé de réception au Ministère chargé des mines ou déposée directement auprès de ses services contre décharge et ce, six mois au plus tard à partir de la date d'entrée en vigueur du Code Minier. A défaut de l'exercice de l'option mentionnée à l'alinéa premier du présent article par le titulaire d'un titre minier, ledit titre demeure, jusqu'à son expiration, régi par les dispositions législatives en vertu desquelles il a été institué.*'
>
> [Free translation into English: '[o]n the date that the mining code comes into force, the holders of valid exploration permits and/or instituted exploitation concessions have the ability to opt relatively to these licences and concessions for the application of the provisions of the Mining Code. The exercising of the option stipulated in the first paragraph of this article must be subject to notification signed by the holder of the licence and/or the exploitation concession or by their legal representative. Each mining right must be subject to separate notification which must be sent by registered post with acknowledgement of receipt to the Ministry responsible for mines or filed directly with its departments against a signed receipt and this must be done six months at the latest from the date the Mining Code comes into force. Failing the exercising of the option mentioned in the first paragraph of this article by the holder of a mining right, the aforesaid right shall remain, until its expiry, governed by the legislative provisions by virtue of which it was instituted'].

[20] See, in particular, in South Africa, special relative legislation applicable to certain mining activities, such as the Diamonds Act 56 of 1986 or the Precious Metals Act 37 of 2005. Besides, Morocco enacted two different laws to govern the exploitation of minerals and quarries on this territory.

6.29 African Governments are increasingly committed to establish a legal framework that encourages foreign investment and promotes exploration while ensuring that the contribution of the extractive industries to local development grows.[21] The wish to put an attractive framework in place sometimes leads to producing states competing through the adoption of new legislation. Besides, regulations governing the oil and mining sectors are often amended in order to attract new investors or increase the state's profits.

6.30 Finally, the rapid change in legislation is a constant in numerous African states. Apart from the laws adopted in 2015,[22] 2016[23] and 2017,[24] numerous reforms have been announced for the years to come.[25]

---

[21] See Magic 'Africa: A future Center of Gravity for Global Mining' at 6.

[22] For instance, in Burkina Faso, the National Transition Council adopted a new Mining Code on 26 June 2015. Similarly, in Morocco, July 2015 was marked by the publication of Law No. 33-13 On Mines in the *Journal officiel* (a reform which entered into force in April 2016 following the publication of its implementation decree in the *Journal officiel*). See also the reforms that have taken place in the oil sector in the Democratic Republic of Congo (Law No. 15/012 of 1 August 2015 Enacting the Petroleum Code), reform of mining legislation which took place in Gabon (Law No. 17/2014 of 30 January 2015 On Regulation of the Mining Sector in the Gabonese Republic) and the new Hydrocarbon Code adopted in Mali (Law No. 2015-035 of 16 July 2015 On Organisation of Exploration, Exploitation and Transportation of Hydrocarbons).

[23] The year 2016 has been marked by the adoption of numerous pieces of legislation, in particular, in Kenya (adoption of the Mining Act 2016), in Djibouti (adoption of Law No. 138/AM/16/7thL of 23 July 2016 Enacting the Mining Code) and in the Republic of Congo (adoption of Decree No. 16/010 of 19 April 2016 Enacting the Petroleum Code).

[24] In Tanzania (the Government enacted several regulations such as: the Petroleum (Local Content) Regulations, 2017, the Natural Wealth and Resources (Permanent Sovereignty) Act, 2017; the Natural Wealth and Resources (Review and Re-Negotiation of Unconscionable Terms) Act, 2017), in Burkina Faso, Senegal and Kenya (enactment of the regulations implementing the new mining laws), in Niger (Ordinance No. 2017-03 of 30 June 2017 Amending and Supplementing Ordinance No. 93-16 of 2 March 1993 Enacting the Mining Law.

[25] For instance, in the Oil and Gas sectors, reforms of the legal framework had been announced in Algeria, in Gabon, in Madagascar, in Nigeria, and in Tunisia. In the mining sector, ongoing reforms concern countries such as Mali, Mauritania, Madagascar and in Malawi.

7

THE INTERNATIONALISATION OF APPLICABLE STANDARDS

7.1 The exploitation of mining and oil resources is mainly a national concern. Nevertheless, their strategic importance means that the issues connected with the extractive industries very often go beyond the national context. Taking these issues into account, both at regional and international level, results from sometimes antagonistic wishes. On the one hand, the desire to come together with a view to regional cooperation—this is the case in the mining sector with the creation of supranational standard—and, on the other hand, the wish to come together in order to avoid a conflict or to put an end to a dispute—this is especially the case in the petroleum sector through agreements which tend to create a joint development zone.

Therefore, in the mining sector, the first supranational standards appeared within the context of regional organisations. The standards, which are particularly developed in French-speaking Africa, currently tend towards a harmonisation of mining legislation in the areas concerned.

7.2 Unlike the mining sector, which is subject to harmonisation at regional level, the hydrocarbon sector for its part has seen standards multiply at an international level. These supranational standards specifically concern the exploitation of offshore hydrocarbons, which presents specific issues.[1]

1 MOVEMENTS OF REGIONAL HARMONISATION

7.3 States have not been the only ones to legislate in the area of natural resources. Africa distinguished itself by a codification or regional unification movement, spearheaded by West Africa.[2] Indeed, regional organisations enacted standards which are designed to govern mining activities over their entire territory; this is the case of the West African Economic and Monetary Union (WAEMU)[3] and the Economic Community of West African States (ECOWAS).[4]

[1] The specific issues are taken into consideration by national oil legislation which contain specific provisions related to the exploitation of offshore hydrocarbons. See, to that effect, Bandoki op cit 48 et seq.

[2] See on the multiplication of regional standards in Africa: *La concurrence des organisations régionales en Afrique, sous la direction de Matthieu Fau-Nougaret* (L'Harmattan 2012).

[3] Member states of the WAEMU are: Benin, Burkina Faso, the Côte d'Ivoire, Guinea-Bissau, Mali, Niger, Senegal and Togo, all remaining members of OHADA. See Thierry Lauriol & Sophie Da Cunha 'La règlementation minière en Afrique de l'Ouest: une multiplicité des normes' in *Études à la mémoire de Fernand-Charles Jeantet* (LexisNexis Litec 2010) 261.

[4] Member states of ECOWAS are: Benin, Burkina Faso, Cape Verde, Côte d'Ivoire, Gambia, Ghana, Guinea, Guinea-Bissau, Liberia, Mali, Niger, Nigeria, Senegal, Sierra Leone and Togo, namely nine out of the fifteen OHADA member states.

This movement is still on-going, not just because it is continuing within these organisations, particularly the WAEMU, but also because others would also like to follow these examples.[5] The movement tending towards harmonisation of mining standards, launched in West Africa, is currently interesting central Africa within the context of the Central African Economic and Monetary Community (CEMAC), where the enactment of a community mining code is expected.[6]

It must be acknowledged that through these supranational standards, the states waived part of their national sovereignty over their rights to control the exploitation of their natural resources, more specifically their mining resources, with petroleum resources and quarries clearly being excluded from the scope of these texts by referring to national regulations.[7]

a The WAEMU mining code

7.4 It was through a text dated 23 December 2003 that the WAEMU equipped itself with a legal framework[8] regulating all operations relating to the prospecting, exploration, exploitation, holding, circulation, processing, transport, and marketing of mining substances over its entire territory.[9]

7.5 This community mining regulation falls under the prerogatives and objectives laid down by the WAEMU, namely the strengthening of the economic and financial activities of the states, the convergence of national economic policies, the achievement of a common market, the setting up of joint sectoral policies and the harmonisation of legislation as far as necessary to achieve the common market.[10]

7.6 The effect of this text on national mining laws is significant, because of the principle of primacy of the WAEMU standard over domestic law.[11] Furthermore, article 41 of the UEMOA mining code reinforces the supremacy

---

[5] For the creation of a Uniform Mining Act within the OHADA see Thierry Lauriol 'L'OHADA, le temps pour un acte uniforme minier' July-September 2015 at 281.

[6] See, in particular, the official website for the Central African Economic and Monetary Community Commission, 'Harmonisation des législations minières: les premiers pas'.

[7] Therefore, Article 2 of Regulation No. 18/2003/CM/UEMOA of 23 December 2003 Enacting the Community Mining Code of the WAEMU excludes liquid or gaseous hydrocarbons from its scope. With regard to substances from quarries, in the absence of any definition, national legislation can in practice regrade quarry substances as mine substances.

[8] Regulation No. 18/2003/CM/UEMOA of 23 December 2003 Enacting the Community Mining Code. See Thierry Lauriol & Sophie Da Cunha op cit 261.

[9] Article 2 of Regulation No. 18/2003/CM/UEMOA of 23 December 2003 Enacting the Community Mining Code.

[10] See, in particular, Articles 4 and 5 of the WAEMU Treaty of 10 January 1995.

[11] Article 43 of the WAEMU Treaty of 10 January 1995:

'[l]es règlements ont une portée générale. Ils sont obligatoires dans tous leurs éléments et sont directement applicables dans tout État membre. Les directives lient tout État membre quant aux résultats à atteindre. Les décisions sont obligatoires dans tous leurs éléments pour les destinataires qu'elles désignent. Les recommandations et les avis n'ont pas de force exécutoire.'

[Free translation into English: '[r]egulations have a general scope. They are mandatory in all their elements and are directly applicable in all member States. The directives are binding upon every member State with regard to the results to be achieved. The decisions

of the community standard over national mining legislation by abrogating and replacing all prior contrary provisions over the entire extent of the Union's territory; from now on, no state can enact standards which are contrary to it. The scope of the UEMOA mining code is all the more important because its adoption in the Council of Ministers by means of a regulation has made it directly applicable in member states. The direct applicability of the regulations is stipulated in article 43 of the WAEMU Treaty, which establishes the direct applicability without the need for vertical or horizontal transposition measures as well as a general and mandatory scope in all its elements.

Among the objectives of the WAEMU mining code, harmonisation, modernisation and clarification of mining legislation should be mentioned. From an economic point of view, it aims to create the conditions to make a substantial contribution to the mining sector in the economies of member states by accelerating the development of mineral substances.

As things stand, the WAEMU mining code states important principles which often leave it up to the states to supplement or implement them.

In this regard, articles 13–17 of the WAEMU mining code establish the guarantees to which mining investors are entitled, including protection of ownership,[12] the free import of goods, materials and equipment, the free choice of suppliers, the free transfer of currency and profits or the stability of tax and customs regime. Articles 25–33 of the WAEMU mining code deal with the tax and customs benefits granted both during the exploration and the production stages. We can mention the principle of temporary admission during the entire period of validity of the mining rights, or exemption from VAT, tax on profits, minimum flat-rate tax, employers' contributions on salaries and wages and occupational taxes. Furthermore, article 12 of the UEMOA mining code provides for the possibility for member states to obtain a free 10% interest in the share capital of operating companies.

7.7 Finally, it should be emphasised implementing regulations for the application of the WAEMU mining code are expected.[13] The WAEMU Commission recently launched a call-for-tenders in order to carry out an overhaul of the mining code and to draw up execution rules concerning the economic, tax, customs, environmental and mining rights management areas, in order to reduce disparities in the mining sector. The contribution of this sector to the development of member states, the efficient and sustainable exploitation of mining resources, the promotion of local preference as well as the improvement of governments within the Union number among the main objectives of the current reform.

---

are mandatory in all their elements for those to whom they are addressed. Recommendations and opinions are not enforceable.']

[12] This protection of ownership is provided for in particular by the free provision of movable and immovable, tangible or intangible property, protection against nationalisation, expropriation or requisition.

[13] The Draft Implementing Regulation of the Community Mining Code drawn up in 2005/2006 has never been adopted.

b The ECOWAS mining directive

7.8 More recently, in May 2009, ECOWAS also equipped itself with a legal framework in order to govern mining operations on its territory, by making a directive on the harmonisation of guiding principles and policies in the mining sector.[14] This directive was supplemented in February 2012 through an additional act,[15] enacting ECOWAS' mineral resource development policy and its action plan.

These directives come directly within ECOWAS's area of jurisdiction, as its objective is to promote the integration of all areas of economic activity, particularly energy and natural resources.[16]

7.9 The directives are binding with regard to member states of ECOWAS, the conditions for achieving the directives' objectives being left to the discretion of the member states. The new article 9 of the revised ECOWAS Treaty provides that:

> 'Directives are acts by which the Conference or the Council fixes the objectives to be reached for member States, allowing each of them the freedom to decide on the procedures for achieving these objectives. Directives are binding Acts for the States'.[17]

From now on, contrary to regulations which are directly enforceable and applicable, member states were under an obligation to adopt the necessary laws and regulations in order to comply with the provisions of the directives. Thus, the ECOWAS directive specifies in its article 22 that the 'State shall make all provisions needed to comply with this directive by 1 July 2014 at the latest'.[18] In this respect, it shall be mentioned that when a member state does not comply with its obligations with regard to the Community, such as the

---

[14] 62nd ordinary session of the Council of Ministers of 26-27 May 2009.

[15] 40th ordinary session of the Conference of Chiefs of State and Government of 16-17 February 2012. Additional Act A/S A.18/02/12 On the Adoption of ECOWAS's Mineral Resource Development Policy (PDRMC).

[16] See, in particular, Article 3(1) and 3(2.a) of ECOWAS Treaty dated 24 July 1993, which respectively provide that:

> '[t]he aims of the Community are to promote co-operation and integration, leading to the establishment of an economic union in West Africa in order to raise the living standards of its peoples, and to maintain and enhance economic stability, foster relations among Member States and contribute to the progress and development of the African Continent.' and '[i]n order to achieve the aims set out in the paragraph above, and in accordance with the relevant provisions of this Treaty, the Community shall, by stages, ensure; a) the harmonisation and co-ordination of national policies and the promotion of integration programmes, projects and activities, particularly in food, agriculture and natural resources, industry, transport and communications, energy, trade, money and finance, taxation, economic reform policies, human resources, education, information, culture, science, technology, services, health, tourism, legal matters'.

[17] Additional Act A/SA.3/01/10 Amending Article 9 of the ECOWAS Treaty as amended by Additional Protocol A/SP.1/06/06.

[18] In addition, the Directive specifies in its article 22(4) that the Chairman of the Commission of ECOWAS shall appoint an *ad hoc* committee to monitor the application of the directive by the member States.

objectives fixed by directive, the Conference can take sanctions against this member state.[19]

The ECOWAS directive aims to provide member states with a mining policy and a harmonised legal framework while taking account of the state of development of their mining sector. It lays out the guiding principles mainly concerning ownership of the management of natural resources, the protection of the environment, the protection of the national interest and the access to mining information.

Therefore, it is reiterated that natural resources are the property of the State and are managed for the benefit of the population of the member state. Their exploitation should take account of the preservation of the environment and the interests of local communities. The ECOWAS directive makes tax and customs exemptions subject to the compliance with social and environmental obligations. It is also specified that the state may participate to the share capital of mining companies.[20]

7.10 It is interesting to note, at different levels of development, the similarity of the subjects dealt within the WAEMU and ECOWAS. Mining matters are subject to codification on identical bases starting with that of the ownership of the natural resource.[21] The risks of overlap or even contradictions between the various texts should therefore be studied with the greatest of attention.

---

[19] ECOWAS Treaty, Article 77.

[20] Therefore, Article 10 of Directive C/DIR 3/05/09 of 27 May 2009 On the Harmonisation of the Guiding Principles and Policies in the Mining Sector provides that:

> '[a] competent authority in a Member State may by notice in writing require a mining right holder which is a corporate body to issue to that Member State special shares, by whatever name called in the company for no consideration. 2. The special shares shall constitute a separate class of shares and shall have rights agreed on between the competent authority of the Member State and the Mining Right Holder. 3. A Member State may also participate in the equity of mining operation within their territorial area on terms to be mutually agreed.'

[21] A reminder of this principle is given in Article 4 of Regulation No. 18/2003/CM/UEMOA of 23 December 2003 Enacting the Community Mining Code of the WAEMU, which provides that

> *'les substances minérales contenues dans le sol et le sous-sol, les eaux territoriales et sur le plateau continental d'un État membre sont propriété de cet État. Toutefois, les titulaires des titres miniers d'exploitation acquièrent la propriété des substances minérales qu'ils extraient.'*
>
> [Free translation into English:
>
> '[t]he mineral substances contained in the soil and subsoil, territorial waters and on the continental shelf of a member State are the property of this State. However, the holders of mining operating rights shall acquire ownership of the mineral substances which they extract.']

With regard to Directive C/DIR 3/05/09 of 27 May 2009 On the Harmonisation of the Guiding Principles and Policies in the Mining Sector, it provides in its Article 3(1) that:

> '[e]very mineral in its natural state in, under or upon land of a Member State, rivers, streams, water courses throughout the sub-region, exclusive economic zone, and an area covered by the territorial seas or continental shelves is the property of the respective Member State.'

7.11 According to practitioners, the existence of a supranational regional standard constitutes an undeniable element of security and stability. However, the existence of regional standards alongside national text is also a source of complexity as knowledge of the mining standard is no longer uniquely local but must include a subregional approach.

2 MOVEMENTS TOWARDS THE ADOPTION OF INTERNATIONAL REGULATIONS

7.12 The petroleum sector has for several decades been experiencing a multiplication of international standards.

The creation of international standards, which apply more specifically to the offshore exploitation of hydrocarbons, has two principal aims: on the one hand, the protection of the environment on an international scale and, on the other hand, the joint exploitation of trans-border deposits.

a International standards applicable to offshore exploitation of hydrocarbons

7.13 The reduction of world reserves of natural resources, the growing demand for energy and the increase in technological know-how has led investors to exploit new, less accessible and deeper deposits.

Currently, one third of hydrocarbon resources and one quarter of gas resources consumed come from submerged zones.[22] The exploitation of these deposits will not decrease, as operators are always looking for new deposits in order to build up their reserves.

The exploitation of hydrocarbons in submerged zones is not without raising some difficulties, particularly with regard to the environmental risk generated by such activity.

7.14 These concerns are particularly taken into account at the international level by the United Nations Convention on the Law of the Sea, adopted in Montego Bay on 10 December 1982. Even though it is not specific as to the offshore exploitation of natural resources, this instrument contains provisions designed to be applied indirectly to offshore operations. Therefore, this convention particularly specifies a positive obligation where these states are responsible for adopting standards which allow pollution to be prevented, reduced and controlled in the marine environment[23] but which also allow measures to be adopted to obtain compensation for damage resulting from pollution of the marine environment.[24]

7.15 With regard more specifically to the oil sector, compensation for damages due to pollution resulting from the spillage of hydrocarbons by tankers is governed by an international scheme drawn up under the auspices

---

[22] See to that effect, Julien Rochette 'Activité pétrolière et gazières en offshore et protection de l'environnement' in *Le contentieux extractif* (ICC 2015) 125.

[23] Article 208 of the United Nations Convention on the Law of the Sea of 1982.

[24] Article 235 of the United Nations Convention on the Law of the Sea of 1982.

of the international maritime organisation: the 1969 International Convention on Civil Liability for Oil Pollution Damage.

7.16 More specifically with regard to Africa, mention should be made of the work in progress to prepare a protocol on regional environmental standards for offshore drilling activities.[25] With regard to blockages encountered at international level, the regional scale appears particularly favourable to strengthening the regulation of offshore activities.[26]

7.17 As this is in spite of the international texts mentioned, the legal arsenal relating to liability for environmental damage connected to the extraction of offshore hydrocarbon deposits is still deemed to be insufficient.[27]

b The joint exploitation of a deposit at international level: example of joint development zones

7.18 It frequently happens that a deposit is located straddling several zones which may or may not coincide with political boundaries.[28] Now, contrary to solid minerals, fluid natural resources are likely to be moved in the subsoil, thus risking, in the case of trans-border deposits, passing from one border to another.

This phenomenon is likely to generate tension between the states concerned. Examples of political conflicts connected to the existence of trans-border deposits are numerous, particularly in the African continent. These include conflicts involving or having involved Nigeria and Cameroon with regard to the Bakassi peninsula, the case of the movement towards the east of the border river Semkili between the Democratic Republic of the Congo and Uganda, the disputes between Ghana and the Côte d'Ivoire or the border dispute which gave rise to the independence of the South Sudan.[29]

---

[25] See Rochette op cit 128. See also Julien Rochette 'Forage en eaux troubles—Pour une régulation internationale des activités pétrolières et gazières offshore' *Institut du développement durable et des relations internationales* Policy Brief 14 February 2014.

[26] See Julien Rochette, Matthieu Wemaëre, Lucien Chabason & Sarah Callet 'En finir avec le beau pétrole—Pour une meilleure régulation des activités pétrolières et gazières offshore' *Institut du développement durable et des relations internationales* Policy Brief 14 February 2014.

[27] See Rochette 'Activité pétrolière et gazières en offshore et protection de l'environnement' loc cit.

[28] It is interesting to mention South Sudan's Petroleum Code (Petroleum Resources Act 1972), Article 34(4) of which obliges the State to enter into negotiation through the authorities of the neighbouring state:

> '[w]here a petroleum accumulation extends beyond the jurisdiction of the Republic, the Government shall endeavour to reach an agreement with the other country with a view to ensure the correct apportionment of the accumulation and the most efficient co-ordination of petroleum activities.'

All the more attention is directed at this provision given the existing conflict between South Sudan and the Sudan.

[29] For a detailed analysis of the various disputes in question, see Philippe Sebille-Lopez 'Origines et règlements politiques des conflits dans le secteur des hydrocarbures' in *Le contentieux extractif* (ICC 2015) 181.

7.19 The determination of borders often being impossible in such a context, the states have been able to find alternative solutions which allow joint exploitation of trans-border deposits. Therefore, numerous unitisation treaties have been signed and joint development zones have been created in order to allow states to jointly exploit their natural resources. This trend is of particular importance concerning the exploitation of offshore trans-border hydrocarbons deposits. For instance, in 2017, Mauritania and Senegal were in discussions for the signature of bilateral treaties for the joint exploitation of natural resources.[30]

7.20 On the African continent, the creation of a joint development zone between Nigeria and São Tome and Principe is an example of joint management of trans-border deposits.[31] The operation of this zone has included the creation of a Joint Development Authority given, in accordance with the treaty, an international legal personality,[32] as well as the creation of a Joint Council of Ministers, a joint representative political body for the party states.[33]

The jurisdiction allocated to the Joint Development Authority is considerable. It is up to it to prepare a specific tax scheme that applies to oil operations in the Joint Development Zone for the approval of the Joint Council of Ministers.[34] Furthermore, no oil activity can be carried out in the

[30] These discussions were initiated following the discovery of the gas deposit of Grand Tortue/Ahmeyim.

[31] This zone was instituted in 2001 by the Treaty between the Republic of São Tome and Principe and the Federal Republic of Nigeria on the Joint Exploitation of Oil Resources, signed in Abuja on 21 February 2001. For more information relating to the background leading led to the signature of this Treaty, see ebille-Lopez op cit 181.

[32] See in this regard Article 9(2) of the Treaty between the Republic of São Tome and Principe and the Federal Republic of Nigeria on the Joint Exploitation of Oil Resources, signed in Abuja on 21 February 2001, which provides that:

> '*[a] autoridade tem personalidade jurídica própria, nos termos do Direito Internacional e das leis de cada um dos Estados Partes, e, bem assim, a competência, no âmbito das leis de ambos os Estados Partes, necessária para o exercício dos seus poderes e a execuç ão das suas funções. Em particular, a Autoridade tem poderes para contratar, adquirir e alienar bens móveis e imóveis, intentar acções e ser parte em processos contenciosos*.'
>
> [Free translation into English: '[a]n authority shall have judicial personality in international law and under the law of each of the States Parties and such legal capacities under the law of both States Parties as are necessary for the exercise of its powers and performance of its functions. In particular, the Authority shall have the capacity to contact, to acquire and to dispose of movable and immovable property and to institute and be party to legal proceedings'].

[33] See to that effect Article 6 of the Treaty between the Republic of São Tome and Principe and the Federal Republic of Nigeria on the Joint Exploitation of Oil Resources, signed in Abuja on 21 February 2001.

[34] See to that effect Article 21 of the Treaty between the Republic of São Tome and Principe and the Federal Republic of Nigeria on the Joint Exploitation of Oil Resources, signed in Abuja on 21 February 2001.

zone unless it is based on an oil agreement concluded between the operator or operators and the Authority.[35]

7.21 This treaty constitutes a particularly interesting example of the development of oil standards at an international level. The rules instituted under this joint development zone may almost appear to be a 'mining code' under international law, enacted at inter-state level.[36]

[35] See to that 'effect Article 23(1) of the Treaty between the Republic of São Tome and Principe and the Federal Republic of Nigeria On the Joint Exploitation of Oil Resources, signed in Abuja on 21 February 2001.

[36] Indeed, just as in an Petroleum Code, exploitation in this joint development zone is subject to specific rules in terms of health and safety (Article 37 of the Treaty between the Republic of São Tome and Principe and the Federal Republic of Nigeria on the Joint Exploitation of Oil Resources, signed in Abuja on 21 February 2001), the employment of local staff (Article 36 of the said Treaty), the rights and obligations of private operators (Articles 25 of the Treaty) or the customs scheme (Article 22 of the said Treaty).

PART 2

Legal basis of the exploitation of natural resources

8

THE CONCEPT OF MINING RIGHTS

8.1 The two traditional legal bases for the exploitation of mining and petroleum resources comprise, on the one hand, the granting to the private operator by the licensing authority of mining titles and, on the other hand, the signing of an agreement between the host state and the private operator.

Originally, the first exploitation contracts, regardless of the resource in question, were based on a mining title. Subsequently, these bases have evolved, particularly in the oil sector. The exploitation agreements have been signed without granting mining rights to the operator.

Nowadays private operators may or may not hold mining rights in connection with the exploitation of petroleum resources. In the mining sector, the granting of mining rights remains the standard and the absence thereof the exception.[1]

A Mining rights

8.2 A mining right or mining title is an administrative act granting its holder rights within a designated perimeter and for certain substances.

The granting of mining rights is most often made mandatory by the mining and petroleum codes themselves.[2]

There are, however, many exceptions to this principle,[3] in particular when national laws allow their citizens under conditions (such as in the absence of prior rights) to exploit natural resources without a mining title.[4] In addition, more often in the petroleum sector, extractive activities can be performed on

[1] This is particularly true of Equatorial Guinea, where Article 11 of Law No. 9/2006 of 3 November 2006 Enacting the Mining Code provides as follows:

> *'[e]l Estado, à través del Ministerio de Minas, Industria y Energía, otorgara concesiones mineras mediante Contratos de Exploración Minera conforme al modelo del Anexo A que forma parte integrante de la presente Ley.'*
>
> [Free translation into English: '[t]he State, through the Ministry of Mines, Industry and Energy, will grant mining concessions through Mining Exploration Contracts in accordance with the model of Annex A, which forms an integral part of the present Law'].

[2] Among many applications, we can mention: in Ethiopia, Article 7(1) of Proclamation No. 678/2010 Promoting Sustainable Development of Mineral Resources, which provides: '[n]o person may undertake mining operations without having obtained the relevant licence under this Proclamation'. In Congo, Article 10 of Law No. 4-2005 of 11 April 2005 Enacting the Mining Code; in Côte d'Ivoire, Article 5 of Law No. 2014-138 of 24 March 2014 Enacting the Mining Code.

[3] As mentioned above, the granting of mining economic rights may result from operating contracts entered into between the host state and the private operator.

[4] As an illustration, in Ethiopia, Article 7(5) of Proclamation No. 678/2010 Promoting Sustainable Development of Mineral Resources provides: '[a]ny Ethiopian may conduct reconnaissance without having licence provided that he does not interfere with the rights of a licensee or any other person'.

the basis of economic rights resulting from the signature of an exploitation agreement with the host state.

Since the granting of mining rights remains the standard, particularly in the mining sector, it is necessary to successively analyse the concept of mining rights and the procedures for granting such title.

8.3 There is a great diversity of mining rights depending of the different stages of the project. Depending of the phase concerned, the mining and petroleum legislation will provide for the granting of prospecting authorisation, exploration permit and exploitation or production permit.

The scope and content of mining rights vary according to the phase in question but also according to the country concerned.

1 CLASSIFICATION OF MINING RIGHTS

8.4 Mining and petroleum operations give rise in principle to the granting of three separate mining rights, each of which corresponds to the different phases of exploitation operations.[5] Generally, according to the designations used, these include:

- prospecting or reconnaissance permits or authorisations,
- exploration or research permits, and
- production or exploitation permits and concessions.

This statement must be balanced considering that some national laws provide for the granting of additional mining titles. This is the case for instance of retention permits or licences which are granted by the licensing authority to the applicant where a deposit with a potential commercial significance has been discovered and where the deposit in question cannot be developed immediately for temporary reasons (such as adverse market conditions, other economic factors, or unavailable processing technologies).[6]

a Prospecting permits or authorisations

8.5 The prospecting phase, also known as 'reconnaissance',[7] is generally considered to be the first stage of mining and petroleum operations. It makes it possible to carry out the preliminary works of surface observation or manual surveys in order to highlight sites likely to present a mining interest.[8]

[5] For more details on the different phases of operation see above para 3.12 et seq.

[6] See, for example, in Ethiopia, article 23 of the Proclamation No. 678/2010, A Proclamation to Promote Sustainable Development of Mineral Resources.

[7] See, for example, in Guinea, whose Mining and Petroleum Codes provide for the granting of reconnaissance authorisations (Articles 42 et seq of Law No. 2011/006/CNT of 9 September 2011 Enacting the Mining Code and Articles 12 et seq of Law No. L/2014/n°034/AN Enacting the Petroleum Code).

[8] Many mining and petroleum codes define the concept of prospecting. See, in particular, in Algeria, Article 14 of Law No. 01-10 of 3 July 2001 Enacting the Mining Law; in Benin, Article 1 of Law No. 2006-17 of 17 October 2006 Enacting Mining and Mining Taxation Code; in Burundi, Article 2(26) of Law No. 1-21 of 15 October 2013 Enacting the Mining Code; in the Central African Republic, Article 1 of Law No. 09-005 of 29 April 2009 Enacting the Mining Code; in Gabon, Article 22 of Law No. 05-2000 Enacting the Mining Code; in the Seychelles, Article 2 of the Minerals Act; in Chad, Article 13 of Law No. 011-PR-1995 of 20 June 1995

The prospecting activity in practice corresponds to the first stage of the exploration phase.[9] The following are considered to be prospecting activities: topographic, geological and geophysical studies, site reconnaissance or any other preliminary investigation work. The preliminary nature of prospecting operations presupposes the completion of surface or subsurface work, work involving deep drilling being excluded from the prospecting phase.[10] The depth of drilling may be the subject of clarifications and may vary according to the states and perimeters concerned. Thus, the mining and petroleum codes may provide that prospecting works of a certain depth are subject to prior declaration.[11]

8.6 In principle, the right to prospect can only be acquired by virtue of a prospecting licence or authorisation issued by the licensing authority.[12] However, certain national laws may, on the contrary, provide that the right to conduct prospecting within the national territory is free, subject to a prior declaration by the private operator[13] and the necessary authorisations from the private owners of the areas concerned.

8.7 Thus, no one can in principle undertake prospecting operations except on the basis of an authorisation. This authorisation is not generally considered as constituting a mining title.[14] This observation must, how-

Enacting the Mining Code; in Tunisia, Article 1 of Law No. 2003-30 of 28 April 2003 Enacting the Mining Code.

[9] Thus, in Algeria, Article 13 of Law No. 1-10 of 3 July 2001 Enacting Mining Law provides that mining exploration is subdivided into two stages: prospecting and exploration.

[10] In Niger, for example, Article 28 of Law No. 2007-01 of 31 January 2007 Enacting the Petroleum Code excludes from prospecting operations drilling at depth of more than 300 meters.

[11] For instance, in Malawi, Article 23(2) of the Mines and Minerals Act, 1981, which provides: '[t]he holder of a reconnaissance licence shall not engage in drilling, excavation or other subsurface techniques, except where and to the extent that he is authorised by his licence to use any of those techniques'.

[12] See, in Algeria, Article 94 of Law No. 01-10 of 3 July 2001 On Mining Law; in Benin, Articles 4 and 14 of Law No. 2006-17 of 17 October 2006 On the Mining Code and Mining Taxation in the Republic of Benin and Article 5 of Law No. 2006-18 of 17 October 2006 On the Petroleum Code; in Niger, Article 29 of Law No. 2007-01 of 31 January 2007 On the Petroleum Code; in Togo, Article 9 of Law No. 99-003 of 18 February 1999 On the Hydrocarbons Code.

[13] See, in particular, in the Democratic Republic of Congo, Article 17 of Law No. 007-2002 of 11 July 2002 On the Mining Code. In this case, the licensing authority will issue a certificate of prospecting to the private operator (Article 18 of the said Law). See also, in Madagascar, Article 20 of Law No. 99-022 of 19 August 1999 On the Mining code amended by Law No. 2005-021 of 17 October 2005. In this case, although the prospecting is not subject to authorisation the operator may nevertheless request that it be granted exclusive authorisation to reserve the perimeter (Article 21 et seq of the said Law).

[14] See, in particular, in Benin, Article 10 of Law No. 2006-18 of 17 October 2006 On the Petroleum Code; in Cameroon, Article 23(2) of Law No. 99-013 of 22 December 1999 Enacting the Petroleum Code; in the Central African Republic, Article 10 of Ordinance No. 93.007 of 25 May 1993 On the Petroleum Code; in Côte d'Ivoire, Article 11 of Law No. 96-669 of 29 August 1996 On the Petroleum Code and Article 1 of Law No. 2014-138 of 24 March 2014 On the Mining Code.

ever, be qualified as prospecting might be authorised by a 'prospecting licence'.[15]

8.8 It should also be noted that, in the case of artisanal mining, prospecting operations may take place under specific authorisations.[16] Furthermore and as an exception, certain laws provide for the signing of a prospecting agreement between the State and the applicant for a prospecting licence.[17] Indeed, the signing of such agreement usually takes place once the exploration permit is granted.

8.9 The prospecting authorisation is generally given in a precarious and revocable manner;[18] any possible withdrawal or the restrictions which may be placed on it do not give rise to compensation;[19] this is particularly the case when the legislation provides for the lapse of the prospecting authorisation in the event of the award of mining rights or the conclusion of an exploitation agreement on the perimeter to another operator,[20] the former thus losing its rights without compensation.

The right to carry out prospecting work is not exclusive[21] and can be

[15] See, in particular, in Eritrea, Article 8 of Proclamation No. 68/1995. Promoting the Development of Mineral Resources; in Ghana, Article 37 of the Minerals and Mining Act, 2006; Lesotho, Article 27 of the Mines and Minerals Act, 2005; in Liberia, Article 5.2 of the Minerals and Mining Law, 2000; in Malawi, Article 22 of the Mines and Minerals Act, 1981; in Namibia, Article 16 of the Minerals Prospecting and Mining Act, 1992; in Uganda, Article 23 of the Mining Act, 2003; in Sierra Leone, Article 64 of the Mines and Minerals Act, 2009; in Swaziland, Article 46 of the Mines and Minerals Act, 2011.

[16] This is particularly the case in Guinea, where prospecting in the context of artisan exploitation is authorised by the issuance of a '*Prospector's Individual Map*' (Article 43 of Law No. 2011/006/CNT of 9 September 2011 Enacting the Mining Code).

[17] See in Chad, Article 26 (5) of Law No. 006-PR-2007 of 2 May 2007 On Hydrocarbons.

[18] See, in particular in Togo, Article 11(3) of Law No. 99-003 of 18 February 1999 On the Hydrocarbons Code.

[19] See, in particular, in Benin, Article 17 of Law No. 2006-18 of 17 October 2006 On the Petroleum Code; in Cameroon, Article 25(2) of Law No. 99-013 of 22 December 1999 Enacting the Petroleum Code.

[20] See, in particular, in the Central African Republic, Article 11 of Ordinance No. 93.007 of 25 May 1993 On the Petroleum Code; in Côte d'Ivoire, Article 12 of Law No. 96-669 of 29 August 1996 On the Petroleum Code; in Djibouti, Article 13 of Law No. 117-AN-05 of 16 October 2005 On the Petroleum Code; in Guinea, Article 12 of Law L/2014/N° 034/AN of 23 December 2014 Enacting the Petroleum Code; in Niger, Article 32 of Law No. 2007-01 of 31 January 2007 Enacting the Petroleum Code; in Senegal, Article 13 of Law No. 98-05 of 8 January 1998 On the Petroleum Code.

[21] See, in particular, to that effect, in Benin, Article 16 of Law No. 2006-18 of 17 October 2006 On the Petroleum Code and Article 15 of Law No. 2006-17 of 17 October 2006 On the Mining Code and Mining Taxation in the Republic of Benin; in Cameroon, Article 23(2) of Law No. 99-013 of 22 December 1999 Enacting the Petroleum Code; in the Central African Republic, Article 11 of Ordinance No. 93.007 of 25 May 1993 Enacting the Petroleum Code; in Côte d'Ivoire, Article 12 of Law No. 96-669 of 29 August 1996 Enacting the Petroleum Code and Article 46 of Law No. 2014-138 of 24 March 2014 On the Mining Code; in Djibouti, Article 12 of Law No. 117-AN-05 of 16 October 2005 On the Petroleum Code; in Ethiopia, Article 16 of Proclamation No. 678/2010 Promoting Sustainable Development of Mineral Resources; in Guinea, Article 12 of Law L/2014/N° 034/AN of 23 December 2014 On the Petroleum Code; in Mauritania, Article 13 of Law No. 2010-033 of 20 July 2010 On the Code of Crude Hydrocarbons modified by Law No. 2011-044 of 25 November 2011; in

exercised concurrently on a same perimeter.[22] Exclusivity of the right to prospect is, however, admitted by certain national countries, both in French-speaking[23] and in English-speaking[24] Africa.

8.10 Notwithstanding some exceptions,[25] the right to prospect is valid throughout the perimeter defined by the licence. Similarly, in the context of mining exploitation, the right to prospect applies to the substances mentioned in the licence or authorisation. Since the rights of the beneficiary of the prospecting licence are limited to the substances mentioned by it, national laws may offer the beneficiary the possibility of applying for the extension of the permit to cover substances discovered and not included in the permit.[26]

8.11 In principle, the granting of a prospecting licence does not confer any right or preference in for the granting of mining rights or the conclusion of an exploitation agreement.[27] Although not afforded any priority, the holder

---

Namibia, Article 16 of the Minerals Prospecting and Mining Act, 1992; in Niger, Article 31 of Law No. 2007-01 of 31 January 2007 On the Petroleum Code; in Uganda, Article 23(1) of the Mining Act, 2003; in Senegal, and Article 12 of Law No. 98-05 of 8 January 1998 On the Petroleum Code; in Swaziland, Article 38(1) of the Mines and Minerals Act, 2011; in Chad, Article 26(1) of Law No. 006-PR-2007 of 2 May 2007 On Hydrocarbons and Article 15 of Law No. 011-PR-1995 of 20 June 1995 On the Mining Code; in Togo, Article 10(1) of Law No. 99-003 of 18 February 1999 Enacting the Hydrocarbon Code and Article 11 of Law No. 2003-012 of 4 October 2003 Amending the Mining Code.

[22] See, in particular, in Cameroon, Article 25(1) of Law No. 99-013 of 22 December 1999 Enacting the Petroleum Code; in the Republic of Congo, Article 19 of Law No. 4-2005 of 11 April 2005 On the Mining Code; in Djibouti, Article 13 of Law No. 117-AN-05 of 16 October 2005 On the Petroleum Code; in Togo, Article 10(1) of Law No. 99-003 of 18 February 1999 On the Hydrocarbon Code; in Tunisia, Article 15 of Law No. 2003-30 of 28 April 2003 Enacting the Mining Code.

[23] See, in particular, in Mali, Article 27 of Law No. 2012-015 of 27 February 2012 On the Mining Code; in Algeria, Article 96 of Law No. 01-10 of 3 July 2001 On the Mining Law.

[24] See, in particular, in Malawi, Article 20(4) of the Mines and Minerals Act, 1981.

[25] See, in particular, in Benin, Article 15 of Law No. 2006-17 of 17 October 2006 On the Mining Code and Mining Taxation Taxes in the Republic of Benin, which provides:

'[s]ous réserve de l'article 7 de la présente loi, l'autorisation de prospection confère à son titulaire, sur tout le territoire de la République du Bénin ne faisant pas déjà l'objet d'un permis de recherche ou d'exploitation et pour toutes les substances minières ou de carrière, le droit non exclusif de procéder concurremment avec les autres titulaires d'autorisation de prospection valable, à des activités de prospection.'

[Free translation into English: '[s]ubject to provisions of Article 7 of this Law, a prospecting authorisation shall entitle its holder, throughout the territory of the Republic of Benin not already covered by an exploration or exploitation permit, and for all mining or quarrying substances, to the non-exclusive right to engage in prospecting activities concurrently with other holders of valid prospecting authorisation.']

[26] See, in particular, in this regard, in the Republic of Congo, Article 24 of Law No. 4-2005 of 11 April 2005 On the Mining Codein Lesotho, Article 25(1) of the Mines and Minerals Act, 2005; in Mali, Article 28 of Law No. 2012-015 of 27 May 2012 on the Mining Code.

[27] See, in particular, in Benin, Article 17 of Law No. 2006-18 of 17 October 2006 On the Petroleum Code, which provides that: '*[l]'obtention de l'autorisation de prospection ne confère pas d'office au titulaire le droit au permis de recherche*.'

[Free translation into English: '[t]he acquisition of the prospecting authorisation does not automatically confer on the holder the right to an investigation permit.'] See also, in

of a prospecting licence may be invited to tender, in the event that the signing of an exploitation contract is envisaged, on all or part of the perimeter covered by the permit.[28] Besides, some regulations grant a pre-emption[29] or preferential right[30] to the holder of a prospecting licence for the granting of mining rights or the signing of an exploitation agreement.

8.12 The rights granted by the prospecting authorisation on discovered substances are also limited. Thus, barring exceptions,[31] the prospecting

---

Cameroon, Article 23(3) of Law No. 99-013 of 22 December 1999 Enacting the Petroleum Code; in the Central African Republic, Article 11 of Ordinance No. 93.007 of 25 May 1993 On the Petroleum Code; in Côte d'Ivoire, Article 11 of Law No. 96-669 of 29 August 1996 On the Petroleum Code and Article 46 of Law No. 2014-138 of 24 March 2014 On the Mining Code; in Djibouti, Article 13 of Law No. 117-AN-05 of 16 October 2005 On the Petroleum Code; in Senegal, Article 13 of Law No. 98-05 of 8 January 1998 On the Petroleum Code; in Chad, Article 15 of Law No. 011-PR-1995 of 20 June 1995 On the Mining Code.

[28] Such provisions are sporadically found in mining and petroleum codes. See, in particular, in Mauritania, Article 13 of Law No. 2010-033 of 20 July 2010 Enacting the Code of Crude Hydrocarbons modified by Law No. 2011-044 of 25 November 2011.

[29] See, to that effect, in Niger, Article 15 of Ordinance No. 93-16 of 2 March 1993 On the Amended Mining Law, which provides:

'*[l]'autorisation de prospection confère à son titulaire un droit de préemption sur le périmètre sollicité, dans les limites et la durée de l'autorisation. Elle n'a aucun caractère exclusif.*'

[Free translation into English: '[t]he prospecting authorisation confers on its holder a pre-emption right on the requested perimeter, within the limit of the term of the licence. It is not of an exclusive nature.'].

[30] See, in particular, to that effect, in Cameroon, Article 23(4) of Law No. 99-013 of 22 December 1999 Enacting the Petroleum Code, which provides:

'*[n]onobstant ce qui précède et si des circonstances exceptionnelles le justifient, notamment pour les Zones d'Opérations Pétrolières Particulières, l'Autorisation de Prospection peut prévoir, pendant sa durée de validité, en faveur de son Titulaire, soit un droit de préférence, à conditions équivalentes en cas de conclusion éventuelle d'un Contrat Pétrolier sur tout ou partie du même périmètre, soit une exclusivité de durée limitée pour conclure un Contrat Pétrolier sur tout ou partie du perimeter.*'

[Free translation into English: '[n]otwithstanding the foregoing and if exceptional circumstances so warrant, in particular for Private Oil Operations Areas, the Prospecting Authorisation may provide, during its term of validity, in favour of its holder, either a preferential right, on equivalent terms in the event of possible conclusion of a Petroleum Agreement on all or part of the same perimeter, or exclusivity of limited duration to conclude a Petroleum Agreement on all or part of the perimeter'].

See also, in the Central African Republic, Article 11 of Ordinance No. 93.007 of 25 May 1993 On the Petroleum Code; in Côte d'Ivoire, Article 12 of Law No. 96-669 of 29 August 1996 On the Petroleum Code.

[31] See, in particular, in Guinea, Article 44 of Law No. 2011/006/CNT of 9 September 2011 On the Mining Code, which provides:

'*[l]e titulaire d'une autorisation de reconnaissance a droit à la libre disposition des produits extraits à l'occasion de ses recherches et essais, à condition que ces travaux ne revêtent pas le caractère de travaux d'exploitation, sous réserve d'en faire la déclaration à la Direction Nationale des Mines et de respecter toutes les dispositions prises par la réglementation minière concernant les substances minières extraites.*'

[Free translation into English: '[t]he holders of a prospecting authorisation have the right to dispose freely of all products extracted in the course of their exploration and assays, provided that these works do not fall within mining activities and provided that the said licence or authorisation does not confer on its holder any right to exploit or dispose of discovered substances.[32] However, prospecting operations require that samples be taken for analysis.[33] In the context of the analysis of discovered substances, the beneficiary of the licence or authorisation may then move the extracted substances within and, under certain conditions, outside the national territory.[34]

8.13 The ownership of the data obtained constitutes a considerable challenge in the framework of prospecting operations. In this regard, national laws may grant the state ownership of data obtained in connection with prospecting operations. The interests of the state will also be protected by a series of provisions requiring the holder of the licence to transfer to the licensing authority, free of charge, the information gathered and the results obtained during the prospecting phase,[35] in particular in the form of a report.[36] Exceptionally, national legislation may provide that the data obtained during the prospecting phase may be assigned by the holder of the

---

disposal is declared to the National Mining Authority and provided that all the provisions of the mining regulations are observed with respect to the extracted mining substances'].

[32] See, in particular, in Benin, Article 18 of Law No. 2006-17 of 17 October 2006 On the Mining Code and Mining Taxation in the Republic of Benin; in the Central African Republic, Article 11 of Ordinance No. 93.007 of 25 May 1993 On the Petroleum Code; in Côte d'Ivoire, Article 12 of Law No. 96-669 of 29 August 1996 Enacting the Petroleum Code and Article 46 of Law No. 2014-138 of 24 March 2014 On the Mining Code; in Chad, Article 15 of Law No. 011-PR-1995 of 20 June 1995 On the Mining Code.

[33] See, in particular, to that effect, in Sierra Leone, Article 64 (2) of the Mines and Minerals Act, 2009.

[34] See, in particular, in the Republic of Congo, Article 19 of Law No. 4-2005 of 11 April 2005 On the Mining Code; in the Democratic Republic of Congo, Article 21 of Law No. 007-2002 of 11 July 2002 On the Mining Code.

[35] For example, in Algeria, Article 22 of Law No. 13-01 of 20 February 2013 Amending and Supplementing Law No. 05-07 of 28 April 2005 On Hydrocarbons provides:

> *'[t]outes données et résultats issus des travaux de prospection doivent être mis à la disposition de l'agence nationale pour la valorisation des ressources en hydrocarbures (ALNAFT).'*
>
> [Free translation into English: '[a]ll data and results obtained from prospecting work must be made available to the National Agency for the Exploitation of Hydrocarbon Resources (ALNAFT)'].

See also, in Algeria, Article 101 of Law No. 01-10 of 3 July 2001 Enacting the Mining Law; in the Central African Republic, Article 11 of Ordinance No. 93.007 of 25 May 1993 Enacting the Petroleum Code; in Côte d'Ivoire, Article 12 of Law No. 96-669 of 29 August 1996 Enacting the Petroleum Code; in Guinea, Article 42 of Law No. 2011/006/CNT of 9 September 2011 Enacting the Mining Code and Article 12 of Law No. L/2014/N°034/AN of 23 December 2014 Enacting the Petroleum Code; in Lesotho, Article 28(3) and (4) of the Mines and Minerals Act, 2005; in Morocco, Article 21 of Law No. 21-90 of 1 April 1992 On the Exploration and Exploitation of Hydrocarbon Deposits amended and supplemented by Law No. 27-99 of 15 February 2000; in Mauritania, Article 14 of Law No. 2010-033 of 20 July 2010 Enacting the Petroleum Code modified by Law No. 2011-044 of 25 November 2011; in Tunisia, Article 18 of Law No. 2003-30 of 28 April 2003 Enacting the Mining code; in Chad, Article 26 (6) of Law No. 006-PR-2007 of 2 May 2007 On Hydrocarbons.

[36] See in particular, in Algeria, Article 101 of Law No. 01-10 of 3 July 2001 Enacting the Mining Law; in Benin, Article 17 of Law No. 2006-17 of 17 October 2006 Enacting the Mining and Mining Taxation Code; in Uganda, Article 24 of the Mining Act, 2003.

permit, in which case the profits derived from such assignment must be distributed between the host state and the holder of the licence.[37]

8.14 Perimeters covered by an exploration agreement[38] or production mining rights[39] are in principle excluded from the areas covered by the prospecting authorisation. This being so, some laws authorise the granting of a prospecting authorisation and exploration mining rights on the same perimeter, while the rights of the holder of the mining rights prevail over the rights conferred on the holder of a prospecting authorisation.[40] In such case, the lack of clearly defined operating rules is likely to create tension and practical difficulties between the holders.

8.15 Among the obligations incumbent upon the beneficiary of a prospecting licence or authorisation, mention should be made in particular to the obligation of reporting a possible discovery to the competent authority[41] or to the obligation to start prospecting works within a certain period of time following the granting of the right to prospect.[42] Failure to fulfil the obligations incumbent upon the holder of the prospecting licence or authorisation may in certain circumstances lead to the withdrawal of the said right by the licensing authority.[43]

8.16 In general, it should be emphasised that the right to prospect is granted for a relatively short period of time.[44] This right may be

---

[37] See in Guinea, Article 12 of Law No. L/2014/N°034/AN of 23 December 2014 Enacting the Petroleum Code.

[38] See, in particular, in Algeria, Article 20 of Law No. 13-01 of 20 February 2013 Amending and Supplementing Law No. 05-07 of 28 April 2005 On Hydrocarbons; in Cameroon, Article 23(1) of Law No. 99-013 of 22 December 1999 Enacting the Petroleum Code.

[39] See, in particular, in Benin, Article 16 of Law No. 2006-18 of 17 October 2006 Enacting the Petroleum Code.

[40] See, in particular, in Benin, Article 16 of Law No. 2006-18 of 17 October 2006 Enacting the Petroleum Code; in Niger, Article 30 of Law No. 2007-01 of 31 January 2007 Enacting the Petroleum Code.

[41] See, in particular, to that effect, in Kenya, Article 67(1) of the Mining Act, 2016; in Lesotho, Article 28(1) of the Mines and Minerals Act, 2005; in Uganda, Article 24 of the Mining Act, 2003; in Swaziland, Article 39 of the Mines and Minerals Act, 2011.

[42] See, in particular, in Ghana, Article 37(2) of the Minerals and Mining Act, 2006; in Kenya, Article 67(1) of the Mining Act, 2016; in Lesotho, Article 28(1) of the Mines and Minerals Act, 2005; in Sierra Leone, Article 65(1) of the Mines and Minerals Act, 2009.

[43] See, in particular, in Guinea, Article 15 of Law No. L/2014/n°034/AN of 23 December 2014 Enacting the Petroleum Code.

[44] For prospecting authorisation with a term of three years, see, in particular: in Benin, Article 16 of Law No. 2006-17 of 17 October 2006 Enacting the Mining and Mining Taxation Code; in Ghana, Article 34(2) of the Minerals and Mining Act, 2006; in Mali, Article 30 of Law No. 2012-015 of 27 February 2012 Enacting the Mining Code.

For prospecting authorisations with a term of two years, see, in particular: in Benin, Article 17 of Law No. 2006-18 of 17 October 2006 Enacting the Petroleum Code; in Lesotho, Article 24(1) of the Mines and Minerals Act, 2005; in Senegal, Article 12 of Law No. 98-05 of 8 January 1998 Enacting the Petroleum Code; in Togo, Article 11(3) of Law No. 99-003 of 18 February 1999 Enacting the Hydrocarbon Code and Article 11 of Law No. 2003-012 of 4 October 2003 Amending the Mining Code; in Zimbabwe, Article 23 of the Mines and Minerals Act.

extended[45] or renewed.[46] Finally, it should be noted that some African legislation does not authorise the renewal of prospecting licences or authorisations.[47]

b Exploration rights

8.17 The research or exploration phase comprises all the activities conducted with the goal of discovering or determining the existence of deposits, delimiting them and evaluating the scope thereof and the possibili-

---

For prospecting authorisations with a term of 18 months, see, in particular: in Ethiopia, Article 17 of Proclamation No. 678/2010 Promoting sustainable development of mineral resources.

For prospecting authorisations with a term of one year, see, in particular: in Algeria, Article 97 of Law No. 01-10 of 3 July 2001 Enacting the Mining Law; in the Central African Republic, Article 10 of Ordinance No. 93.007 of 25 May 1993 Enacting the Petroleum Code; in Côte d'Ivoire, Article 11 of Law No. 96-669 of 29 August 1996 Enacting the Petroleum Code and Article 47 of Law No. 2014-138 of 24 March 2014 Enacting the Mining Code; in Eritrea, Article 8(2) of Proclamation No. 68/1995 Promoting Development of Mineral Resources; in Guinea, Article 14 of Law No. L/2014/n° 034/AN of 23 December 2014 Enacting the Petroleum Code; in Malawi, Article 21 of the Mines and Minerals Act, 1981; in Morocco, Article 20 of Law No. 21-90 of 1 April 1992 On to the Exploration and Exploitation of Hydrocarbon Deposits amended and supplemented by Law No. 27-99 of 15 February 2000; in Mauritania, Article 13 of Law No. 2010-033 of 20 July 2010 Enacting the Code of Crude Hydrocarbons amended by Law No. 2011-044 of 25 November 2011; in Niger, Article 17 of Ordinance No. 93-16 of 2 March 1993 Amending the Mining Act; in Namibia, Article 22 of the Minerals Prospecting and Mining Act, 1992; in South Sudan, Article 17(4) of the Petroleum Act, 2012; in Chad, Article 15 of Law No. 011-PR-1995 of 20 June 1995 Enacting the Mining Code; in Tunisia, Article 14 of Law No. 2003-30 of 28 April 2003 Enacting the Mining Code.

For prospecting authorisations with a term of six months, see, in particular: in Guinea, Article 45 of Law No. 2011/006/CNT of 9 September 2011 Enacting the Mining Code.

[45] See, in particular, in Algeria, Article 97 of Law No. 01-10 of 3 July 2001 Enacting the Mining Law; in Benin, article 17 of Law No. 2006-18 of 17 October 2006 Enacting the Petroleum Code; in Morocco, Article 20 of Law No. 21-90 of 1 April 1992 On the Exploration and Exploitation of Hydrocarbon Deposits amended and supplemented by Law No. 27-99 of 15 February 2000.

[46] See, in particular, in Benin, Article 15 of Law No. 2006-17 of 17 October 2006 Enacting the Mining and Mining Taxation Code; in the Central African Republic, Article 10 of Ordinance No. 93.007 of 25 May 1993 Enacting the Petroleum Code; in Côte d'Ivoire, Article 11 of Law No. 96-669 of 29 August 1996 Enacting the Petroleum Code and Article 47 of Law No. 2014-138 of 24 March 2014 Enacting the Mining Code; in Guinea, Article 14 of Law No. L/2014/n°034/AN of 23 December 2014 Enacting the Petroleum Code; in Lesotho, Article 24(2) of the Mines and Minerals Act, 2005; in Mali, Article 30 of Law No. 2012-015 of 27 February 2012 Enacting the Mining Code; in Mauritania, Article 13 of Law No. 2010-033 of 20 July 2010 Enacting the Code of Crude Hydrocarbons modified by Law No. 2011-044 of 25 November 2011; in Niger, Article 17 of Ordinance No. 93-16 of 2 March 1993 Amending the Mining Act; in Chad, Article 15 of Law No. 011-PR-1995 of 20 June 1995 Enacting the Mining Code; in Togo, Article 11(3) of Act No. 99-003 d of 18 February 1999 Enacting the Hydrocarbon Code; in Tunisia, Article 14 of Law No. 2003-30 of 28 April 2003 Enacting the Mining Code.

[47] See in particular, in Eritrea, Article 8(2) of Proclamation No. 68/1995 Promoting Development of Mineral Resources; in Ethiopia, Article 17 of Proclamation No. 678/2010 Promoting the Sustainable Development of Mineral Resources.

ties of commercial exploitation.[48] The exploration phase is subsequent to and different from prospecting operations, in particular because it generally involves carrying out works not only on the surface but also at depths.

8.18 Exploration activities are in principle reserved for those holding a mining right: the exploration permit.[49] However, exploration activities do not necessarily give rise to the granting of a mining right and may lead to the granting of simple authorisations by administrative means, in particular in the case of artisanal mining or quarrying.

8.19 The exploration permit will most often be accompanied by an exploitation agreement.[50] The signature of this agreement is sometimes a prerequisite to the granting of the exploitation permit.[51] Moreover, the agreement will in most cases cover the phases subsequent to exploration activities, namely development and production following a commercial discovery.

[48] For examples of definitions of exploration activities, see, in particular: in Benin, Article 1 of Law No. 2006-17 of 17 October 2006 Enacting the Mining Code and Mining Taxation.

For other definitions of exploration activities, see also; in Burundi, Article 2(27) of Law No. 1-21 of 15 October 2013 Enacting the Mining Code; in the Central African Republic, Article 1 of Law No. 09-005 of 29 April 2009 Enacting the Mining Code; in Guinea, Article 1 of Law No. L/2014/n°034/AN of 23 December 2014 Enacting the Petroleum Code; in Madagascar, Article 2 of Law No. 99-022 of 19 August 1999 Enacting the Mining Code amended by Law No. 2005-021 of 17 October 2005; in Mali, Article 1(44) of Law No. 2012-015 of 27 February 2012 Enacting the Mining Code; in Morocco, Article 2(5) of Law No. 21-90 of 1 April 1992 On the Exploration and Exploitation of Hydrocarbon Deposits amended and supplemented by Law No. 27-99 of 15 February 2000; in Niger, Article 20 of Ordinance No. 93-16 of 2 March 1993 on the amended Mining Law; in Uganda, Article 2 of the Mining Act, 2003; in Chad, Article 17 of Law No. 011-PR-1995 of 20 June 1995 Enacting the Mining Code; in Togo, Article 13 of Law No. 96-004 of 26 February 1996 Enacting the Mining Code.

[49] See, in particular, in Algeria, Article 102 of Law No. 01-10 of 3 July 2001 Enacting the Mining Law, which provides: '*[n]ul ne peut entreprendre des travaux d'exploration minière s'il n'est titulaire d'un permis d'exploration.*'

[Free translation into English: '[n]o person may engage in mining exploration works unless such person holds an exploration permit'.]

See also in Benin Articles 4 and 19 of Law No. 2006-17 of 17 October 2006 Enacting the Mining and Mining Taxation Code and Article 6 of Law No. 2006-18 of 17 October 2006 Enacting the Petroleum Code; in Burundi, Article 39 of Law No. 1-21 of 15 October 2013 Enacting the Mining Code; in Gabon, Article 31 of Law No. 05-2000 of 12 October Enacting the Mining Code; in Morocco, Article 31 of Law No. 33-13 On Mines; in Niger, Article 34 of Law No. 2007-01 of 31 January 2007 Enacting the Petroleum Code; in Togo, Article 12(1) of Law No. 99-003 of 18 February 1999 Enacting the Hydrocarbons Code.

[50] See, in particular, in Gabon, Article 35 of Law No. 05-2000 of 12 October Enacting the Mining Code; in Senegal, Article 17 of Law No. 98-05 of 8 January 1998 Enacting the Petroleum Code; in Benin, Article 23 of Law No. 2006-17 of 17 October 2006 Enacting the Mining and Mining Taxation Code.

[51] See, in particular, in Togo, Article 15 of Law No. 99-003 of 18 February 1999 Enacting the Hydrocarbons Code.

8.20 The exploration permit confers on its holder, within the perimeter, the right to search for the substances specifically defined in such permit. Unlike prospecting licences or authorisations, the exploration permit confers on the holder an exclusive right.[52] Thus, no mining right can cover a perimeter already subject to an exploration permit.[53] However, the existence of an exploration permit does not preclude the granting of another exploration permit for different substances, such as quarry substances[54] or

[52] See, in particular, in Algeria, Article 105 of Law No. 01-10 of 3 July 2001 Enacting the Mining Law; in Benin, Articles 1 and 18 of Law No. 2006-18 of 17 October 2006 Enacting the Petroleum Code; in Cameroon, Article 27 of Law No. 99-013 of 22 December 1999 Enacting the Petroleum Code; in the Central African Republic, Article 12 of Ordinance No. 93.007 of 25 May 1993 Enacting the Petroleum Code and Article 20 of Law No. 09-005 of 29 April 2009 Enacting the Mining Code; in the Comoros, Article 8 of Decree-Law No. 54-1110 of 13 November 1954 On the Reform of the Regime of Mineral Substances; in the Republic of Congo, Article 27 of Law No. 4-2005 of 11 April 2005 On the Mining Code; in Côte d'Ivoire, Article 20 of Law No. 2014-138 of 24 March 2014 Enacting the Mining Code and Article 21 of Law No. 96-669 of 29 August 1996 Enacting the Petroleum Code; in Djibouti, Article 14 of Law No. 117-AN-05 of 16 October 2005 Enacting the Petroleum Code; in Ethiopia, Article 18(1) of Proclamation No. 678/2010 Promoting Sustainable Development of Mineral Resources; in Guinea, Article 19 of Law No. L/2013/n°053/CNT of 8 April 2013 Amending Law No. 2011/006/CNT of 9 September 2011 Enacting the Mining Code; in Madagascar, Article 33 of Law No. 99-022 of 19 August 1999 Enacting the Mining Code amended by Law No. 2005-021 of 17 October 2005; in Morocco, Article 23 of Law No. 21-90 of 1 April 1992 On the Exploration and Exploitation of Hydrocarbon Deposits amended and supplemented by Law No. 27-99 of 15 February 2000; in Niger, Article 21 of Ordinance No. 93-16 of 2 March 1993 On the Amended Mining Law and Article 35 of Law No. 2007-01 of 31 January 2007 Enacting the Petroleum Code; in Uganda, Section 31(1) of the Mining Act, 2003 and Section 60 of the Petroleum (Exploration, Development and Production) Act, 2013; in Senegal, Article 14 of Law No. 98-05 of 8 January 1998 Enacting the Petroleum Code; in Sierra Leone, Article 77(1) of the Mines and Minerals Act, 2009; in Chad, Article 19 of Law No. 011-PR-1995 of 20 June 1995 Enacting the Mining Code and Article 27(5) of Law No. 006-PR-2007 of 2 May 2007 Enacting the Hydrocarbons Code; in Togo, Article 14 of Law No. 2003-012 of 4 October 2003 Amending and Supplementing Law No. 96-004 of 26 February 1996 Enacting the Mining Code and Article 13(1) of Law No. 99 -003 of 18 February 1999 Enacting the Hydrocarbon Code; in Tunisia, Article 18(1) of Law No. 99-93 of 17 August 1999 Enacting the Code of Hydrocarbons and Article 31 of Law No. 2003-30 of 28 April 2003 Enacting the Mining Code.

[53] See, in particular, in Madagascar, Article 36 of Law No. 99-022 of 19 August 1999 Enacting the Mining Code amended by Law No. 2005-021 of 17 October 2005, which provides:

> '*[t]ant qu'un périmètre est couvert par un permis de recherche, aucun permis minier, aucune autorisation exclusive de réservation de périmètre ne peuvent y être octroyés, hormis le permis d'exploitation sollicité par le titulaire dudit permis de recherche*.'
>
> [Free translation into English: '[i]nsofar as a perimeter is covered by an exploration permit, no mining permit, no exclusive authorisation to reserve the perimeter can be granted thereupon, except the exploitation permit requested by the holder of the exploration permit'].

See also, in Tunisia, Article 28 of Law No. 2003-30 of 28 April 2003 Enacting the Mining Code; in Benin, Article 25 of Law No. 2006-17 of 17 October 2006 Enacting the Mining and Mining Taxation Code.

[54] See, in particular, in Burundi, Article 40 of Law No. 1-21 of 15 October 2013 Enacting the Mining Code; in Benin, Article 25 of Law No. 2006-17 of 17 October 2006 Enacting the Mining and Mining Taxation Code; in the Central African Republic, Article 20 of Law No. 09-005 of 29 April 2009 Enacting the Mining Code; in Côte d'Ivoire, Article 21 of Law

hydrocarbons, when the exploration permit covers mineral substances.[55] The granting of such rights, on a perimeter covered by a mining permit, presupposes that the works will not interfere with the exploration activities and may require the prior consent of the permit holder. A general agreement either between each permit holder and the mining authority or between the mining authority and all permits holders, is useful in practice but is often poorly designed or simply overlooked.

8.21 Exploration activities are carried out by the holder of an exploration permit (within the limits of the perimeter) on the surface but also in depth.[56]

In order to avoid the numerous abuses found that freeze surface areas without carrying out exploration works; national legislation increasingly establishes a maximum or minimum surface[57] and specifies the form of the perimeter covered by the exploration mining right.[58]

No. 2014-138 of 24 March 2014 Enacting the Mining Code; in Niger, Article 26 of Ordinance No. 93-16 of 2 March 1993 Amending the Mining Law.

[55] See, in particular, Article 26 of Ordinance No. 93-16 of 2 March 1993 Amending the Mining Law.

[56] See, in particular, in Benin, Article 20 of Law No. 2006-17 of 17 October 2006 Enacting the Mining and Mining Taxation Code, which provides:

'*[l]e permis de recherche confère, dans les limites de son périmètre et indéfiniment en profondeur, le droit exclusif de prospection et de recherche de toutes les substances minières ou de carrière qui en font l'objet.*'

[Free translation into English: '[t]he exploration permit grants, within the limits of its perimeter and indefinitely in depth, the exclusive right to prospect and search for all mining or quarrying substances that are subject thereto.']

See also, in Benin, Article 18 of Law No. 2006-18 of 17 October 2006 Enacting the Petroleum Code; in Burundi, Article 49 of Law No. 1-21 of 15 October 2013 Enacting the Mining Code; in Cameroon, Article 27 of Law No. 99-013 of 22 December 1999 Enacting the Petroleum Code; in the Central African Republic, Article 12 of Ordinance No. 93.007 of 25 May 1993 Enacting the Petroleum Code and Article 20 of Law No. 09-005 of 29 April 2009 Enacting the Mining Code; in the Republic of Congo, Article 27 of Law No. 4-2005 of 11 April 2005 Enacting the Mining Code; in Guinea, Article 19 of Law No. L/2013/n°053/CNT of 8 April 2013 Amending Law No. 2011/006/CNT of 9 September 2011 Enacting the Mining Code; in Mali, Article 35 of Law No. 2012-015 of 27 February 2012 Enacting the Mining Code; in Niger, Article 21 of Ordinance No. 93-16 of 2 March 1993 Enacting the Amended Mining Law; in Togo, Article 13(1) of Law No. 99-003 of 18 February 1999 Enacting the Hydrocarbons Code.

[57] See, in particular, in Benin, Article 22 of Law No. 2006-17 of 17 October 2006 On Mining and Mining Taxation Code; in the Central African Republic, Article 22 of law No. 09-005 of 29 April 2009 Enacting the Mining Code; in the Democratic Republic of Congo, Article 53 of Law No. 007-2002 of 11 July 2002 Enacting the Mining Code; in Côte d'Ivoire, Article 23 of Law No. 2014-138 of 24 March 2014 Enacting the Mining Code; in Guinea, Article 19 of Law No. 2011/006/CNT of 9 September 2011 Enacting the Mining Code amended by Law No. L/2013/N°053/CNT of 8 April 2013 Amending Law No. L/2011/006 /CNT of 9 September 2011 Enacting the Mining Code; in Morocco, Article 25 of Law No. 21-90 of 1 April 1992 On the Exploration and Exploitation of Hydrocarbon Deposits amended and supplemented by Law No. 27-99 of 15 February 2000; In Niger, Article 24 of Ordinance No. 93-16 of 2 March 1993 Amending the Mining Law; in Uganda, Article 27(2) of the Mining Act, 2003; in Sierra Leone, Article 75 of the Mines and Minerals Act, 2009; In South Sudan, Article 45(1) of the Mining Act, 2012; in Chad, Article 21 of Law No. 011-PR-1995 of 20 June 1995 Enacting the Mining Code; in Togo, Article14 of Law No. 2003-012 of 4 October 2003 Amending and

8.22 The exploration operations may only involve the substances covered by the exploration permit. However, the holder of the permit may request extension of the exploration permit to other substances should the work reveal the presence of substances not covered by the exploration permit.

The extension which may be requested by the holder of an exploration permit may cover the substances concerned but also the perimeter covered by the mining right.[59] Certain regulations provide that, in the event of discovery of substances not covered by the mining right, the extension of the permit to these new substances is *de jure*.[60]

8.23 Exploration permits generally confer on the holder the right to dispose of the resources extracted during exploration works,[61] sometimes under certain conditions. For instance, the right to dispose of such substances

---

Supplementing Law No. 96-004 of 26 February 1996 Enacting the Mining Code; in Tunisia, Article 23 of Law No. 2003-30 of 28 April 2003 Enacting the Mining Code.

[58] See, in particular, in Burundi, Article 44 of Law No. 1-21 of 15 October 2013 Enacting the Mining Code; in the Republic of Congo, Article 28 of Law No. 4-2005 of 11 April 2005 Enacting the Mining Code; in Côte d'Ivoire, Article 23 of Law No. 2014-138 of 24 March 2014 Enacting the Mining Code; in Niger, Article 24 of Ordinance No. 93-16 of 2 March 1993 On the Amended Mining Law; in Togo, Article 17(1) of Law No. 99-003 of 18 February 1999 Enacting the Hydrocarbons Code; in Tunisia, Article 23 of Law No. 2003-30 of 28 April 2003 Enacting the Mining Code.

However, some national laws leave it to the holders of mining rights to delimit the surface and the shape of the perimeter without stipulating special conditions. In particular, in Benin, this is the case with Article 25 of Law No. 2006-18 of 17 October 2006 Enacting the Petroleum Code, which provides that: '*[l]es permis de recherche d'hydrocarbures ou permis H peuvent concerner un périmètre présentant une forme et une superficie quelconques qui sont définies dans l'acte qui les a institutes*'. [Free translation into English: '*[t]he licences or H permits for hydrocarbon exploration may relate to a perimeter having any shape and surface area as defined in the act that institutes them*'].

[59] See, in particular, in Algeria, Article 111 of Law No. 01-10 of 3 July 2001 On Mining Law which provides that:

> '*[s]i les travaux entrepris dans le périmètre d'un permis d'exploration font apparaître la nécessité de les étendre à d'autres substances, autres que celles visées par le titre minier, et/ou de les étendre à l'extérieur du périmètre attribué, le titulaire peut demander l'extension du titre à ces substances, et/ou l'intégration de zones contiguës au périmètre initial.*'
>
> [Free translation into English: '[i]f the works undertaken within the scope of an exploration permit highlights the need to extend it to other substances, other than those covered by the mining right, and/or to extend them outside the allocated perimeter, the holder may request extension of the right to these substances, and/or the integration of adjacent areas to the initial perimeter'].

For other examples of provisions relating to the extension of the right to other substances or to the perimeter, see, in particular, in the Central African Republic, Article 20 of Law No. 09-005 of 29 April 2009 Enacting the Mining Code; in the Democratic Republic of Congo, Article 59 of Law No. 007-2002 of 11 July 2002 Enacting the Mining Code; in Mali, Article 36 of Law No. 2012-015 of 27 February 2012 Enacting the Mining Code.

[60] See, in particular, in Burundi, Article 39 of Law No. 1-21 of 15 October 2013 Enacting the Mining Code.

[61] See, in particular, in the Central African Republic, Article 20 of Law No. 09-005 of 29 April 2009 Enacting the Mining Code; in Côte d'Ivoire, Article 20 of Law No. 2014-138 of 24 March 2014 Enacting the Mining Code; in Mali, Article 35 of Law No. 2012-015 of

may be subject to a prior declaration to the authority.[62] In any event, prospecting activities does not constitute exploitation works[63] and cannot be of a commercial nature.[64]

8.24 The exploration permit also allows the holder to dispose of samples for investigation work,[65] in particular for the purpose of industrial testing and analysis.

8.25 The activities of the holder of an exploration permit may not take the form of production works as stated in certain legislation.[66] This principle, however, is subject to exceptions when the holder of an exploration permit is authorised to carry out exploitation works on a provisional basis. This authorisation is, as a matter of principle, provisional and allows the private operator to undertake exploitation works within the perimeter concerned by the exploration permit. In addition, the holder of an exploration permit could

---

27 February 2012 Enacting the Mining Code; in Senegal, Article 21 of Law No. 98-05 of 8 January 1998 Enacting the Petroleum Code.

[62] See, in particular, in Algeria, Article 110 of Law No. 01-10 of 3 July 2001 Enacting the Mining Law; in Cameroon, Article 27 of Law No. 98-013 of 22 December 1999 Enacting the Petroleum; in the Central African Republic, Article 16 of Ordinance No. 93.007 of 25 May 1993 Enacting the Petroleum Code and Article 24 of Law No. 09-005 of 29 April 2009 Enacting the Mining Code; in the Republic of Congo, Article 35 of Law No. 4-2005 of 11 April 2005 Enacting the Mining Code; in Côte d'Ivoire, Article 26 of Law No. 2014-138 of 24 March 2014 Enacting the Mining Code and Article 25 of Law No. 96-669 of 29 August 1996 Enacting the Petroleum Code; in Guinea, Article 27 of Law No. 2011/006/CNT of 9 September 2011 Enacting the Mining Code; Article 36 of Law No. 2007-01 of 31 January 2007 Enacting the Petroleum Code and Article 28 of Ordinance No. 93-16 of 2 March 1993 On the Amended Mining Law; in Chad, Article 24 of Law No. 011-PR-1995 of 20 June 1995 Enacting the Mining Code.

[63] See, in particular, in Benin, Article 27 of Law No. 2006-17 of 17 October 2006 Enacting the Mining and Mining Taxation Code; in the Central African Republic, Article 24 of Law No. 09-005 of 29 April 2009 Enacting the Mining Code; in Côte d'Ivoire, Article 26 of Law No. 2014-138 of 24 March 2014 Enacting the Mining Code; in Guinea, Article 27 of Law No. 2011/006/CNT of 9 September 2011 Enacting the Mining Code; in Niger, Article 28 of Ordinance No. 93-16 of 2 March 1993 Enacting the Amended Mining Law; in Chad, Article 24 of Law No. 011-PR-1995 of 20 June 1995 Enacting the Mining Code.

[64] See, in particular, in Algeria, Article 110 of Law No. 01-10 of 3 July 2001 Enacting the Mining Law.

[65] See, in particular, in Burundi, Article 52 of Law No. 1-21 of 15 October 2013 Enacting the Mining code:

> '*[l]e permis de recherche permet à son titulaire de disposer des échantillons provenant de ses travaux en vue de procéder à toute étude ou à tout essai nécessaires et à l'exécution de tous les travaux d'exploitation.*'
>
> [Free translation into English: '[t]he exploration permit allows the holder to dispose of samples resulting from its works in order to conduct any necessary study or test and to perform any exploitation works.']

[66] See, in particular, in Benin, Article 18 of Law No. 2006-18 of 17 October 2006 Enacting the Petroleum Code:

> '*[i]l est interdit de transformer les travaux de recherche en travaux d'exploitation.*'
>
> [Free translation into English: '[i]t prohibited to transform exploration works into exploitation works']

apply for the partial conversion of the exploration permit into an exploitation permit[67] .

8.26 The main obligations of the holder of an exploration permit vis-à-vis the host state and its administration are first and foremost detailed in the mining and petroleum codes. Under these obligations, the holder of the exploration mining right is, in particular, required to begin the exploration works within a certain period of time following the granting of the permit[68] and to comply with the established exploration schedule.[69] The holder of an exploration permit has to comply with the minimum work programme and the financial commitments defined in the mining title. Failure by the operator to comply with the work programme or the minimum financial commitments may trigger serious consequences and lead to a refusal to renew the exploration permit, withdrawal of the mining right, automatic reduction of the surface area of the permit and sometimes payment of a penalty.[70]

8.27 The holder of an exploration permit is also required to inform the granting authority of the results obtained over the course of the exploration works, usually in the form of a report. Failure to communicate the results could justify withdrawal of the exploration permit.[71] In order to protect the rights of the state to the data collected during the exploration works, the applicable regulations may also provide that studies carried out during exploration works automatically fall into the public domain.[72]

8.28 In addition, the mining and petroleum codes prescribe the rights and obligations of the holder of the exploration permit in the event of a discovery. Firstly, the holder of an exploration mining right must notify the licensing authority of any discovery,[73] which is in its best interest. Similarly,

---

[67] See, in particular, in the Democratic Republic of Congo, Article 63 of Law No. 007-2002 of 11 July 2002 Enacting the Mining Code.

[68] See, in particular, in Burundi, Article 53 of Law No. 1-21 of 15 October 2013 Enacting the Mining Code; in the Republic of Congo, Article 36 of Law No. 4-2005 of 11 April 2005 Enacting the Mining Code; in Ethiopia, Article 20(2) of Proclamation No. 678/2010 Promoting Sustainable Development of Mineral Resources; in Guinea, Article 26 of Law No. 2011/006/CNT of 9 September 2011 Enacting the Mining Code; in Uganda, Article 32(1) of the Mining Act, 2003; in South Sudan, Article 52(1) of the Mining Act, 2012; in Tunisia, Article 42 of Law No. 2003-30 of 28 April 2003 Enacting the Mining Code; in Niger, Article 27 of Ordinance No. 93-16 of 2 March 1993 On the Amended Mining Law.

[69] See, in particular, in the Central African Republic, Article 23 of Law No. 09-005 of 29 April 2009 Enacting the Mining Code; in Côte d'Ivoire, Article 25 of Law No. 2014-138 of 24 March 2014 Enacting the Mining Code; in Uganda, Article 32(1) of the Mining Act, 2003; in South Sudan, Article 52(1) of the Mining Act, 2012; in Chad, Article 23 of Law No. 011-PR-1995 Enacting the Mining Code.

[70] See, in particular, Article 112 of Law No. 017/2014 of 30 January 2015 Regulating the Mining Sector in Gabon.

[71] See, in particular, in Guinea, Article 50 of Law No. 2011/006/CNT of 9 September 2011 Enacting the Mining Code.

[72] See, in this regard, in Gabon, Article 195 of Law No. 17/2014 of 30 January 2015 Regulating the Mining Sector in Gabon.

[73] See, in particular, in Cameroon, Article 31(1) of Law No. 99-013 of 22 December 1999 Enacting the Petroleum Code; in the Central African Republic, Article 15 of Ordinance No. 93.007 of 25 May 1993 Enacting the Petroleum Code; in Côte d'Ivoire, Article 24 of Law No.

national laws may require the holder of an exploration permit to carry out an assessment of the deposit[74] and to delimit it[75] in order to assess feasibility of exploitation.

8.29 Besides these obligations, the holder of an exploration permit is entitled to the granting of a production mining right.[76] Thus, the holder of an exploration permit will benefit from priority,[77] or even exclusi-

96-669 of 29 August 1996 Enacting the Petroleum Code; in Senegal, Article 20 of Law No. 98-05 of 8 January 1998 Enacting the Petroleum Code; in Togo, Article 14(1) of Law No. 99-003 of 18 February 1999 Enacting the Hydrocarbons Code.

[74] See in particular, in Cameroon, Article 31(2) of Law No. 99-013 of 22 December 1999 Enacting the Petroleum Code, which provides:

'*[l]orsque la découverte d'Hydrocarbures permet de présumer de l'existence d'un gisement commercialement exploitable, le Titulaire de l'Autorisation de Recherche est tenu d'effectuer, avec diligence, les travaux nécessaires à l'évaluation du caractère commercial dudit gisement. Cette évaluation peut consister à délimiter le gisement en question et/ou à évaluer les structures et prospects avoisinants à l'intérieur du périmètre contractuel. A l'issue de ces travaux, le Titulaire établit le caractère commercial ou non de la découverte.*'

[Free translation into English: '[w]hen the discovery of Hydrocarbons leads to the presumption of the existence of a commercially exploitable deposit, the Holder of the Exploration Authorisation is required to conduct, with diligence, the work necessary for the assessment of the commercial nature of the said deposit. This assessment may involve delineating the deposit and/or assessing neighbouring structures and prospects within the contractual perimeter. At the end of this work, the Holder establishes the commercial or non-commercial nature of the discovery'.]

See also in Senegal, Article 20 of Law No. 98-05 of 8 January 1998 Enacting the Petroleum Code.

[75] See, in particular, in Benin, Article 21 of Law No. 2006-18 of 17 October 2006 Enacting the Petroleum Code; in Burundi, Article 55 of Law No. 1-21 of 15 October 2013 Enacting the Mining Code; in the Central African Republic, Article 15 of Ordinance No. 93.007 of 25 May 1993 Enacting the Petroleum Code.

[76] See, in particular, in Algeria, Article 109 of Law No. 01-10 of 3 July 2001 On the Mining Law which provides:

'*[a]u cours des travaux d'exploration, la découverte de substances minérales confère au titulaire du permis en cours de validité, le droit à l'obtention d'un titre minier d'exploitation.*'

[Free translation into English: '[d]uring the exploration work, the discovery of mineral substances gives the holder of the valid permit the right to obtain an exploitation mining right'].

See also, in Burundi, Article 51 of Law No. 1-21 of 15 October 2013 Enacting the Mining Code.

[77] See, in particular, in the Republic of Congo, Article 36 of Law No. 4-2005 of 11 April 2005 Enacting the Mining Code, which provides:

'*[l]e permis de recherches minières confère en outre à son titulaire, dans le cas des résultats fructueux des recherches, la priorité dans l'octroi des titres d'exploitation de ressources découvertes, notamment celles visées par le permis de recherches minières, dans le périmètre concerné.*'

[Free translation into English: '[t]he mining exploration permit further confers on its holder, in the event of successful exploration results, priority in the granting of exploitation rights on discovered resources, in particular those covered by the mining exploration permit, within the relevant perimeter'].

See also, in Madagascar, Article 35 of Law No. 99-022 of 19 August 1999 Enacting the Mining Code amended by Law No. 2005-021 of 17 October 2005.

vity,[78] in the granting of a production/exploitation mining licence or permit. Some jurisdictions provide that the holder of an exploration permit is required, in the event of a commercial discovery, to apply for the granting of a production permit.[79] It should be noted that the granting of an exploitation or production permit results in the cancellation of the exploration permit for the area concerned,[80] since two licences for the same substance cannot compete in the exploration and production phases. However, an exploitation permit will allow its holder to continue exploration work within its perimeter.

8.30 The exploration permit is granted for a limited term, which may be extended under conditions defined by national laws. The holder of an exploration permit may also request renewal thereof. Regulations may make

---

[78] See, in particular, in Benin, Article 20 of Law No. 2006-17 of 17 October 2006 Enacting the Mining and Mining Taxation Code:

> '*[p]endant la durée de validité du permis de recherche, seul son titulaire, sous réserve de l'article 57 de la présente loi ou d'une prise de participation par l'État si elle est prévue dans la convention minière, peut demander et obtenir un permis d'exploitation pour le ou les gisements se trouvant à l'intérieur du périmètre du permis de recherche.*'
>
> [Free translation into English: '[d]uring the term of validity of the exploration permit, only its holder, subject to section 57 of this Law or an acquisition of a stake by the State if so provided for in the mining agreement, may apply for and obtain an exploitation permit for the deposit(s) within the perimeter of the exploration permit'.]

See also, in the Central African Republic, Article 20 of Law No. 09-005 of 29 April 2009 Enacting the Mining Code; in Côte d'Ivoire, Article 20 of Law No. 2014-138 of 24 March 2014 Enacting the Mining Code; in Guinea, Article 19 of Law No. L/2013/n°053/CNT of 8 April 2013 Amending Law No. 2011/006/CNT of 9 September 2011 Enacting the Mining Code; in South Sudan, Article 50(1) of the Mining Act, 2012; in Chad, Article 22 of Law No. 011-PR-1995 of 20 June 1995 On the Mining Code.

This exclusivity also applies in the context of granting mining concessions. See, in particular, in Tunisia, Article 31 of Law No. 2003-30 of 28 April 2003 Enacting the Mining Code and Article 18(2) Law No. 99-93 of 17 August 1999 Enacting the Hydrocarbons Code.

[79] See, in particular, in the Central African Republic, Article 17 of Ordinance No. 93.007 of 25 May 1993 Enacting the Petroleum Code, which provides:

> '*[d]ès que l'existence d'un gisement d'hydrocarbures commercialement exploitable est établie, le titulaire du permis de recherche est tenu de demander l'octroi d'une concession d'exploitation et d'entreprendre les travaux de développement*'.
>
> [Free translation into English: '[a]s soon as the existence of a commercially exploitable hydrocarbon deposit is established, the holder of the exploration permit is required to apply for the granting of an exploitation concession and to undertake the development works'].

[80] See, in particular, in Côte d'Ivoire, Article 26 of Law No. 96-669 of 29 August 1996 Enacting the Petroleum Code, which provides:

> '*[l]'octroi d'une autorisation d'exploitation entraîne l'annulation de l'autorisation de recherche à l'intérieur du périmètre d'exploitation, mais la laisse subsister à l'extérieur de ce périmètre jusqu'à sa date d'expiration, sans modifier le programme minimum de travaux de recherche souscrit par le titulaire.*'
>
> [Free translation into English: '[t]he granting of an exploitation authorisation results in the cancellation of the exploration authorisation within the exploitation perimeter but leaves it intact outside this perimeter until its date of expiration, without modifying the minimum exploration works programme subscribed by the holder'.]

the renewal subject to compliance by the licencee with its obligations.[81] In the event of non-compliance with the minimum expenditure commitments or work programme, the renewal of the exploration permit may be subject to the payment of compensation generally corresponding to the difference between the minimum expenditure commitments to be made and the amount actually incurred or the amount needed to complete the exploration.[82] In the event that the requirements under the exploration permit have been met, renewal will generally be granted automatically to the permit holder.[83]

As a general rule, the renewal of an exploration permit gives rise to the relinquishment of a part of the perimeter to the host state. The return of such areas is provided for under numerous regulations.[84] Less common, the regulations may specify that the reduction of the area in the event of renewal will take place at the request of the holder of the exploration permit.[85]

---

[81] See, in particular, in Benin, Article 25 of Law No. 2006-18 of 17 October 2006 Enacting the Petroleum Code; in the Central African Republic, Article 13 of Ordinance No. 93.007 of 25 May 1993 Enacting the Petroleum Code and Article 21 of Law No. 09-005 of 29 April 2009 Enacting the Mining Code; in Côte d'Ivoire, Article 22 of Law No. 96-669 of 29 August 1996 Enacting the Petroleum Code; in Uganda, Article 64(2) of the Petroleum (Exploration, Development and Production) Act, 2013.

[82] See in Tunisia, Article 27 of Law No. 99-93 of 17 August 1999 Enacting the Hydrocarbons Code.

[83] See, in particular, in Benin, Article 21 of Law No. 2006-17 of 17 October 2006 Enacting the Mining and Mining Taxation Code; in Mali, Article 38 of Law No. 2012-015 of 27 February 2012 Enacting the Mining Code; in Niger, Article 25 of Ordinance No. 93-16 of 2 March 1993 on the Amended Mining Law; in Togo, Article 15 of Law No. 96-004 of 26 February 1996 Enacting the Mining Code and Article 17(4) of Law No. 99-003 of 18 February 1999 Enacting the Hydrocarbons Code.

[84] See, in particular, in Cameroon, Article 28(4) of Law No 99-013 of 22 December 1999 Enacting the Petroleum Code; in the Central African Republic, Article 13 of Ordinance No. 93.007 of 25 May 1993 Enacting the Petroleum Code and Article 22 of Law No. 09-005 of 29 April 2009 Enacting the Mining Code; in the Democratic Republic of Congo, Article 62 of Law No. 18/001 of 9 March 2018 Amending and Supplementing Law No. 007/2002 of 11 July 2002 Enacting the Mining Code; in Côte d'Ivoire, Article 22 of Law No. 96-669 of 29 August 1996 Enacting the Petroleum Code and Article 24 of Law No. 2014-138 of 24 March 2014 Enacting the Mining Code; in Guinea, Article 24 of Law No. 2011/006/CNT of 9 September 2011 Enacting the Mining Code; in Niger, Article 25 of Ordinance No. 93-16 of 2 March 1993 On the Amended Mining Law and Article 41 of Law No. 2007-01 of 31 January 2007 Enacting the Petroleum Code; in Senegal, Article 16 of Law No. 98-05 of 8 January 1998 Enacting the Petroleum Code; in South Sudan, Article 45(2) of the Mining Act, 2012; in Togo, Article 15 of Law No. 96-004 of 26 February 1996 Enacting the Mining Code and Article 17(3) of Law No. 99-003 of 18 February 1999 Enacting the Hydrocarbons Code.

[85] See, in particular, in Benin, Article 21 of Law No. 2006-17 of 17 October 2006 Enacting the Mining and Mining Taxation Code, which provides:

> '*[l]ors de chaque renouvellement du permis de recherche, le titulaire du permis peut decider ou non de garder toute ou partie de la superficie couverte par le permis.*'
>
> [Free translation into English: '[a]t the time of each renewal of an exploration permit, the holder of the permit may decide whether to keep all or part of the area covered by the permit'].

8.31 Finally, the holder of an exploration permit may relinquish the exploration permit,[86] subject to compliance with the legal procedures established by the mining and petroleum codes, in particular with regard to advance notice. Relinquishment of the exploration permit must be the subject of a detailed report.[87] This relinquishment may be partial or total and involves cancellation of the permit on the perimeter to which it relates. In the event of partial relinquishment, the holder of an exploration permit shall remain liable for all of its obligations, which usually cannot be reduced.[88] The relinquishment may involve the liability of the private operator relinquishing the mining right; the financial consequences of such liability will generally correspond to the minimum financial commitments it did not incur.

c Production or exploitation mining rights

8.32 The works undertaken during the exploration phase may reveal the existence of a commercial deposit and thus lead to the exploitation phase of the project.

This phase refers to development works and extraction of substances for utilitarian or commercial purposes.[89] More broadly, this phase is also intended to include activities such as the first processing operations,[90] or even downstream-related activities such as decommissioning.[91] This phase is divided into two stages: the development phase[92] and the production phase.

---

[86] See, in particular, in Benin, Article 26 of Law No. 2006-18 of 17 October 2006 Enacting the Petroleum Code; in the Central African Republic, Article 27 of Law No. 09-005 of 29 April 2009 Enacting the Mining Code; in Côte d'Ivoire, Article 28 of Law No. 96-669 of 29 August 1996 Enacting the Petroleum Code; in Guinea, Article 49 of Law No. 2011/006/CNT of 9 September 2011 Enacting the Mining Code; in Mali, Article 42 of Law No. 2012-015 of 27 February 2012 Enacting the Mining Code; in Niger, Article 30 of Ordinance No. 93-16 of 2 March 1993 On the Amended Mining Law and Article 52 of Law No. 2007-01 of 31 January 2007 Enacting the Petroleum Code; in Tunisia, Article 40 of Law No. 2003-30 of 28 April 2003 Enacting the Mining Code.

[87] See in Burundi, Article 57 of Law No. 1-21 of 15 October 2013 Enacting the Mining Code.

[88] See, in particular, in the Central African Republic, Article 19 of Ordinance No. 93.007 of 25 May 1993 Enacting the Petroleum Code; in Côte d'Ivoire, Article 28 of Law No. 96-669 of 29 August 1996 Enacting the Petroleum Code; in Niger, Article 54 of Law No. 2007-01 of 31 January 2007 Enacting the Petroleum Code.

[89] For definitions of the exploitation phase, see for example: in the Central African Republic, Article 1 of Law No. 09-005 of 29 April 2009 Enacting the Mining Code; in Côte d'Ivoire, Article 1 of Law No. 2014-138 of 24 March 2014 Enacting the Mining Code; in Morocco, Article 2(6) of Law No. 21-90 of 1 January 1992 On the Exploration and Exploitation of Hydrocarbon Deposits amended and supplemented by Law No. 27-99 of 15 February 2000.

[90] See to that effect, in Algeria, Article 16 of Law No. 01-10 of 3 July 2001 Enacting the Mining Law.

[91] See to that effect, in Niger, Article 59 of Law No. 2007-01 of 31 January 2007 Enacting the Hydrocarbons Code.

[92] The development phase includes all preparatory works, such as geological and geophysical work or the drilling of development wells. For examples of definitions of the development phase, see, in particular, in Morocco, Article 2(6) of Law No. 21-90 of 1 April 1992 On the Exploration and Exploitation of Hydrocarbon Deposits amended and supplemented by Law No. 27-99 of 15 February 2000.

8.33 The production of a deposit takes various forms, depending in particular on the volume of activity, the amounts invested, the method of carrying out the mining operations or the resource concerned. Thus, a distinction must be made between large- or small-scale exploitation.[93] Regarding the methods of undertaking the activity, the mining codes distinguish, in particular, between the operations of collecting, semi-mechanised exploitation, artisan exploitation,[94] semi-industrial exploitation[95] or industrial exploitation.[96] Finally, in the case of mineral resources, a distinction should be drawn between mining and quarrying and also the exploitation of mining tailing. These various types of exploitation result in a wide variety of mining rights. This analysis is therefore not intended to be

---

[93] See to that effect, in the mining sector, in the Democratic Republic of Congo, Article 98 of Law No. 007-2002 of 11 July 2002 Enacting the Mining Code, which provides:

'*[l]orsque les conditions techniques caractérisant certains gisements des substances minérales ne permettent pas d'en faire une exploitation à grande échelle économiquement rentable, mais permettent une exploitation minière de petite taille avec un minimum d'installations fixes utilisant des procédés semi-industriels ou industriels, ceux-ci sont considérés comme gisements d'exploitation minière à petite échelle.*'

[Free translation into English: '[w]hen the technical conditions characterising certain deposits of mineral substances do not permit economically profitable large-scale mining but permit small-scale mining with a minimum of fixed installations using semi-industrial or industrial processes, these are considered small-scale mining deposits'].

[94] Artisanal exploitation refers to '*toute activité par laquelle une personne physique de nationalité centrafricaine d'origine se livre, dans une zone d'exploitation artisanale délimitée en surface et en profondeur jusqu'h trente mètres au maximum, à extraire et à concentrer des substances minérales en utilisant des outils, des méthodes et des procédés non industriels, manuels et peu mécanisés.*'

[Free translation into English: 'any activity whereby a natural person of Central African nationality of origin sets out, in an artisan zone delimited on the surface and up to a maximum depth of 30 meters, to extract and concentrate mineral substances using non-industrial, manual and mechanised tools, methods and processes'] (Article 1 of the Central African Law No. 09-005 of 29 April 2009 Enacting the Mining Code).

[95] Semi-industrial exploitation refers to '*tout exploitation minière de petite taille, permanente, fondée sur la justification de l'existence d'un Gisement, utilisant selon les règles de l'art, des procédés semi-industriels et font la production annuel en régime de croisière n'excède pas un certain tonnage du produit commercialisable (minerai concentré ou métal) fixé par substance et par la réglementation minière.*'

[Free translation into English: 'any small, permanent mining operation, started based on the justification of the existence of a Deposit, using, according to the rules of the art, semi-industrial processes and making annual production as business as usual and not exceeding a certain tonnage of the marketable product (concentrated ore or metal) set per substance by the mining regulations']

(in Guinea, Article 1 of Law L/2013/n° 053/CNT of 8 April 2013 Amending Law No 2011/006/CNT of 9 September 2011 Enacting the Mining Code).

[96] Industrial exploitation refers to '*toute exploitation fondée sur la mise en évidence préalable d'un gisement, possédant des installations fixes, nécessaires pour la récupération dans les règles de l'art, de minerais exploités par des précédés industriels.*'

[Free translation into English: 'any exploitation based on the prior identification of a deposit, possessing fixed installations, necessary for the proper recovery, according to rules of the art, of minerals exploited by industrial process']

(in the Central African Republic, Article 1 of Law No. 09-005 of 29 April 2009 Enacting the Mining Code).

exhaustive and does not specify the regime applicable to each mining right. It merely involves stating their main principles.

8.34 As a preliminary point, a distinction must be made in the so-called exploitation mining rights between production or exploitation permits and licences, on the one hand, and concessions on the other.[97]

Exploitation operations may only be carried out on the basis of a mining right.[98] This general principle also applies to the exploitation of quarries[99] and artisan mining[100] which, in principle, can only be carried out on the basis of an authorisation from the licensing authority.

8.35 The granting of an exploitation mining right follows a discovery, generally carried out by the holder of an exploration permit. Thus, following a discovery in commercial quantities,[101] a mining right for production may be granted to the person who proves the existence thereof.[102] In the event that the mining right is not granted to the person who discovered the deposit, the 'discoverer', the latter may, in accordance with certain regulations, be entitled to compensation.[103]

---

[97] See, in particular, in the Central African Republic, Articles 23 et seq of Law No. 93.007 of 25 May 1993 Enacting the Petroleum Code; in Djibouti, Article 23 of Law No. 117-AN-05 of 16 October 2005 Enacting the Petroleum Code; in Guinea, Articles 35 et seq of Law No. 2011/006/CNT of 9 September 2011 Enacting the Mining Code; in Morocco, Article 27 of Law No. 21-90 of 1 January 1992 On the Exploration and Exploitation of Hydrocarbon Deposits amended and supplemented by Law No. 27-99 of 15 February 2000; in Mozambique, Articles 42 et seq of Mining Law No. 20-2014 of 18 August 2014; in Togo, Articles 20 et seq of Law No. 99-003 of 18 February 1999 Enacting the Hydrocarbons Code.

[98] See, in particular, in Niger, Article 60 of Law No. 2007-01 of 31 January 2007 Enacting the Petroleum Code.

[99] See, in particular, in Burundi, Article 106 of Law No. 1-21 of 15 October 2013 Enacting the Mining Code.

[100] See, in particular, in Benin, Article 51 of Law No. 2006-17 of 17 October 2006 Enacting the Mining and Mining Taxation Code; in Mali, Article 43 of Law No. 2012-015 as of 27 February 2012 Enacting the Mining Code.

[101] For an example of a definition of commercial discovery see, in Morocco, Article 28 of Law No. 21-90 of 1 April 1992 On the Exploration and Exploitation of Hydrocarbon Deposits amended and supplemented by Law No. 27-99 of 15 February 2000, which provides:

> '*[u]ne découverte d'hydrocarbures est réputée commerciale lorsqu'après l'exécution d'un programme de forages d'appréciation adéquat en conformité avec les dispositions particulières de l'accord pétrolier décrit au chapitre V ci-après, le titulaire du permis aura démontré que cette découverte recèle un potentiel de réserves d'hydrocarbures récupérables pouvant aboutir à une exploitation économiquement rentable et qu'il s'engage à la développer.*'
>
> [Free translation into English: '[a] discovery of hydrocarbons is deemed to be commercial when, following the performance of an adequate appraisal drilling programme in accordance with the special provisions of the petroleum agreement described in Chapter V below, the permit holder has demonstrated that this discovery has the potential of recoverable hydrocarbon reserves that can lead to economically viable exploitation and that it is committed to developing it'].

[102] See, in particular, in Niger, Article 33 of Ordinance No. 93-16 of 2 March 1993 Enacting the Amended Mining Law.

[103] See, in particular, in Guinea, Article 31 of Law No. 2011/006/CNT of 9 September 2011 Enacting the Mining Code:

8.36 Like an exploration permit, an exploitation agreement generally accompanies the granting of an exploitation permit.[104] the duration of which coincides with the term of the production mining right.[105] Either this agreement will already exist in the exploration phase and it will be pursued with some adjustments depending on the specificities of the discovery made, or a special agreement will be negotiated and signed.

8.37 The granting of an exploitation/production permit or concession entails the cancellation of the exploration permit within the perimeter granted,[106] subject to the rights of other exploration permit holders for other substances.

The rights conferred by the exploitation/production mining right must be exercised within the perimeter defined by the permit. Generally, the perimeter will be determined on the basis of the area of the deposit for which the mining right is granted.[107] The shape and orientation of the perimeter

---

'[s]i le permis d'exploitation est délivré à une personne autre que l'inventeur du gisement, le titulaire du permis doit verser à ce dernier une juste indemnité fixée dans le cadre d'une transaction commerciale privée. L'indemnité est destinée à compenser le montant des frais effectivement engagés par l'inventeur pour les travaux de recherche proprement dits effectués sur le gisement en vertu d'un permis de recherche. L'inventeur ne peut se prévaloir de cette disposition s'il n'a pas satisfait à toutes ses obligations selon le présent Code.'

[Free translation into English code: '[i]f the Exploitation Permit is issued to any person other than the Discoverer of the Deposit, the holder of the Permit must pay the latter fair compensation, the amount of which is determined by a private commercial transaction. The compensation is intended to cover the costs which the Discoverer effectively incurred for exploration works carried out on the Deposit pursuant to an Exploration Permit. The Discoverer cannot rely on this provision unless it has met all of its obligations under this Code'.

104 See, in particular, in Burundi, Article 61 of Law No. 1-21 of 15 October 2013 Enacting the Mining Code, which provides: *'[l]e titre minier d'exploitation est constitué du permis d'exploitation auquel est annexé obligatoirement une convention minière.'*

[Free translation into English: '[t]he exploitation mining right consists of the exploitation permit to which a mining agreement is mandatorily attached'].

See also, in Niger, Article 51 of Ordinance No. 93-16 of 2 March 1993 Enacting the Amended Mining Law.

105 See, in particular, in Burundi, Article 71 of Law No. 1-21 of 15 October 2013 Enacting the Mining Code.

106 See in particular, in Tunisia, Article 51 of Law No. 99-93 of 17 August 1999 Enacting the Hydrocarbons Code and Article 53 of Law No. 2003-30 of 28 April 2003 Enacting the Mining Code; in the Central African Republic, Article 35 of Law No. 09-005 of 29 April 2009 Enacting the Mining Code.

107 See, in particular, in Benin, Article 33 of Law No. 2006-17 of 17 October 2006 Enacting the Mining and Mining Taxation Code, which provides:

'[l]a superficie du permis d'exploitation est délimitée en fonction de la taille du gisement telle qu'elle a été définie dans l'étude de faisabilité.'

[Free translation into English: '[t]he surface area of the exploitation permit shall be delineated on the basis of the size of the deposit as defined in the feasibility study'].

See also, in the Central African Republic, Article 40 of Law No. 09-005 of 29 April 2009 Enacting the Mining Code; in Mali, Article 27 of Law No. 04-037 of 2 August 2004 On the Organization of Exploration, Exploitation, Transport and Refining of Hydrocarbons; in Niger, Article 37 of Ordinance No. 93-16 of 2 March 1993 On the Amended Mining Law; in

may be specified by the mining and petroleum codes.[108] The surface area of the perimeter granted to the holder of a so-called exploitation mining right may be limited by national legislation, which may provide for a maximum area.[109] Finally, the perimeter will be encompassed within the scope of the exploration permit on the basis of which the exploitation permit is granted.[110]

8.38 Within the defined perimeter, the rights conferred on the holder of a production/exploitation mining right extend to the surface but also in depth.[111] The rights conferred on the holder of an artisanal production mining right shall, for the most part, extend to a limited depth.[112]

8.39 The exploitation mining rights confer on the holder the exclusive right to conduct, within the perimeter concerned, the development and exploitation of the deposits located within the perimeter. The granting of exclusive rights is widely accepted and applies in particular to the holder

Chad, Article 29 of Law No. 011-PR-1995 of 20 June 1995 Enacting the Mining Code; in Togo, Article 18 of Law No. 2003-012 of 4 October 2003 Amending and Supplementing Law No. 96-004 of 26 February 1996 Enacting the Mining Code.

[108] See, in particular, in Burundi, Article 64 of Law No. 1-21 of 15 October 2013 Enacting the Mining Code, which provides: '*[l]e permis d'exploitation porte sur un polygone compris dans l'intérieur de celui-ci. L'un des côtés au moins du polygone doit être orienté Nord-Sud géographique*'. [Free translation into English: '[t]he exploitation permit covers a polygon included within it. At least one side of the polygon must be oriented north-south geographically'.] For other examples see also, in Niger, Article 37 of Ordinance No. 93-16 of 2 March 1993 Enacting the Amended Mining Law.

[109] With regard to exploitation permits, see, in particular, in Madagascar, Article 28 of Law No. 99-022 of 19 August 1999 Enacting the mining code amended by Law No. 2005-021 of 17 October 2005.

[110] See in particular, in Benin, Article 33 of Law No. 2006-17 of 17 October 2006 Enacting the Mining Code and Mining Taxation, which provides: '*[l]e périmètre du permis d'exploitation doit être entièrement situé à l'intérieur du permis de recherche dont il dérive*'. [Free translation into English: '[t]he perimeter of the exploitation permit must be fully within the exploration permit from which it is derived.'] See also, in Guinea, Article 29 of Law No. 2011/006/CNT of 9 September 2011 Enacting the Mining Code.

[111] See, in particular, in Burundi, Article 75 of Law No. 1-21 of 15 October 2013 Enacting the Mining Code; in Cameroon, Article 37(1) of Law No. 99-013 of 22 December 1999 Enacting the Petroleum Code; in Guinea, Article 28 of Law No. 2011/006/CNT of 9 September 2011 Enacting the Mining Code; in Mali, Article 63 of Law No. 2012-015 of 27 February 2012 Enacting the Mining Code; in the Republic of Congo, Article 58 of Law No. 4-2005 of 11 April 2005 Enacting the Mining Code; in the Central African Republic, Article 37 of Law No. 09-005 of 29 April 2009 Enacting the Mining Code.

[112] See, in particular in Niger, Article 45 of Ordinance No. 93-16 of 2 March 1993 Amending the Mining Law; in Mali, Article 48 of Law No. 2012-015 of 27 February 2012 Enacting the Mining Code.

of an exploitation or production permit,[113] a concession,[114] an artisanal authorisation,[115] a quarry authorisation,[116] a small-scale mining authorisation or a semi-industrial mining authorisation.[117] The exclusivity granted to the holder of a mining right of production excludes any competing exploitation of the deposit and therefore presupposes that no other mining right is granted on the perimeter. The holder of a production mining right may, in parallel with the exploitation operations, pursue exploration and prospecting activities.[118]

---

[113] See, in particular, in Burundi, Article 75 of Law No. 1-21 of 15 October 2013 Enacting the Mining Code; in Côte d'Ivoire, Article 31 of Law No. 2014-138 of 24 March 2014 Enacting the Mining Code; in Guinea, Article 28 of Law No. 2011/006/CNT of 9 September 2011 Enacting the Mining Code; in Madagascar, Article 37 of Law No. 99-022 of 19 August 1999 Enacting the Mining Code amended by Law No. 2005-021 of 17 October 2005; in Mali, Article 63 of Law No. 2012-015 of 27 February 2012 Enacting the Mining Code; in Niger, Article 68 of Law No. 2007-01 of 31 January 2007 Enacting the Petroleum Code; in Chad, Article 27 of Law No. 011-PR-1995 of 20 June 1995 Enacting the Mining Code; in Togo, Article 18 of Law No. 2003-012 of 4 October 2003 Amending and Supplementing Law No. 96-004 of 26 February 1996 Enacting the Mining Code; in the Central African Republic, Article 37 of Law No. 09-005 of 29 April 2009 Enacting the Mining Code.

[114] See, in particular, in Tunisia, Article 54 of Law No. 2003-30 of 28 April 2003 Enacting the Mining Code.

[115] See, in particular, in Burundi, Article 100 of Law No. 1-21 of 15 October 2013 Enacting the Mining Code; in the Republic of Congo, Article 41 of Law No. 4-2005 of 11 April 2005 Enacting the Mining Code; in Côte d'Ivoire, Article 66 of Law No. 2014-138 of 24 March 2014 Enacting the Mining Code.

[116] See, in Côte d'Ivoire, Article 77 of Law No. 2014-138 of 24 March 2014 Enacting the Mining Code.

[117] See, in Côte d'Ivoire, Article 54 of Law No. 2014-138 of 24 March 2014 Enacting the Mining Code.

[118] See, in Burundi, Article 75 of Law No. 1-21 of 15 October 2013 Enacting the Mining Code, which provides:

> '*[l]e titre minier d'exploitation confère à son titulaire, dans les limites de son périmètre et indéfiniment en profondeur, un droit exclusif de prospection, de recherche et d'exploitation des substances minérales pour lesquelles il a été délivré, sous réserve des lois et de la réglementation en vigueur, notamment en matière d'environnement, de santé publique et de sécurité sociale.*'
>
> [Free translation into English: '[t]he exploitation mining right confers on its holder, within the limits of its perimeter and indefinitely in depth, an exclusive right of prospecting, exploration and exploitation of the mineral substances for which it is issued, subject to the laws and regulations in force, in particular with regard to the environment, public health and public safety'].

See also, in Guinea, Article 28 of Law No. 2011/006/CNT of 9 September 2011 Enacting the Mining Code; in Madagascar, Article 37 of Law No. 99-022 of 19 August 1999 Enacting the Mining Code amended by Law No. 2005-021 of 17 October 2005; in Mali, Article 63 of Law No. 2012-015 of 27 February 2012 Enacting the Mining Code; in Togo, Article 18 of Law No. 2003-012 of 4 October 2003 Amending and Supplementing Law No. 96-004 of 26 February 1996 Enacting the Mining Code; in the Central African Republic, Article 37 of Law No. 09-005 as of 29 April 2009 Enacting the Mining Code.

8.40 In addition, the exploitation mining right confers the right to dispose of the extracted resources,[119] and in particular to market and export them out of the territory of the host state.[120] In order to support the national economy and the development of integrated projects, national laws may limit the export of extracted natural resources by making it subject to proof that it is impossible to process the extracted materials within the national territory at an economically viable cost.[121]

The rights conferred by the exploitation mining right usually include the right to transport the extracted products,[122] to proceed with the transformation thereof,[123] and thus to establish within the national territory facilities for packaging, processing and transformation of extracted substances.[124]

---

[119] See, in particular, in Cameroon, Article 37(1) of Law No. 99-013 of 22 December 1999 Enacting the Petroleum Code.

[120] See, in particular, in the Central African Republic, Article 38 of Law No. 09-005 of 29 April 2009 Enacting the Mining Code.

[121] For example, in the Democratic Republic of Congo, Article 85 of Law No. 2007-2002 of 11 July 2002 Enacting the Mining Code provides:

'*[s]ous réserve des dispositions de l'alinéa suivant, la commercialisation des produits miniers qui proviennent des Périmètres d'exploitation est libre. Le titulaire d'un Permis d'Exploitation peut vendre ses produits aux clients de son choix à des prix librement négociés. Toutefois, l'autorisation du Ministre est requise pour l'exportation des minerais à l'état brut pour traitement à l'extérieur du Territoire National. Cette autorisation ne sera accordée que si le titulaire qui la demande démontre à la fois: a) l'inexistence d'une possibilité de traitement dans le Territoire National à un coût économiquement rentable pour le projet minier; b) les avantages pour la République Démocratique du Congo au cas où l'autorisation d'exportation est accordée.*'

[Free translation into English: '[s]ubject to the provisions of the following paragraph, the marketing of mining products originating from the exploitation perimeters shall be free. The holder of an Exploitation Permit may sell its products to the customers of its choice at freely negotiated prices. However, the Minister's authorisation is required for the export of raw minerals for processing outside the National Territory. Such authorisation shall be granted only if the holder demonstrates both: (a) the non-existence of processing capabilities within the National Territory at an economically viable cost for the mining project; (b) the advantages for the Democratic Republic of Congo in the event that the export authorisation is granted'].

[122] See, in particular, in the Democratic Republic of Congo, Articles 84 of Law No. 007-2002 of 11 July 2002 Enacting the Mining Code; in the Central African Republic, Article 38 of Law No. 09-005 of 29 April 2009 Enacting the Mining Code.

[123] See, in particular, in the Democratic Republic of Congo, Article 81 of Law No. 007-2002 of 11 July 2002 Enacting the Mining Code:

'*[s]ous réserve des dispositions prévues à l'article 10, litera j, le traitement ou la transformation des substances minérales peut être réalisée soit par le titulaire d'un Permis d'Exploitation, soit par une Entité de traitement ou une Entité de transformation.*'

[Free translation into English: '[s]ubject to the provisions of Article 10(j), processing or transformation of mineral substances may be carried out either by the holder of an Exploitation Permit or by a processing Entity or a transformation Entity'].

[124] See, in particular, in the Central African Republic, Article 38 of Law No. 09-005 of 29 April 2009 Enacting the Mining code; in Côte d'Ivoire, Article 31 of Law No. 2014-138 of 24 March 2014 Enacting the Mining Code.

8.41 The rights conferred upon a private operator only cover substances covered by the exploitation mining right. If in the course of the exploitation works the private operator discovers substances not covered by the mining right, the latter is in principle required to inform the licensing authority,[125] sometimes under penalty of withdrawal of the production permit and legal action.[126] In the event of discovery of other substances within the perimeter, the holder of the production permit may apply for the extension of its mining right to the discovered substances, without the granting thereof being automatic most of the time.[127]

8.42 The exploitation mining right, which can be extended to other substances, may also be subject to transformation. Thus, a small-scale mining right can be transformed into a large-scale production permit[128] under the conditions defined by regulations.

8.43 Development and exploitation of the deposit must begin within the period prescribed in the mining and petroleum codes,[129] exploitation agreements, or work programmes validated by the licensing authority.[130]

---

[125] See, in particular, in Madagascar, Article 30 of Law No. 99-022 of 19 August 1999 Enacting the Mining Code amended by Law No. 2005-021 of 17 October 2005.

[126] See, in particular, in Burundi, Article 70 of Law No. 1-21 of 15 October 2013 Enacting the Mining Code.

[127] See, in particular, in Burundi, Article 70 of Law No. 1-21 of 15 October 2013 Enacting the Mining Code; in the Democratic Republic of Congo, Article 77 of Law No. 18/001 of 9 March 2018 Amending and Supplementing Law No. 007/2002 of 11 July 2002 Enacting the Mining Code; in Madagascar, Articles 30 and 31 of Law No. 99-022 of 19 August 1999 Enacting the Mining Code amended by Law No. 2005-021 of 17 October 2005.

[128] See, in particular, in Niger, article 42 of Ordinance No. 93-16 of 2 March 1993 Amending the Mining Law; in Togo, article 20 of Law No. 96-004 of 26 February 1996 Enacting the Mining Code.

[129] See in particular, in Cameroon, article 57(3) of Law No. 2016/017 of 14 December 2016 Enacting the Mining Code which provides:

> '*[l]e titulaire d'un permis d'exploitation de la mine industrielle doit commencer les travaux de développement dans un délai de deux (02) ans, à compter de la date de notification du permis, faute de quoi, il s'expose au retrait du permis, après une mise en demeure restée sans suite*.'
>
> [Free translation into English: '[t]he holder of an exploitation permit for an industrial mine must begin the development works within two (2) years from the date of notification of the permit, otherwise, the permit may be withdrawn following a formal notice which remained unanswered'].

See also in the Central African Republic, Article 41 of Law No. 09-005 of 29 April 2009 Enacting the Mining Code; in Côte d'Ivoire, Article 34 of Law No. 2014-138 of 24 March 2014 Enacting the Mining Code; in Guinea, Article 34 of Law No. L/2013/n°053/CNT of 8 April 2013 Amending Law No. 2011/006/CNT of 9 September 2011 Enacting the Mining Code; in Niger, Article 39 of Ordinance No. 93-16 of 2 March 1993 Enacting the Amended Mining Law; in Chad, Article 29(5) of Law No. 006-PR-2007 of 2 May 2007 On Hydrocarbons; in Tunisia, Article 57 of Law No. 2003-30 of 28 April 2003 Enacting the Mining Code.

[130] See in particular, in South Sudan, article 61(1) of the Mining Act, 2012, which provides: '[i]n addition to any terms and conditions as may be prescribed by regulation, every Small-Scale Mining Licence is granted subject to the terms and conditions that the Mining Holder shall: (a) commerce Mine Development or processing plant development, where the Mining

The timeframe in which the work must begin is likely to vary from one production mining right to another.[131] If the work has not commenced within the specified timeframe, the holder of the exploitation/production permit is liable to penalties for delay[132] or even withdrawal of the mining right.[133]

Some jurisdictions, however, offer the holder of an exploitation/production permit the opportunity to apply for an exemption from commencing the work or continuing the exploitation of the deposit. The applicable laws may provide that the exemption will be granted if the reason invoked is related to unfavourable market conditions for the products concerned, as demonstrated by an economic study.[134] The interest of such a provision in the current economic context is far from negligible and may make it possible to avoid many disputes. In fact, when facing such changes in circumstances, operators will sometimes be tempted to postpone the start of operations or to suspend projects whose profitability would be suddenly called into question.[135] However, exemption from commencing development and exploitation works within the specified time frames does not prevent the host state from

---

Licence Area is for Mineral processing operations in accordance with the approved program'. See also, in the case of a 'Large-Scale Mining Licence', section 67(1) of the said law.

[131] For example, in Guinea, a holder of a semi-industrial exploitation permit is required to begin development and exploitation works within a maximum of six months from the date of granting the permit compared to one year for the holder of an industrial exploitation permit (article 34 of Law No. L/2013/n°053/CNT of 8 April 2013 Amending Law No 2011/006/CNT of 9 September 2011 Enacting the Mining Code).

[132] See, in particular, in Guinea, article 34 of Law No. L/2013/n°053/CNT of 8 April 2013 Amending Law No. 2011/006/CNT of 9 September 2011 Enacting the Mining Code.

[133] See, in particular, in the Republic of Congo, Article 62 of Law No. 4-2005 of 11 April 2005 Enacting the Mining Code.

[134] See, in particular, in Burkina Faso, Article 53 of Law No. 036-2015/CNT of 16 June 2015 Enacting the Mining Code of Burkina Faso concerning the industrial exploitation permit:

> '*[u]ne dispense de commencer les travaux de développement et de mise en exploitation ou de continuer l'exploitation du gisement peut être obtenue par arrêté conjoint des ministres chargés des mines et des finances.* [. . .] *Elle est toujours accordée lorsque le motif invoqué est la conjoncture défavorable du marché des produits concernés au moment de la demande de dispense, telle que démontrée par une étude économique produite par l'opérateur.*'
>
> [Free translation into English: '[a]n exemption from commencing development works and exploitation or continuing exploitation of the deposit may be obtained by joint order of the ministers responsible for mining and finance. [. . .] It is always granted when the reason given is the unfavourable market situation of the products concerned at the time of the exemption request, as demonstrated by an economic study produced by the operator'.]

With regard to the semi-mechanised exploitation permit, see Article 62 of the said law. See also, in the Central African Republic, Article 41 of Law No. 09-005 of 29 April 2009 Enacting the Mining Code.

[135] Reference is made to the decisions taken by some companies in 2014 to suspend iron ore mining projects in Mauritania, notably the decision taken by Sphere Minerals, a subsidiary of the Glencore group, in relation to the Askaf mining project, or Rio Tinto's decision in July 2016 to suspend the Simandou project in Guinea, presented as a 'mega project' with the consequences involved for the host state.

proceeding with the withdrawal of the mining right in the event that the suspension of works is for a prolonged period of time.[136] Similarly, the prolonged suspension of exploitation operations is often likely to lead to the withdrawal of the production mining right.[137]

8.44 Always in order to ensure the speed and smooth operation of exploitation operations, the mining and petroleum codes may limit the development phase[138] to a certain length of time, in particular by requiring the operator to achieve the production phase within a specified period of time.[139]

8.45 Mining rights are granted for a specific period, which may vary from one mining right to another, in particular depending on the substances concerned.[140] Furthermore, the exploitation of deep offshore deposits may

---

[136] See also, in the Central African Republic, Article 41 of Law No. 09-005 of 29 April 2009 Enacting the Mining Code.

[137] See, in particular, in Cameroon, Article 43 of Law No. 99-013 of 22 December 1999 Enacting the Petroleum Code; in Côte d'Ivoire, Article 36 of Law No. 96-669 of 29 August 1996 Enacting the Petroleum Code.

[138] See, in particular, in Algeria, Article 8(2) of Presidential Decree of 1 March 2003 On the Model Mining Convention, which provides:

> '*[l]a Période de Développement et de Construction sera celle prévue dans l'Étude de Faisabilité, mais elle ne devra en aucun cas excéder une période de cinq (5) ans à compter de la date d'octroi de la Concession Minière.*'
>
> [Free translation into English: '[t]he Development and Construction Period shall be that provided in the Feasibility Study but shall in no case exceed a period of five (5) years from the date of granting of the Mining Concession'].

[139] See, in particular, in Guinea, Article 34 of Law No. L/2013/n°053/CNT of 8 April 2013 Amending Law No. 2011/006/CNT of 9 September 2011 Enacting the Mining Code, which provides:

> '*[l]e titulaire d'un Permis d'exploitation minière est tenu d'atteindre la phase d'exploitation, telle que définie à l'article 168, dans la durée prévue dans l'étude de faisabilité et, ce, dans la limite maximale de quatre (4) ans à compter de la date de l'octroi du Titre minier pour les Permis d'exploitation destinés à l'extraction et à l'exportation du minerai brut et cinq (5) ans pour ceux destinés à la transformation de matières premières sur le territoire guinéen. Passé ce délai une pénalité de retard correspondant au solde non exécuté des dépenses prévues sur une année civile, lui sera appliquée. Cette pénalité ne sera pas due si l'écart non exécuté des dépenses ainsi constaté est inférieur à 10% des dépenses pour l'année civile concernée et/ou résulte d'un ajustement du programme des travaux validé par le Ministre, après avis favorable de la Commission Nationale des Mines.*'
>
> [Free translation into English: '[t]he holder of a mining Exploitation Permit is required to begin the Exploitation Phase, as defined in Article 168, within the timeframe stipulated in the feasibility study, and this within the grant a maximum period of four (4) years from date of the Mining Permit for the Mining Exploitation Permit for the extraction and export of unprocessed ore and five (5) years for those dedicated to the processing of raw materials within Guinea. After this time period, a penalty for delay corresponding to the unused balance of expenditures contemplated for the calendar year will be applied. This penalty will not be due if the established unused balance of expenditures is less than 10% of the expenditure for the relevant calendar year and/or is the result of an adjustment to the works programme validated by the Minister, following the approval by the National Mining Commission'.]

[140] For example, in Benin, while the production permit is valid for a period of 20 years (Article 32 of Law No. 2006-17 of 17 October 2006 Enacting the Mining Code and Mining Taxation

justify the granting of a production permit for a longer duration.[141] Once the mining right has expired, the holder may request a possible extension or renewal.

8.46 As with exploration mining rights, the holders of a production mining right may in principle be able to renounce it,[142] under the conditions prescribed by the mining and petroleum codes.

8.47 Upon expiry of the mining right or in the event of renunciation, the holder must fulfil the obligations thereupon regarding abandonment of the deposit and rehabilitation of the site.[143]

2 THE LEGAL NATURE OF MINING RIGHTS

8.48 The legal nature of mining rights is far from uniform and varies considerably from one jurisdiction to another, which in practice has important consequences on the content of these rights.

a The distinction related to the movable or immovable nature of the mining rights

8.49 One of the main difficulties in this area is the determination of the movable or immovable nature of the mining right. This distinction is a physical distinction; thus, as opposed to movable or personal property, immovable or real property refers to goods which by their nature cannot be moved.

While certain jurisdictions unequivocally determine the nature of each mining right, others provide sporadic responses for certain mining rights only. Finally, some mining and petroleum codes remain silent on the matter, which in practice creates certain difficulties.[144] The failure to address the issue is particularly notable in the context of English-speaking countries.

8.50 Faced with the lack of precision as to the nature of the mining right, one should proceed *mutatis mutandis* by referring to other goods whose

---

in the Republic of Benin), the quarry exploitation and commencing licence is only valid for a period of 5 years (Article 44 of the said law).

[141] See, in particular, in Benin, Article 37(2) of Law No. 2006-18 of 17 October 2006 Enacting the Petroleum Code.

[142] See, in particular, in Côte d'Ivoire, Article 35 of Law No. 96-669 of 29 August 1996 Enacting the Petroleum Code; in Mali, Article 70 of Law No. 2012-015 of 27 February 2012 Enacting the Mining Code; in Niger, Article 41 of Ordinance No. 93-16 of 2 March 1993 On the Amended Mining Law and Article 77 of Law No. 2007-01 of 31 January 2007 Enacting the Petroleum Code.

[143] See, in particular, in Cameroon, Article 44(1) of Law No. 99-013 of 22 December 1999 Enacting the Petroleum Code; in Côte d'Ivoire, Article 37 of Ordinance No. 2012-369 of 18 April 2012 Amending Law No. 96-669 of 29 August 1996 Enacting the Petroleum Code; in Niger, Article 74 of Law No. 2007-01 of 31 January 2007 Enacting the Petroleum Code.

[144] See to that effect, Thierry Lauriol 'Financement de projets miniers dans les États parties à l'OHADA: l'apport des titres miniers en garantie' in *Les Grands articles de doctrine de l'OHADA parus au Recueil Penant de 2000 à 2013* (Juris Africa 2013) 221.

legal nature is defined.[145] Thus, the movable or immovable nature of the mining right can sometimes be deduced from the type of security provided for in the mining or petroleum code, depending on whether it is a pledge or a mortgage.[146]

8.51 As for legislation that specifies the movable or immovable nature of mining rights, they vary considerably from one state to another and depending on the right concerned.

Thus, the qualification of movable or personal property is generally retained for prospecting licences[147] and exploration permits,[148] although many exceptions are to be mentioned.[149] Contrarily, exploitation permits generally receive the qualification of immovable or real property,[150] again with certain exceptions.[151] The same applies to concessions, which in principle constitute real property rights.[152]

---

145 See to that effect, Lauriol 'Financement de projets miniers dans les États parties à l'OHADA: l'apport des titres miniers en garantie' at 222.

146 See, in Madagascar, Articles 58 and 59 of Law No. 99-022 of 19 August 1999 On the Mining Code amended by Law No. 2005-021 of 17 October 2005, which provide, respectively:
'*[l]e permis de recherche minière ou permis "R" constitue un droit cessible, transmissible, amodiable et susceptible de gage.*'
[Free translation into English: '[t]he mining exploration permit or "R" permit constitutes a right that can be assigned, transferred, farmed-out and pledged'] and
'*Le permis d'exploitation minière ou permis "E" constitue un droit cessible, transmissible, amodiable et susceptible d'hypothèque.*'
[Free translation into English: 'The mining exploitation permit or "E" permit constitutes a right that can be assigned, transferred, farmed-out and mortgaged.']
Based on these provisions, it is possible to conclude that the nature of the exploration permit is movable, which may be pledged, and the immovable nature of the exploitation permit, which, in turn, may be mortgaged.

147 See in Tunisia, Article33 of Law No. 99-93 of 17 August 1999 Enacting the Hydrocarbons Code.

148 See in Algeria, Article 8 of Law No. 01-10 of 3 July 2001 On Enacting the Mining Law; in Benin, Article 59 of Law No. 2006-17 of 17 October 2006 Enacting the Mining and Mining Taxation Code; in the Central African Republic, Article 25 of Law No. 09-005 of 29 April 2009 Enacting the Mining Code; in Côte d'Ivoire, Article 20 paragraph 3 of Law No. 2014-138 of 24 March 2014 Enacting the Mining Code; in Mali, Article 40 of Law No. 2012-015 of 27 February 2012 Enacting the Mining Code; in Togo, Article 13(3) of Law No. 99-003 of 18 February 1999 Enacting the Hydrocarbons Code; in Tunisia, Article 38 of Law No. 2003-30 of 28 April 2003 Enacting the Mining Code and Article 33 of Law No. 99-93 of 17 August 1999 Enacting the Hydrocarbons Code.

149 See in the Republic of Congo, Article 29 of Law No. 4-2005 of 11 April 2005 Enacting the Mining Code; in the Democratic Republic of Congo, Article 51 of Law No. 007-2002 of 11 July 2002 Enacting the Mining Code.

150 See in Algeria, Article 8 of Law No. 01-10 of 3 July 2001 Enacting the Mining Law; in Benin, Article 33 of Law No. 2006-18 of 17 October 2006 Enacting the Petroleum Code; in Côte d'Ivoire, Article 31(6) of Law No. 2014-138 of 24 March 2014 Enacting the Mining Code; in Mali, Article 68 of Law No. 2012-015 of 27 February 2012 Enacting the Mining Code.

151 See in Benin, Article 59 of Law No. 2006-17 of 17 October 2006 Enacting the Mining and Mining Taxation Code.

152 See in Togo, Article 22(1) of Law No. 99-003 of 18 February 1999 Enacting the Hydrocarbons Code.

8.52 As for mining rights constituting real property rights, the mining and petroleum codes often specify that these real property rights are distinct from the ownership of the land;[153] such precision makes it possible in practice to avoid major disputes with the owners of the areas concerned by the mining rights and the taking of random security.

8.53 To illustrate this, the diversities encountered among the mining codes is illustrated below.

Nature of the main mining rights within the OHADA zone Analysis of the mining codes						
Country	**Exploration permit**		**Exploitation permit**		**Mining concession**	
Benin	Personal property right (article 59)		Personal property right (article 59)		**No provision relative to concessions**	
Burkina Faso	**No specification of the nature of the exploration permit**			Real property right (articles 47 and 59)	**No provision relative to concessions**	
Cameroon	**No specification of the nature of the mining rights**				**No provision relative to concessions**	
Central Africa	Personal property right (article 25)			Real property right (article 38)	**No provision relative to concessions**	
Comoros	Personal property right (article 16)		Personal property right (article 16)			Real property right (article 16)
Congo		Real property right (article 29)		Real property right (article 16)	**No provision relative to concessions**	
Côte d'Ivoire	Personal property right (article 20)			Real property right (article 31)	**No provision relative to concessions**	
Gabon	**No specification of the nature of the exploration permit**			Real property right (article 150)		Real property right (article 150)
Guinea-Bissau	**No specification of the nature of the mining rights**					
Guinea	Personal property right (article 19)		Personal property right (article 28)			Real property right (article 35)
Equatorial Guinea	**Exploitation on the basis of mining economic rights**					
Mali	Personal property right (article 40)			Real property right (article 68)	**No provision relative to concessions**	

[153] See in Algeria, Article 8 of Law No. 01-10 of 3 July 2001 Enacting the Mining Law; in Angola, Article 47 of Mining Law No. 31/11 of 23 September 2011; in Benin, Article 33 of Law No. 2006-18 of 17 October 2006 Enacting the Petroleum Code; in Togo, Article 22(1) of Law No. 99-003 of 18 February 1999 Enacting the Hydrocarbons Code.

Nature of the main mining rights within the OHADA zone Analysis of the mining codes						
Niger	Personal property right (article 57)			Real property right (article 57)	**No provision relative to concessions**	
Democratic Republic of Congo		Real property right (article 51)		Real property right (article 65)	**No provision relative to concessions**	
Senegal	Personal property right (article 15)			Real property right (article 28)		Real property right (article 24)
Chad	Personal property right (article 50)		Personal property right (article 15		**No provision relative to concessions**	
Togo	**No specification of the nature of the mining rights**				**No provision relative to concessions**	

8.54 This diversity is found in the petroleum codes enacted by OHADA member states.

Nature of the main mining rights within the OHADA zone Analysis of the petroleum codes						
Country	**Exploration permit**		**Exploitation permit**		**Concession**	
Benin	Personal property right (article 24)			Real property right (article 33)	**No provisions relative to concessions**	
Burkina Faso	**No text applicable to the upstream oil sector**					
Cameroon	**No specification of the nature of the mining rights**					
Central Africa	**No specification of the nature of the exploration permit**		**No provisions relative to the exploitation permits**		**Real property right (article 27)**	
Comoros	**Text not available**					
Côte d'Ivoire	**No specification of the nature of the mining rights**				**No provisions relative to concessions**	
Gabon	**No specification of the nature of the mining rights**					
Guinea-Bissau	**Text not available**					
Guinea	**No specification of the nature of the mining rights**					
Equatorial Guinea	**Text not available**					
Mali		Real property right (article 37)		Real property right (article 54)	**No provisions relative to concessions**	
Niger	**No specification of the nature of the mining rights**					
Democratic Republic of Congo	**No provisions relating to mining rights since the exploitation and production operations are now conducted on the basis of a petroleum agreement**					
Senegal	**No specification of the nature of the mining rights**					
Chad	Personal property right (article 50)		Personal property right (article 50)		**No provisions relative to concessions**	

Nature of the main mining rights within the OHADA zone Analysis of the petroleum codes					
Country	**Exploration permit**		**Exploitation permit**	**Concession**	
Togo	Personal property right (article 13)		**No provisions relative to the exploitation permits**		Real property right (article 22)

b The distinction related to the divisibility and indivisibility of mining rights

8.55 In addition to the fundamental distinction between personal and real property, national legislation sometimes provides certain indications as to the divisibility or indivisibility of mining rights.

As for the countries of French-speaking Africa, many jurisdictions provide for the indivisibility of mining rights on the basis of which exploitation operations are undertaken,[154] although certain exceptions must be mentioned.[155] The indivisibility of the mining right is not surprising in the Francophone African countries that are part of the civil-law system, as

---

[154] As regards mining rights in general, this principle is affirmed in particular by Article 28 of Law No. 1-21 of 15 October 2013 Enacting the Mining Code in Burundi, which provides: '*[l]es titres miniers sont personnels et indivisibles.*' [Free translation into English: '[t]he mining rights are personal and indivisible'].

Regarding exploration permits, this principle is expressed by the following laws: in Benin, Article 59 of Law No. 2006-17 of 17 October 2006 Enacting the Mining and Mining Taxation Code in the Republic of Benin, Article 24 of Law No. 2006-18 of 17 October 2006 Enacting the Petroleum Code; in the Central African Republic, Article 25 of Law No. 09-005 of 29 April 2009 Enacting the Mining Code; in the Republic of Congo, Article 29 of Law No. 4-2005 of 11 April 2005 Enacting the Mining Code; in Mali, Article 40 of Law No. 2012-015 of 27 February 2012 Enacting the Mining Code; in Tunisia, Article 38 of Law No. 2003-30 of 28 April 2003 Enacting the Mining Code and Article 33 of Law No. 99-93 of 17 August 1999 Enacting the Hydrocarbons Code.

With regard to exploitation permits, this principle is affirmed, in particular, by the following laws: in Benin, Article 59 of Law No. 2006-17 of 17 October 2006 Enacting the Mining and Mining Taxation Code; in Côte d'Ivoire, Article 31(4) of Law No. 2014-138 of 24 March 2014 Enacting the Mining Code; in Niger, Article 67 of Law No. 2007-01 of 31 January 2007 Enacting the Petroleum Code; in Togo, Article 13 of Law No. 99-003 of 18 February 1999 Enacting the Hydrocarbons Code.

[155] See, in particular, in Madagascar, Article 58 of Law No. 99-022 of 19 August 1999 On the Mining Code amended by Law No. 2005-021 of 17 October 2005 (On the Mining Exploration Permit): '*[l]e permis de recherche minière ou permis "R" constitue un droit cessible, transmissible, amodiable et susceptible de gage. Il est divisible seulement en carrés entiers.*' [Free translation into English: '[t]he mining exploration permit or "R" permit constitutes a right that can be assigned, transferred, farmed-out and pledged. It is divisible only in whole squares'. and Article 59 of the said Law (Relating to Mining Exploitation Permits); in Mali, Article 58 of Law No. 2012-015 of 27 February 2012 Enacting the Mining Code (relating to small mine exploitation permits):

> '*[u]ne autorisation d'exploitation de petite mine est cessible, amodiable, transmissible, mais divisible seulement sous certaines conditions qui sont précisées dans le décret d'application du présent Code minier.*'
>
> [Free translation into English: '[a]n operating licence for a small-scale mine can be assigned, farmed-out, transferred, but divisible only under certain conditions specified in the implementing decree of this Mining Code.']

unlike common-law countries, they do not provide for the division of ownership of a mining right.[156]

c The consequences arising from the nature of mining rights

8.56 The nature of mining rights has far-reaching consequences on the operations that may be carried out thereunder, in particular when mining rights are required as security or to transfer ownership thereof.

i *Mining rights provided as security*

8.57 The legal nature of mining rights—personal or real property—determines the type of security that may be created thereupon.

The principle in this regard is that real property may be mortgaged, while personal property may be used as pledge or collateral. The distinction between real and personal property securities is also taken up in OHADA law by the Uniform Act of 15 December 2010 on the organisation of security rights, which separates collateral and pledge from mortgage.[157]

8.58 The issue of the use of mining rights as security may be addressed in national legislation and the answers provided are still far from uniform.

Thus, in the case of mining titles creating real property rights, certain jurisdictions provide for the impossibility of establishing a mortgage,[158] whereas this possibility is left open by certain laws.[159]

---

[156] See to that effect, Thierry Lauriol 'Financement de projets miniers dans les États parties à l'OHADA l'apport des titres miniers en garantie' at 223.

[157] Regarding the application of the Uniform Act concerning the organisation of security interests on mining rights, see Lauriol 'Financement de projets miniers dans les États parties à l'OHADA: l'apport des titres miniers en garantie' at 224.

[158] As for jurisdictions that prohibit the establishment of a mortgage on exploitation permits, see, in particular, in Niger, Article 67 of Law No. 2007-01 of 31 January 2007 Enacting the Petroleum Code.
For jurisdictions that prohibit mortgages on temporary exploitation permits for quarries (Article 92) and permanent exploitation permits for quarries (Article 146), see Law No. 17/2014 of 30 January 2015 Regulating the Mining Sector in the Gabonese Republic.
For jurisdictions that prohibit the creation of a mortgage on artisanal exploitation permits, see, in particular: in the Central African Republic, Article 72 of Law No. 09-005 of 29 April 2009 Enacting the Mining Code; in Niger, Article 45(2) of Ordinance No. 93-16 of 2 March 1993 Enacting the Amended Mining Law.

[159] As for exploitation permits, the establishment of a mortgage is, in particular, permitted by the following provisions: in Algeria, Article 8 of Law No. 01-10 of 3 July 2001 Enacting Mining Law; in Benin, Article 33 of Law No. 2006-18 of 17 October 2006 Enacting the Petroleum Code; in the Democratic Republic of Congo, Article 168 of Law No. 007-2002 of 11 July 2002 Enacting the Mining Code; in Côte d'Ivoire, Article 31(4) of Law No. 2014-138 of 24 March 2014 Enacting the Mining Code; in Madagascar, Article 59 of Law No. 99-022 of 19 August 1999 Relating to the Mining Code modified by Law No. 2005-021 of 17 October 2005.
As for concessions, the establishment of a mortgage is in particular permitted by the following provisions: in Burundi, Article 68 of Decree-Law No. 1/138 of 17 July 1976 On the Mining and Petroleum Code; in Togo, Article 22(1) of Law No. 99-003 of 18 February 1999 Enacting the Hydrocarbons Code.

The same applies to pledges and collateral, which, may be allowed[160] or, on the contrary, prohibited,[161] as the case may be.

8.59 Understanding of the matter of the contribution of mining rights as security is sometimes complicated by the existence of contradictions in national mining and petroleum laws. This is for instance the case in Mali, where the mining code which qualifies the exploitation/production permit as real property admits the possibility of establishing on the latter a pledge or mortgage.[162] However, as mentioned above, the pledge is reserved for movable property not subject to a mortgage.[163] This clarification—unnecessary because personal property cannot be the subject of a mortgage—illustrates the lack of clarity in the matter.

8.60 In order to preserve the smooth running of exploitation operations, certain laws are careful to specify that the establishment of a security on a mining right cannot prevent further exploration and exploitation works.[164]

It should also be noted that the end of validity of a mining right must in principle entail the termination of the security.[165]

---

[160] For jurisdictions authorising the establishment of pledges and/or collateral with mining exploration rights, see, in particular: in Madagascar, Article 58 of Law No. 99-022 of 19 August 1999 On the Mining Code amended by Law No. 2005-021 of 17 October 2005.

[161] For jurisdictions prohibiting the establishment of pledges and/or collateral with mining exploration rights, see, in particular: in Algeria, Article 8 of Law No. 01-10 of 3 July 2001 Enacting Mining Law; in Benin, Article 59 of Law No. 2006-17 of 17 October 2006 Enacting the Mining and Mining Taxation Code; in Côte d'Ivoire, Article 20(3) of Law No. 2014-138 of 24 March 2014 Enacting the Mining Code.

[162] See to that effect Article 68 of law 2012-015 of 27 February 2012 Enacting the Mining Code in Mali, which provides:

> '*[l]e permis d'exploitation constitue un droit immobilier de durée limitée, distinct de la propriété du sol, susceptible d'hypothèque ou de nantissement.*'
>
> [Free translation into English: '[t]he exploitation permit constitutes a real property right of limited duration, distinct from the ownership of the land, which is subject to mortgage or pledge.']

Another contradiction can be found in the new Burkina Faso Mining Code, which provides that a large-scale or small-scale industrial exploitation permit constitutes a real property right that can be pledged (Article 47 of Law No. 036-2015/CNT of 16 June 2015 Enacting the Mining Code of Burkina Faso). The same confusion is found with regard to permits for the semi-mechanised exploitation of mining substances (Article 59 of the said Law).

[163] See, in particular, in Burundi, Article 68 of Order No. 1/138 of 17 July 1976 Enacting the Mining and Petroleum Code; in Côte d'Ivoire, Article 20(3) of Law No. 2014-138 of 24 March 2014 Enacting the Mining Code.

[164] See, in particular, in Algeria, Article 9 of Law No. 01-10 of 3 July 2001 Enacting the Mining Law which provides:

> '*[l]es mutations de propriété, les hypothèques, les sûretés immobilières consenties par les propriétaires du sol en surface ou des ayant droits, sur les immeubles par nature ou par destination et sur les droits immobiliers, ne font pas obstacle à la poursuite des activités de recherche ou d'exploitation au sens de la présente loi.*'
>
> [Free translation into English: '[t]he transfer of ownership, mortgages, real property security rights granted by surface landowners or rights-holders, on real property by nature or by destination and on the real property rights, shall not prevent the continuation of exploration or exploitation activities within the meaning of this Law.']

[165] See, in particular, in Algeria, Article 8 of Law No. 01-10 of 3 July 2001 On Mining Law, which provides: '*[l]es autorisations de ramassage, d'exploitation des carrières et des sablières*

8.61 The possibility of constituting a security on a mining right is intrinsically linked to the possibility of transferring the said right.[166] In fact, only transferable property can be the object of security interests, the latter involving the possibility of a forced transfer of the mining right if the security is enforced.

ii *Transfer of mining rights*

8.62 Many transactions may concern mining rights, such as assignments, company contributions, farm-in and farm-out agreements[167] or, more broadly, transfers.

8.63 In general, these operations raise the question of the transferability of mining rights. The answers given to this question by African mining and petroleum laws are quite diverse and vary according to the mining right concerned. For example, the farm-in/farm-out of mining rights, generally accepted for exploitation permits, is in many cases prohibited for exploration permits.[168]

---

considérées comme bien meubles, ne constituent pas des titres miniers.' [Free translation into English: '[t]he authorisation to collect, from the exploitation of quarries and sand pits considered as personal property, shall not constitute mining rights'.]

[166] Only transferable property can be the object of security interests. On the other hand, it is possible to provide for the transferability of a mining right even though the law does not authorise the establishment of security rights. See, in particular, in Algeria, Article 8 of Law No. 01-10 of 3 July 2001 On Mining Law which provides:

> '*[l]es titres miniers relatifs aux activités de recherche minière constituent des biens meubles, transmissibles et cessibles dans les conditions fixées par la présente loi et les dispositions du Code Civil et du Code de Commerce. Ils ne sont pas susceptibles d'amodiation ou de gage ou de nantissement.*'
>
> [Free translation into English: '[t]he mining rights relating to mining exploration activities constitute personal property, transferable and assignable under the conditions laid down by this law and the provisions of the Civil Code and the Commercial Code. They cannot be the object of a "farm-in/farm-out" agreement, collateral or pledge.']

[167] The concept of a 'farm-in/farm-out' agreement is defined by Article 177 of Law No. 18/001 of 9 March 2018 Amending and Supplementing Law No. 007/2002 of 11 July 2002 Enacting the Mining Code in the Democratic Republic of Congo, which provides:

> '*[l]'amodiation consiste en un louage pour une durée fixe ou indéterminée, sans faculté de sous louage, de tout ou partie d'un droit minier d'exploitation ou d'Autorisation d'exploitation de carrières permanente, moyennant une rémunération fixée par accord entre l'amodiant et l'amodiataire.*'
>
> [Free translation into English: '[a] "farm-in/farm-out" agreement consists of a lease for a fixed or indefinite period, without the possibility of sub-letting, of all or part of a mining right or a permanent authorisation to exploit quarries, the rights attached to a mining right or a quarry authorisation, in return of a composentation determined by an agreement between the farmme and the farmor by way of remuneration established by agreement between the lessee and lessor'].

See also in Guinea, Article 1 of Law No. L/2013/n°053/CNT of 8 April 2013 Amending Certain Provisions of Law No. 2011/006/CNT of 9 September 2011 Enacting the Mining Code or in Morocco, Article 1 of Law 33-13 On Mines.

[168] As regards codes prohibiting the 'farm-in/farm-out' of exploration mining rights, see, in particular: in Algeria, Article 8 of Law No. 01-10 of 3 July 2001 Enacting Mining Law; in Benin, Article 24 of Law No. 2006-18 of 17 October 2006 Enacting the Petroleum Code; in Burundi, Article 50 of Law No. 1-21 of 15 October 2013 Enacting the Mining Code and

8.64 Similarly, the transferability of a mining right depends largely on the mining right in question. Thus, the transferability of mining rights, generally accepted for exploration[169] and exploitation[170] permits, is in principle restricted in the case of artisanal production permits[171] and prospecting

Article 36 of Decree-Law No. 1/138 of 17 July 1976 Enacting the Mining and Petroleum Code; in the Central African Republic, Article 25 of Law No. 09-005 of 29 April 2009 Enacting the Mining Code; in the Comoros, Article 16 of Order No. 54-1110 of 13 November 1954 Reforming the Regime of Mineral Substances; in Côte d'Ivoire, Article 20(3) of Law No. 2014-138 of 24 March 2014 Enacting the Mining Code; in Mali, Article 40 of Law No. 2012-015 of 27 February 2012 Enacting the Mining Code.

As for the codes authorising the "farm-in/farm-out" of mining exploration rights, see, in particular: in Morocco, Article 9 of Law No. 21-90 of 1 April 1992 On the Exploration and Exploitation of Hydrocarbon Deposits, amended and supplemented by Law No. 27-99 of 15 February 2000 and Articles 11 and 14 of Law No. 33-13 On Mines; in Tunisia, Article 38 of Law No. 2003-30 of 28 April 2003 Enacting the Mining Code; in Madagascar, Article 58 of Law No. 99-022 of 19 August 1999 Enacting the Mining Code amended by Law No. 2005-021 of 17 October 2005.

[169] As regards the codes admitting the assignability and transferability of mining exploration rights, see, in particular: in Algeria, Article 8 of Law No. 01-10 of 3 July 2001 Enacting Mining Law; in Benin, Article 24 of Law No. 2006-18 of 17 October 2006 Enacting the Petroleum Code; in Burundi, Article 36 of Decree-Law No. 1/138 of 17 July 1976 Enacting the Mining and Petroleum Code; in the Central African Republic, Article 25 of Law No. 09-005 of 29 April 2009 Enacting the Mining Code; in the Comoros, Article 16 of Decree-Law No. 54-1110 of 13 November 1954 Reforming the Regime of Mineral Substances; in the Republic of Congo, Article 29 of Law No. 4-2005 of 11 April 2005 Enacting the Mining Code; in Madagascar, Article 58 of Law No. 99-022 of 19 August 1999 Enacting the Mining Code as amended by Law No. 2005-021 of 17 October 2005; in Mali, Article 40 of Law No. 2012-015 of 27 February 2012 Enacting the Mining Code; in Niger, Article 50 of Law No. 2007-01 of 31 January 2007 Enacting the Petroleum Code; in Togo, Article 13(3) of Law No. 99-003 of 18 February 1999 Enacting the Hydrocarbons Code; in Tunisia, Article 38 of Law No. 2003-30 of 28 April 2003 Enacting the Mining Code.

[170] With regard to the codes enacting the assignability and the transferability of exploitation permits, see in Benin, Article 33 of Law No. 2006-18 of 17 October 2006 Enacting the Petroleum Code; in Burundi, Article 68 of Decree-Law No. 1/138 of 17 July 1976 Enacting the Mining and Petroleum Code; in the Comoros, Article 16 of Order No. 54-1110 of 13 November 1954 Reforming the Regime of Mineral Substances; in the Republic of Congo, Articles 52 and 64 of Law No. 4-2005 of 11 April 2005 Enacting the Mining Code; in the Democratic Republic of Congo, Article 65 of Law No. 007-2002 Enacting the Mining Code; in Madagascar, Article 59 of Law No. 99-022 of August 19, 1999 Enacting the Mining Code amended by Law No. 2005-021 of 17 October 2005; in Mali, Article 68 of Law No. 2012-015 of 27 February 2012 Enacting the Mining Code; in Togo, Article 18(5) of Law No. 2003-012 of 4 October 2003 Amending and Supplementing Law No. 96-004 of 26 February 1996 Enacting the Mining Code.

[171] As for the codes enacting the non-assignable nature of mining rights for artisanal production, see in Burundi, Article 101 of Law No. 1-21 of 15 October 2013 Enacting the Mining Code; in the Central African Republic, Article 72 of Law No. 09-005 of 29 April 2009 Enacting the Mining Code; in Niger, Article 45(2) of Ordinance No. 93-16 of 2 March 1993 Enacting the Amended Mining Law.

See, to the contrary, in the Republic of Congo, Article 43 of Law No. 4-2005 of 11 April 2005 Enacting the Mining Code; in Togo, Article 22(4) of Law No. 2003-012 of 4 October 2003 Amending and Supplementing Law No. 96-004 of 26 February 1996 Enacting the Mining Code.

permits[172] or authorisaitons.[173] The non-transferability of mining rights is generally due to the personal nature of the mining right concerned.[174]

8.65 Although the principle of transfer, assignment or farm-in and farm-out of a mining title is widely accepted, such operations assume, by virtue of their importance, a prior control by the licensing authority. Indeed, these transactions involve a change in ownership of the right and are therefore subject to the prior authorisation of the licensing authority.[175] In

---

[172] With regard to the codes establishing the non-assignable nature of prospecting licences: in the Republic of Congo, Article 23 of Law No. 4-2005 of 11 April 2005 Enacting the Mining Code; in Côte d'Ivoire, Article 49 of Law No. 2014-138 of 24 March 2014 Enacting the Mining Code; in Niger, Article 31 of Law No. 2007-01 of 31 January 2007 Enacting the Petroleum Code; in Chad, Article 45 of Law No. 011-PR-1995 of 20 June 1995 Enacting the Mining Code; in Togo, Article 11(3) of Law No. 2003-012 of 4 October 2003 Amending and Supplementing Law No. 96-004 of 26 February 1996 Enacting the Mining Code; in Tunisia, Article 19 of Law No. 2003-30 of 28 April 2003 Enacting the Mining Code.

See, to the contrary, in Mali, Article 32 of Law No. 2012-015 of 27 February 2012 Enacting the Mining Code.

[173] With regard to the codes establishing the non-assignable nature of prospecting licences: in the Republic of Congo, article 23 of Law No. 4-2005 of 11 April 2005 Enacting the Mining Code; in Côte d'Ivoire, Article 49 of Law No. 2014-138 of 24 March 2014 Enacting the Mining Code; in Niger, article 31 of Law No. 2007-01 as 31 January 2007 Enacting the Petroleum Code; in Chad, Article 45 of Law No. 011-PR-1995 of 20 June 1995 Enacting the Mining Code; in Togo, article 11(3) of Law No. 2003-012 as 4 October 2003 Amending and Supplementing Law No. 96-004 of 26 February 1996 Enacting the Mining Code; in Tunisia, Article 19 of Law No. 2003-30 of 28 April 2003 Enacting the Mining Code.

See, to the contrary, in Mali, Article 32 of Law No. 2012-015 of 27 February 2012 Enacting the Mining Code.

[174] For example, in Burundi, Article 101 of Law No. 1-21 of 15 October 2013 Enacting the Mining Code provides:

'*[l]e permis d'exploitation artisanale est un droit conféré à titre personnel. Ce droit est incessible et non transmissible par succession.*'

[Free translation into English: '[t]he artisanal exploitation permit is granted as a personal right. This right is not assignable and not transferable by succession'.]

[175] The prior consent of the licensing authority in connection with the transfer of a mining right is in particular provided for in section 11(1) of the Mineral and Petroleum Resources Development Act 28 of 2002; in Benin, Article 24 of Law No. 2006-18 of 17 October 2006 Enacting the Petroleum Code; in Botswana, Articles 23, 36, 50 and 59 of the Mines and Minerals Act and Article 51 of the Petroleum (Exploration and Production) Act; in Burundi, Articles 25, 36 and 68 of Order No. 1/138 of 17 July 1976 Enacting the Mining and Petroleum Code; in the Republic of Congo, Articles 43, 52 and 64 of Law No. 4-2005 of 11 April 2005 Enacting the Mining Code; in Côte d'Ivoire, Articles 41 and 93 of Law No. 2014-138 of 24 March 2014 Enacting the Mining Code; in Eritrea, Articles 9(3), 14(3) and 16(2) of Proclamation No. 68/1995 Promoting Development of Mineral Resources; in Ethiopia, Article 38 of Proclamation No. 678/2010 Promoting Sustainable Development of Mineral Resources; in Ghana, Article 14 of the Minerals and Mining Act, 2006; in Guinea, Article 90 of Law No. 2011/006/CNT of 9 September 2011 Enacting the Mining Code and Article 107 of Law No. L/2014/n°034/AN of 23 December 2014 Enacting the Petroleum Code; in Libya, Article 17 of Petroleum Law No. 25/55 of 11 May 2012; in Malawi, Article 60 of the Mines and Minerals Act, 1981; in Mauritania, Article 47 of Law No. 2010-033 of 20 July 2010 Enacting the Code of Crude Hydrocarbons amended by Law No. 2011-044 of 25 November 2011 and Article 43 of Law No. 2008-011 Enacting the Mining Code; in Mozambique, Article 62 of Mining Law No. 20-2014 of 18 August 2014 and Article 53 of the Petroleum Law No. 21-2014 of 18 August 2014; in Niger, Article 58 of Ordinance No. 93-16

this regard, certain mining or petroleum codes specify that any agreement resulting in the assignment, transfer, farm-in, or farm-out of a mining right must be made conditional on the authorisation of the licensing authority.[176] Exceptionally, some jurisdictions provide that the prior consent of the licensing authority is not required if the transaction takes place for the benefit of an affiliate.

8.66 The licensing authority must ensure that the operator for whose benefit the transfer of the mining rights is performed, has the necessary technical and financial skills[177] to carry out mining and oil operations.

However, the requirement of obtaining the prior authorisation of the granting authority is not without some difficulty, in particular in the context

---

of 2 March 1993 Enacting the Amended Mining Law and Article 50 of Law No. 2007-01 of 31 January 2007 Enacting the Petroleum Code; in Nigeria, Article 147 of the Minerals and Mining Act, 2007; in Uganda, Article 6 of the Mining Act, 2003 and Article 87 of the Petroleum (Exploration, Development and Production) Act, 2013; in Senegal, Article 56 of Law No. 98-05 of 8 January 1998 Enacting the Petroleum Code; in Sierra Leone, Articles 83 and 119 of the Mines and Minerals Act, 2009; in Swaziland, Article 30 of the Mines and Minerals Act, 2011; in Chad, Article 45 of Law No. 011-PR-1995 of 20 June 1995 Enacting the Mining Code and Article 21.1 of Law No. 006-PR-2007 of 2 May 2007 On Hydrocarbons; in Togo, Articles 13(3) and 22(1) of Law No. 99-003 of 18 February 1999 Enacting the Hydrocarbons Code and Articles 14, 18 and 22 of Law No. 2003-012 of 4 October 2003 Amending and Supplementing Law No. 96-004 of 26 February 1996 On the Mining Code; in Tunisia, Articles 38 and 61 of Law No. 2003-30 of 28 April 2003 Enacting the Mining Code and Articles 64 and 55 of Law No. 99-93 of 17 August 1999 Enacting the Hydrocarbon Code.

[176] Thus, Article 41 of Law No. 2014-138 of 24 March 2014 Enacting the Ivorian Mining Code provides:

> '*[l]e titre minier est cessible ou transmissible sous réserve de l'approbation préalable du Ministre chargé des Mines et dans les conditions prévues par décret. Tout accord ainsi conclu ne peut être passé que sous condition suspensive de cette autorisation.*'
>
> [Free translation into English: '[t]he mining right is assignable or transferable subject to the prior approval of the Minister of Mines and under the conditions prescribed by decree. Any agreement thus concluded may only be executed under the condition precedent of this authorisation'].

See also, in Burundi, Article 25 of Decree-Law No. 1/138 of 17 July 1976 Enacting the Mining and Petroleum Code, which provides:

> '*[l]es contrats de cession entre vifs ou d'amodiation de titres miniers doivent, à peine de nullité de plein droit, contenir la clause suspensive d'autorisation préalable du Ministre chargé des mines.*'
>
> [Free translation into English: '[t]he contracts of *inter vivos* transfer or 'farming-in/farming-out' of mining rights must, on pain of nullity by operation of law, contain the condition precedent clause of prior authorisation of the Minister in charge of mines'].

In the Central African Republic, Article 25 of Law No. 09-005 of 29 April 2009 Enacting the Mining Code.

[177] To that effect, Article 67 of Law No. 2012-015 of 27 February 2012 Enacting the Mining Code in Mali states:

> '*[n]ul ne peut se voir transférer un permis d'exploitation, s'il ne possède les capacités techniques et financières nécessaires pour mener à bien les travaux d'exploitation et répondre aux obligations mentionnées aux articles 92 et 93 ci-dessous.*'
>
> [Free translation into English: '[n]o person shall benefit from the transfer of an exploitation permit unless it possesses the technical and financial capacity to carry out the work and fulfil the obligations referred to in Articles 92 and 93 below'].

On the technical and financial capacities of the operators, see below para 9.3 et seq.

of the contribution of a mining right as security.[178] In fact, the beneficiaries, very often banks, do not possess the necessary technical capacity and do not have the vocation to undertake mining or petroleum operations. Thus, in practice, these guarantees may have the effect of blocking the defaulting borrower.

8.67 Finally, it should be pointed out that assignment or transfer operations generally have financial consequences[179] giving rise to taxation of realized capital gains[180] (capital gains are understood as the gain resulting from the difference between the transfer price or the transfer value of the security and the cost of the investments made on the mining right).[181] In practice, this subject raises great difficulties.

It is sometimes difficult for the private operator and the authorities of the host state (the tax authorities) to agree on the amount of capital gains realised, in particular when the investor asserts that the profits of the transfer would have been absorbed by the costs incurred in the operations.

Lastly, it shall be pointed out that mining and petroleum codes will generally impose the obligation to pay fixed duties in the event of the transfer of mining rights.[182]

8.68 We shall illustrate our developments with a table containing the provisions relating to the assignment, transfer and farm-in and farm-out of mining rights in the OHADA zone.

Assignment, transfer and farm-in/farm-out of the main mining rights Analysis of the mining codes enacted by OHADA member states						
	Exploration permit			**Exploitation permit**		
Country	Assignment	Transfer	Farm in/ Farm out	Assignment	Transfer	Farm in/ Farm out
Benin	Yes (article 60)	Yes (article 60)	Yes (article 60)	Yes (article 60)	Yes (article 60)	Yes (article 60)
Burkina Faso	Yes (article 106)	Yes (article 106)		Yes (article 106)	Yes (article 106)	

[178] To that effect, see Lauriol 'Financement de projets miniers dans les États parties à l'OHADA: l'apport des titres miniers en garantie' at 227.

[179] For more details on the methods of taxing capital gains in connection with the sale of a mining right, see Alain Charlet, Bertrand Laporte & Grégoire Rota-Graziosi 'La fiscalité minière en Afrique de l'Ouest et du Centre'(2013) 48 *Revue de droit fiscal* 27 et seq.

[180] See, in particular, in the Central African Republic, Article 58 of Law No. 09-005 of 29 April 2009 Enacting the Mining Code; in Guinea, Article 91-II of Law No. L/2013/N°053/CNT of 8 April 2013 Amending Law No. 2011/006/CNT of 9 September 2011 Enacting the Mining Code; in Chad, Article 21(7) of Law No. 006-PR-2007 of 2 May 2007 On Hydrocarbons; in the Democratic Republic of Congo, Article 253 of Law No. 18/001 of 9 March 2018 Amending and Supplementing Law No. 007/2002 of 11 July 2002 Enacting the Mining Code; in Burkina Faso, Article 107 of Law No. 036-2015/CNT of 16 June 2015 Enacting the Mining Code of Burkina Faso; in the Democratic Republic of Congo, Article 84 of Law No. 15/012 Enacting the General Hydrocarbon Regime.

[181] See to that effect, in Guinea, Article 1 of Law L/2013/n°053/CNT of 8 April 2013 Amending Law No. 2011/006/CNT of 9 September 2011 Enacting the Mining Code.

[182] See to that effect, in Gabon, Articles 275 et seq of Law No. 17/2014 of 30 January 2015 Regulating the Mining Sector in the Gabonese Republic.

Assignment, transfer and farm-in/farm-out of the main mining rights Analysis of the mining codes enacted by OHADA member states						
Cameroon	Yes (article 20)	Yes (article 20)		Yes (article 20)	Yes (article 20)	
Central Africa	Yes (article 25)	Yes (article 25)	No (article 25)	Yes (article 58)	Yes (article 58)	Yes (article 58)
Comoros	Yes (article 16)	Yes (article 16)	No (article 16)	Yes (article 16)	Yes (article 16)	Yes (article 16)
Congo	Yes (article 29)	Yes (article 29)		Yes (article 64)	Yes (article 64)	Yes (article 64)
Côte d'Ivoire	Yes (article 41)	Yes (article 41)	No (article 20)	Yes (article 41)	Yes (article 41)	Yes (article 1)
Gabon	Yes (article 276)	Yes (article 276)	Yes (article 276)	Yes (article 147)	Yes (article 147)	Yes (article 147)
Guinea	No (article 90)	No (article 90)		Yes (article 90)	Yes (article 90)	Yes (article 28)
Guinea-Bissau	Yes (article 127)	Yes (article 126)		Yes (article 127)	Yes (article 126)	
Equatorial Guinea	**Exploration and exploitation activities take place on the basis of mining economic rights**					
Mali	Yes (article 40)	Yes (article 40)	No (article 40)	Yes (article 68)		Yes (article 68)
Niger	Yes (article 58)	Yes (article 58)		Yes (article 58)	Yes (article 58)	Yes (article 58)
Democratic Republic of Congo	Yes (article 51)	Yes (article 51)		Yes (article 65)	Yes (article 65)	Yes (article 65)
Senegal	Yes (article 19)	Yes (article 19)	}	Yes (article 28)	Yes (article 28)	Yes (article 28)
Chad	Yes (article 45)	Yes (article 45)	Yes (article 45)	Yes (article 45)	Yes (article 45)	Yes (article 45)
Togo	Yes (article 14)	No (article 14)	No (article 14)	Yes (article 18)	Yes (article 18)	No (article 18)

3 THE LEGAL CONSEQUENCES OF MINING RIGHTS VIS-À-VIS THIRD PARTIES

8.69 Mining and petroleum laws are intended to regulate the relations between the owners of the land[183] and the holders of mining rights.[184] Quite often, reference should also be made to national land legislation in order to accurately determine all rules pertaining to the rights of the operator on the land. It should be noted, however, that there could be discrepancies between property laws, sometimes established long ago,[185] and the ever-changing mining and petroleum legislation.

---

183 The mining and petroleum codes make reference to the rights of the owners but also more generally to the rights of the legitimate occupants or even beneficiaries of customary rights.

184 See, in particular, on this subject, Bandoki op cit 52 et seq.

185 However, this issue should not be generalised, since certain countries have recently reformed their land laws (see, in particular, Senegal, Law No. 2011-07 of 30 March 2011 On Land Ownership).

8.70 A mining right does not provide any right to the surface property and in no case extinguishes the rights of the landowner.[186] However, the exercise of mining rights over a given perimeter may impact the property rights within the said area. Having mining rights grants the holders thereof the right to exploit the natural resources present in the subsoil; however, the exercise of this right requires the holder to access the surface in order to carry out the mining or petroleum operations. The holder of mining rights is therefore authorised to access and occupy the land necessary to carry on its activities.

This right of occupation will be subject to consideration for the owner of the land, in particular in the form of the payment of compensation for the occupation, tied to the loss of use, or the payment of financial compensation in order to repair the damage caused to surface property.

a The right of occupation

8.71 The rights enjoyed by the landowner are not an obstacle to the granting of surface rights to the holder of mining rights. The standard is that the holder of mining rights may, under certain conditions, access and occupy the land necessary for the performance of mining and petroleum operations.[187] Generally speaking, this occupation involves both land within and outside the mining perimeter, as access to and occupation of these lands is necessary for operations.[188]

---

[186] Thus, Article 123 of Law No. 2011/006/CNT of 9 September 2011 Enacting the Guinean Mining Code provides that: '*[l]e droit minier n'éteint pas le droit de propriété.*' [Free translation into English: '[t]he mining right does not extinguish the right of ownership'].

[187] The right of the holder of mining rights to occupy the land is referenced by numerous national laws. Thus, for example, Article 71 of Law No. 2006-18 of 17 October 2006 Enacting the Petroleum Code in Benin provides that:

'*[l]e titulaire est autorisé à occuper sur sa demande, les terrains qui sont nécessaires à son activité de recherches ou d'exploitation.*'

[Free translation into English: '[t]he holder [of mining rights] is authorised to occupy, upon request, the land that is necessary for its exploration or exploitation activities'].

This standard is also referenced in Algeria, in Article 134 of Law No. 01-10 of 3 July 2001 Enacting the Mining Law; in Benin, in Article 115 of Law No. 2006-17 of 17 October 2006 Enacting the Mining and Mining Taxation Code in the Republic of Benin; in Guinea, in Article 49 of Law No. L/2014/n°034/AN Enacting the Petroleum Code; in Mauritania, in Article 54 of Law No. 2010-033 of 20 July 2010 Enacting the Code of Crude Hydrocarbons amended by Law No. 2011-044 of 25 November 2011; in Namibia, in Article 54 of the Petroleum (Exploration and Production) Act, 1991; in Niger, in Article 13 of Law No. 2007-01 of 31 January 2007 Enacting the Petroleum Code.

[188] For example, in Algeria, Article 19(1) of the Presidential Decree of 1 March 2003 On the Model Mining Convention provides that:

'*[p]our l'exploitation minière du gisement, L'ÉTAT garantit, dans le cadre de la législation en vigueur notamment les dispositions des articles 133 à 148 de la Loi Minière, à l'Investisseur l'occupation et l'utilisation de tous les terrains nécessaires, qu'ils soient situés à l'intérieur ou à l'extérieur du Périmètre.*'

[Free translation into English: '[f]or the exploitation of the deposit, the STATE shall guarantee to the Investor, within the framework of the legislation in force, in particular

8.72 In addition to the right to occupy the land, the holder of mining rights may generally benefit from legal easements for access and passage necessary for its installations or the conduct of mining and petroleum operations.[189]

The occupation of land necessary to carry out the exploitation activities entails in principle the right to cut the wood necessary for this activity or to use surface and underground water,[190] subject in most cases to compensation or payment of the fees and charges provided for by the laws or regulations in force.

8.73 However, this right of occupation is not absolute, as the holder of mining rights will generally have to ensure compliance with certain formalities imposed by national laws. Thus, in certain cases, the holder of mining rights must first inform the landowner of its intention to carry out the work necessary for the exploitation of the deposit.[191] In certain geographical areas where the publishing of information is not guaranteed, it is also incumbent upon the holder to provide the surface owner with proof of the mining rights.[192] Some laws go further in protecting the surface owner by providing that no right of exploration or exploitation is valid without the consent of the landowner.[193]

> the provisions of Articles 133 to 148 of the Mining Act, the occupation and use of all land necessary, whether located inside or outside the Perimeter'].

See also to that effect, in Benin, Article 115 of Law No. 2006-17 of 17 October 2006 Enacting the Mining and Mining Taxation Code.

189 See, in particular, to that effect, in Algeria, Article 140 of Law No. 01-10 of 3 July 2001 Enacting the Mining Law.

190 See, in particular, in Benin, Article 121 of Law No. 2006-17 of 17 October 2006 Enacting the Mining and Mining Taxation Code:

> '*[l]'arrêté du ministre chargé des mines prévu à l'article 115 de la présente loi autorise en outre le titulaire après avis des autorités compétentes: à couper les bois nécessaires à ses travaux, à utiliser les chutes d'eau non utilisées ni réservées et à les aménager pour le besoin de ses travaux à l'intérieur du périmètre* [. . .]'.
>
> [Free translation into English: '[t]he order of the Minister responsible for mines as provided for in Article 115 of this law also authorises the holder, after consulting with the competent authorities: to cut the woods required for its work, to utilise the waterfalls that are not in use or reserved and to arrange them according to the needs of its work within the perimeter [. . .]'].

191 See, in particular, to that effect, in South Africa, section 5(4) of the Minerals and Petroleum Resources Development Act 28 of 2002; in Botswana, Article 60(3) of the Mines and Minerals Act; in Malawi, Article 66 of the Petroleum (Exploration and Production) Act, 1983 and Article 106 of the Mines and Minerals Act, 1981.

192 See in this regard, in Botswana, Article 60(2) of the Mines and Minerals Act, which provides that:

> '[a] person exercising any right under a mineral concession shall produce evidence of the possession of such concession to the owner or lawful occupier of any land upon which such right is to be exercised upon demand being made to him and, in default of such production, such person may be treated as a trespasser.'

See also in Swaziland, Article 118 of the Mines and Mineral Act, 2011.

193 See, in particular, in Mali, Article 73 of Law 2012-015 of 27 February 2012 Enacting the Mining Code, which provides that: '*[n]ul droit de recherche ou d'exploitation découlant des titres miniers ne vaut sans le consentement du (des) propriétaire(s) foncier(s), et de*

8.74 Numerous national laws also provide that the occupation of land by the holder of mining rights must be the object of an agreement between the said holder and the owner of the land, the consent of the landowner being therefore required.[194] The search for an agreement on the occupation of the land proceeds *a priori* with the identification of the owner or the lawful occupier by the holder of the mining rights, which is not always easy. Certain legislation provides for the possibility for the operator to apply for a land survey in order to identify the status of the parcels requested and to identify the beneficiaries on the parcels in question.[195]

8.75 Moreover, the agreement on the occupation of the land by the holder of mining rights may take the form of a lease agreement[196] or 'land lease',[197] which will allow the landowner to receive compensation in the form of a rent in consideration for the occupation of the land by the holder of the mining rights.

ses (leurs) ayant droit, en ce qui concerne les activités impliquant la surface ou ayant un effet sur celle-ci.' [Free translation into English: '[n]o right of exploration or exploitation deriving from mining rights shall be valid without the consent of the landowner(s) and their beneficiaries, with regard to the activities involving the surface or having an effect thereon]. See also, in Guinea, Article 123 of Law No. 2011/006/CNT of 9 September 2011 On the Mining Code.

194 See, in particular, in this regard, in Algeria, Article 135 of Law No. 01-10 of 3 July 2001 Enacting the Mining Law; in Côte d'Ivoire, Article, Article 61 of Law No. 96-669 of 29 August 1996 Enacting the Petroleum Code; in Guinea, Article 50 of Law No. L/2014/N°034/AN of 23 December 2014 Enacting the Petroleum Code; in Mali, Article 73 of Law No. 2012-015 of 27 February 2012 Enacting the Mining Code; in Mauritania, Article 55 of Law No. 2010-033 of 20 July 2010 Enacting the Code of Crude Hydrocarbons amended by Law No. 2011-044 of 25 November 2011; in Tunisia, Article 79 of Act No. 2003-30 of 28 April 2003 Enacting the Mining Code and Article 85 of Law No. 99-93 of 17 August 1999 Enacting the Hydrocarbon Code.

195 See to this effect, in Guinea, Article 50 of Law No. L/2014/N°034/AN of 23 December 2014 Enacting the Petroleum Code.

196 See, in particular, in Madagascar, Article 125 of Law No. 99-022 of 19 August 1999 Enacting the Mining Code amended by Law No. 2005-021 of 17 October 2005, which provides:

> '*[l]e titulaire du permis minier et le propriétaire du sol conviennent par contrat de bail, de leurs droits et obligations respectifs.*'
>
> [Free translation into English: '[t]he holder of the mining permit and the owner of the land agree under a lease agreement, on their respective rights and obligations'].

197 See, in particular, in Uganda, Article 81(1) of the Mining Act, 2003, which provides:

> '[t]he holder of a mining lease may, if he or she requires the exclusive use of the whole or any part of the mining area concerned, and if so requested by the owner or lawful occupier of any part of such area, obtain a land lease or other rights to use the area upon such terms as to duration or the extent of the land to which the lease shall relate, as may be agreed between the holder and the owner or lawful occupier of the land in question, or failing an agreement, as may be determined by arbitration.'

8.76 In the absence of agreement, the granting authority generally has the right to intervene, after consultation with the parties, in order to grant the holder of mining rights authorisation for occupation,[198] thereby enabling the holder to gain access to private land. Such intervention is essential in order not to delay the development of the operations.

8.77 The second possibility is that the occupation of the land by the holder of the mining rights may be subject to the issuance of administrative authorisation for occupation,[199] the procedure for which is described in the applicable regulations.

In the event of absolute necessity, and in particular when an agreement between the landowner and the holder of mining rights proves impossible, an expropriation procedure may be carried out,[200] if appropriate, at the request of the holder of the mining rights. The expropriated land will then be made available to the holder of the mining rights in accordance with the methods provided for by land and public property legislation. The expropriation procedure may also be requested by the owner of the land itself, in the event that the occupation granted to the holder of mining rights renders the use of

---

[198] See, in particular, in Ivory Coast, Article 61 of Law No. 96-669 of 29 August 1996 Enacting the Mining Code; in Guinea, Article 50 of the Law No. L/2014/No. 034/AN of 23 December 2014 Enacting the Mining Code; In Moroco, Article 14 of the Law No. 21-90 as 1 April 1992 On Exploration and Exploitation of Hydrocarbon Deposits Amended and Supplemented by the Law No. 27-99 of 15 February 2000; In Mauritius, Article 55 of the Law No. 2010-033 of 20 July 2010 Enacting the Hydrocarbons Code modified by the Law No. 2011-044 of 25 November 2011; in Togo, Article 42(2) of Law No. 99-003 of 18 February 1999 On Hydrocarbons Code; In Tunisia, Article 79 of Law No. 2003-30 of 28 April 2003 Enacting the Mining Code and Article 85 of Law No. 99-93 of 17 August 1999 Enacting the Hydrocarbons Code.

[199] See, in particular, in Benin, Articles 115 et seq. of Law No. 2006-17 of 17 October 2006 Enacting the Mining and Mining Taxation Code and Article 69 of Law No. 2006-18 of 17 October 2006 Enacting the Petroleum Code; in Cameroon, Article 54 of Law No. 99-013 of 22 December 1999 Enacting the Petroleum Code; in the Comoros, Article 33 of Decree-Law No 54-1110 of 13 November 1984 Reforming the Regime of Mineral Substances; in the Republic of Congo, Article 104 of Law No. 4-2005 of 11 April 2005 Enacting the Mining Code; in Niger, Article 113 of Ordinance No. 93-16 of 2 March 1993 Enacting the Amended Mining Law; in Togo, Article 42 of Law No. 99-003 of 18 February 1999 Enacting the Hydrocarbons Code.

[200] See, in particular, to that effect, in Algeria, Article 148 of Law No. 01-10 of 3 July 2001 Enacting the Mining Law; in Cameroon, Article 54(4) of Law No. 99-013 of 22 December 1999 Enacting the Petroleum Code; in the Central African Republic, Article 48 of Ordinance No. 93.007 of 25 May 1993 Enacting the Petroleum Code; in Côte d'Ivoire, Article 62 of Law No. 96-669 of 29 August 1996 Enacting the Petroleum Code; in Guinea, Article 125 of Law No. L/2013/N°053/CNT of 8 April 2013 Amending Law No. 2011/006/CNT of 9 September 2011 Enacting the Mining Code and Article 50 of Law L/2014/n°034/AN of 23 December 2014 Enacting the Petroleum Code; in Morocco, Article 16 of Law No. 21-90 of 1 April 1992 On the Investigation and Exploitation of Hydrocarbon Deposits, amended and supplemented by Law No. 27-99 of 15 February 2000; in Mauritania, Article 57 of Law No. 2010-033 of 20 July 2010 Enacting the Code of Hydrocarbons amended by Law No. 2011-044 of 25 November 2011 and Article 77(2) of Law No. 2008-011 of 27 April 2008 Enacting the Mining Code; in Niger, Article 15 of Law No. 2007-01 of 31 January 2007 Enacting the Petroleum Code; in South Sudan, Articles 140 and 141 of the Mining Act, 2012.

the land impossible.[201] Expropriation of the owner of the land will give rise to compensation, the methods of calculation of which may be stipulated in the mining and petroleum codes in force.[202] It should be emphasised that the costs and fees resulting from the expropriation proceedings will, in principle, be borne by the holder of the mining rights.[203]

8.78 The trend, evident in the evolution of mining and petroleum legislation, is that of better protection of land ownership. This trend is found in particular in the provisions relating to the compensation attached to the exercise by the operator of its right of occupancy.

b Compensation attached to the exercise of the right of occupancy

8.79 Access to private lands and the occupation thereof for the purposes of mining and petroleum operations are subject to consideration for the owner of the land. The disruption caused by the exploitation operations is considerable and the surface property is greatly impacted by mining and petroleum projects, which modify the landscapes and can cause a lot of damage to the depths and therefore to the owner of the surface.

Thus, the mining and petroleum codes establish a set of mechanisms aimed at protecting the landowners affected by the occupation of their land.

First of all, it should be emphasised that many laws, mainly in English-speaking Africa, reserve the right of the holder to cultivate the land on which the mining and petroleum operations take place, so long as the exercise of this right does not impair the proper conduct of operations.[204]

---

[201] See, in particular, the Republic of Congo, Article 106 of Law No. 4-2005 of 11 April 2005 Enacting the Mining Code; in Mali, Article 74 of Law No. 2012-015 of 27 February 2012 Enacting the Mining Code.

[202] As an example, in the Central African Republic, Article 48 of Ordinance No. 93.007 of 25 May 1993 Enacting the Petroleum Code provides that:

> '*[l]'indemnité d'expropriation est égale à la valeur du terrain exproprié, cette valeur étant déterminée par l'usage du terrain avant l'expropriation ou, le cas échéant, avant l'occupation. Les contestations relatives à cette indemnité sont soumises aux tribunaux civils.*'
>
> [Free translation into English: '[t]he compensation for expropriation shall be equal to the value of the land expropriated, the value being determined by the use of the land before expropriation or, if applicable, before the occupation. Disputes relating to this compensation are submitted to the civil courts'].

See also, in this regard, in Guinea, Article 125 of Law No. L/2013/N°053/CNT of 8 April 2013 Amending Law No. 2011/006/CNT of 9 September 2011 Enacting the Mining Code and Article 50 of Law L/2014/N°034/AN of 23 December 2014 Enacting the Petroleum Code.

[203] See, in particular, in Côte d'Ivoire, Article, article 62 of Law No. 96-669 of 29 August 1996 Enacting the Petroleum Code; in Guinea, Article 50 of Law No. L/2014/N°034/AN of 23 December 2014 Enacting the Petroleum Code; in Mauritania, Article 57 of Law No. 2010-033 of 20 July 2010 Enacting the Code of Hydrocarbons amended by Law No. 2011-044 of 25 November 2011.

[204] See, in particular, in Botswana, Article 61 of the Mines and Minerals Act and Article 69 of the Petroleum (Exploration and Production) Act; in Ghana, Article 72(3) of the Minerals and Mining Act, 2006; in Malawi, Article 104(1) of the Mines and Minerals Act, 1981 and Article 64 of the Petroleum (Exploration and Production) Act, 1983; in Nigeria, Article 101(3) of the Minerals and Mining Act, 2007; in Uganda, Article 80 of the Mining Act, 2003; in

8.80 The protection of the owner of the land also involves the obligation of the holder of the authorisation to minimise the consequences that may arise from the exploitation of the subsoil.[205]

8.81 Moreover, the occupation of private land generally involves financial compensation for the benefit of the landowner. Such consideration may, as applicable, take the form of compensation or reparation in the event of damage caused to land ownership. In some areas, part of the compensation paid to the local population is calculated on the basis of income from the exploitation of the natural resources contained under their land.

In fact, the occupation of the land by the holder of mining rights is such that it deprives the owner of the land of his right of enjoyment. In order to compensate the landowner's loss of enjoyment, the mining and petroleum laws will generally impose the payment of compensation on the holder of the permit.[206] The payment of the occupancy compensation is sometimes a prerequisite for the occupation of the land by the holder of the mining rights.[207] The applicable regulations generally provide that the amount of this

---

Swaziland, Article 119 of the Mines and Mineral Act, 2011; in Zimbabwe, Article 179 of the Mines and Minerals Act.

[205] See, in particular, in this regard, in Botswana, Article 61(3) of the Mines and Minerals Act, which provides that: '[t]he rights conferred by a mineral concession shall be exercised reasonably and so as to affect as little as possible the interests of any owner or lawful occupier of the land on which such rights are exercised consistent with the reasonable and proper conduct of the operations concerned'.

[206] See, in particular, in Algeria, Article 136 of the Law No. 01-10 of 3 July 2001 Enacting the Mining Law; In Burundi, Article 28 of Law No. 1-21 of 15 October 2013 Enacting the Mining Code; In Cameroun, Article 73(1) of Law No. 001 of 16 April 2001 Enacting the Mining Code; In Central African Republic, Article 47 of the Ordinance No. 93.007 du 25 May 1993 Enacting the Petroleum Code and Article 93 of the Law No. 09-005 of 29 April 2009 Enacting the Mining Code; in the Democratic Republic of Congo, Article 281 of Law No. 007-2002 Enacting the Mining Code; in Côte d'Ivoire Article 127 of Law No 2014-138 of 24 March 2014 Enacting Mining Code and Article 61 of the Law No 96-669 of 29 August 1996 Enacting Petroleum Code; in Ghana, Article 73 of Minerals and Mining Act, 2006; in Guinea, Article 124 of the Law No. 2011/006/CNT of 9 September 2011 Enacting the Mining Code and Article 50 of Law No. L/2014/No. 034/AN of 23 December 2014 Enacting the Petroleum Code; in Mauritania, Article 56 of the Law No. 2010-033 as 20 July 2010 Enacting the Hydrocarbons Code modified by the Law No. 2011-044 of 25 November 2011 and Article 77(1) of the Law No. 2008-011 of 27 April 2008 Enacting the Mining Code; in Namibia, Article 54 of Petroleum (Exploration and Production) Act, 1991; in Tchad, Article 78 of the Law No. 011-PR-1995 of 20 June 1995 Enacting the Mining Code.

[207] See to that effect, in the Democratic Republic of Congo, Article 108 of Law No. 4-2005 of 11 April 2005 Enacting the Mining Code, which provides:

> '*[l]e bénéficiaire ne peut occuper une parcelle de terrain visée par l'autorisation préfectorale qu'après présentation à l'autorité administrative des mines de l'acte attestant le paiement ou caution de payer l'indemnité d'occupation fixée dans les conditions prévues à l'article 112 ci-dessous.*'
>
> [Free translation into English: '[t]he beneficiary may occupy a parcel of land covered by the prefectoral authorisation only after filing with the administrative authority of mines the document evidencing the payment or security to pay the occupancy compensation established in the conditions laid out in Article 112 below'].

compensation be the subject of an agreement between the parties,[208] an amount which may in particular take into account the damage caused to the crops or the loss of earnings which would have been generated from the use of the land.[209] In the absence of agreement between the parties, mining and petroleum legislations may designate the court having jurisdiction, generally the national courts, to decide disputes concerning the amount of such compensation.[210] Quite often, determining the amount of compensation is a difficult task, especially when it involves taking into account the resources present on the surface. This difficulty considerably complicates the search for a compromise between the landowner and the holder of mining rights.

8.82 In a more original way, certain laws may also provide that the owner of the land may be compensated for the disruption caused to the enjoyment of his property by the payment of a portion of the royalties paid by the holder of the permit.[211]

8.83 It should be emphasised that the holder of the mining rights may be relieved of the payment of this occupancy compensation if the occupation concerns public land,[212] in particular the private domain of the State and

See also, in this regard, in Tunisia, Article 86(1) of Law No. 99-93 of 17 August 1999 Enacting the Hydrocarbons Code and Article 80 of Law No. 2003-30 of 28 April 2003 Enacting the Mining Code.

[208] See in Algeria, Article 136 of Law No. 01-10 of 3 July 2001 Enacting the Mining Law.

[209] See in Guinea, Article 50 of Law No. L/2014/N°034/AN of 23 December 2014 Enacting the Petroleum Code.

[210] For example, Article 281 of Law No. 18/001 of 9 March 2018 Amending and Supplementing Law No. 007/2002 of 11 July 2002 Enacting the Mining Code provides:

'*[l]e règlement à l'amiable du litige s'effectue par toutes voies de droit non juridictionnelles, notamment la transaction, le compromis, l'arbitrage ou devant un Officier de Police Judiciaire ou un Officier du Ministère Public. Faute d'arrangement à l'amiable entre les parties dans les trois mois à compter de la date de la survenance du litige, les indemnités seront allouées par le tribunal compétent en vertu des règles de l'organisation et de la compétence judiciaires en vigueur en République Démocratique du Congo.*'

[Free translation into English: '[t]he amicable resolution of the dispute shall be arranged by any non-judicial legal remedies, notably by settlement, compromise, arbitration or before an Officer of the Investigation Department or Officer of the Prosecution Department. Barring an amicable settlement between the parties within three months from the date of the occurrence of the dispute, compensation shall be awarded by the competent court in accordance with the judicial rules of jurisdiction and organisation in force in the Democratic Republic of Congo'].

[211] Thus, Article 83 of the Ugandan Mining Act, 2003 provides:

'[t]he owner or lawful occupier of any land subject to a mining right is entitled to compensation under either section 82 of this Act or to a share of royalties under section 98 of this Act.' Article 98 of the same Code provides for the following in this regard: '(1) Subject to section 100 of this Act all minerals obtained or mined in the course of prospecting, exploration, mining or mineral beneficiation operations shall be subject to the payment of royalties on the gross value of the minerals based on the prevailing market price of the minerals at such rates as shall be prescribed. (2) Royalty shall be shared by the Government, Local Governments and owners or lawful occupiers of land subject to mining rights as specified in the Second Schedule to this Act.'

[212] See in this regard, in Côte d'Ivoire, Article 61 of Law No. 96-669 of 29 August 1996 Enacting the Petroleum Code; in Mauritania, Article 56 of Law No. 2010-033 of 20 July 2010 Enacting the Code of Hydrocarbons amended by Law No. 2011-044 of 25 November 2011; in the

public authorities in general, as catching-up is involved with the payment of the surface royalties established for the petroleum and mining laws.

8.84 In addition to compensation for loss of enjoyment, the landowner may also receive financial compensation in the event that mining or petroleum operations cause damage to the surface property.[213] The amount of this compensation will generally be equal to the amount of the damage caused.[214]

8.85 Finally, mining and petroleum legislation may provide for an obligation on the part of the operator to purchase parcels damaged by the exploitation of natural resources. This obligation to purchase the land from the landowner is implemented if the owner is deprived of the use of the land for a certain period of time or when the occupied plots are no longer suitable for their previous use.[215] This is particularly the case where the land in

---

Central African Republic, Article 47 of Ordinance No. 93.007 of 25 May 1993 Enacting the Petroleum Code; in Guinea, Article 51 of Law No. L/2014/n°034/AN of 23 December 2014 Enacting the Petroleum Code.

[213] See, in particular, in Benin, Article 68 of Law No. 2006-18 of 17 October 2006 Enacting the Petroleum Code; in Cameroon, Article 61 of Law No. 99-013 of 22 December 1999 Enacting the Petroleum Code; in the Central African Republic, Article 50 of Ordinance No. 93.007 of 25 May 1993 Enacting the Petroleum Code; in the Comoros, Article 34 of Decree-Law No. 54-1110 of 13 November 1954 On the Reform of the Mineral Substances Regime; in the Democratic Republic of Congo, Article 280 of Law No. 007-2002 of 11 July 2002 Enacting the Mining Code; in Guinea, Article 126 of Law No. 2011/006/CNT of 9 September 2011 Enacting the Mining Code; in Equatorial Guinea, Article 51 of Mining Law No. 9/2006 of 3 November 2006; in Niger, Article 117 of Ordinance No. 93-16 of 2 March 1993 Enacting the Amended Mining Law; in Tunisia, Article 86(3) of Law No. 99-93 of 17 August 1999 Enacting the Code of Hydrocarbons and Article 82 of Law No. 2003-30 of 28 April 2003 Enacting the Mining Code.

[214] See, in particular, in Benin, Article 68 of Law No. 2006-18 of 17 October 2006 Enacting the Petroleum Code and Article 120 of Law No. 2006-17 of 17 October 2006 Enacting the Mining and Mining Taxation Code; in the Comoros, Article 34 of Decree-Law No. 54-1110 of 13 November 1954 Reforming the Mineral Substances Regime. It should be noted that certain laws establish a ceiling for this compensation. In particular, this includes, in Tunisia, Article 83(3) of Law No. 99-93 of 17 August 1999 Enacting the Hydrocarbons Code, which provides:

> '*[l]orsque à la fin des travaux, il appert que les terrains occupés ont été trop endommagés ou dégradés et ne sont plus propres à leur usage d'origine, le Titulaire est tenu soit de réparer le dommage, soit de payer une indemnité au propriétaire du sol pour la réparation du préjudice résultant de ce dommage. Cette indemnité ne peut dépasser le double de la valeur vénale des terrains concernés.*'
>
> [Free translation into English: '[w]hen upon completion of the works, it is apparent that the occupied land has been damaged or degraded and is no longer suitable for its original use, the Holder is required either to repair the damage or to pay compensation to the landowner to repair the harm resulting from such damage. This compensation shall not exceed twice the market value of the land concerned'].

See also Article 82 of Law No. 2003-30 of 28 April 2003 Enacting the Tunisian Mining Code.

[215] See, in particular, in Algeria, Article 139 of Law No. 01-10 of 3 July 2001 Enacting the Mining Law; in Benin, Article 73 of Law No. 2006-18 of 17 October 2006 Enacting the Petroleum Code; in the Central African Republic, Article 47 of Ordinance No. 93.007 of 25 May 1993 Enacting the Petroleum Code; in the Comoros, Article 33(B) of Decree Law No 54-1110 of 13 November 1954 On the Reform of the Mineral Substances Regime; in the Republic of

question can no longer be farmed.[216] The procedures for determining the price of the transfer of such land may, depending on the case, be provided for by law.[217] Similarly, the mining and petroleum laws may specify the appropriate jurisdiction in the event that the establishment of the sale price of the land poses difficulties.[218]

Congo, Article 108 of Law No. 4-2005 of 11 April 2005 Enacting the Mining Code; in Guinea, Article 50 of Law L/2014/N°034/AN of 23 December 2014 Enacting the Petroleum Code; in Mauritania, Article 55 of Law No. 2010-033 of 20 July 2010 Enacting the Code of Hydrocarbons amended by Law No 2011-044 of 25 November 2011; in Niger, Article 116 of Ordinance No. 93-16 of 2 March 1993 Enacting the Amended Mining Law; in Tunisia, Article 81 of Law No. 2003-30 of 28 April 2003 Enacting the Mining Code.

216 See in this regard, in Benin, Article 73 of Law No. 2006-18 of 17 October 2006 Enacting the Petroleum Code.

217 The legal criteria established to determine the selling price of occupied land vary. For example, in Algeria (Article 139 of Law No. 01-10 of 3 July 2001 Enacting the Mining Law), in Guinea (Article 50 of Law No. L/2014/N°034/AN of 23 December 2014 Enacting the Petroleum Code) and in Mauritania (Article 56 of Law No. 2010-033 of 20 July 2010 Enacting the Code of Hydrocarbons amended by Law No. 2011-044 of 25 November 2011), it is specified that the estimate of the value of the land to be purchased dates back to the date of the occupation. On the contrary, in Benin (Article 73 of Law No. 2006-18 of 17 October 2006 On the Petroleum Code), in the Comoros (Article 33(B) of Decree-Law No. 54-1110 of 13 November 1954 Reforming the Regime of Mineral Substances), in Niger (Article 116 of Ordinance No. 93-16 of 2 March 1993 Enacting the Amended Mining Law) and in Tunisia (Article 81 of Law No. 2003-30 of 28 April 2003 Enacting the Mining Code), the price of the land to be purchased is estimated to be twice the value it was before the occupation.

218 See, in particular, in Benin, Article 73 of Law No. 2006-18 of 17 October 2006 Enacting the Petroleum Code; in Guinea, Article 50 of Law No. L/2014/N°034/AN of 23 December 2014 Enacting the Petroleum Code, in Mauritania, Article 56 of Law No. 2010-033 of 20 July 2010 Enacting the Crude Hydrocarbons Code amended by Law No. 2011-044 of 25 November 2011 and in Tunisia Article 82 of Law No. 2003-30 of 28 April 2003 Enacting the Mining Code, which grants jurisdiction to ordinary courts.

9

THE GRANTING OF MINING RIGHTS

9.1 The conditions for the granting of mining rights are more or less detailed depending on the regulation at stake, thus allowing more or less space for a discretionary assessment by the licensing authority.

1 ELIGIBILITY CONDITIONS

9.2 Mining and petroleum codes define the rules for allocating mining and petroleum rights and determine the persons who can apply for such rights.

The eligibility conditions enacted by national legislation are of two kinds: conditions relating to the capacities of the private operator, on the one hand, and conditions relating to persons who could be granted mining rights, on the other hand.

a Conditions relating to the capacities of the private operator

9.3 Extractive projects are, by nature, complex and costly. A lack of technical knowledge or a potential financing deficit can call into question the overall project. To ensure that the private operator has the financial resources and technical knowledge needed to conduct mining and oil operations, national legislation lays down a set of conditions relating to the capacities of the private operator.

Almost all African mining and petroleum legislation compel a private operator to have the technical and financial capacities needed for operations to run smoothly.[1] Justification by the private operator of its technical and

[1] See, in particular, section 23(1) of the Mineral and Petroleum Resources Development Act 28 of 2002; in Algeria, Article 103 of Law No. 01-10 of 3 July 2001 On Mining Law; in Botswana, Article 39 of the Mines and Minerals Act and Article 37 of the Petroleum (Exploration and Production) Act; in Burundi, Article 42 of Law No. 1-21 of 15 October 2013 Enacting the Mining Code and Articles 33 and 46 of Order No. 1/138 of 17 July 1976 Enacting the Mining and Petroleum Code; in the Republic of Congo, and Articles 10, 51 and 59 of Law No. 4-2005 of 11 April 2005 Enacting the Mining Code; in the Democratic Republic of Congo, Article 56 of Law No. 007-2002 of 11 July 2002 Enacting the Mining Code; in Djibouti, Article 8 of Law No. 117-AN-05 of 16 October 2005 Enacting the Petroleum Code; in Guinea, Article 5 of Law No. L/2014/n°.034/AN of 23 December 2014 Enacting the Petroleum Code and Articles 15 and 22 of Law No. 2011/006/CNT of 9 September 2011 Enacting the Mining Code; in Malawi, Article 27(4) of the Mines and Minerals Act 1981 and Article 29 of the Petroleum (Exploration and Production) Act 1983; in Mali, Articles 16, 18 and 39 of Law No. 2012-015 of 27 February 2012 Enacting the Mining Code; in Mauritania, Article 9 of Law No. 2010-033 of 20 July 2010 Enacting the Hydrocarbon Code amended by Law No. 2011-044 of 25 November 2011 and Article 41 of Law No. 2008-011 of 27 April 2008; in Mozambique, Articles 39, 42, 45 and 50 of Mining Law No. 20-2014 of 18 August 2014 and Articles 26 and 46 of Law No. 21-2014 of 18 August 2014; in Niger, Article 22 and 46 of Ordinance No. 93-16 of 2 March 1993 Enacting the

financial capacities is also required as part of the signing of an exploitation agreement.[2]

The private operator's obligation to have the financial capacity needed to conduct operations sometimes undergoes the insertion of legal provisions which require a private operator have a minimum share capital of which the amount is determined by the application regulation.[3] Other provisions require the private operator to set up a bank reserve within a local financial institution.[4] Mining and petroleum codes may lay down the procedures for calculating the minimum financial capacities of the operator.[5] Generally speaking, the operator's financial capacities may result from its shareholders' equity, borrow funds or a bank guarantee.

9.4 With regard to the technical capacities of the operator, these will be assessed with regard to the number of years of experience within the mining or petroleum sector, the number of projects carried out prior to the licence application or with regard to the skills of the technical works manager.

9.5 Generally speaking, it is up to the private operator to provide a statement to the licensing authority giving details of its technical and financial resources.[6] Sometimes, in order to assess the required capacities, the host state may require to be provided with the elements enabling it to make a ruling on the technical and financial capacities, such as the *curriculum vitae* of the experts, certified copies of diplomas, bank statements or the private operator's balance sheet and profit and loss statements.[7] The amount

---

amended Mining Law; in São Tomé and Principe, Article 8(2) of the Framework Law on Oil Operations; in Senegal, Article 8 of Law No. 98-05 of 8 January 1998 Enacting the Petroleum Code; in Sudan, Articles 9 and 11 of the Mineral Resources and Mining Development Act 2007; in South Sudan, Articles 55 and 62 of the Mining Act 2012; in Chad, Articles 4, 8, 25 and 26 of Law No. 006-PR-2007 of 2 May 2007 On Hydrocarbons; in Togo, Article 3(11) of Law No. 99-003 of 18 February 1999 Enacting the Hydrocarbon Code and Article 6 of Law No. 96-004 of 26 February 1996 Enacting the Mining Code; in Tunisia, Article 11 of Law No. 2003-30 of 28 April 2003 Enacting the Mining Code and Article 7, 15 and 39 of Law No. 99-93 of 17 August 1999 Enacting the Hydrocarbon Code; in Zambia, Articles 12 and 34 of the Petroleum (Exploration and Production) Act 2008.

[2] See, in particular, in Central Africa, Article 7 of Ordinance No. 93.007 of 25 May 1993 Enacting the Petroleum Code; in Côte d'Ivoire, Article 8 of Law No. 96-669 of 29 August 1996 Enacting the Petroleum Code.

[3] See, in particular, in Central Africa, Article 19 of Law No. 09-005 of 29 April 2009 Enacting the Mining Code; in Nigeria, Article 54 of the Minerals and Mining Act 2007; and in Uganda, Article 168 of the Petroleum (Exploration, Development and Production) Act 2013.

[4] See, in particular, in Côte d'Ivoire, Article 19 of Law No. 2014-138 of 24 March 2014 Enacting the Mining Code.

[5] See, in particular, in Mali, Article 8 of Decree No. 2012-311/P-RM of 21 June 2012 Laying Down the Terms and Conditions for Application of the Mining Code.

[6] See, in particular in Tanzania, Article 28(3) of the Mining Act 2010.

[7] See, in particular, in Burundi, Article 43 of Law No. 1-21 of 15 October 2013 Enacting the Mining Code; and in Mali, Article 8 of Decree No. 2012-311/P-RM of 21 June 2012 Laying Down the Terms and and Conditions for Implementation of the Mining Code.

corresponding to the minimum financial commitments may also be deposited on an escrow account.[8]

The licensing authority may also appoint a committee of experts or a regulation body in order to assess the technical and financial capacities of the private operator.[9]

9.6 Some legislation furthermore specifies that the holder of artisanal mining rights is not obliged to justify the technical and financial capacity that is normally required for the exploitation of a deposit.[10]

b Conditions relating to persons

9.7 Apart from the conditions relating to technical and financial capacities, mining and petroleum codes require the holders of mining rights to satisfy a set of conditions, particularly concerning their nationality and their domiciliation.

Often, these conditions are supplemented by a list of persons who may not carry out mining and oil activities.

9.8 A distinction should be made between legal entities, on the one hand, and natural persons, on the other hand. While some countries reserved the granting of mining rights, such as artisanal production mining rights, to natural persons,[11] others conversely require that a legal entity hold the

[8] See, in particular, in Democratic Republic of Congo, Law No. 18/001 of 9 March 2018 Amending and Supplementing Law No. 007/2002 of 11 July 2002 Enacting the Mining Code.

[9] See, in particular, in Burundi, Article 42 of Law No. 1-21 of 15 October 2013 Enacting the Mining Code.

[10] See, in particular, in Eritrea, Article 5(2) of Proclamation No. 68/1995 Promoting the Development of Mineral Resources; in Ethiopia, Article 1(2) of Proclamation No. 678/2010 Promoting Sustainable Development of Mineral Resources.

[11] In particular, it concerns Guinea, Article 53 of Law No. 2011/006/CNT of 9 September 2011 Enacting the Mining Code which provides that:

> '*[l]'octroi de l'autorisation d'exploitation artisanale est réservé aux seules personnes physiques de nationalité guinéenne, aux personnes morales dont les capitaux sont entièrement détenus par des guinéens ou aux ressortissants des pays accordant la réciprocité aux Guinéens. L'exploitation artisanale est interdite aux actionnaires et employés des sociétés minières, des comptoirs d'achat et des bureaux d'achat d'or.*'
>
> [Free translation into English: '[t]he granting of an authorisation for artisanal exploitation is reserved solely for natural persons of Guinean nationality, legal entities, the share capital of which is entirely held by Guineans or by nationals of countries granting reciprocity to Guineans. Artisanal exploitation is prohibited for shareholders and employees of mining companies, purchasing agencies and gold purchasing offices'].

mining rights.[12] Similarly, some mining and petroleum codes reserve the allocation of mining rights to commercial companies.[13]

9.9 Furthermore, national legislation may require that the company object relates to the exploitation of mineral or petroleum resources.[14]

i *Conditions relating to nationality*

9.10 In order to promote access of nationals to natural resources, mining and petroleum laws often lay down conditions relating to the nationality of the holders of mining rights.

Frequently, the nationality requirement only applies during the exploitation phase. Whilst prospecting and exploration works can generally be carried out by foreign companies, exploitation operations will be reserved to companies governed by local law,[15] thus requiring the establishment of an operating company in the host state.[16] The foreign operator's obligation to

---

[12] In particular, in Benin, Article 11 of Law No. 2006-18 of 17 October 2006 Enacting the Mining Code, which provides that: '*[l]es titres pétroliers ne peuvent être accordés qu'à une personne morale ou conjointement et solidairement à plusieurs personnes morales, qu'elles soient nationales ou étrangère.*' [Free translation into English: '[p]etroleum rights can only be granted to a legal entity or jointly and severally to several legal entities, whether they are nationals or foreign']. Similarly, in Burundi, Article 27 of Law No. 1-21 of 15 October 2013 Enacting the Mining Code which provides that: '*[s]eules les personnes morales peuvent être titulaires d'un titre minier.*' [Free translation into English: 'Only legal entities can hold mining rights'].

[13] In particular, it concerns Burundi, Article 46 of Order No. 1/138 of 17 July 1976 Enacting the Mining and Petroleum Code, which provides that the 'H' permit can only be granted to one or more commercial companies.

[14] See, in particular, in the Democratic Republic of Congo, Article 23 of Law No. 18/001 of 9 March 2018 Amending and Supplementing Law No. 007/2002 of 11 July 2002 Enacting the Mining Code.

[15] See, in particular, in Chad Article 8 of Law No. 011-PR-1995 of 20 June 1995 Enacting the Mining Code which provides that:

> '*[u]ne société ne peut obtenir ou détenir un permis d'exploitation si elle n'est pas constituée conformément aux lois régissant le statut des sociétés en République du Tchad.*'
>
> [Free translation into English: '[a] company cannot obtain or hold an exploitation permit if it is not incorporated in accordance with the laws governing the status of companies in the Republic of Chad'].

See also, Article 11 of Ordinance No. 93-16 of 2 March 1993 Amending the Mining Code; and in Guinea, Article 69 of Law No. 2011/006/CNT of 9 September 2011 Enacting the Mining Code.

[16] See, in particular, in Algeria, Article 78 of Law No. 01-10 of 3 July 2001 Enacting the Mining Law, which provides that:

> '*[l]e demandeur d'un titre minier doit au préalable remplir les conditions suivantes: 1) Pour l'autorisation de prospection et le permis d'exploration: être une société commerciale de droit Algérien ou étranger régulièrement constituée et immatriculée dans son pays d'origine, 2) Pour la concession minière et le permis d'exploitation de petite ou moyenne exploitation minière: être une société commerciale de droit Algérien régulièrement constituée.*'
>
> [Free translation into English: '[a]n applicant for mining rights must first fulfil the following conditions: 1) For a prospecting licence and exploration licence: be a commercial company governed by Algerian or foreign law incorporated in accordance

create an operating company on the territory of the host state is reiterated by legislation of numerous other states.[17] This operating company can be granted mining rights or obtain the transfer of these, once it is set up.[18]

9.11 Sometimes, the condition relating to the nationality of the holder of mining rights is more widely applied, in the prospecting[19] or exploration[20] phases.

9.12 The nationality requirement can, in some cases, be replaced by a reciprocity condition. Therefore, mining rights will be granted to nationals of the host state or to nationals of countries granting reciprocity to nationals of this state.[21]

9.13 It should be emphasised that many mining rights remain reserved for local interests, particularly in the mining sector. Several African mining laws provide that the carrying out of artisanal (or small-scale) mining activities is reserved for nationals of the host state.[22] Therefore, with regard to legal

---

with the law and registered in its country of origin, 2) For mining concessions and small or medium mining operation exploitation permits: be a commercial company governed by Algerian law and incorporated in accordance with the law'].

[17] See, in particular, in Algeria, Article 4 of Presidential Decree of 1 March 2003 On Standard Mining Agreements; in Côte d'Ivoire, Article 7 of Law No. 2014-138 of 24 March 2014 Enacting the Mining Code; and in Mali, Article 65 of Law No. 2012-015 of 27 February 2012 Enacting the Mining Code.

[18] See in Côte d'Ivoire, Article 7 of Law No. 2014-138 of 24 March 2014 Enacting the Mining Code.

[19] In particular, it concerns Mali, where Article 28 of Law No. 2012-015 of 27 February 2012 Enacting the Mining Code provides that:

'*[l]'autorisation de prospection est attribuée par arrêté du ministre chargé des Mines à toute personne morale de droit malien, dont au moins un actionnaire est malien.*'

[Free translation into English: '[a] prospecting authorisation is granted through an order of the minister responsible for Mines to any legal entity governed by Malian law, of which at least one shareholder is Malian'].

[20] In particular, it concerns the Central African Republic, Law No. 09-005 of 29 April 2009 Enacting the Mining Code, Article 19 of which provides that:

'*[l]e permis de recherché est attribute, sous reserve des droits antérieurs* [. . .] *à toute personne morale de droit centrafricain*'.

[Free translation into English: '[a]n exploration permit is granted, subject to prior rights [. . .] to any legal entity governed by Central African law'.

See also in Côte d'Ivoire Article 18 of Law No. 2014-138 of 24 March 2014 Enacting the Mining Code.

[21] In particular, it concerns Article 45 of Law No. 2012-015 of 27 February 2012 Enacting the Mining Code of Mali, which provides that:

'*[l]'autorisation d'exploitation artisanale est attribué à des personnes physiques de nationalité malienne ou morales de droit malien ou aux ressortissants des pays accordant la réciprocité aux maliens.*'

[Free translation into English: '[a]n artisanal exploitation authorisation is granted to natural persons of Malian nationality or legal entities governed by Malian law or to nationals of countries granting reciprocity to Malians'].

[22] See, in particularin Central Africa, Article 62 of Law No. 09-005 of 29 April 2009 Enacting the Mining Code; in the Republic of Congo, Article 40 of Law No. 4-2005 of 11 April 2005 Enacting the Mining Code, in the Democratic Republic of Congo, Articles 5 and 26 of Law No. 007-2002 of 11 July 2002 Enacting the Mining Code; in Ghana, Article 83 of the Minerals and Mining Act 2006; in Madagascar, Article 11 of Law No. 99-002 of 19 August 1999

entities, legislation can provide that mining rights are reserved for cooperative companies with local majority shareholders[23] or legal entities (the share capital of which is entirely[24] or partially[25] held by nationals).

9.14 The protection of national interests in the granting of mining rights may also involve the setting up of a preference right for the benefit of local companies and foreign operators associated with local companies.[26]

9.15 Apart from the nationality requirement, some categories of persons may benefit from priority in terms of the granting of mining rights. This is the case in South Africa where legislation provides that persons who were 'historically disadvantaged' may benefit from special assistance regarding access to the resources.[27]

ii *Conditions relating to domiciliation*

9.16 The granting of mining rights is not just reserved for companies governed by local laws or nationals. Some codes grant foreign private operators the benefit of being able to seek the allocation of mining rights. However, the granting of such rights often requires the private operator to satisfy a domiciliation requirement.

Enacting the Mining Code as amended by Law No. 2005-021 of 17 October 2005; in Mali, Articles 45 and 46 paragraph 2 of Law No. 2012-015 of 27 February 2012 Enacting the Mining Code; and in Sierra Leone, Article 84 (for artisanal mining exploitation) and Article 95 (for small-scale mining exploitation) of the Mines and Minerals Act 2009.

[23] See, in particular, in the Côte d'Ivoire, for semi-industrial mining operations, Article 53 of Law No. 2014-138 of 24 March 2014 Enacting the Mining Code; and for artisanal exploitation, Article 65 of the same law.

[24] See, in particular, in Mali, Article 46 paragraph 2 of Law No. 2012-015 of 27 February 2012 Enacting the Mining Code; and in Sierra Leone, Article 84 of the Mines and Minerals Act 2009.

[25] With respect to small-scale mining exploitation, see, in particular, in Sierra Leone, Article 95 of the Mines and Minerals Act 2009.

[26] See, in particular, in Mozambique, Article 26(4) of Oil Law No. 21-2014 of 18 August 2014.

[27] Therefore, section 12 of the Mineral and Petroleum Resources Development Act 28 of 2002 provides that:

> '(1) The Minister may facilitate assistance to any historically disadvantaged person to conduct prospecting or mining operations. (2) The assistance referred to in subsection (1) may be provided subject to such terms and conditions as the Minister may determine. (3) Before facilitating the assistance contemplated in subsection (1), the Minister must take into account all relevant factors, including *(a)* the need to promote equitable access to the nation's mineral resources; *(b)* the financial position of the applicant; *(c)* the need to transform the ownership structure of the minerals and mining industry; and *(d)* the extent to which the proposed prospecting or mining project meets the objects referred to in section 2*(c)*, *(d)*, *(e)*, *(f)* and *(i)*. (4) When considering the assistance referred to in subsection (1), the Minister may request any relevant organ of State to assist the applicant concerned in the development of his or her prospecting or mining project.'

Therefore, regulations can provide that foreign natural persons[28] or foreign legal entities[29] can only be granted mining rights on condition that they have elected domicile within the host state. To this end, mining and petroleum codes can provide that a natural person or legal entity not residing in the host state can elect domicile at a representative[30] or an agent established on the national territory.[31]

9.17 The domiciliation requirement meets the need for the private operator to have a stable place of business[32] on the territory of the host state, either from the commencement of the exploration phase, or only from the

---

[28] See in particular, in Benin, Article 5 of Law No. 2006-17 Enacting the Mining and Mining Taxation Code; and in Burundi, Article 19 paragraph 2 of Order No. 1/138 of 17 July 1976 Enacting the Mining and Petroleum Code.

[29] See, in particular, in Burundi, Article 19 paragraph 2 of Order No. 1/138 of 17 July 1976 Enacting the Mining and Petroleum Code; in Madagascar, Article 11 of Law No. 99-022 of 19 August 1999 Enacting the Mining Code amended by Law No. 2005-021 of 17 October 2005; in South Sudan, Article 7(2) of the Mining Act 2012; in Swaziland, Article 27 of the Mines and Mineral Act 2011; in Tunisia, Article 12 of Law No. 99-93 of 17 August 1999 Enacting the Hydrocarbon Code; and in Zambia, Article 62 of the Petroleum (Exploration and Production) Act, 2008.

[30] This solution remains uncommon. For instance, this solution was admitted in Cameroon before the enactement of the new mining law (Article 10 of Law No. 001 of 16 April 2001 Enacting the Mining Code, which provided that:

> '*[t]out titulaire d'un titre minier émis en vertu de la présente loi doit justifier d'un domicile au Cameroun. S'il ne réside pas lui-même au Cameroun, il doit y élire domicile chez son représentant dont il fait connaître l'identité et les qualifications à l'administration chargée des mines.*'
>
> [Free translation into English: '[a]ny holder of mining rights issued by virtue of this law must prove domicile in Cameroon. If it does not reside itself in Cameroon, it must elect domicile with its representative, whose identity and qualifications it shall make known to the administration responsible for mines']).

The new mining law provides that only legal persons incorporated under the law of Cameroon may hold a mining right (Article 15(5) of Law No. 2016/ 017 of 14 December 2016 Enacting the Mining Code).

[31] See, in particular, in the Democratic Republic of Congo, Article 23 of Law No. 007-002 of 11 July 2002 Enacting the Mining Code; in Côte d'Ivoire, Article 9 of Law No. 2014-138 of 24 March 2014 Enacting the Mining Code; and in Madagascar, Article 11 of Law No. 99-022 of 19 August 1999 Enacting the Mining Code amended by Law No. 2005-021 of 17 October 2005.

[32] In particular, in Côte d'Ivoire, this requirement is established by Article 8 of Law No. 96-669 of 29 August 1996 Enacting the Petroleum Code, which provides that:

> '*[u]n contrat pétrolier et les autorisations et, s'il y a lieu, les titres miniers d'hydrocarbures en dérivant, ainsi qu'une autorisation de reconnaissance, ne peuvent être attribués qu'à une société commerciale ou, conjointement, à plusieurs sociétés commerciales, de droit ivoirien ou étranger. Ces sociétés, lorsqu'elles sont de droit étranger, doivent justifier pendant toute la durée du contrat pétrolier d'un établissement stable en République de Côte d'Ivoire inscrit au registre du commerce, qui peut être une société de droit ivoirien ou une succursale.*'
>
> [Free translation into English: '[a] petroleum agreement and licences and, if there are any, mining rights for hydrocarbons derived therefrom, as well as a prospecting licence, can only be granted to a commercial company or, jointly, to several commercial companies, governed by Côte d'Ivoire or foreign law. These companies, when they are governed by foreign law, must prove throughout the term of the oil agreement that they have a stable

discovery of a natural resource in marketable quantities. The obligation for a foreign operator to elect domicile on the territory of the host state may result in the establishment of a company governed by local law or by the creation of a branch office.[33] It should be mentioned that the creation of a branch office by a foreign private operator may be subject to certain constraints; therefore, pursuant to OHADA law (and more specifically article 120 of the OHADA Uniform Act on Commercial Companies and Economic Interest Groups of 30 January 2014, which provides that the operation of a branch office should not exceed two years).[34]

9.18 Lastly, it shall be mentioned that a private operator which is a legal entity may have to register with the host state's register of companies.[35]

iii *Incompatibilities affecting certain persons*

9.19 Not all persons, regardless of whether they are natural persons or legal entities, can carry out extractive activities.

For instance, public officials, elected members or, more generally speaking, employees of the host state are often those who cannot carry out mining and petroleum activities on the territory of the host state[36] . As a result, public officials of the host state cannot be granted mining rights or hold any interest,

place of business in the Republic of the Côte d'Ivoire which is registered on the trade register, which can be a company governed by Côte d'Ivoire law or a branch office'].

See also, in Togo, Article 8(2) of Law No. 99-003 of 18 February 1999 Enacting the Hydrocarbon Code.

[33] See, in particular, in Côte d'Ivoire, Article 8 of Law No. 96-669 of 29 August 1996 Enactingthe Petroleum Code; and in Togo, Article 8(2) of Law No. 99-003 of 18 February 1999 Enacting the Hydrocarbons Code.

[34] See to that effect Alain Feneon *Droit des sociétés en Afrique: OHADA* (LGDJ 2015) 856-857; Deloitte 'La réforme du régime juridique des succursales des sociétés étrangères en droit des sociétés commerciales OHADA' (2014) 2 *Revue Juridique et fiscale d'Afrique Francophone* 8.

[35] As an example, Article 10 of Law No. 2014-138 of 24 March 2014 Enacting the Mining Code provides that a legal entity holding mining rights should be registered on the trade and property lending register. See also in Niger, Article 10 of Law No. 2007-01 of 31 January 2007 Enacting the Petroleum Code; and in Togo, Article 8(2) of Law No. 99-003 of 18 February 1999 Enacting the Hydrocarbon Code.

[36] See, in particular, in Algeria, Article 80 of Law No. 01-10 of 3 July 2001 Enacting the Mining Law, which provides that: '*[l]es fonctionnaires de l'État et des Collectivités territoriales, les élus et les employés des établissements et Organismes publics en poste, ne peuvent exercer des activités minières, durant l'exercice de leur fonction.*'

[Free translation into English: '[p]ublic officers of the State and Territorial Local Authorities, elected members and employees of public establishments and bodies in post may not carry out mining activities while they are carrying out their duties'].

See also Article 27 of Law No. 007-2002 of 11 July 2002 Enacting the Mining Code of the Democratic Republic of Congo, which states:

'*[n]e sont pas éligibles pour solliciter et obtenir les droits miniers et/ou de carrières, les cartes d'exploitant artisanal, de négociants ainsi que l'agrément au titre de comptoir d'achat et de vente des substances minérales d'exploitation artisanale: a) les agents et fonctionnaires de l'État, les Magistrats, les membres des Forces Armées, la Police et les Services de Sécurité, les employés des organismes publics habilités à procéder aux opérations minières.*'

directly or indirectly, in mining rights,[37] under penalty of seeing the rights acquired in breach of this rule cancelled.[38] If such persons would benefit from interests in mining rights, these interests should generally be subject to a statement to the competent authorities.[39] Public officials are generally not able to carry out such activities throughout their entire period of office. Some legislation, however, provides that the ban is extended but only for a certain period of time following the end of their period of office.[40] Furthermore, the ban made on public officials from carrying out mining and petroleum activities may be extended to the spouses, ascending or descending relatives of these persons. This rule is justified as part of the fight against corruption of public officials in the host state and allows any conflict of interest when granting mining rights to be avoided.

9.20 In addition, as far as natural persons are concerned, minors cannot in general be granted mining rights. The condition of being of age is largely taken up by African mining and petroleum legislation.[41] This requirement therefore assumes that the holders of mining rights have full and entire legal capacity.

[Free translation into English: '[t]he following are not eligible to seek and obtain mineral and/or quarrying interests, artisanal operator and trader cards, as well as the approval for a purchasing and sales agency for mineral substances extracted by artisanal mining methods: a) State officers and public officials, Magistrates, members of the Armed Forces, the Police and Security Services, employees of public bodies authorised to carry out mining operations'].

This principle is also affirmed: in Côte d'Ivoire, in Articles 10 and 11 of Law No. 2014-138 of 24 March 2014 Enacting the Mining Code and Article 9 of Law No. 96-669 of 29 August 1996 Enacting the Petroleum Code; in Madagascar, Article 10 of Law No. 99-022 of 19 August 1999 Enacting the Mining Code amended by Law No. 2005-021 of 17 October 2005; in Mauritania, Article 11 of Law No. 2010-033 of 20 July 2010 Enacting the Hydrocarbons Code amended by Law No. 2011-044 of 25 November 2011; and in Togo, Article 6 of Law No. 96-004 of 26 February 1996 Enacting the Mining Code.

[37] See, in particular, in Côte d'Ivoire, Article 10 of Law No. 2014-138 of 24 March 2014 Enacting the Mining Code; in Mauritania, Article 11 of Law No. 2010-033 of 20 July 2010 Enacting the Hydrocarbons Code amended by Law No. 2011-044 of 25 November 2011. *A contrario*, some codes provide that these persons may take an equity participation in the share capital of mining companies (Article 27 of Law No. 007-2002 of 11 July 2002 Enacting the Mining Code in the Democratic Republic of Congo).

[38] See, in particular, in Algeria, Article 82 of Law No. 01-10 of 3 July 2001 Enacting the Mining Law.

[39] See, in particular, in Côte d'Ivoire, Article 11 of Law No. 2014-138 of 24 March 2014 Enacting the Mining Code.

[40] See, in particular, in the Côte d'Ivoire, Article 11 of Law No. 2014-138 of 24 March 2014 Enacting the Mining Code.

[41] See, in particular, in Botswana, Article 6 of the Mines and Minerals Act and Article 11 of the Petroleum (Exploration and Production) Act; in the Republic of Congo, Articles 21, 31 and 40 of Law No. 4-2005 of 11 April 2005 Enacting the Mining Code; in the Democratic Republic of Congo, Articles 23 and 26 of Law No. 007-2002 of 11 July 2002 Enacting the Mining Code; in Lesotho, Article 5(1) of the Mines and Minerals Act 2005; in Uganda, Article 5 of the Mining Act 2003; in South Sudan, Article 7(2) of the Mining Act 2012; in Tanzania, Article 8(1) of the Mining Act 2010; in Zambia, Article 7(2) of the Petroleum (Exploration and Production) Act 2008; and in Zimbabwe, Article 24 of the Mines and Minerals Act.

Mining and petroleum codes can also provide that mining rights can only be granted to natural persons who have full enjoyment of their rights[42] and are not being affected by legal incapacity.[43]

9.21 Most mining and petroleum legislation provide that legal entities which are or have been subject to insolvency proceedings (most often liquidation) cannot apply for the granting of mining rights.[44] The same principle applies to natural persons who have been declared bankrupt.[45]

Legal entities or natural persons also cannot apply for mining rights if, in the past, they have been convicted, or sometimes even simply been investigated, for infringements of mining and petroleum legislation or for other offences, particularly for fraud, money laundering, corruption or breach of environmental, social or security rules.[46]

9.22 Finally, persons who, in the past, have had their mining rights withdrawn by the licensing authority cannot generally apply for new mining rights.[47]

[42] See, in particular, in Côte d'Ivoire, Article 10 of Law No. 2014-138 of 24 March 2014 Enacting the Mining Code.

[43] See in particular, in the Democratic Republic of Congo, Article 27 of Law No. 007-2002 of 11 July 2002 Enacting the Mining Code.

[44] See, in particular, in Botswana, Article 6 of the Mines and Minerals Act and Article 11 of the Petroleum (Exploration and Production) Act; in Côte d'Ivoire, Article 10 of Law No. 2014-138 of 24 March 2014 Enacting the Mining Code; in Ethiopia, Article 11 of Proclamation No. 678/2010 Promoting Sustainable Development of Mineral Resources; in Uganda, Article 5 of the Mining Act 2003; in South Sudan, Article 7(2) of the Mining Act 2012; in Tanzania, Article 8(1) of the Mining Act 2010; and in Zambia, Article 7(2) of the Petroleum (Exploration and Production) Act 2008.

[45] See, in particular, in Botswana, Article 6 of the Mines and Minerals Act and Article 11 of the Petroleum (Exploration and Production) Act; in Lesotho, Article 5(1) of the Mines and Minerals Act 2005; in Uganda, Article 5 of the Mining Act 2003; in South Sudan, Article 7(2) of the Mining Act 2012; in Tanzania, Article 8(1) of the Mining Act 2010; and in Zambia, Article 7(2) of the Petroleum (Exploration and Production) Act 2008.

[46] See, in particular, in Botswana, Article 11 of the Petroleum (Exploration and Production) Act; in the Democratic Republic of Congo, Article 27 of Law No. 007-2002 of 11 July 2002 Enacting the Mining Code; in Côte d'Ivoire, Article 10 of Law No. 2014-138 of 24 March 2014 Enacting the Mining Code as well as Article 5 of Decree No. 2014-397 Determining the Terms and Conditions for Implementation of Law No. 2014-138 of 24 March 2014 Enacting the Mining Code; in Guinea, Article 15 of Law No. 2011/006/CNT of 9 September 2011 Enacting the Mining Code; in Niger, Article 10 of Ordinance No. 93-16 of 2 March 1993 Enacting the Mining Law; in South Sudan, Article 7(2) of the Mining Act, 2012; and in Zambia, Article 7(2) of the Petroleum (Exploration and Production) Act 2008.

[47] See, in particular, in the Democratic Republic of Congo, Article 27 of Law No. 007-2002 of 11 July 2002 Enacting the Mining Code; in Ethiopia, Article 11(4) of Proclamation No. 678/2010 Promoting Sustainable Development of Mineral Resources; and in Madagascar, Article 10 of Law No. 99-022 of 19 August 1999 Enacting the Mining Code amended by Law No. 2005-021 of 17 October 2005.

2 THE PROCEDURES FOR GRANTING MINING RIGHTS

9.23 The administrative procedures—which aim to select private operators—are subject to detailed provisions, most often contained in the regulations implementing the mining and petroleum codes.[48]

The selection of the private operator for the allocation of mining rights generally undergoes two distinct procedures. The granting of mining rights can result from a bidding process or an application submitted by the private operator.

a The granting of mining rights as part of a tender procedure

9.24 Recently, a new pattern emerged in Africa in favour of the development of tender procedures as part of the granting of mining rights[49] and the signature of exploitation agreements.[50] This contractual and legislative unity is recent and tends to be widespread.

The selection of operators through a bidding process promotes transparency as part of the granting of mining rights and also meets good governance requirements. The wish to promote transparency in the extractive industries[51] includes clear selection procedures[52] and the implementation of fair

---

[48] In many cases, mining and petroleum codes refer to the relevant regulatory provisions. See, in particular, in Algeria, Article 73 of Law No. 01-10 of 3 July 2001 Enacting the Mining Law; and in Benin, Article 4 of Decree No. 2008-804 of 31 December 2008 Implementing the Mining and Mining Taxation Code.

[49] For a definition of the bidding process in the mining and oil sectors, see, in particular, in the Republic of Congo, Article 4 of Decree No. 2008-15 of 11 February 2008 Establishing the Procedure for Granting Liquid or Gaseous Hydrocarbon Mining Rights, which provides that:

> '*[a]u sens du présent décret, l'appel d'offres est une procédure de sélection par laquelle l'administration des hydrocarbures met en concurrence les société visées à l'article 31 du code des hydrocarbures, pour attribuer un titre minier d'hydrocarbures liquides ou gazeux dans chaque zone libre du domaine minier congolais.*'
>
> [Free translation into English: '[w]ithin the meaning of this Decree, a call-for-tenders is a selection procedure by which the hydrocarbon administration makes the companies specified in Article 31 of the Hydrocarbon Code compete for the granting of liquid or gaseous hydrocarbon mining rights in each free zone of the Congolese mining sector'].

[50] See, in particular, to that effect, in Mauritania, Article 18 of Law No. 2010-033 of 20 July 2010 Enacting the Hydrocarbon Code amended by law no. 2011-044 of 25 November 2011, which provides that: '*[l]e Contrat d'exploration-production est, en principe, conclu suite à un appel à la concurrence conformément aux procédures établies par voie réglementaire.*' [Free translation into English: '[t]he exploration and production agreement is, as a rule, concluded following a call-for-tender in accordance with the procedures established by regulations'].

[51] On transparency in extractive industries, see above para 11.28.

[52] See to that effect, in the Democratic Republic of Congo, Article 32 of Law No. 007-2002 of 11 July 2002 Enacting the Mining Code:

> '*[e]n vue d'assurer la transparence, l'objectivité, l'efficacité et la rapidité dans les processus de réception, d'instruction, de décision et de notification relatifs aux demandes d'octroi des droits miniers ou de carrières ainsi que dans la délivrance des titres octroyés y afférents, la procédure prévue dans le présent chapitre, s'applique, sous réserve des dispositions particulières à chaque droit minier et de carrières, à l'octroi des droits miniers et de carrières organisés dans le présent Code. La procédure d'octroi*

competition. The selection of operators through tender procedures does satisfy these requirements.

9.25 The use of tender procedures may be made compulsory or left up to the discretion of the host state's competent authorities.

Therefore, while some mining and petroleum codes provide that the use of a tender procedure is the principle,[53] others make an exception to this. Such is the case for national legislation which specifies that the licensing authority has the ability to use the tender procedure in the case where the public interest so requires,[54] thus leaving a wide margin of discretion regarding the possibility of using such a procedure. Similarly, the licensing authority may decide to use a call-for-tenders procedure for unallocated sites on which work

---

des droits miniers ou de carrières et de délivrance des titres y afférents est de stricte application.'

[Free translation into English: '[i]n order to ensure transparency, objectivity, efficiency and speed in the processes for receiving, reviewing, decision-making and notification relating to applications for the granting of mining or quarrying rights as well as in the issuing of the granted rights relating thereto, the procedure stipulated in this section applies, subject to special provisions for each mining and quarrying right, to the granting of mining and quarrying rights organised under this Code. The procedure for granting mining or quarrying rights and for issuing the rights relating thereto must be strictly followed'].

[53] See, in particular, in the Republic of Congo, Article 3 of Decree No. 2008-15 of 11 February 2008 Establishing the Procedure for Granting Liquid or Gaseous Hydrocarbon Mining Rights, which provides that:

'[h]ormis les cas exceptionnels régis par des accords cadres entre États ou pour des raisons de souveraineté, l'attribution des titres miniers d'hydrocarbures liquides ou gazeux se fait après une procédure d'appel d'offres.' [Free translation into English: '[e]xcept for exceptional cases governed by framework agreements between States or for reasons of sovereignty, the granting of liquid or gaseous hydrocarbon mining rights shall be executed after a call-for-tenders procedure'].

See also, in Mozambique, Article 21 of Petroleum Law No. 21-2014 of 18 August 2014.

[54] See, in particular, in the Democratic Republic of Congo, Article 33 of Law No. 007-2002 of 11 July 2002 Enacting the Mining Code, which provides that:

'[s]i l'intérêt public l'exige, le Ministre soumet exceptionnellement à un appel d'offres, ouvert ou restreint, les droits miniers et de carrières portant sur un gisement étudié, documenté ou éventuellement travaillé par l'État ou ses services, qui est considéré comme un actif d'une valeur importante connue.'

[Free translation into English: 'If the public interest so requires, the Minister shall exceptionally submit the mining and quarrying rights concerning a deposit explored, documented or potentially worked upon by the State or its departments which is considered to be an asset of known important value, to a call-for-tenders, open or restricted'].

See also, in the Central African Republic, Article 32 of Law No. 09-005 of 29 April 2009 Enacting the Mining Code; and in Mozambique, Article 10 of Mining Law No. 20-2014 of 18 August 2014.

has proved the existence of a mining or oil potential,[55] for areas already surveyed[56] or following the revocation of a mining right.[57]

9.26 The call-for-tenders will be open or restricted,[58] as appropriate. Legislation specifies, as appropriate, the content of the call-for-tenders notice[59] and the procedures for publication of the notice, particularly in local and international newspapers.[60] Besides, tenderers may be obliged to pay record disposal fees.[61]

9.27 In order to ensure that the procedure runs smoothly, the main selection criteria may be established by the mining and petroleum codes.[62] Similarly, legislation may specify that the selection of bids will take place in accordance with procedures generally accepted or recognised by international practice.[63]

Tenderers are mainly obliged to provide a bid security, the amount of which is equal to a percentage of the value of the contract, indicated in the call-for-tenders documents.[64] The purpose of this security is to encourage the

---

[55] See, in particular, in the Côte d'Ivoire, Article 37 of Law No. 2014-138 of 24 March 2014 Enacting the Petroleum Code; and in Guinea, Article 22 of Law No. 2011/006/CNT of 9 September 2011 Enacting the Mining Code.

[56] See, in particular, in Guinea, Article 22 of Law No. 2011/006/CNT of 9 September 2011 Enacting the Mining Code.

[57] See, in particular, in Burundi, Article 74 of Law No. 1-21 of 15 October 2013 Enacting the Mining Code, which provides that:

> *'[u]ne procédure d'adjudication d'un permis d'exploitation est ouverte lorsque l'ancien titre minier d'exploitation a été révoqué pour l'un des motifs prévus à l'article 37.'*
>
> [Free translation into English: '[a]n adjudication procedure for an exploitation licence opens when the old exploitation mining rights have been revoked in accordance with one of the grounds provided for in Article 37'].

[58] See, in particular, in the Republic of Congo, Article 5 of Law No. 2008-15 of 11 February 2008 Establishing the Procedure for Allocating Liquid or Gaseous Hydrocarbon Mining Rights; in the Central African Republic, Article 32 of Law No. 09-005 of 29 April 2009 Enacting the Mining Code; and in the Democratic Republic of Congo, Article 33 of Law No. 007-2002 of 11 July 2002 Enacting the Mining Code.

[59] See, in particular, in the Republic of Congo, Article 5 of Decree No. 2008-15 of 11 February 2008 Establishing the Procedure for Granting Liquid or Gaseous Hydrocarbon Mining Rights.

[60] See, in particular, in Central Africa, Article 32 of Law No. 09-005 of 29 April 2009 Enacting the Mining Code; in the Republic of Congo, Article 5 Decree No. 2008-15 of 11 February 2008 Establishing the Procedure for Allocating Liquid or Gaseous Hydrocarbon Mining Rights; and in Zambia, Article 9 of the Petroleum (Exploration and Production) Act 2008.

[61] See, in particular, in the Democratic Republic of Congo, Article 15 of Decree No. 2008-15 of 11 February 2008 Establishing the Procedure for Granting Liquid or Gaseous Hydrocarbon Mining Rights.

[62] See, in particular, in the Central African Republic, Article 33 of Law No. 09-005 of 29 April 2009 Enacting the Petroleum Code; and the Democratic Republic of Congo, Article 33 of Law No. 007-2002 of 11 July 2002 Enacting the Mining Code.

[63] See, in particular, in the Democratic Republic of Congo, Article 33 of Law No. 007-2002 of 11 July 2002 Enacting the Mining Code.

[64] See, in particular, in the Republic of Congo, Article 14 of decree No. 2008-15 of 11 February 2008 Laying Down the Procedure for Granting Liquid or Gaseous Hydrocarbon Mining Rights.

tenderer not to withdraw its bid as the adjudication procedure is still pending.

Finally, with a view to ensuring the efficiency of the bidding process, regulations usually specify the period in which the decision to grant the mining rights should take place.[65]

b The granting of mining rights following an application

9.28 In many cases, the granting of mining rights does not result from a tender procedure but an application for mining rights submitted to the granting authority by the private operator.

i *Submission of applications for mining rights*

9.29 National laws, and more precisely implementation decrees, contain numerous rules detailing the procedures for submitting applications for the granting of mining rights.

An application for mining rights should, as a general rule, be made based on a form provided by the mining administration,[66] attached to applicable legislation[67] or drawn up in an implementation decree.[68]

Similarly, the language in which the application should be drafted is specified by the legislation,[69] and the use of a language other than the official language of the host state may be tolerated.

---

[65] See, in particular, in the Republic of Congo, Article 26 of Decree No. 2008-15 of 11 February 2008 Establishing the Procedure for Granting Liquid or Gaseous Hydrocarbon Mining Rights, which provides that:

> '*[l]e Décret d'attribution du permis de recherche, pris en conseil des ministres sur la base du rapport du ministre des hydrocarbures est signé au plus tard 30 jours après dépouillement.*'
>
> [Free translation into English: '[t]he Decree for granting an exploration licence, passed by the Council of Ministers based on the report from the Minister of Hydrocarbons, shall be signed at latest 30 days from its reception'].

See also, in the Democratic Republic of Congo, Article 33 of Law No. 007-2002 of 11 July 2002 Enacting the Mining Code, which provides that:

> '*[l]'appel d'offres est conclu dans un délai d'un an à compter de l'entrée en vigueur de l'Arrêté portant reservation du gisement à soumettre à l'appel d'offres.*'
>
> [Free translation into English: '[t]he call-for-tenders shall be concluded within a period of one year from the date the order regarding reservation of the deposit to be subjected to the call-for-tenders came into force'].

[66] See, in particular, in the Democratic Republic of Congo, Article 35 of Law No. 007-2002 of 11 July 2002 Enacting the Mining Code.

[67] See, in particular, in Algeria, Article 2 of Executive Decree No. 02-65 of 6 February 2002 Defining the Conditions and Procedures for Allocating Mining Rights; and in Lesotho, Articles 20(2), 32 and 46(2) of the Mines and Minerals Act 2005.

[68] See, in particular, in Madagascar, Article 43 of Law No. 99-022 of 19 August 1999 Enacting the Mining Code amended by Law No. 2005-021 of 17 October 2005.

[69] See, in particular, in the Democratic Republic of Congo, Article 36 of Law No. 007-2002 of 11 July 2002 Enacting the Mining Code.

The authority to which the application for mining rights should be sent[70] as well as the number of copies which must be submitted are generally specified by legislation.

A list of data, information and documents provided in support of the application is specified by the legislation. In particular, it should indicate the substances for which the right is sought and define the area concerned.[71]

In the event of joint and several requests presented by several operators, the regulation may provide that information concerning the applicants should be supplied by each of them.[72] If the granting authority deems it necessary, additional information may be requested.[73]

Following submission of the application, the granting authority will rule on the admissibility of the application[74] (usually in a specific timeframe) and will notify the applicant of its decision.

9.30 In the event of an irregularity in the application, the legislation may provide for the possibility of seeking correction of the application,[75] within the prescribed periods.[76]

9.31 The submission of the application may give rise to the payment of a fee covering the costs of submission of the application.[77]

ii *Review of applications for mining rights*

9.32 The review of the application for mining rights often requires preliminary investigations to be made, so that the granting authority has all the information needed to rule on the application. The review procedure,

---

[70] As an example, in Algeria, an application for the granting of mining rights should be sent to the *Agence Nationale du Patrimoine Minier* (Article 2 of Executive Decree No. 02-65 of 6 February 2002 Defining the Conditions and Procedures for Granting Mining Rights).

[71] See, in particular, in Benin, Article 30 of Law No. 2008-804 of 31 December 2008 Implementing the Mining and Mining Taxation Code.

[72] See, in particular, in the Central African Republic, Article 10 of the Decree Establishing Conditions for Implementation of Ordinance No. 93.007 of 25 May 1993 Enacting the Petroleum Code.

[73] See, in particular, in Malawi, Article 12 of the Mines and Minerals Act 1981 and Article 12 of the Petroleum (Exploration and Production) Act 1983; and, in Mali, Article 12 of Decree No. 2012-311/P-RM of 21 June 2012 Establishing the Terms and Conditions for Implementation of the Mining Code.

[74] See, in particular, in Algeria, Article 2 of Executive Decree No. 02-65 of 6 February 2002 Defining the Conditions and Procedures for Allocating Mining Rights.

[75] See, in particular, in Algeria, Article 2 of Executive Decree No. 02-65 of 6 February 2002 Defining the Conditions and Procedures for Allocating Mining Rights; in Central Africa, Article 73 of Decree No. 09-126 of 30 April 2009 Establishing the Implementation Conditions for Law No. 09-005 of 29 April 2009 Enacting the Mining Code; in Madagascar, Article 7 of Decree No. 2004-669 of 29 June 2004 Implementing Law No. 99-10; and in Niger, Articles 164, 161, 123 and 111 of Decree No. 2007-082/PRN/MME of 28 March 2007 Laying down the Procedures for Implementing Law No. 2007-01 of 31 January 2007 Enacting the Petroleum Code.

[76] See in particular in Algeria, Article 4 of Executive Decree No. 02-65 of 6 February 2002 Defining the Conditions and Procedures for Allocating Mining Rights.

[77] See, in particular, in the Democratic Republic of Congo, Article 37 of Law No. 007-2002 of 11 July 2002 Enacting the Petroleum Code; and in Eritrea, Article 6 of Legal Notice No. 19/1995, Regulations on Mining Operations.

most often time-limited, may be governed by principles governing the choice of the applicants.

(1) Preliminary investigations

9.33 It is during the review of the application for granting mining rights that the public and administrative investigations will be conducted.

As a general rule, a public investigation will precede the granting of mining rights,[78] the aim of which is to evaluate the consequences of the proposed project, particularly on the environment and the local populations concerned.[79]

Mining and petroleum codes may also provide for the conduct of administrative investigations,[80] or the conduct of cadastral, technical and environmental enquiries.[81]

Besides, the licensing authority may, if the circumstances so require, carry out all investigations which it deems necessary,[82] in order to obtain information on the applicant's moral or technical and financial guarantees.[83]

In order to ensure the promptness of the procedure for granting mining rights, the investigation period may be limited.[84]

The applicable legislation may provide that the applicant bears the investigation costs.[85]

9.34 Parallel to these investigations, the licensing authority may consult

[78] See, in particular, in Benin, Articles 23 and 29 of Law No. 2006-18 of 17 October 2006 Enacting the Petroleum Code; in Central Africa, Article 42 of Law No. 09-005 of 29 April 2009 Enacting the Mining Code; in the Republic of Congo, Article 26 of Decree No. 2008-15 of 11 February 2008 Establishing the Procedure for Granting Liquid or Gaseous Hydrocarbon Mining Rights; in the Republic of Congo, Article 40 of Decree No. 2007-274 of 21 May 2007 Establishing Conditions for Prospecting, Exploration and Exploitation of Mineral Substances and Carrying out Administrative Monitoring; and, in Chad, Article 26 of Law No. 0011-PR-1995 of 20 June 1995 Enacting the Mining Code.

[79] See, in particular, in Chad, Article 26 of Law No. 0011-PR-1995 of 20 June 1995 Enacting the Mining Code.

[80] See in Algeria, Article 7 of Executive Decree No. 02-65 of 6 February 2002 Defining Conditions and Procedures for Allocating Mining Rights.

[81] See in the Democratic Republic of Congo, Articles 39, 40, 41 and 42 of Law No. 007-2002 of 11 July 2002 Enacting the Mining Code.

[82] See in Benin, Article 21 of Decree No. 2008-804 of 31 December 2008 Implementing the Mining and Mining Taxation Code; in the Republic of Congo, Articles 14 and 20 of Decree No. 2007-274 of 21 May 2007 Establishing Conditions for Prospecting, Exploring and Exploiting of Mineral Substances and Carrying out Administrative Monitoring.

[83] See in the Republic of Congo, Article 20 of Decree No. 2007-274 of 21 May 2007 Establishing Conditions for Prospecting, Exploring and Exploiting of Mineral Substances and Carrying out Administrative Monitoring.

[84] See in particular, in the Republic of Congo, Articles 14 and 21 of Decree No. 2007-274 of 21 May 2007 Establishing the Conditions for Prospecting, Exploring and Exploiting Mineral Substances and Carrying out Administrative Monitoring.

[85] See in the Republic of Congo, Article 21 of Decree No. 2007-274 of 21 May 2007 Establishing the Conditions for Prospecting, Exploring and Exploiting Mineral Substances and Carrying out Administrative Monitoring.

local communities[86] and gather the objections received from third parties who are opposed to the granting of the mining rights.[87] These investigations are now increasing, as the impact of mining projects on local populations is often considerable.

(2) Periods for processing the applications

9.35 The speed of the administrative procedures for granting mining rights is taken into consideration by the host states.[88]

Therefore, in order to ensure the swiftness of administrative procedures, national legislation may provide for time-limits in which applications for the granting of rights must be dealt with by the granting authority.[89] These time-limits may vary according to the mining rights concerned. If the application for mining rights is not processed within the specified time-limits, the mining rights may be deemed to be granted to the applicant.[90]

Besides concerning the renewal, some legislation provides that the mining title will be tacitly extended if the administration has not ruled on an application to renew the mining rights before its period of validity expires; the extension will be valid up to the time the decision to grant or refuse the permit is made by the licensing authority.[91]

(3) Selection of the operator

9.36 In order to provide the greatest transparency possible in the selection of operators, mining and petroleum codes more and more often insert rules relating to competitive applications presented in the same area by different operators.

Except for exploitation or mining licences, for the granting of which the holders of an exploration licence usually take priority,[92] mining and petroleum codes may provide that the mining rights will be granted

---

[86] See in Central Africa, Article 42 of Law No. 09-005 of 29 April 2009 Enacting the Mining Code.

[87] See in Eritrea, Article 8 of Legal Notice No. 19/1995, Regulations on Mining Operations; and in Ethiopia, Article 7 of Regulation No. 182/1994 On Mining Operations.

[88] See in this regard, in the Democratic Republic of Congo, Article 32 of Law No. 007-2002 of 11 July 2002 Enacting the Mining Code.

[89] See, in particular, in the Central African Republic, Articles 29 and 72 of Decree No. 09-126 of 30 April 2009 Establishing the Conditions for Implementing Law No. 09-005 of 29 April 2009 Enacting the Mining Code; in Mali, Article 13 of Decree No. 2012-311/P-RM of 21 June 2012 Establishing the Terms and Conditions for Implementing the Mining Code; and in Madagascar, Article 44 and 45 of Law No. 99-022 of 19 August 1999 Enacting the Mining Code amended by Law No. 2005-021 of 17 October 2005.

[90] This solution remains uncommon. The administrative slowness of the mining administrations usully refrain the legislator to enact this kind of provisions. For instance, in Cameroon prior to the adoption of the new mining law, Article 22(2) of Decree No. 2002-648-PM of 26 March 2002 Establishing the Procedures for Implementing Law No. 001 of 16 April 2001 Enacting the Mining Code as amended by Decree No. 2014/1882/PM of 4 July 2014.

[91] See in Tunisia Article 29 of Law No. 99-93 of 17 August 1999 Enacting the Hydrocarbon Code.

[92] On so-called exploitation licences, see above para 8.32 et seq.

according to the precedence of applications, ie the 'first come, first served' rule.[93] In some cases national legislation, on the contrary, gives the granting authority discretionary power in the event of competitive applications.[94]

9.37 The priority rule may also be implemented by some legislation. According to this principle, applications for the granting of mining rights are recorded in chronological order of their submission and as long as one application is pending, no other competitive application can be reviewed.[95]

9.38 In order to facilitate the review and selection of operators, mining and petroleum codes may enable the licensing authority to use the services of local or international consultants for examining applications for mining rights. In this case, the costs connected with these services may be invoiced to the applicant.[96]

iii *The granting of mining rights*

9.39 The mining rights are generally granted by an order of the Minister in charge of mining or petroleum activities or by decree. Sometimes, the decree will be made in the Council Ministers based on an issue raised by the Minister in charge of mining or petroleum activities.

National laws may provide that the decision to grant the mining rights will be subject to publication in the official gazette of the host state.[97] The rights granted will generally be recorded in a register and numbered.[98]

Besides, the granting of mining rights generally gives rise to the payment of fixed duties,[99] which may be considered as operating charges.[100]

9.40 In the event that the mining rights are not granted, legislation may provide that grounds must be provided for the decisions.[101] The rejection,

---

[93] See in Burundi, Article 41 of Law No. 1-21 of 15 October 2013 Enacting the Mining Code; in Madagascar, Article 41 of Law No. 99-022 of 19 August 1999 Enacting the Mining Code amended by Law No. 2005-021 of 17 October 2005; and in Mozambique, Article 9 of Mining Law No. 20-2014 of 18 August 2014.

[94] See also, in Libya, Article 8 of Petroleum Law No. 25/55 of 11 May 2012.

[95] See, in particular, in the Democratic Republic of Congo, Article 34 of Law No. 007-2002 of 11 July 2002 Enacting the Mining Code.

[96] Id.

[97] See in Mauritania, Article 18 of Law No. 2010-033 of 20 July 2010 Enacting the Hydrocarbons Code as amended by Law No. 2011-044; and in Tunisia, Articles 29 and 51 of Law No. 2003-30 of 28 April 2003 Enacting the Mining Code and Article 17(1) of Law No. 99-93 of 17 August 1999 Enacting the Hydrocarbon Code.

[98] See in Benin, Article 10 of Law No. 2008-804 of 31 January 2008 Implementing the Mining and Mining Taxation Code.

[99] See in Algeria, Article 156 of Law No. 01-10 of 3 July 2001 Enacting the Mining Code; in Madagascar, Article 21 of Law No. 99-022 of 19 August 1999 Enacting the Mining Code amended by Law No. 2005-021 of 17 October 2005.

[100] See in Madagascar, Article 21 of Law No. 99-022 of 19 August 1999 Enacting the Mining Code amended by Law No. 2005-021 of 17 October 2005.

[101] See in the Democratic Republic of Congo, Article 44 of Law No. 007-2002 of 11 July 2002 Enacting the Mining Code; and in Chad, Articles 14, 36, 18 and 32 of Law No. 011-PR-1995 of 20 June 1995 Enacting the Mining Code.

duly justified, of an application does not give rise to compensation to the benefit of the unsuccessful applicant.[102]

9.41 Mining rights are administrative authorisations which are essential for the exploitation of natural resources. This being said, the existence of contractual elements should not be ignored; therefore, parallel to the mining rights, the legal environment applicable to exploitation activities is also governed by contractual elements resulting from exploitation or investment agreements.

[102] See in Chad, Articles 14, 36, 18 and 32 of Law No. 011-PR-1995 of 20 June 1995 Enacting the Mining Code; and in Togo, Articles 11, 14 and 22 of Law No. 2003-012 of 4 October 2003 Amending and Supplementing Law No. 96-004 of 26 February 1996 Enacting the Mining Code.

10

Agreements with the Host State

B Contractual Agreements for the Exploitation of Natural Resources

10.1 The signature of contracts for the exploitation of natural resources is crucial to take into consideration the specifics of each extraction project as well as the peculiarities of the host state. In practice, these specific details have led to a great deal of diversity in exploitation agreements.

Exploitation agreements between the private operator and the host state, or its agency, can be based on mining rights assigned to the private operator or, in the absence of mining rights, confer economic mining rights on this operator.[1]

10.2 Alongside exploitation agreements, we are seeing an increase in contracts associated with natural resources projects. In many cases, the complexity of mining and petroleum projects call for a pooling of technical knowledge and financing means. For the same project, there may be a whole host of companies and the proliferation of a contractual framework governing the relations between operators and their contractors, as well as contractual relations between the various operators involved in the same extraction project.

10.3 The private operator may access the natural resources within the state through mining rights but can also benefit from economic mining rights granted through the signature of an agreement with the host state.

Depending on the applicable regulation, two different systems can generally be found. In the first scenario, the exploitation agreement will be based on a mining title granted in parallel. The exploitation agreement will usually be signed with the state once an exploration permit is allocated to the private operator and will be enforceable throughout the duration of the exploitation permit (this is the solution implemented in most of African jurisdictions in the mining sector).[2] In the second scenario, once the exploration phase which was covered by an exploration permit comes to an

[1] Some national legislation provides for economic mining rights to be granted in the form of exploration contracts, without the private operator obtaining any mining rights. This solution has, for example, been adopted in the New Petroleum Code of the Democratic Republic of Congo, which oil suppressed the references to petroleum licences in favour of a single production sharing agreement (Law No. 15/012 of 1 August 2015 Enacting the General Regime for Hydrocarbons).

[2] The granting of a mining title and the concomitant signature of a mining convention (usually during the exploration phase) constitutes the general rule prevailing in French-speaking African countries. In the process of reforming the legal regime applicable to mining operations, some countries are in the process of suppressing the reference to mining conventions in their legislation considering that such agreements may result in the application of a derogatory regime (for instance, Niger—which is currently in the process of reforming its mining code—is envisaging to do so).

end, the private operator will enter into an exploitation agreement with the host state. Such agreement will grant the private operator with economic mining rights giving access to the natural resource.

The following developments refer to exploitation agreements through which the private operator accesses the natural resource of the host state.

10.4 Contracts granting economic mining rights all constitute political window-dressing associated with the four main local factors, ie politics, the level of development, the economic importance of the natural resources and the progress made with exploratory works. Naturally, at a subsidiary level, international external factors also have an influence.

10.5 A great variety of contractual forms and terminology is used in reference with these contractual arrangements. Charles Leben simplified matters on the subject of hydrocarbons by differentiating between two main contract categories:

> '[t]he first generation has a precise, limited goal: crude oil exploration and production contracts, still referred to as oil contracts. The second generation, on the other hand, is pursuing much more ambitious objectives by aiming to achieve mass transfer of industrial and commercial control between two partners from a developed country and a developing country respectively: industrial cooperation contracts.'[3]

Care must be taken with the terminology used.[4] For example, 'petroleum operating contracts' (*contrats d'opérations pétrolières*), the term used by Professor Leben, are a specific form of exploitation agreement, in the past used by some Latin American countries.[5] It is not possible, at least not without creating confusion, to use it generically. Industrial cooperation contracts also establish a well-defined framework for the development of contractual structures for exploiting natural resources; however, they have not innovated in terms of 'mass transfer of industrial and commercial control'. Moreover, contextualising this transfer within the context of these 'North-South' relations by stating exclusively that they bring together people from 'developed countries' and others from 'developing countries' is no longer accurate. The legal conditions for exploitation of hydrocarbons in the North Sea[6] or mineral resources in Canada are a very clear counterexample. Besides,

---

[3] Charles Leben 'Les modes de coopération entre pays en développement et entreprises multinationales dans le secteur de la production des matières premières minérales' 1980 *JDI* 576.

[4] Such terminology errors are common. For instance, Philippe Leboulanger equates with '*Le régime juridique des investissements pétroliers en Égypte*' 1972 *Rev Égypt Dr Int* 112. See also Jean-Flavien Lalive, who places the emergence of production-sharing agreements between the two World Wars and within the context of contracts signed with the Soviet Union. Jean-Flavien Lalive '*Contrats entre États ou entreprises étatiques et personnes privées—Développements récents*' (1983) 181 *RCADI* 190.

[5] Particularly, in Bolivia (Law of 28 March 1972) and Peru (Law of 12 March 1952).

[6] To that effect, see, in particular, Commission on Transnational Corporations 'Main Features and Trends in Petroleum and Mining Agreements' (1983 ST/CTC/29) at 28 et seq; Alexandre Stärker *Des concessions aux concentrations pétrolières dans les pays du Moyen-Orient* (PhD Thesis: Université de Lausanne, 1974) 1; Alan C Page 'Transnational Mining Contracts' in *The transnational law of international commercial transactions* (Kluwer 1982) 224.

on the African continent, the emergence of Pan-African companies active in the mining and petroleum sectors has to be mentioned.

Furthermore, such a distinction overlooks the real motivations of mining and oil companies who all want to gain access to the natural resources extracted. Such industrial cooperation contracts are not in themselves a type of exploitation agreement, but rather a way of obtaining, following this cooperation, access to natural resources through preferential purchasing agreements or even exploitation agreements. They involve keeping the company's human technological capital active and, at the same time, demonstrating technical capacities and a desire to cooperate with the host state. Lastly, making a differentiation based on the burden of financial risk and stating that only the 'industrial cooperation contracts category' protects the private operator is illusory. Indeed, by adopting such a position, we once again ignore the fact that financial risk can be not only posed by the commitment of key investors in areas where the presence of natural resources has not been proven, but also by the mobilisation of people and equipment often by accepting random remuneration based wholly or partly on profit sharing, even in this kind of contract.

10.6 If a distinction has to be made, the main difference (as we already mentioned) is whether or not the private operator holds mining rights. In other words, it depends on whether the exploitation agreement grants economic mining rights or not.

10.7 For the natural resources we are interested in, we must recognise greater diversity of contractual forms within the context of petroleum operations, whereas the exploitation of minerals has greater unity.

10.8 It should also be noted that there is no direct correlation between the type of exploitation agreement chosen and the financial benefits or the control granted by the host state. These aspects will also depend on the actual structure of the contractual provisions.

10.9 Lastly, care should be taken not to give too much credit to the legal qualifications given in contracts by the operators themselves. These may result in many whims, such as the petroleum production-sharing agreement qualified as a service contract, or even production-sharing agreements qualified as concession contracts.

10.10 It therefore seems to us wiser and more realistic to present in turn the main contractual arrangements granting economic mining rights in favour of the private operator, rather than attempting to classify them, since it is true that there are as many contract types as there are contracts. We will look at the historical aspects of these contracts, which can sometimes take us outside of the African continent.

10.11 It is the history of natural resources that has dictated and continues to dictate the main changes to contracts. Within this context, Africa did not lead the way but has managed to set up innovative contract forms.

1 SERVICE CONTRACTS

10.12 In international trade, the term 'service contract' is generic and the content of such agreements is not unified. Service contracts applied to natural resources are profoundly specific and the origin for this goes a long way back.

a A historical approach to service contracts in the extractive industry

10.13 Service contracts have their origins in Mexico, where, in accordance with the Law of 18 March 1938 on nationalisation of petroleum and mining resources, a new type of contractual arrangement was introduced in response to the many difficulties faced by the state-owned company PEMEX. Under this new contractual form, the involvement of private companies is permitted but in a different framework than the one for concession contracts.[7]

These contractual forms have always been presented as workarounds to the state being unable to lead exploitation operations on its own following refusal to grant or withdrawal of mining rights held by national or foreign private companies.[8]

10.14 At the present time, service contracts are widely used in the extractive industry, especially on the African continent. In this way, legislation in many countries refers to the opportunity given to the local State operator to partner with private operators by way of service contracts.[9]

---

[7] The exact origin of service contracts is still the subject of debate, with some belief that this contractual model takes its origins from the oil contracts signed in Argentina at the end of the 1950s, while others see the precedent of this contractual model in the Contract signed in 1966 between the National Iranian Oil Company and the French company ERAP (*Entreprise de Recherches et d'Activités Pétrolières*). On this matter, see Zhiguo Gao *International Petroleum Contracts: Current Trends and New Directions* (Graham & Trotman/Martinus Nijhoff 1994) 115.

[8] See Bret-Rouzaut & Favennec op cit 188–189 and 212–213.

[9] See, in particular, in Angola, Article 14(3) of Petroleum Law No. 10/04 of 12 November 2004: '*[t]he National Concessionaire shall also be permitted to carry out petroleum operations by means of risk services agreements*'; in Benin, Article 1 of Law No. 2006-18 of 17 October 2006 Enacting the Petroleum Code:

> '*[u]n contrat pétrolier peut être un contrat de concession, un contrat de partage de production, un contrat de service ou tous les autres types de contrat en usage dans l'industrie pétrolière internationale admis et autorisés par décrets pris en Conseil des Ministres*'.
>
> [Free translation into English: '[a]n oil contract can be a concession contract, a production-sharing agreement, a service contract or any other type of contract used in the international petroleum industry permitted and authorised by Decrees adopted by the Council of Ministers'].

See also, in the Central African Republic, Article 32 of Ordinance No. 93.007 of 25 May 1993 Enacting the Petroleum Code; in Côte d'Ivoire, Article 13 of Law No. 96-669 of 29 August 1996 Enacting the Petroleum Code.

b An analysis of service contracts

10.15 The service contract is based on a traditional mechanism used in work agreements, ie the remuneration of a company for services rendered under the contract.[10]

Thus, the host appoints the private operator state, or the local State operator, to undertake and perform the operations associated with the various phases of extractive activities. Subsequently, the private operator may intervene to cover all of the exploitation operations, from exploration to transportation of the production. The private operator therefore becomes a 'contractor' of services. In general, it must finance the exploitation operations and therefore provide capital, personnel, machinery and equipment.

The services rendered by the private operator are paid for in local or foreign currency.[11] Indeed, the aim of a foreign oil company entering into a contract to exploit the natural resources of the producing state is, as service contract terminology suggests, to get paid for services rendered within the context of these oil operations.

Therefore, a private operator will have several roles at the same time: contractor and funder. In the preliminary phase of the exploitation agreement, the role is that of a contractor. It will be responsible for managing and carrying out the exploitation work. The private operator also has the role of a funder as it provides the funds necessary to carry out the work. Only if a commercial deposit is discovered, the local State operator will remunerate the private operator and reimburse all of the sums invested in exploration and those spent on development and production.

10.16 Service contracts are designed to not involve any transfer of ownership of the oil resources extracted to the private operator; in this way, the producing state remains the sole owner of the extracted resources. Unlike with the concession contract, the private operator will not be granted any direct ownership of the quantities of oil produced. Similarly, unlike with other contractual models, such as the production-sharing agreement, the

---

[10] For examples defining the concept of service contracts, see, in particular, in Benin, Article 1 of Law No. 2006-18 of 17 October 2006 Enacting the Petroleum Code:

> '*[t]out contrat pétrolier dans lequel le titulaire assure la direction des opérations pour le compte de l'État propriétaire, moyennant une rémunération financière, fixe ou variable.*'
>
> [Free translation into English: '[a]ny oil contract in which the holder manages operations on behalf of the State owner, in return for fixed or variable financial remuneration'].

For a definition of the service contract, see Bret-Rouzaut & Favennec op cit 213:

> '[t]he service contract is a contract whereby a contractor must search for deposits, at its own risk and expense, and on behalf of a national company, in return for reimbursement of expenses incurred and monetary remuneration in line with the success of the exploration works. All production goes to the national company and the contractor may sometimes purchase a fraction of this production under the conditions agreed'.

[11] The principle of cash payment of the private operator is reiterated, in particular, in Liberia by Article 6(1) of the 2002 Petroleum Law: '[a] petroleum agreement shall be called a "services and risks contract" when it provides for the reimbursement of petroleum costs and the remuneration to the holder shall be paid in cash.'

private operator will not, under a service contract, be remunerated in the form of allocation of a share of production.[12]

However, under certain service contracts, the private operator may have a preferential right to purchase any production surplus under a buy-back clause included in the service contract.

2 CONTRACTS OF ASSOCIATION

10.17 Alongside the development of service contracts in the oil industry, hydrocarbon exploitation contractual arrangements in the Middle East underwent massive change, with 1957 being the pivotal date. These 'contracts of association' served as the basis and as legal support for implementation of host state involvement in the extractive operations.

10.18 There are two distinct categories of contracts of association: *joint ventures* and *joint-operating agreements*.

However, it is worth pointing out that this distinction can sometimes seem rather blurred; some legislation confuses the concepts of joint-operating agreements and joint ventures. This is the case with the Malagasy Petroleum Code, which adopts the concept of 'contrat d'association en joint-venture'.[13]

a Joint ventures

10.19 Joint ventures, which appeared in the extractive industry in the second half of the 20th century, are now widespread, especially on the African continent.

i *Historical approach to joint ventures in the extractive industry*

10.20 Historically, two oil companies were behind the development of joint ventures in the extractive industry, the Italian company ENI and the French company ERAP.

---

[12] Thus, in the Central African Republic, Article 34 of Ordinance No. 93.007 of 25 May 1993 Enacting the Petroleum Code reiterates the difference between the service contract and the production sharing agreement:

> '*[e]n cas de production commerciale, l'État ou la société d'État, suivant les cas, en est propriétaire et, à partir de cette production, conformément aux dispositions du contrat de services à risques, rembourse le titulaire du contrat de ses investissements et coûts, et lui verse une rémunération, en espèces ou en nature. Dans le cas d'un remboursement et d'une rémunération en nature, le titulaire reçoit une part de la production obtenue et le contrat est dit "contrat de partage de production"*.'

> [Free translation into English: '[i]n the event of commercial production, the State, or where appropriate the state-owned company, shall own this production and, based on this production, in accordance with the provisions of the services and risks contract, shall reimburse the contractor for its investment and costs and pay it remuneration, in cash or in kind. If reimbursement and remuneration are done in kind, the contractor receives a share of the production obtained and the contract is considered to be a "production sharing agreement"'].

[13] See, to that effect, in Madagascar, Article 11 of Law No. 96-018 of 4 September 1996 Enacting the Petroleum Code.

(1) Transformations dictated by ENI

10.21 On 31 July 1957, Iran signed a contract with AGIP, a subsidiary of the Italian company ENI,[14] which established a new allocation of responsibilities, powers and benefits between the operators.[15]

AGIP had no ownership rights either on the surface or on the natural resources contained in the subsoil. It was acting as the operator of a joint enterprise, SIRIP, in which the operators were equal. SIRIP, a company under Iranian law, was invested with all the rights and became the holder of the mining rights. AGIP bore all of the exploration costs, for which it would only be reimbursed in the event of it discovering commercial quantities before sharing the benefits with its Iranian partner.

Pursuant to this policy, ENI then signed other contracts of this type in other countries, for instance in Morocco.[16]

(2) Transformations dictated by ERAP

10.22 ERAP went one step further in 1965, specifically in Saudi Arabia within the context of the AUXIRAP contract.[17]

This contract was signed on 4 April 1965 between the Saudi Government and AUXIRAP, a French subsidiary of ERAP.

Preliminary exploration permits covering three zones in the Red Sea with a total surface area of 26,800 km^2 were issued to AUXIRAP for a two-year period; it had to finance the work on its own, with a minimum expenditure of five million dollars over the first two years. The exploration costs were not reimbursable. At the end of these two years, the Saudi Government could extend the term of the permit by a further three years. Following a discovery of hydrocarbons in commercial quantities, a legal framework in the form of a joint venture was created for a period of 30 years.

A local public company, governed by Saudi laws, was then set up,[18] in which the State held a 40% share and the French company the remaining 60%. The great innovation in comparison to the 1957 SIRIP contract was the disappearance of an equal partnership between the operators in favour of the private operator.

---

[14] See, on this matter, Dow Votaw *The Six-Legged Dog, Mattei and ENI—A Study of Power* (University of California Press 1964) 17 et seq and 77 et seq; Kamal Hossain *Law and Policy in Petroleum Development—Changing Relations Between Transnationals and Governments* (Nichols Publishing Company 1979) 121 et seq; E H Wall 'The Iranian Italian Oil Agreement of 1957' (1958) 7 *Int'l & Comp LQ* 736.

[15] Contract of Association between NIOC and AGIP Mineraria, 31 July 1957 (Source: *The Petroleum Times* 1957 at 1031 (F)).

[16] See Contract of Association for the Creation of a Joint Enterprise (Société mixte marocaine italienne des pétroles—SOMIP) between the Moroccan Government and AGIP Mineraria, dated 26 July 1958.

[17] See, in particular, the Joint Operating Agreement in the Event of Commercial Discovery between PETROMIN and AUXIRAP, April 1965 (Source: BOLACC—PMO Supplement 7, C-1 (A)).

[18] See the Joint Operating Agreement in the Event of Commercial Discovery between PETROMIN and AUXIRAP, April 1965, Article 4 (Source: BOLACC—PMO Supplement 7, C-1 (A)).

It was the first time that a producing state in the Middle East had held such a high share in the capital of an operating company.

ii *The main characteristics of joint ventures*

10.23 Joint ventures have developed in the extractive industry[19] and the use of this type of association is envisaged by some legislation throughout the African continent.[20]

As a general rule, the joint venture takes the form of a joint-stock company with legal personality, governed by the law of the host state. The company created under local law will generally hold mining rights. Thus, the agreement of joint ventures may result in the State operator assigning mining rights to the joint company, as well as permits for exploitation of the deposit.[21]

---

[19] For a recent example of a joint venture contract in the mining sector, see, in particular, in the Democratic Republic of Congo, the Joint Venture Agreement Between Congolaise d'Exploitation Minière (COMINIERE) and DBB Resources Corporation (DBB CORP) relating to the exploitation of mineral deposits in the Manomo region (29 October 2010), which provides for the establishment of a partnership between the two companies and the creation of a joint operating company.

[20] In this way, in Angola, Article 14 of Petroleum Law No. 10/04 of 12 November 2004 provides: '[s]ubject to the prior consent of the Government, the National Concessionaire may associate with Angolan or foreign entities of recognised capacity, technical knowledge and financial capability. 2. Such association may take the following forms: (a) Corporation; (b) Consortium; (c) Production-Sharing Agreement.'

[21] See, in particular, in the Democratic Republic of Congo, the joint venture between Congolaise d'Exploitation Minière (COMINIERE) and DBB Resources Corporation (DBB CORP) relating to the exploitation of mineral deposits in the Manomo region (29 October 2010), which, in Article 8(2) 'Assignment of exclusive exploitation rights to MANOMIN', provides:

> '*[l]a COMINIERE s'engage par la présente à céder à MANOMIN dès la date de constitution de cette dernière, tous les droits exclusifs sur le Bien* [c'est-à-dire les gisements] *ainsi que les droits exclusifs d'occuper, d'avoir pleine jouissance, d'utiliser, de maintenir, d'améliorer, de développer, d'exploiter les rejets (tailings), et tous autres droits et propriété sur les Installations à l'intérieur du Périmètre. La COMINIERE s'engage à obtenir conformément à la législation de la RDC, toute approbation relative auxdites cessions auprès des autorités habilitées.*'
>
> [Free translation into English: 'COMINIERE [a private company 90% owned by the State of DRC and 10% by the Institut National de Sécurité Sociale] hereby undertakes to assign to MANOMIN, from the date of its constitution, all of the exclusive rights to the Goods [namely the deposits] as well as the exclusive rights to occupy, have full enjoyment, use, maintain, improve, develop, exploit the tailings, and any other rights and property on the Installations inside the Area. COMINIERE undertakes to obtain, in accordance with DRC legislation, any approvals regarding the said assignments from the relevant authorities'].

b Joint operating agreements

10.24 These contracts were originally limited geographically to Iran under the name 'joint structure'[22] and then were adopted in other states.

In April 1958, the first joint structure contract was signed with Pan American Petroleum. It was followed by others of the same type with most of the large oil companies. This new contractual form originates from the fact that US oil companies, in order to benefit from certain tax incentives in the USA, had to provide evidence that they directly held title deeds to the production, which was not the case with previous contractual formulas.

10.25 Unlike joint ventures, joint-operating agreements do not take the form of a company with legal personality.[23]

Joint operating agreements are sometimes mentioned in legislation in African countries, which provide in particular for joint-operating agreements in the form of a consortium between the local State operator and the private operators.[24]

3 PRODUCTION SHARING AGREEMENTS

10.26 In 1966, with the emergence in Indonesia of production sharing agreements, there was a real turning point in contractual history for exploitation of natural resources. This emergence took almost six years. Since Indonesia's rise to political independence, in the wake of World War II, the old concessionary system inherited from an 1899 Law needed to be modified.[25] The interests of the operating companies and the holders of exclusive mining rights had to be reconciled with those of a young

---

[22] We have chosen to use the Anglo-Saxon term rather than its approximate French translation of 'entreprise en participation' or even the term *'joint operating association' in* Robert Brasseur *Législation et fiscalité internationales des hydrocarbures—Exploration et Production* (Éd Technip 1975) 225 et seq.

[23] This principle is reiterated in Mali in Article 1 of Law No. 2015-035 of 16 July 2015 Organising the Exploration, Exploitation and Transportation of Hydrocarbons, which defines the consortium as:

> '*tout groupement de sociétés ou autres entités juridiques, dépourvu de la personnalité juridique et constitué en vue d'effectuer des Opérations Pétrolières* [. . .]'.
>
> [Free translation into English: 'any group of companies or other legal entities, devoid of legal personality and formed with a view to performing Petroleum Operations [. . .]'].

[24] See, in particular, in Angola, Article 14 of Petroleum Law No. 10/04 of 12 November 2004: '[s]ubject to the prior consent of the Government, the National Concessionaire may associate with Angolan or foreign entities of recognised capacity, technical knowledge and financial capability. 2. Such association may take the following forms: (a) Corporation; (b) Consortium; (c) Production-Sharing Agreement.' See also, in Mali, in Article 1 of Law No. 2015-035 of 16 July 2015 Organising the Exploration, Exploitation and Transportation of Hydrocarbons.

[25] Indonesia has a one-hundred-year-old oil industry. The first barrel was extracted in 1885 by a Dutch tobacco plantation owner on the north coast of Sumatra. Its field was then acquired by Shell. The exploitation of coal began in 1849 in the east of Kalimantan. On the issue of Indonesia's oil potential, see Allan Cree 'Status of Petroleum Exploration in Indonesia' 10 *Exp Eco Pet Ind* 123.

Indonesian nation that wanted to assume responsibility for and take control of its natural resources.

a Production-sharing agreements

10.27 Nowadays, the production-sharing agreement is undoubtedly the most widespread contractual form in the oil industry on the African continent. In this way, legislation in many countries provides for the option of using this contractual form within the context of exploitation of natural resources.[26]

374 Although, production-sharing agreements primarily refer to the exploitation of hydrocarbon resources, such contracts can also be applied in the mining sector as demonstrated by the new Senegalese mining code, which provides for the signature of production-sharing agreements within the context of the exploration and exploitation of mineral substances.[27]

b The main components of production-sharing contracts

10.28 The production-sharing contract is a service contract whereby the contractor is entitled to a share of the hydrocarbon production as payment for its services.[28]

10.29 With regard to financing, the principle is that the private operator finances the operations and assumes all the risks.[29] If a commercial discovery

---

[26] See, in particular, in Côte d'Ivoire, Article 13 of Law No. 96-669 of 29 August 1996 Enacting the Petroleum Code; in Benin, Article 1 of Law No. 2006-18 of 17 October 2006 Enacting the Petroleum Code; in Angola, Article 14 of Petroleum Law No. 10/04 of 12 November 2004; in Cameroon, Article 13(1) of Law No. 99-013 of 22 December 1999 Enacting the Petroleum Code; in Guinea, Article 22 of Law No. L/2014/N°034/AN of 23 December 2014 Enacting the Petroleum Code; in Niger, Article 105 of Law No. 2007-01 of 31 January 2007 Enacting the Petroleum Code.

[27] See Article 33 of Law No. 2016-32 of November 2016.

[28] See in Niger, Article 2 of Law No. 2007-01 of 31 January 2007 Enacting the Petroleum Code provides:

> '*[c]ontrat de partage de production: le contrat pétrolier dans lequel le titulaire s'engage à effectuer les opérations pétrolières, à ses frais et risques, pour le compte de l'État moyennant une part des hydrocarbures produits sur la zone contractuelle liée comme rémunération en cas d'exploitation.*'
>
> [Free translation into English: '[p]roduction-sharing agreement: the oil contract in which the contractor undertakes to perform petroleum operations, at its own risk and expense, on behalf of the State for a share of the hydrocarbons produced in the contractual zone as remuneration in the event of exploitation'].

For other definitions, see, in particular, in Cameroon, Article 2(h) of Law No. 99-013 of 22 December 1999 Enacting the Petroleum Code; in Djibouti, Article 2(i) of Law No. 117-AN-05 of 16 October 2005 Enacting the Petroleum Code; in Liberia, Article 7(1) of the 2002 Petroleum Law.

[29] See, in particular, in Cameroon, Article 15(1) of Law No. 99-013 of 22 December 1999 Enacting the Petroleum Code:

> '[p]*ar le Contrat de Partage de Production, l'État directement ou par l'entremise d'un établissement public dûment mandaté à cet effet, contracte les services d'un Titulaire en vue d'effectuer pour son compte et de façon exclusive, à l'intérieur du périmètre défini, les activités de Recherche et, en cas de découverte d'un gisement*

is made and result in a subsequent production, investment costs will be recovered in the form of a right to collect part of this production.

Besides, some petroleum legislation provides for state participation in financing of all the development and exploitation costs from the start of the deposit's production, although the private operator, which finances the exploration work itself, continues to assume the risks.[30]

Under a production-sharing contract, only the State retains ownership of the natural resource; the private operator receives remuneration in the form of a share of the production. The production-sharing agreement is therefore based on the principle of sharing the production between the private operator and the host state, which alone holds the mining rights.

c Conditions for production sharing

10.30 Under production-sharing contracts, the private operator alone assumes the risk of non-discovery. Indeed, it is only after the discovery of a commercial deposit that the operator can obtain any reimbursement of the costs incurred during the exploration and be awarded a share of production as remuneration.

10.31 As such, the private operator may, in the event of discovery, be reimbursed for the petroleum costs incurred for the exploitation of the deposit. This share of production constitutes the '*cost oil*'.[31]

---

d'Hydrocarbures commercialement exploitable, les activités d'Exploitation. Le Titulaire assure le financement de ces Opérations Pétrolières'.

[Free translation into English: '[t]hrough the Production-Sharing Agreement, the State, either directly or through the intermediary of a public establishment duly appointed for this purpose, contracts the services of a Contractor to carry out on its behalf and exclusively, and within the defined area, the Exploration activities and, in the event that a commercially viable deposit of hydrocarbons is discovered, the Exploitation activities. The Contractor finances these Petroleum Operations'].

[30] See, to that effect, in Gabon, Article 63 of Law No. 011/2014 of 28 August 2014 Enacting Regulations in the Hydrocarbons Sector, which provides:

'*[d]ès la mise en production d'un gisement, l'État participe à concurrence du taux défini dans le contrat de production et de partage de production, au financement de tous les coûts de développement et d'exploitation, à l'exclusion des nouveaux coûts d'exploration dans la zone délimitée.*'

[Free translation into English: '[f]rom the start of production of a deposit, the State participates, up to the rate defined in the production and production-sharing agreement, in the financing of all of the development and exploitation costs, excluding new exploration costs in the area marked out'].

[31] See, to that effect, in Côte d'Ivoire, Article 15 of Law No. 96-669 of 29 August 1996 Enacting the Petroleum Code:

'[. . .] *une part de la production totale d'hydrocarbures est affectée au remboursement des coûts pétroliers effectivement supportés par le titulaire au titre du contrat pour la réalisation des opérations pétrolières. Cette part de production, couramment appelée dans l'industrie "cost oil", ne peut être supérieure au pourcentage de la production fixé dans le contrat, lequel définit les coûts pétroliers récupérables ainsi que les conditions et modalités de leur récupération par prélèvement sur la production* [. . .]'.

[Free translation into English: '[. . .] a share of the total production of hydrocarbons is set aside for reimbursing the petroleum costs incurred by the contractor under the contract for performance of petroleum operations. This share of production, known in the

10.32 This share of production is almost always limited to the 'cost stop',[32] which constitutes the limit of production affected to the reimbursement of the costs incurred by the operator.

10.33 If the costs to be recovered ('cost oil') are greater than the 'cost stop', the balance can be carried over to subsequent financial years. If the 'cost oil' is lower than the 'cost stop', the difference (known as 'excess cost') may be either paid to the host state or added to the 'profit oil' depending on the agreement.

10.34 Once the production costs have been recovered by the private operator, the remaining share of production (the '*profit oil*') will be split between the State and the private operator depending on the conditions laid down in the exploitation agreement.[33]

10.35 The '*profit oil*' can be split between the operators following two different methods: to determine the 'share allocated to the State and the operator there may be, on the one hand, a fixed percentage unrelated to the level of production or, on the other hand, and more commonly, a variable percentage. As a general rule, the greater the production, the higher the share of 'profit oil' that goes to the State.[34]

---

industry as "cost oil", may not be greater than the percentage of production established in the agreement, which defines the recoverable petroleum costs and the terms and conditions for recovering them in the form of production [. . .]'.]

[32] So, in Niger, Article 108 of Law No. 2007-01 of 31 January 2007 Enacting the Petroleum Code provides:

'[. . .] *une part de la production totale d'hydrocarbures, nette de la redevance ad valorem définie à l'article 113 ci-dessous, est affeectée au remboursement des coûts pétroliers effectivement supportés par le titulaire au titre du contrat pour la realisation des operations pétrolières. Cette part, couramment appelée "cost oil" ou "production pour la recuperation des coûts", ne peut être supérieure à un pourcentage de la production appelé "cost stop"* [. . .]'.

[Free translation into English: '[. . .] a share of the total production of hydrocarbons, net of the *ad valorem* tax defined in Article 113 below, is set aside for reimbursing the petroleum costs incurred by the contractor under the contract, for performance of petroleum operations. This share, commonly called "cost oil", may not be greater than a percentage of production commonly called "cost stop" set aside for recovery of costs [. . .]'.

[33] See, in particular, in Côte d'Ivoire, Article 15 of Law No. 96-669 of 29 August 1996 Enacting the Petroleum Code:

'[. . .] *le solde de la production total d'hydrocarbures, après déduction de la part prélevée au titre de l'alinéa a) ci-dessus, couramment appelé dans l'industrie « profit oil fl, est partagé entre l'État et le titulaire, selon modalités de partage fixées dans le contrat, lequel précise si le partage est effectué avant ou après impôt sur les bénéfices industriel et commerciaux.*'

[Free translation into English: '[. . .] the balance of the total production of hydrocarbons, after the share referred to in paragraph (a) above has been deducted, known in the industry as "profit oil", is shared between the State and the contractor, depending on the sharing conditions laid down in the agreement, which indicated whether sharing is done before or after tax on industrial and commercial profits'].

[34] See, in particular, in the Democratic Republic of Congo, Article 12.1 of the Production-Sharing Agreement between the Democratic Republic of Congo and the Association: Dominion Petroleum Congo & Soco Exploration-Production RDC & La Congolaise des

10.36 National legislation can sometimes lay down conditions for specific production sharing as circumstances dictate. It is usually about encouraging the exploitation of unassociated gas deposits and deep offshore deposits or deposits that call for the use of laborious extraction techniques.[35] Thus, petroleum codes may in particular set a '*cost stop*' that is higher than the one generally applied[36] or provide for a different production-sharing

---

Hydrocarbures sur le Bloc V du Graben Albertine (November 2007), which provides for the profit oil to be distributed as shown below:

- net cumulative production less than 50,000,000 barrels: 60% for the contractor, 40% for the State;
- net cumulative production between 50,000,001 and 75,000,000 barrels: 55% for the contractor, 45% for the State;
- net cumulative production between 75,000,001 and 100,000,000 barrels: 50% for the contractor, 50% for the State;
- net cumulative production between 100,000,001 and 125,000,000 barrels: 45% for the contractor, 55% for the State;
- net cumulative production greater than 125,000,001 barrels: 40% for the contractor, 60% for the State.

[35] See, in particular, in Côte d'Ivoire, Article 16 of Law No. 96-669 as 29 August 1996 Enacting the Petroleum Code:

> '[l]es règles de partage de la production visées aux alinéas a) et b) de l'article 15 ci-dessus peuvent être différentes pour les hydrocarbures liquides et les hydrocarbures gazeux, en vue de promouvoir l'exploitation des gisements de gaz naturel notamment non associé. En outre, pour encourager les opérations pétrolières dans les zones marines profondes, le contrat de partage de production prévoit des règles de remboursement des coûts pétroliers et de rémunération du titulaire qui prendront en compte, directement ou indirectement, l'incidence de la profondeur d'eau des gisements concernés. Il peut également prévoir pour ces zones marines profondes la possibilité d'inclure dans les coûts pétroliers récupérables au titre de l'alinéa a) de l'article 15 ci-dessus un montant supplémentaire égal à une fraction des investissements de développement appelé "crédit d'investissement en mer profonde", qui est défini dans le contrat.'
>
> [Free translation into English: '[t]he rules for sharing production given in paragraphs (a) and (b) of Article 15 above may differ for liquid hydrocarbons and gaseous hydrocarbons, with a view to promoting the exploitation of natural gas deposits, especially those that are not associated. Moreover, to encourage petroleum operations in deep sea zones, the production-sharing agreement lays down rules for reimbursement of petroleum costs and remuneration of the contractor that will take into account, directly or indirectly, the impact of the water depth of the deposits in question. It may also provide, for these deep sea zones, the option to include in the petroleum costs that are recoverable under paragraph (a) of Article 15 above an additional amount equal to a fraction of the development investments called "deep sea investment credit", which is defined in the agreement.'

See also, in Liberia, Article 7(2) of the 2002 Petroleum Law: '[t]he rules for production sharing set forth may be different for gaseous hydrocarbon, in order to promote the development of natural gas field, particularly those not associated with other reserves'].

[36] See, in particular the former petroleum law applied in Republic of Congo (Article 35 of Law No. 24-94 as 23 August 1994 Enacting the Hydrocarbons Code):

> '*[u]ne part de la production d'hydrocarbures est affectée au remboursement des coûts pétroliers encourus par la société ("cost oil"). Cette part de production qui est fixée dans le contrat ne peut être supérieure à 60% de la production annuelle provenant de l'ensemble des permis d'exploitation découlant d'un même permis de recherche. Toutefois, lorsque l'importance des travaux de recherche ou de développement, l'utilisation d'une technologie particulièrement onéreuse ou la difficulté exceptionnelle*

percentage depending on whether the operations are onshore or deep offshore deposits.[37]

4 INDUSTRIAL COOPERATION CONTRACTS

10.37 The concept of industrial cooperation is generally incredibly extensive. The United Nations Economic Commission for Europe has adopted the following definition and:

> 'considers as industrial cooperation contracts operations which go beyond the straightforward sale on purchase of goods and services and which involve the creation between parties belonging to different countries of a lasting community of interests designed to provide the parties with mutual advantages. Such contracts relate to, inter alia, the following: 1. transfer of technology and of technical experience; 2. cooperation in the field of production including, where appropriate, cooperation in research and development in the specialization of production; 3. cooperation in developing natural resources; and 4. joint marketing or marketing on joint account, in the countries of parties to the contract or in the third markets, of the products of international cooperation.'[38]

---

de la zone comme par exemple les zones marines très profondes et le bassin intérieur dit Bassin de la Cuvette Congolaise le justifie, cette part de production peut être portée à un niveau supérieur n'excédant pas soixante dix pourtant (70%) de la production annuelle provenant de l'ensemble des permis d'exploitation découlant d'un même permis de recherche.'

[Free translation into English: '[a] share of the production of hydrocarbons is allocated to reimbursement of the petroleum costs incurred by the company ("cost oil"). This share of production, which is established in the agreement, may not be greater than 60% of the annual production from all of the exploitation permits arising from the same exploration permit. However, where the magnitude of this exploration or development work, the use of a particularly laborious technology or the exceptional difficulty of the zone, for instance very deep sea zones and the inner basin known as the Bassin de la Cuvette Congolaise so justifies, this share of production can be increased to a level not in excess of seventy per cent (70%) of the annual production from all of the exploitation permits arising from the same exploration permit']).

37 See, to that effect, in Gabon, Article 68 of Law No. 011/2014 of 28 August 2014 Enacting the Hydrocarbons Code:

'*[l]e mode de calcul de tout partage de production est basé sur la production totale disponible cumulée de l'ensemble des zones d'exploitation d'un même contrat. Pour la part revenant à l'État, le taux de la première tranche ne peut être inférieur à: 55% pour la zone conventionnelle; 50% pour la zone off-shore profond et très profond.*'

[Free translation into English: '[t]he method for calculating any production sharing is based on the total available cumulative production of all of the exploitation zones for the same agreement. For the share that goes to the State, the rate of the first portion may not be lower than: 55% for the contractual zone; 50% for the deep and very deep offshore zone.'].

38 Paragraph 4 of the Guide published on 21 May 1976 by the United Nations with reference ECE/TRADE/124, Sales No.: F.76.II.E.14. It is also stated in the notes that this listing does not specifically include the more integrated forms of industrial cooperation such as mixed companies. This text was also reproduced in the 1976 review *Droit et pratique du commerce international* 649 et seq.

The exploitation of natural resources features prominently in this listing due to their economic and political importance, as well as the highly advanced technologies used.

10.38 Industrial cooperation contracts differ from service contracts in that their purpose is mainly limited to technology transfer aspects. They also differ from production-sharing agreements and services contracts as the private operator does not assume the financial risks in terms of investments.

The operator will be remunerated like any other service provider, in cash or in kind,[39] but the local State operator holds full and permanent ownership rights over all of the production.

10.39 By only keeping to these contracts, the private operator does not in the short term improve its actual presence in the exploitation cycle of the natural resource and will not therefore have any control over it. However, it will attempt to strengthen its brand image and it is at this level that it takes a risk that could be financially quantifiable; indeed, fulfilment of an industrial cooperation contract will tie up some of its human capital, and maybe its material resources, but above all will lead to the transmission of a larger or smaller fraction of its technological capital.

10.40 There is one difficulty associated with these agreements that should be mentioned: how is fair remuneration for the services provided by the private operator calculated? It is difficult to determine[40] . Indeed, the private operator provides intellectual and intangible services, the results of which will be identical in nature. How do we translate into financial terms technological experience acquired over a number of years, in many countries and from multiple contracts? Besides, a project can be more or less risky depending on the technical level of the local State operator and the amount of effort the private operator has to put in. As a general rule, two factors are taken into consideration: the scarcity of the technological knowledge required, and the benefits expected by the local State operator.

10.41 The terminology relating to the different types of industrial cooperation contract is extensive. We will look at it under two main headings depending on whether or not their scope is more than just a simple transfer of technology.

a Agreements making up a limited industrial cooperation

10.42 These agreements have limited scopes: licence, patent, *ad hoc* management advice, miscellaneous construction or maintenance services. They only meet occasional needs that are completely separate from a consistent set of operations.

[39] It may be that the private operator sets up a joint management company with the local State operator and then provides its share of the capital itself, for a very reduced amount.

[40] See on this issue, the work of the symposium on 'Les contrats internationaux de coopération industrielle et le nouvel ordre économique international', Jean D'Herbes & Jean Touscoz (Presses Universitaires de France 1980).

In terms of exploitation of natural resources, they mainly concern management contracts and technical support contracts.

i *Management contracts*

10.43 Under management contracts, the private operator is entrusted with the management of exploitation operations for a given period.

10.44 However, the purpose of the management contract may be broader. The host state can charge the private operator with managing one of its companies that dedicates its activity to the exploitation of one or more natural resources, or even with managing a stake it holds in an operating company.

Remuneration of the private operator is variable and it can be based on fixed fees, be profit-based or depend on the amount of investments made.

ii *Technical support contracts*

10.45 Within the context of a technical support contract, the private operator, for a given period, assists with the exploitation of a natural resource from a technical point of view.

10.46 Nonetheless, it will assume full responsibility for management and performance. This responsibility is limited in practice as certain areas elude the private operator, such as subcontractor selection.

However, the intervention of the private operator is usually remunerated regardless of the actual result. The contracts signed by the operator are identical to service contracts commonly used in international economic exchanges.

b Agreements establishing an extended industrial cooperation

10.47 Extended industrial cooperation contracts differ from the previous contracts in the size of the cooperation scope they establish. Indeed, it will involve the private operator passing on general operating experience.

10.48 Another difference is that in principle the private operator is not seeking as a priority or at the very least is not only seeking, immediate and maximum profit. The central point of an industrial cooperation contract relies on the absolute transfer of knowledge in such a way that at the end of the private operator's intervention, the local State operator can, alone, successfully conduct the exploitation operations. Nonetheless, it is not certain that the knowledge acquired may be applied by the State entity in other circumstances.

10.49 Industrial cooperation contracts are signed for a relatively short term, but they may be supplemented by agreements relating in particular to the long-term purchase of a larger or smaller share of production.

10.50 Analysis of contractual arrangements for exploitation of natural resources effectively illustrates the difficulties of organising a typology. It is from this contractual diversity that agreements relating to the exploitation of natural resources in Africa draw their inspiration. These contracts are not, except in very rare circumstances, innovative, but they have adopted and reconfigured the contractual practice that exists within the context of exploitation of natural resources.

11

The organisation of exploitation between operators

11.1 It is in the common interest of operators to join forces with a view of carrying out the exploitation operations together.[1] This phenomenon is not specific to natural resources projects.

In the context of national resources, the intervention of several operators may be subject to different modalities.

Besides, the way in which operators are organised is far from static and evolves throughout the performance of the exploration works. These evolutions can result in the intervention of new operators or the departure of others.

1 METHODS OF JOINT INTERVENTION OF OPERATORS

11.2 It is in the common interest of operators to join forces with a view to carrying out the exploitation operations.[2]

11.3 The operators' decisions to intervene jointly may intervene prior to the signature of the exploitation agreement. This situation refers to the existence of a 'horizontal' grouping whereby the operators present the bid and sign the exploitation agreement jointly and severally.

In other circumstances, the various companies interested in a project may manifest their interest individually; however, the licensing authority may decide to sign a single agreement with several bidders.

a The intervention of several operators with a view to signing the exploitation agreement

11.4 In this scenario, the operators join together before the exploitation agreement is signed. We will consider the general framework that motivates and justifies this partnership, then the contractual structures used to implement this decision to partner.

i *Explanation of the phenomenon*

11.5 From the beginning of the 20th century, large companies based their exploration works for new mineral reserves and oil deposits on separate actions, each one zealously defending its own sphere of influence that it wanted to be exclusive. Thereafter, and faced with a changing international context, they were forced to accept other companies being involved in their operations.

---

[1] See, in particular, Claude Champaud 'Les méthodes de groupement des sociétés'(1967) 20 *RTD Com* 1004.

[2] Ibid.

These transformations marked the beginning of international cooperation agreements in the mining and oil sector. They differ depending on whether or not the expression of a third-party desire creates them, generally a political one, and therefore irrespective of the operators.

(1) Private operator cooperation dictated by an outside decision

11.6 Private operators may be brought together by a political desire of the state of origin of the private operators, but also sometimes by the host state itself for which this cooperation will constitute a condition *sine qua non* for awarding mining rights or for signing an exploitation agreement.

(2) Private operator cooperation independent of an outside decision

11.7 In terms of exploitation of natural resources, determining the motivations for bringing together several companies, often of different nationalities, involves searching for evidence.[3]

11.8 The dimension and nature of mining and petroleum projects involve a pooling of resources and effort. From a financial point of view, it brings together more availability, not just of equity, but also in terms of credit facilities. Bringing different stakeholders together for a given project is often motivated by the financial advantages of performing a specific project as a group.

From a technical point of view, the joining of several highly qualified teams will enable priceless exchanges of expertise, far beyond the use of patented technologies by each member of the union. Incidentally, this cooperation may lead to the filing of joint patents.

11.9 In practice, this cooperation may take on a variety of forms; they may be general but limited in time or dedicated to a specific project.[4]

ii *Cooperation agreement typology*

11.10 The contractual structures that sanction the union of private operators with a view to cooperating have common characteristics regardless of the natural resource at stake. However, some forms may be more specific to the solid minerals sector and others to the hydrocarbons sector.

(1) The characteristics of contractual cooperation structures

11.11 For a long time, the vast majority of authors only looked at contractual structures for private operator cooperation in the form of

[3] See on this issue, Joseph Russell Milton 'International Consortia: Definition, Purpose and the Consortium Agreement', (1979-1980) 3 (2) *FILJ* 125 et seq; Jean-Albert Boon 'Problèmes juridiques soulevés par la conclusion de consortiums internationaux' 1964 *Ann Fac Dt Liège* 374 et seq.

[4] The first example of private operator cooperation not obligated by external forces, is given by the three participants to the Turkish Petroleum Company, which, in March 1914, had undertaken to carry out together exploration and production operations only in the then Ottoman Empire, with the exception of Kuwait and Egypt.

associations.[5] And yet, this term, which incidentally is very vague and poorly defined, hardly fits with the legal practice materialising the exploitation of natural resources, which largely favours the signature of consortium agreements.

11.12 In the vast majority of situations, a contractual cooperation platform is established before work begins and even before the exploitation agreement is signed. Therefore, a consortium may only have a fleeting existence, with a considerable proportion of mining and oil projects collapsing during negotiations compared to those that succeed.

11.13 However, it may be decided to either create a consortium during performance of the exploitation operations, particularly if financing is needed, or following encouraging exploration results.

(2) The content of contractual cooperation structures

11.14 As a general rule, the scope of application of the contractual cooperation structures will be determined in their recitals, which generally state whether these structures are intended for one or more projects.

The members of the cooperation structure decide to act jointly and state their desire to collaborate exclusively with one another to achieve objectives defined in the recitals of the agreement. A confidentiality clause, inherent to this provision, is also inserted.

(3) The consortium or partnership agreement ('contrat d'association')

11.15 The diversity of situations and of the parties involved in a consortium encourages the development of contractual structures.

11.16 The consortium is a partnership whereby several operators establish a common organisation with a view to determining and implementing given operations performed by their respective companies.[6] The consortium or partnership agreement will specify the conditions of this cooperation.

---

[5] Milton op cit 121 et seq.

[6] For a definition of a consortium, see, in particular, in Niger, Article 2 of Law No. 2007-01 of 31 January 2007 Enacting the Petroleum Code:

'*[c]onsortium: tout groupement de sociétés ou autres entités juridiques constitué en vue d'effectuer des opérations pétrolières de quelques natures que ce soit, dont les membres sont conjointement titulaires d'un permis ou d'une autorisation. Un consortium peut être créé postérieurement à la conclusion d'un contrat pétrolier. Le terme consortium n'est utilisé dans la présente loi que dans un souci de commodité et ne saurait en aucun cas indiquer une intention quelconque de la part des sociétés et personnes morales constituant le Consortium, de former entre elles une entité dotée de la personnalité juridique d'après les lois de quelque État ou juridiction que ce soit.*'

[Free translation into English: '[c]onsortium: any grouping of companies or other legal entities formed with a view to performing petroleum operations of any kind, the members of which jointly hold a permit or a licence. A consortium can be created after a petroleum contract has been signed. The term consortium is only used in this Law for ease and may not in any case indicate any intention by the companies and legal entities in the Consortium to form an entity with legal personality according to the laws of any State or jurisdiction'].

The legal nature of the consortium is not unified and is often uncertain. This group of operators may or may not have legal personality.

11.17 One essential aspect involves determining a hierarchy for responsibilities, in particular a 'leading company'.[7]

11.18 As well as the consequences of the failure of one of the members of the consortium and the extent of each member's commitments (often given in the appendices) the definition of the exact mission of the lead partner is a central point of the agreement. Indeed, several elements are to be defined, such as for instance the allocation of tasks and the monitoring of their performance, the content of the coordinator assignment or the control that the other members must have over its work.

11.19 Traditionally, the consortium forbids its members, even the leading partner, from acting on behalf of another member or the group as a whole.[8] However, this is not true in respect of a mining or oil consortium. Indeed, it is the leading partner, assisted or not by its partners, that will negotiate with the state's authorities. In principle, all the members of the consortium will, however, sign the exploitation agreement.

Between the members of the consortium, responsibility is generally limited except in the case of gross misconduct.

11.20 National legislation that anticipates the signature of a consortium agreement or partnership agreement may also lay down minimum clauses that these agreements should include, such as: the term of the agreement, the decision-making process, the stakeholders' obligations in particular in terms of financing or exit clauses.[9]

---

[7] Philippe Janin & Barthélémy Mercadal *Les contrats de coopération inter-entreprise* (Juridiques Lefebvre 1974) 311.

[8] 'In fulfilment of its mission, the leader must ensure in particular that it does not give third parties the impression that it is acting on behalf of an organised group. Presenting itself as speaking for the group must therefore be avoided, especially in correspondence, as should not mentioning the name of the group.' Janin & Mercadal op cit 312.

[9] See in particular, to that effect, in Niger, Article 103 of Decree No. 2007-082/PRN/MME of 28 March 2007 Establishing the Conditions for Implementation of Law No. 2007-01 of 31 January 2007 Enacting the Petroleum Code:

'*[c]onformément à l'article 9, dernier alinéa, du Code Pétrolier,l'ensemble des accords et conventions relatifs au Consortium, notamment le Contrat d'Association, est annexé à la demande mentionnée à l'alinéa 102.4 ci-dessus. Le Contrat d'Association est soumis à l'approbation du Ministre chargé des Hydrocarbures et doit comporter, au minimum, les clauses relatives à: la durée de l'accord; la désignation de l'Opérateur; les obligations de l'Opérateur, notamment dans le cadre de la représentation des membres du Consortium; la répartition des compétences en matière de commercialisation des Hydrocarbures extraits; les droits et obligations des parties notamment en ce qui concerne: leur part dans la production; l'audit des coûts de l'association; le processus des dépenses; le processus de prise de décision et, notamment, à travers la mise en place d'un comité d'association (CA): la direction de l'exécution des travaux; les prérogatives du CA; le suivi des directives du CA; la préparation et l'examen des programmes et budgets par le CA; l'autorisation des dépenses; le processus d'appels de fonds; les obligations des parties prenantes notamment en matière de financement; les stipulations relatives à la tenue de la comptabilité, qui doivent être conformes à l'accord*

11.21 Lastly, another contractual phenomenon should be mentioned in support of the partnership of private operators. This mechanism is not often used since it involves a merger.[10] The merging of parent companies is still very limited, unlike that of their local subsidiaries in a same state.

b The intervention of several operators with a view to fulfilling the exploitation agreement

11.22 Once the exploitation agreement has been signed, the partnership of several operators may intervene at different levels. First, the cooperation may result from the exploitation agreement itself through the implementation of various joint decision-making and management bodies. In addition to these contractual mechanisms, a joint-enterprise can be created; its primary purpose will be to organise exploitation and the allocation of tasks related in particular to performance and control.

i *Cooperation within the exploitation agreement: joint-operating agreement between the State and the private operator*

11.23 Until recently, only private operators were involved in this cooperation. Gradually, a desire for action by the local State operator, both in the performance of the works and in decision-making, emerged.

We will look at the principles and the modalities of this contractual cooperation.

(1) Operating principles

11.24 The exploitation of hydrocarbons has paved the way for contractual cooperation; since then, other examples have followed suit.

To be effective, such cooperation underlies the needs for all operators of a high level of equivalent technological mastery.

---

comptable annexé au Contrat Pétrolier; le processus de séparation (sortie de l'association).'

[Free translation into English: '[i]n accordance with the final paragraph of Article 9 of the Petroleum Code, all of the agreements and conventions relating to the Consortium, in particular, the Contract of Association, are appended to the application referred to in paragraph 102.4 above. The Contract of Association is subject to the approval of the Minister for Hydrocarbons and must, as a minimum, include clauses relating to: the term of the agreement; the name of the Operator; the Operator's obligations, particularly within the context of representation of the Consortium members; the distribution of the skills in terms of marketing of the Hydrocarbons extracted; the rights and obligations of the parties, especially with regard to: their share in the production; the audit of the association's costs; the expenses process; the decision-making process and, in particular, through the creation of an association committee: the works performance department; the prerogatives of the association committee; the monitoring of the association committee's guidelines; preparation and examination of the programmes and budgets by the association committee; authorisation of expenses; the call for funds process; the obligations of the stakeholders, particularly with regard to financing; the provisions relating to accounting, which must comply with the accounting agreement appended to the Oil Contract; the separation process (departure from the association)'].

[10] This was, for example, the origin of the French oil company Elf Aquitaine.

Oil and mining operations are meticulously prepared, week by week, in a 'work programme'. In most cases, it is upon the private operator to draw up this programme, although the local State operator may give its approval or provide its opinion.

During the contract term, joint committees are formed on which representatives of the operators will be sitting, usually in equal numbers or with an advantage in favour of one of them.

(2) Operating conditions

11.25 Joint committees have a variety of functions, the most significant being to act as coordination bodies.

Their practical actions are of various types and concern the provision of information to operators, the performance of the work programme, the preparation of budgets and the determination of operating costs, as well as the setting of the production level and the production prices.

ii *Cooperation outside the main exploitation agreement*

11.26 The creation of a joint company or the institution of an *ad hoc* grouping between the operators can be made mandatory or left optional.

(a) An obligation to partner for the exploitation of national resources

11.27 This obligation can result from the provisions of the mining or petroleum code but also from the exploitation agreement itself.

11.28 The institution of a partnership can be the result of a legal obligation, and in particular can arise from rules relating to unitisation within the context of hydrocarbon exploitation operations. Deposits often straddle several perimeters for which the exploitation rights have been granted to different operators.

11.29 Unlike solid minerals, liquid natural resources are likely to move in the subsoil; they are therefore likely to move from one area to another and compromise exploitation by one of the operators. Common sense tells us there should be an agreement between the operators of the different areas to get the most profitable exploitation of the deposit possible through joint performance of works.[11] As many states prefer not to resort to just the common sense of the operators, or wish to facilitate it, they have laid down very strict measures relating to unitisation.[12]

---

[11] The aims of unitisation are furthermore reiterated by national legislation in some countries. Thus, in Central African Republic, Article 53 of Ordinance No. 93.007 of 25 May 1993 Enacting the Petroleum Code specifies that unitisation agreements aim to exploit the deposit under the best possible technical and economic conditions.

[12] For a definition of the concept of unitisation, see, in particular, in Niger, Article 2 of Law No. 2007-01 of 31 January 2007 Enacting the Petroleum Code:

> *'[a]ccord d'unitisation: l'accord par lequel plusieurs titulaires de permis d'exploitation ou d'autorisations exclusives d'exploitation contigus et portant sur un même gisement commercial, désignent un opérateur unique pour le gisement et s'entendent sur les conditions de financement des dépenses et de partage des produits résultant du développement et de l'exploitation de ce gisement.'*

11.30 Thus, in the event that a deposit extended over several perimeters, subject to different mining rights or exploitation agreements, the holders of the mining rights or the operators may be forced to sign a unitisation agreement[13] or to establish a unitisation plan,[14] which will generally be submitted to the competent administrative authority for approval.

[Free translation into English: '[u]nitisation agreement: the agreement whereby several holders of adjoining exploitation permits or exclusive exploitation licences for the same commercial deposit appoint a single operator for the deposit and agree on the conditions for financing expenses and sharing of products resulting from the development and exploitation of this deposit.']

[13] See, in particular, in Central African Republic, Article 53 of Ordinance No. 93.007 of 25 May 1993 Enacting the Petroleum Code:

'[a]u cas où un gisement d'hydrocarbures s'étend sur plusieurs périmètres contractuels attribués à des titulaires distincts, ceux-ci sont tenus de conclure des accords dits "d'unitisation" afin d'exploiter ce gisement dans les meilleures conditions d'efficacité technique et économique. Les accords ainsi que les programmes d'exploitation commune doivent être communiqués au Ministre.'

[Free translation into English: '[i]n the event that a hydrocarbon deposit extends over several contractual areas allocated to different holders, these have an obligation to sign so-called unitisation agreements in order to exploit this deposit under the best possible technical and economic efficiency conditions. The agreements and the joint exploitation programmes must be communicated to the Minister.']

See also, in Morocco, Article 30 of Law 21-90 of 1 April 1992 On the Exploration and Exploitation of Hydrocarbons Deposits amended and supplemented by Law No. 27-99 of 15 February 2000:

'[l]orsqu'un gisement se prolonge au-delà du périmètre d'un permis de recherche mitoyen d'un ou de plusieurs autres permis son développement et son exploitation doivent se faire, le cas échéant, au moyen d'un accord dit d'unitisation entre les titulaires desdits permis selon des conditions qui doivent être approuvées par l'administration.'

[Free translation into English: '[i]f a deposit extends beyond the area of an exploration permit adjoining one or more other permits, its development and exploitation must be done, where appropriate, by way of a so-called unitisation agreement between the holders of the said permits according to conditions that must be approved by the authorities.']

[14] See, in particular, in Algeria, Article 54 of Law 13-01 of 20 February 2013 Amending and Supplementing Law No. 05-07 of 28 April 2005 On Hydrocarbons:

'[d]ans le cas où un gisement déclaré commercial s'étend sur au moins deux périmètres, objet de contrats distincts, les contractants concernés doivent, après notification par l'agence nationale pour la valorisation des ressources en hydrocarbures (ALNAFT), établir un plan conjoint pour le développement et l'exploitation du gisement. Ce plan est désigné par "plan d'unitisation". Il est soumis à l'approbation de l'agence nationale pour la valorisation des ressources en hydrocarbures (ALNAFT).'

[Free translation into English: '[i]n the event that a deposit declared to be commercial extends over at least two areas, covered by separate agreements, the contracting parties in question, following notification by the national agency in charge of promoting hydrocarbon resources (ALNAFT), must establish a joint plan for development and exploitation of the deposit. This plan is known as the "unitisation plan". It is subject to the approval of the national agency in charge of promoting hydrocarbon resources (ALNAFT).']

See also, in Guinea, Article 37 of Law No. L/2014/N°034/AN of 23 December 2014 Enacting the Petroleum Code:

11.31 If the parties cannot reach an agreement regarding joint exploitation of the deposit, it is normally the responsibility of the petroleum authorities to draw up the agreement or the unitisation plan. Where appropriate, some texts provide that the petroleum authorities may use, sometimes at the expense of the operators,[15] the services of an international expert to write such an agreement.[16]

'[d]ans le cas où un gisement déclaré commercial s'étend sur au moins deux (2) périmètres, objets de Contrats Pétroliers distincts, les Contractants concernés doivent, à la demande du Ministre en charge des Hydrocarbures, établir un plan conjoint pour le développement et l'exploitation du gisement, et le soumettre à l'approbation du Ministre.'

[Free translation into English: '[i]n the event that a deposit declared to be commercial extends over at least two (2) areas, covered by separate Oil Contracts, the Contracting Parties in question, upon request from the Minister in charge of Hydrocarbons, must establish a joint plan for development and exploitation of the deposit, and submit it to the Minister for approval.']

In Chad, Article 101(1) of Law No. 006-PR-2007 of 2 May 2007 On Hydrocarbons:

'[d]ans le cas où la découverte d'un gisement commercialement exploitable s'étend sur un périmètre couvert par deux ou plusieurs Permis de Recherches ou d'Exploitation accordées à deux ou plusieurs titulaires, il est procédé à l'unitisation du développement et de l'exploitation dudit gisement. A cet effet, un plan de développement conjoint du gisement découvert est présenté par les titulaires des Permis au Ministre chargé des Hydrocarbures pour approbation.'

[Free translation into English: '[i]n the event that the discovery of a commercially viable deposit extends over an area covered by two or more Exploration or Exploitation Permits granted to two or more holders, unitisation must be carried out for development and exploitation of the said deposit. To this end, a joint development plan for the deposit discovered is presented by the holders of the Permits to the Minister in charge of Hydrocarbons for approval.']

[15] See, in particular, in Algeria, Article 54(2) of Law No. 13-01 of 20 February 2013 Amending and Supplementing Law No. 05-07 of 28 April 2005 On Hydrocarbons:

'[d]ans le cas où les contractants ne s'accordent pas sur un plan d'unitisation, six (6) mois après la notification de l'agence nationale pour la valorisation des ressources en hydrocarbures (ALNAFT) à l'effet de préparer un plan d'unitisation, ou si l'agence nationale pour la valorisation des ressources en hydrocarbures (ALNAFT) n'approuve pas le plan d'unitisation soumis par les contractants, l'agence nationale pour la valorisation des ressources en hydrocarbures (ALNAFT) engage, à la charge des contractants, un expert indépendant, choisi conformément au contrat, pour établir un plan d'unitisation.'

[Free translation into English: '[i]n the event that the contracting parties have not reached an agreement on a unitisation plan six (6) months after notification by the national agency in charge of promoting hydrocarbon resources (ALNAFT) for the purpose of preparing a unitisation plan, or if ALNAFT does not approve the unitisation plan submitted by the contracting parties, ALNAFT commissions, at the expense of the contracting parties, an independent expert, chosen in accordance with the contract, to produce a unitisation plan.']

[16] See, in particular, in Guinea, Article 37(2) of Law No. L/2014/N°034/AN of 23 December 2014 Enacting the Petroleum Code:

'[à] défaut d'accord des Contractants sur ce plan ou à défaut d'approbation du Ministre, ce dernier engage un expert international pour établir un tel plan.'

[Free translation into English: '[s]hould the Contracting Parties fail to reach an agreement on this plan or if this plan is not approved by the Minister, the Minister commissions an international expert to establish such a plan'].

11.32 It is important to point out that national legislation in some countries lays down specific provisions that take into consideration the consequences of creating a joint development zone between the operators in terms of taxation.[17]

(b) The option to partner for the exploitation of natural resources

11.33 The decision to create a grouping between operators often results in the creation of a joint company; this decision is related to a desire to clarify and simplify relations between the operators.

2 MODIFICATIONS CONCERNING OPERATORS

11.34 During the life time of the project, the operators themselves may change; some of them may disappear or refuse to continue their efforts, others may want to take part in the project. These are the rules of the oil and mining game.

The ensuing modifications and transformations may or may not call the exploitation agreement into question.

a Reconsideration of the exploitation agreement

11.35 The exploitation agreement may be called into question, in particular by a joint decision of the operators to abandon operations or by the unilateral waiver by one of the operators holding a power that is decisive from a technological or financial point of view. We must differentiate between abandonment and waiver.

11.36 Abandonment by an operator of its contractual obligations does not fall within a contractual framework; it would therefore be a breach of the commitments made in the exploitation agreement. However, waiver falls within an assumption made and provided for by the parties at the time of signature of the exploitation agreement in a so-called exit clause.

[17] See, in particular, in Guinea, Article 37(3) of Law No. L/2014/N°034/AN of 23 December 2014 Enacting the Petroleum Code *'[d]ans les cas visés ci-dessus, le régime fiscal applicable et le partage de production sont déterminés à partir des paramètres de calcul applicables à chaque périmètre, au prorata des volumes d'Hydrocarbures contenus dans chaque périmètre'*. [Free translation into English: '[i]n the scenarios mentioned above, the applicable taxation system and the sharing of production are determined based on calculation variables that apply to each area, pro rata to the volumes of hydrocarbons contained in each area']. See also, in Algeria, Article 54(5) of Law No. 13-01 of 20 February 2013 Amending and Supplementing Law No. 05-07 of 28 April 2005 On Hydrocarbons: *'[l]orsque le gisement déclaré commercial s'étend sur deux ou plusieurs zones, le régime fiscal applicable est déterminé à partir des paramètres de calcul applicables à chaque zone, au prorata des volumes d'hydrocarbures contenus originellement dans chaque zone'*. [Free translation into English: '[i]f the deposit declared to be commercial extends over two or more zones, the applicable taxation system is determined based on the calculation variables applicable to each zone, pro rata to the volumes of hydrocarbons originally contained in each zone'].

11.37 Waiver is also widely accepted by mining and petroleum codes, which allow this possibility within the context of waiver of mining rights.[18]

11.38 On the one hand, we will distinguish the conditions for reconsideration of the exploration agreement and, on the other hand, the consequences of this decision.

i *Conditions for the reconsideration of the exploitation agreement*

11.39 There are a wide variety of conditions for reconsideration depending on the agreements and the circumstances at stake.

(1) Time-based conditions

11.40 There are two scenarios, depending on whether or not the unilateral waiver is subject to a time condition.

11.41 A time condition may apply for a unilateral waiver in that the reconsideration of the agreement will be forbidden before a certain period of time that should, generally, enable at least the preliminary exploration work to be performed. However, clauses are more commonly found in which no deadline is indicated; an operator who wishes to waive these rights should first have carried out the specific work. Indirectly, this provision also implies a temporary limit.

11.42 Conversely, some circumstances can result in the termination of the exploitation agreement without the obligation to keep to a certain amount of time between signature of the exploitation agreement and the unilateral waiver decision being applied; these circumstances may include the disappearance of the private operator following merger, absorption or bankruptcy.

(2) Material conditions

11.43 Of course, the waiver by one of the operators should not prevent others from continuing to exploit the deposit; this is why such a waiver is subject to several material conditions that primarily involve compliance with a notice period and communication of information.

11.44 In this way, one of the fundamental conditions is compliance by the operator with a notice period. Furthermore, the waiving operator must generally leave all of the documents and information it has gathered to the operator or operators continuing the exploitation. This obligation relates in particular to maps, geological and geophysical surveys and their interpretations.

ii *The consequences of the reconsideration of the exploitation agreement*

11.45 The consequences associated with the reconsideration of an exploitation agreement are mainly financial. We will differentiate between the abandonment and unilateral waiver.

---

[18] For operator waiver of the exploration permit and the exploitation permit, see paras 8.31 and 8.46 above.

(1) The consequences of abandonment

11.46 As we have already mentioned, abandonment is characterised by the breach of one of the waiver conditions as provided for in the exploitation agreement. Regulations rarely provide for abandonment.

The finding of the existence of abandonment is variable but most of the time comes from arbitral decisions or the use of a third-party expert.

In this case, no guidelines are given in the agreement, unless under the form of a penalty clause. In general, the amount of the penalty is equal to the amount of investment that should have been made throughout the whole exploitation cycle or just during one of its phases.

However, the contentious climate that prevails in an abandonment situation often leads to the seizure of the equipment and documents belonging to the operator in the host state.

In the majority of countries, if the operator has mining rights, it will be automatically deprived of the rights attached.

(2) The consequences of unilateral waiver

11.47 The provisions relating to the consequence of the unilateral waiver are widely developed in exploitation agreements. In this respect, two mechanisms that are generally applied as compensation for damages may be fixed or scalable.

(a) Fixed compensation for damages

11.48 In this case, regardless of when the waiver occurs, the financial consequences that apply to the waiving operator will be the same.

(b) Variable compensation for damages

11.49 If a variable compensation for damages is possible in the exploitation agreement, the solution will generally be as follows:

The closer the waiver to the exploitation agreement is from the term of the agreement, the lower the financial consequences of such a decision will be.

b Continuation of the exploitation agreement

11.50 Considering the long duration of exploitation agreements, change may occur during its term without resulting in a termination of the said agreement.

For instance, changes may occur due to the arrival of new operators, either without the corresponding departure of the initial operators, or by the replacement of some of them.

i *Increase in the number of operators*

11.51 An increase in the number of operators is in large part motivated by financial pressures.

The arrival of new operators, other than the affiliates of the initial operators, is in principle governed by the host state within the context of very strict authorisation procedures. Indeed, the entrance of a new operator must offer the host state complete assimilation of situations and guarantees of proper fulfilment of the exploitation agreement.

11.52 The change in the initial operators may result from a change of control in an operating company or an assignment of mining rights. There are far fewer provisions relating to the change of control than there are relating to assignment of mining rights,[19] but numbers seem to be growing. The definition of control is often tricky, especially when it is not envisaged by national legislation, as is most often the case.[20] If change of control of the operator is anticipated by regulations, the rule is that the said transfer must be notified to the competent authority for approval.[21]

---

[19] On the rules relating to the transfer of mining rights, see para 8.61 et seq above.

[20] It is, however, worth mentioning certain, generally recent, legislation that offers a definition of the concept of control. Thus, in Guinea, Article 107 of Law No. L/2014/N°034/AN of 23 December 2014 Enacting the Mining Code provides:

> '*[a]ux fins du présent Article, "contrôle" signifie le pouvoir de diriger et de dicter les décisions de la société par l'exercice des droits de vote attachés à la détention d'un nombre d'actions suffisant pour obtenir la majorité dans les assemblées générales d'actionnaires, et la détention par une personne, ou par un groupe de personnes agissant ensemble, de trente pour cent (30%) du capital social est présumée assurer le contrôle d'une société, s'il existe pas d'autres personnes, ou groupe de personnes agissant ensemble, détenant un pourcentage supérieur du capital social de la société en question*'.
>
> [Free translation into English: '[f]or the purposes of this Article, "control" means the power to direct and dictate the decisions of the company by exercising the voting rights attached to the holding of a large enough number of shares to obtain the majority in the general shareholders' meetings, and the holding by a person, or by a group of persons acting together, of thirty per cent (30%) of the share capital is assumed to ensure the control of a company, if there are no other persons, or groups of persons acting together, holding a higher percentage of the share capital of the company in question.']

See also, in Gabon, Article 5(14) of Law No. 011/2014 of 28 August 2014 Enacting Regulations in the Hydrocarbons Sector:

> '*[c]hangement de contrôle: toute modification dans le contrôle d'une personne morale, tel que défini par les dispositions de l'Acte Uniforme relatif au droit des sociétés commerciales et du groupement d'intérêt économique de l'OHADA*'.
>
> [Free translation into English: '[c]hange of control: any change in the control of a legal entity, as defined by the provisions of the Uniform Act relating to commercial company law and the economic interest group of OHADA.']

[21] In particular, in Guinea, Article 90 of Law No. L/2011/No.006/CNT of 9 September 2011 Enacting the Mining Code provides:

> '*[t]out changement de contrôle direct ou indirect de tout titulaire d'un intérêt dans un titre sera soumis à l'approbation ou à la validation du Ministre en charge des Mines. Toute acquisition directe ou indirecte, partielle ou cumulée égale ou supérieure à cinq pour cent (5%) du capital de la société détentrice du titre doit être soumise au Ministre en charge des Mines pour sa validation.*'
>
> [Free translation into English: "[a]ny change in the direct or indirect control of any holder of an interest in rights will be submitted to the Minister in charge of Mines for approval or validation. Any direct or indirect, partial or cumulative acquisition greater than or equal to five percent (5%) of the capital of the company holding the rights must be submitted to the Minister in charge of Mines for validation.']

And Article 107 of Law No. L/2014/N°034/AN Enacting the Petroleum Code:

Where this is provided for by regulations, the approval of the change of control may constitute a suspensive condition.[22] National legislation may also provide that any change of control done without prior approval from the competent authorities is null and void; such a situation is also likely to lead to forfeiture of the mining rights.[23]

11.53 Some regulations provide that change of control in favour of an affiliate or another operator are not subject to approval by the host state, although such an operation must be notified.[24] Furthermore, some legislation states that notification to the local authorities of a change of control is not required if the operation is the result of a transaction on a stock exchange.[25]

'*[l]es intérêts, droits et obligations au titre d'un contrat Pétrolier sont cessibles et transmissibles, en totalité ou en partie, sous réserve de l'approbation préalable du Ministre en charge des Hydrocarbures, dans les conditions prévues au Contrat Pétrolier.* [. . .] *Toute cession d'actions ou de parts d'une société constituant le Contrat ou d'une société contrôlant directement ou indirectement une société constituant le Contractant est assimilées à une cession d'intérêts aux fins du présent Article si elle résulte en un changement de contrôle* [. . .]'.

[Free translation into English: '[t]he interests, rights and obligations under an Oil Contract may be assigned and transmitted, in whole or in part, subject to the prior approval of the Minister in charge of Hydrocarbons, under the conditions laid down in the Oil Contract. [. . .] Any assignment of shares in a company constituting the Contracting Party or a company that directly or indirectly controls a company constituting the Contracting Party is treated as an assignment of interests for the purpose of this Article if it results from a change of control [. . .]'

[22] See in particular, in Niger, Article 50(2) of Law No. 2007-01 of 31 January 2007 Enacting the Petroleum Code: '[. . .] *l'approbation de l'opération constitue de droit une condition suspensive de la cession ou du changement de contrôle*'. [Free translation into English: '[. . .] the approval of the operation constitutes a suspensive condition of the assignment or change in control'].

[23] See, in particular, in Niger, Article 51 of Law No. 2007-01 of 31 January 2007 Enacting the Petroleum Code:

'*[t]out changement de contrôle ou cession réalisé en violation de l'article précédent est de nul effet et peut entraîner pour le titulaire le retrait du ou des permis ou autorisation(s) concerné(s) directement ou indirectement par la cession ou le changement de contrôle*'.

[Free translation into English: '[a]ny change in control or assignment that breaches the previous article is null and void and may result in the holder being stripped of the permits or licences affected directly or indirectly by the assignment or the change in control'].

[24] See, in particular, in Guinea, Article 107 of Law No. L/2014/No. 034/AN of 23 December 2014 Enacting the Petroleum Code.

[25] See, in particular, in Guinea, Article 107 of Law No. L/2014/No. 034/AN of 23 December 2014 Enacting the Petroleum Code:

'[t]*oute cession d'actions ou de parts d'une société constituant le Contractant ou d'une société contrôlant directement ou indirectement une société constituant le Contractant est assimilée à une cession d'intérêts aux fins du présent Article si elle résulte en un changement de contrôle, sauf si le changement de contrôle est le résultat direct d'une transaction sur une bourse officielle de valeurs*'.

[Free translation into English: '[a]ny assignment of shares in a company constituting the Contracting Party or a company that directly or indirectly controls a company constituting the Contracting Party is treated as an assignment of interests for the purpose of this Article if it results in a change in control, unless the change in control is the direct result of a transaction on an official stock exchange'].

11.54 Lastly, a change of control may have financial consequences, particularly when local legislation provides that such an operation is subject to registration duty.[26]

ii *Departure of the original operators: the assignments of mining rights*

11.55 Since the beginning of the 20th century, there have been frequent assignments of mining rights from private individuals to large companies.

An assignment of mining rights may intervene during the lifetime of a project, the departure of one of the original operators does not affect continuation of mining and oil operations.[27]

The departure of original operators may fall within the context of preferential rights. Most of the time, a separate contract, the 'farm-in/farm-out' contract, is signed.

(1) Preferential rights

11.56 Preferential rights allow an operator who want to continue operations to have a pre-emptive right over third parties to acquire the share of the outgoing operator.

11.57 Such provisions, which are generally included in the contractual framework agreed between the operators, will enable the operations to run smoothly despite the withdrawal of one of the initial stakeholders.

---

[26] See in particular, in Guinea, Article 91 of Law No. L/2013/N°053/CNT of 8 April 2013 Amending Law No. 2011/006/CNT of 9 September 2011 Enacting the Mining Code.

[27] See, in particular, in the Central African Republic, Article 20 of Ordinance No. 93.007 of 25 May 1993 Enacting the Petroleum Code:

> *'[l]orsque le permis est accordé à plusieurs titulaires conjoints, le retrait d'un ou plusieurs d'entre eux n'entraîne ni l'annulation du permis ni la résiliation du contrat pétrolier, si le ou les autres titulaires reprennent à leur compte les engagements qui avaient été souscrits pour ledit permis. Ce retrait est accepté par arrêté du Ministre'*.
>
> [Free translation into English: '[w]hen the permit is granted to several joint holders, the withdrawal of one or more of them does not result in the cancellation of the permit or the termination of the oil contract, if the other holders continue the commitments made for the said permit. This withdrawal is accepted by Ministerial Order'].

See also, in Cameroon, Article 20 of Law No. 99-013 of 22 December 1999 Enacting the Petroleum Code:

> *'[s]ans préjudice des dispositions de l'article 7 ci-dessus, lorsqu'un Contrat Pétrolier est conclu avec plusieurs co-Titulaires, le retrait d'un ou de plusieurs d'entre eux n'entraîne ni l'annulation des Autorisations dérivant du Contrat, ni la résiliation du Contrat, si le ou les autres co-Titulaires reprennent à leur compte les engagements qui avaient été souscrits pour ledit Contrat. Ce retrait est réputé accepté par le Ministre chargé des hydrocarbures'*.
>
> [Free translation into English: '[w]ithout prejudice to the provisions of Article 7 above, if an Oil Contract is signed with several joint holders, the withdrawal of one or more of them does not result in the cancellation of the Licences arising from the Contract, or the termination of the Contract, if the other joint holders continue the commitments made for the said Contract. This withdrawal is accepted by the Minister in Charge of Hydrocarbons'].

(2) *Farm-in and farm-out agreement*

11.58 Farm-in and farm-out agreements,[28] like many other concepts we have already studied, have their origins in the US private land system.[29]

11.59 From a strictly legal point of view, it is a contract involving the acquisition of an interest in a permit held by a third party. The holder of a working interest wants to assign all or some of its rights in order to share financing of the exploitation. For the holder of the mining rights (the farmor), this operation is referred to as a farm-out agreement, whereas for those taking all or some of the working interest (the farmee), it constitutes a farm-in agreement.

The farming involves the leasing for a fixed or indeterminate term of all or part of the rights attached to mining rights, in return for remuneration fixed by agreement between the farmor and the farmee.

11.60 The possibility offered to the operator to enter into a farm-in/farm-out agreement is generally provided for by mining and petroleum laws. However, this option is not permitted by all laws; indeed, regulations may provide that certain mining rights cannot be farmed out, in particular exploration mining rights[30] or even small-scale exploitation licences.[31]

Like any operation relating to the transfer of mineral interests, the principle is that farm-in/farm-out agreements must be authorised by the competent

---

[28] The concept of farming out is defined by the Mining Code of the Democratic Republic of Congo, which provides:

> '*[l]'amodiation consiste en un louage pour une durée fixe ou indéterminée, sans faculté de sous-louage, de tout ou partie des droits attachés à un droit minier ou une autorisation de carrières, moyennant une rémunération fixée par accord entre l'amodiant et l'amodiataire*'.
>
> [Free translation into English: '[f]arming involves the leasing for a fixed or indeterminate term, with no option to sublet, of all or part of the rights attached to mining rights or a quarry authorisation, in return for remuneration fixed by agreement between the farmor and the farmee.'] (Article 177 of Law No. 007-2002 as of 11 July 2002 Enacting the Mining Code).

[29] For a definition of the farm-out concept, see in particular, in Nigeria, Article 17(4) of the Petroleum Act: 'farm out means an agreement between the holder of an oil mining lease and a third party which permits the third party to explore, prospect, win, work and carry away any petroleum encountered in a specified area during the validity of the lease.'

[30] See, in particular, in the Democratic Republic of Congo, Article 177 of Law No. 007-2002 of 11 July 2002 Enacting the Mining Code: '*[l]es droits miniers et/ou de carrières de recherche ne peuvent pas faire l'objet d'amodiation*'. [Free translation into English: '[m]ineral and/or quarry exploration rights cannot be farmed out']. See, in particular, in Algeria, Article 8 of Law No. 01-10 of 3 July 2001 Enacting the Mining Code: '*[l]es titres miniers relatifs aux activités de recherche minière constituent des biens meubles, transmissibles et cessibles dans les conditions fixées par la présente loi et les dispositions du Code Civil et du Code de Commerce. Ils ne sont pas susceptibles d'amodiation ou de gage ou de nantissement*'. [Free translation into English: '[m]ineral rights relating to mining exploration activities constitute movable property that can be transmitted and assigned under the conditions laid down by this law and the provisions of the Civil Code and the Commercial Code. They are not liable for farming out or as a pledge or collateral'].

[31] See, in particular, in Niger, Article 45(2) of Ordinance No. 93-16 of 2 March 1993 Enacting the Amended Mining Law.

authorities in the host state.[32] Thus, regulations may provide that the farming will only take effect after it has been approved by the competent authority or even require insertion into the farm-in/farm-out agreement of a suspensive clause related to prior approval by the local authorities.[33]

11.61 In some scenarios, applicable regulations specify the content of the farm-in/farm-out agreement, in particular by laying down the clauses that it must contain. In particular, the farm-in/farm-out agreement may contain a termination clause in the event of a failure by the farmee to pay the required duties, taxes and royalties, etc.[34]

Similarly, legislation may establish the liability system that applies in the presence of a farm-in/farm-out agreement. In this way, it may be provided that the farm-in/farm-out agreement carries a joint and several liability for the farmor and the farmee with regard to the State.[35]

---

[32] See, in particular, in Morocco, Articles 9 and 28 of Law No. 21-90 of 1 April 1992 On the Exploration and Exploitation of Hydrocarbons Deposits, amended and supplemented by Law No. 27-99 of 15 February 2000:

'*[l]'amodiation, partielle ou totale, d'un permis de recherche ou d'une concession d'exploitation doit faire l'objet d'une autorisation préalable de l'administration*' and '*L'amodiation totale ou partielle d'une concession d'exploitation doit faire l'objet d'une autorisation préalable suivant la procédure prévue pour le cas de la cession d'une concession*'.

[Free translation into English: '[f]arming out, in whole or in part, of an exploration permit or an exploitation concession is subject to the prior approval of the authorities' and '[f]ull or partial farming of an exploitation concession is subject to the prior approval of the administration according to the procedure laid down in the event of assignment of a concession'].

[33] See, in particular, in Burundi, Article 25 Order No. 1/138 of 17 July 1976 Enacting the Mining and Petroleum Code.

[34] See in particular, in the Democratic Republic of Congo, Article 177 of Law No. 007-2002 of 11 July 2002 Enacting the Mining Code:

'*[t]out contrat d'amodiation doit comporter, sous peine de nullité, des clauses fixant les conditions d'entretien et de réinvestissement nécessaires à l'exploration et au développement raisonnables du gisement*'.

[Free translation into English: '[a]ny farm-in/farm-out agreement must include, under penalty of becoming null and void, clauses establishing the maintenance and reinvestment conditions necessary for reasonable exploration and development of the deposit'].

[35] See, in particular, in the Democratic Republic of Congo, Article 177 of Law No. 007-2002 of 11 July 2002 Enacting the Mining Code:

'*[t]out contrat d'amodiation comporte la responsabilité solidaire et indivisible de l'amodiant et de l'amodiataire vis à vis de l'État. L'amodiataire est, nonobstant toute clause contraire, redevable des impôts, taxes et redevances dus en vertu du titre minier ou de carrières. Toutefois, en cas de défaillance de l'amodiataire, l'amodiant est responsable vis à vis de l'État, sous réserve de son droit de recours contre l'amodiataire défaillant*'.

[Free translation into English: '[a]ny farm-in/farm-out agreement carries a joint and several liability for the farmor and the farmee with regard to the State. The farmee is, notwithstanding any clause to the contrary, liable for the duties, taxes and royalties payable by virtue of the mining or quarrying rights. However, in the event of failure by the farmee, the farmor is responsible with regard to the State, subject to its right to appeal against the defaulting farmee'].

12

MINING RIGHTS OVER NON-EXTRACTED RESOURCES

C Access to the extracted natural resource

12.1 The legal consequences resulting from the granting of mining rights or the allocation of economic mining rights may differ, in particular in terms of access to the natural resource extracted by the private operator.

Remember that there are disparities depending on whether the operator performs its activities based on mining rights or based on a contract granting economic mining rights.[1] The purely contractual nature of economic mining rights has, in practice, consequences in terms of ownership of the natural resource produced. Indeed, some exploitation agreements may lead to a complete and therefore definitive transfer of ownership of the production (or part of it) to the benefit of the private operator, which is not always the case.

The main principle is that the rights of the private operator will defer depending on whether the natural resource to which they apply has already been extracted or not.

12.2 Mining rights over a natural resource that has not yet been extracted (granted by way of a mining licence or an exploitation agreement) are based on two main principles: exclusivity and, generally, the right to set the production rate.

1 EXCLUSIVITY

12.3 The private operator enjoys exclusivity in the perimeter and over the natural resources referred to in the exploitation agreement[2] or in the mining title.[3]

---

[1] See above para 10.1 et seq above.

[2] See, in particular, in the Seychelles, Article 2(1) of the 2013 Seychelles Model Petroleum Agreement:

> '[i]n consideration of the covenants and agreements on the part of the Company hereinafter contained, the Government hereby grants to the Company the exclusive rights to conduct Petroleum Operations in the Agreement Area subject to the provisions of this Agreement.'

See also, in Algeria, Article 24 as Law No. 13-01 as 20 February 201 Amending and Supplementing Law No. 05-07 as 28 April 2005 On Hydrocarbons:

> *'[l]e contrat de recherche et d'exploitation confère au contractant le droit exclusif d'exercer dans le périmètre défini par ledit contrat: des activités de recherche; des activités d'exploitation, en cas de découverte déclarée commerciale par le contractant et après approbation, par l'agence nationale pour la valorisation des ressources en hydrocarbures (ALNAFT), du plan de développement relatif à ladite découverte. Le contrat d'exploitation relatif à un ou plusieurs gisement(s) déjà découvert(s) confère au contractant le droit exclusif d'exercer, dans le périmètre défini par ledit contrat, des*

The exclusivity granted to the operator precludes any exploitation of the deposit by a competitor and consequently assumes that no mining rights are granted for the area and the resources covered by the exploitation agreement or the mining title.

With regard to natural resources not covered by the mining rights, the host state is free to grant other mining rights insofar as these do not affect the first exploitation. In particular, the granting of mining rights can be envisaged for a hydrocarbon deposit when the area in question is covered by mining rights relating to a mineral resource and vice versa. This scenario presupposes that the works will not interfere with the exploration activities conducted by another operator on the same perimeter and may cause difficulties in practice.

2 THE RIGHT TO SET THE PRODUCTION RATE

12.4 Traditionally, only the private operator would determine the production level. The host state will be involved to a very small extent, in particular establishing production lower limits.

12.5 External factors can influence the production rate, such as the quota regulations imposed by intergovernmental organisations of which the host state is a member.

12.6 There are many issues surrounding the setting of the production rate and these highlight the potential conflicts of interest between the host state and the private operator. On the face of it, the interests of the parties to the exploitation agreement converge. Indeed, while the operator wishes to maximise the efficiency of its project, the host state wants to maximise the income resulting from the exploitation of its natural resources, which in practice assumes maximum production from the deposit. However, the State may have different interests in the long term and may in particular wish to implement a policy of rational management of its natural resources; this may in practice lead to a curbing of the production rate.

activités d'exploitation sur le ou les niveau(x) géologique(s) précisé(s) dans le plan de développement approuvé par l'agence nationale pour la valorisation des ressources en hydrocarbures (ALNAFT).'

[Free translation into English: '[t]he exploration and exploitation agreement grants the holder the exclusive right to perform in the area defined by the said agreement: exploration activities; exploitation activities, in the event of a discovery that is declared to be commercial by the holder and following approval, by the national agency in charge of promoting hydrocarbon resources (ALNAFT), of the development plan relating to the said discovery. The exploitation agreement for one or more deposits that have already been discovered grants the holder the exclusive right to perform, in the area defined by the said agreement, the exploitation activities on the geological level(s) indicated on the development plan approved by ALNAFT'].

[3] With regard to the exclusive nature of the rights granted under an exploitation or exploration permit, see nos. 240 and 259 above.

a Unilateral setting of the production rate

12.7 Depending on the countries and the exploitation agreements, the production rate can be established unilaterally by the local State operator or by the private operator.

i *The host state establishes thresholds that operators must respect*

12.8 The host state may intervene by imposing on the private operator minimum and maximum production thresholds that the private operator must respect.

The host state may decide to set maximum production thresholds. In this case, the fear is no longer of operators deliberately limiting production levels in order to affect international prices for natural resources, but rather for the State to limit uncontrolled exploitation that damages the possibilities of recovering natural resources in the medium and long term.

Indeed, although in the past maximum production was the main objective, new guidelines are now in place, ie the rapid depletion of reserves goes against local national interests.

Consistent with this approach, the host state may set aside some of the deposits from contractual operations, qualifying them as national reserves.

The State can unilaterally set production levels, in particular by way of Decrees. Accordingly, national legislation in some countries authorises the Minister in charge of Mines or Hydrocarbons to set, for reasons linked to national energy policy targets, production limitations.[4]

12.9 Besides, for various reasons associated with the nature of the mining rights granted, the host state can implement maximum production thresholds, in particular, in respect of artisanal mining or small-scale exploitation.

12.10 The host state can also intervene by setting minimum production targets. If these are not achieved, serious sanctions up to and including withdrawal of the mining rights or termination of the exploitation agreement can be applied.

---

[4] See, in particular, in Algeria, Article 50 of Law 13-01 of 20 February 2013 Amending and Supplementing Law No. 05-07 of 28 April 2005 On Hydrocarbons:

'*[p]our des raisons liées aux objectifs de la politique nationale énergétique, des limitations de production des gisements d'hydrocarbures liquides ainsi que l'approvisionnement en priorité du marché national en hydrocarbures liquides peuvent être appliqués. Ces limitations font l'objet d'une décision du ministre chargé des hydrocarbures qui fixe les quantités, la date d'intervention de ces limitations et leur durée. La répartition de ces limitations est appliquée, de manière équitable, par l'agence nationale pour la valorisation des ressources en hydrocarbures (ALNAFT) à l'ensemble des contractants, au prorata de leur production respective.*'

[Free translation into English: '[f]or reasons related to national energy policy targets, limitations on production of liquid hydrocarbon deposits and the priority supply of the national liquid hydrocarbons market may be applied. These limitations are covered by a decision by the Minister in charge of hydrocarbons who sets the quantities, the date of application of these limitations and their duration. The distribution of these limitations is applied, in a fair manner, by the National Agency in Charge of Promoting Hydrocarbon Resources (ALNAFT) to all contractors, pro rata to their respective production'].

ii *The private operator determines the production level unilaterally*

12.11 Whether or not mining rights are transferred, contractual unity is achieved on this point: the principle is that of unilateral setting by the private operator of the production level.

Generally, the operator must notify the host state of the production level it intends to implement.

12.12 The production rate may also result from the parties' negotiation when drafting the exploitation agreement. Several aspects are taken, such as the recoverable reserves and the developments expected in the international markets for the raw materials produced from the natural resources extracted. The exploitation agreement may also set maximum production limits, such as the Maximum Efficient Rate (MER), a concept specific to hydrocarbons, which can be supplemented by the calculation of a Maximum Production Rate (MPR) for each producing well.

b Setting of the production rate at intergovernmental level

12.13 The host state can limit the production rate because of an intergovernmental decision. OPEC takes decisions with a view to setting production quotas.

12.14 Setting production quotas within OPEC assumes a unanimous decision of its members with which the states should comply.[5]

12.15 In practice, the introduction of quotas assumes the existence of clauses in exploitation agreements that enable the State to apply the decision taken at intergovernmental level without it leading to potential contestation from private operators that could see it as a change to the original contractual balance.[6]

The setting of quotas within the context of OPEC is still a topical issue for consideration in the general context of global reduction in hydrocarbon prices.

[5] See Gilles Darmois 'Contingentements de l'OPEP et équilibre économique des contrats' in *Le contentieux extractif* (ICC 2015) 74.

[6] Id.

13

MINING RIGHTS OVER THE EXTRACTED RESOURCE

13.1 With regard to the extracted natural resources, a distinction must be made depending on whether the rights arise from a mining title or an exploitation agreement.

If the operator's rights arise from a mining title, the extracted natural resources belong to it,[1] which is not always the case in respect of economic mining rights. Indeed, we must remember that even if the private operator owns part of the production, it does so by virtue of an indirect contractual right and not a direct right attached to the holding of a mining title. Thus, the private operator who benefits from economic mining rights will own the natural resource extracted only if this right is expressly envisaged by legal[2] or contractual provisions.

We will consider the rights of the private operator associated with the sale, evacuation, transformation and use of the extracted products.

[1] See, in particular, to that effect, in Benin, Article 12 of Law No. 2006-17 of 17 October 2006 Enacting the Mining and Mining Taxation Code:

'[l]a propriété des mines et carrières est distincte de la propriété du sol. Les mines et carrières appartiennent à l'État et constituent un domaine public particulier dont la gestion est régie par la présente loi. Les gîtes naturels de substances minières contenues dans le sous-sol ou existant en surface sont, sur le territoire de la République du Bénin, la propriété de l'État et ne peuvent être, sous réserve de la présente loi, susceptibles d'aucune forme d'appropriation privée. Toutefois, le titulaire de titres miniers d'exploitation acquiert la propriété des substances minérales qu'il extrait.'

[Free translation into English: '[t]he property of the mines and quarries is different from the land ownership. The mines and quarries belong to the State and constitute a specific public domain, management of which is governed by this Law. Pursuant to this Law, the natural deposits of mineral substances contained in the subsoil or on the surface are, in the Republic of Benin, the property of the State and cannot be liable to any form of private appropriation. However, the holding of exploitation mining rights gets ownership of the mineral substances it extracts'].

[2] Thus, in Algeria, Article 25 of Law 13-01 of 20 February 2013 Amending and Supplementing Law No. 05-07 of 28 April 2005 On Hydrocarbons: *'[l]es hydrocarbures extraits, dans le cadre d'un contrat de recherche et/ou d'exploitation, sont propriété du contractant au point de mesure et soumis à une redevance selon les termes et conditions établis par ledit contrat.'* [Free translation into English: '[t]he hydrocarbons extracted, within the context of an exploration or exploitation agreement, belong to the contractor at the measurement point and are subject to a royalty in accordance with the terms and conditions laid down in the said agreement'].

1 MARKETING OF EXTRACTED PRODUCTS

13.2 Marketing of the extracted products inevitably involves the issue of the selling price.

13.3 Nowadays, the price of many raw materials is established on the financial markets, which is not without its difficulties, especially given the volatility of international prices.

13.4 With regard to hydrocarbons, concepts that have now disappeared must still be reported. The 1972 New York agreement between OPEC member states and private operators is at the origin of these particularities. Temporarily, private operators had to market the share of production that now goes to the local states. To this end, three categories of crude oil were created:[3]

- 'bridging oil': Operating companies were committed to market 75% of the local state's share in the first year, then 50% in the second, and lastly 25% in the third; this crude had to enable them to keep their previous commitments;
- 'free crude': It represented the share of production going to the local state and that it was supposed to market itself;
- 'phase in crude': While the producing states were implementing the national means to market the production themselves, the operating companies had an obligation to buy a decreasing share of the free crude.

13.5 Besides the question of prices, the private operator, or the host state as the case may be, gets a right to purchase part of the production.

13.6 Furthermore, the private operator will generally be bound by an obligation to satisfy local demand as a priority before exporting.[4] These provisions, introduced to the benefit of the host state, also often represent an advantage for the private operator because of the market opportunities that are then offered.

13.7 Some legal or, more commonly, contractual provisions may provide the private operator or the host state with a right to purchase part of the production.

13.8 When it is the private operator that has this right, it is usually presented as part of the operator's remuneration and can therefore be related to a benefit in kind. This assumption concerns scenarios in which the private operator holds economic mining rights and is not therefore automatically granted ownership of the natural resource extracted.

The right to purchase enables, for a given period, part of the extracted products to be bought, often at a preferential price. This right can either be optional or mandatory.

13.9 With economic mining rights, and particularly in contractual agreements, the private operator may benefit from an option to purchase, during the production period, enabling it to buy a percentage of the

---

[3] These provisions are given in Article 5 of the 1972 New York Agreement Between OPEC Member states.

[4] On the obligation to prioritise supply of the local market, see para 23.48 et seq below.

hydrocarbons extracted, which varies depending on the location of the deposit, generally at a Free on Board (FOB) price.[5]

13.10 Exercise of the option can also work in favour of the local State operator or the host state when ownership of the natural resource extracted reverts to the private operator, which is automatically the case with a mining title.

13.11 In this way, in Guinea, the Mining Code offers the host state the right to purchase and market some of the production up to its participation in the mining project. In this case, the host state will not benefit from preferential purchasing conditions as this right will be exercised under conditions equivalent to those offered by other buyers. The other shareholders of the operating company also have a pre-emptive right over the mineral sold by the State to third parties.[6]

Alongside this right to purchase, the Guinea Mining Code also establishes a pre-emptive right that the host state may exercise over the raw or transformed mineral substances produced by the holders of mining rights when transactions take place within the context of a non-competitive market. This pre-emptive right is limited and cannot cover more than 50% of the production.[7]

---

[5] Free on Board.

[6] See, in Guinea, Article 138 of Law No. L/2013/N°053/CNT of 8 April 2013 Amending Law No. 2011/006/CNT of 9 September 2011 Enacting the Mining Code:

'*[l]'État ou toute entité agissant en son nom se réserve le droit d'acheter et de commercialiser une quantité de la production du titulaire d'un Titre d'exploitation minière à hauteur de sa participation, pour toute offre de prix supérieure au prix FOB en cours. L'exercice de ce droit est notifié par écrit au plus tard à la fin du premier trimestre de l'année en cours, pour la production de l'exercice suivant, ou lors de la conclusion des contrats de vente à long terme de la société titulaire d'un Titre d'exploitation minière. Ce droit est exercé dans des conditions au moins équivalentes à celles offertes par les autres acheteurs. Il ne peut remettre en cause les dispositions des contrats de vente du minerai en cours de validité et ne peut porter sur une quantité supérieure à la part correspondant à la participation de l'État dans la société titulaire du Titre d'exploitation minière. Les autres actionnaires de la société titulaire du Titre d'exploitation minière bénéficient d'un droit de préemption sur le minerai vendu par l'État à des tiers.*'

[Free translation into English: '[t]he State or any entity acting on its behalf reserves the right to purchase and market a quantity of the production of the holder of a mining right up to its participation, for any price offer above the current FOB price. Exercise of this right is notified in writing no later than the end of the first quarter of the current year, for the following year's production, or upon signature of the long-term contracts of sale of the company that holds the mining rights. This right is exercised under conditions at least equivalent to those offered by other buyers. It may not call into question the provisions of the contracts of mineral sale in force and may not relate to a quantity greater than the share corresponding to the State's participation in the company that holds the mining rights. The other shareholders in the company that holds the mining rights have a pre-emptive right over the mineral sold by the State to third parties'].

[7] See on this issue, Article 138-II of Law No. L/2013/No. 053/CNT of 8 April 2013 Amending Law No. 2011/006/CNT of 9 September 2011 Enacting the Mining Code.

13.12 Lastly, some national legislation offers the host state a priority entitlement to purchase a share of the production in order to meet the needs of the national market.[8]

a Obligation to prioritise local demand

13.13 In certain circumstances, the production belonging to the private operator cannot be marketed freely. Sale is often dependent on prioritising the meeting of local needs.[9] Such a provision is generally popular with international oil and mining companies as they benefit from a preference for the sale of their production, with the local authorities almost always getting involved in that respect. However, it may be otherwise if the private operator has to meet local demand by selling a share of the production at a price lower than the international market price.

13.14 Less commonly, the private operator may have an obligation to keep stocks in the host state to satisfy the local market or contribute to a national political and economic strategy.

2 EVACUATION OF THE EXTRACTED PRODUCTS

13.15 Two issues must be resolved with regard to the evacuation of production: first of all, that of the actual content of the private operator's right of ownership, then that of its export.

a Ownership of production reverting to the private operator

13.16 The transmission to the private operator of a right of ownership over all or part of the production should imply the right to move it freely both within and outside the host state territory, but this is not always the case.

Even if the host state and the local State operator recognise the right to freely export of portion of production belonging to the private operator, it

---

[8] See, in particular, in Tunisia, Article 50(1) of Law No. 99-93 as 17 August 1999 Enacting the Hydrocarbons Code:

> '*[p]our couvrir les besoins de la consommation intérieure tunisienne, l'Autorité Concédante a le droit d'acheter, en priorité, une part de la production des Hydrocarbures Liquides extraits par le Titulaire, ou pour son compte, de ses Concessions en Tunisie. Les quantités destinées au marché local au titre de cet achat sont calculées au prorata des quantités produites par chaque Concession jusqu'à concurrence de vingt pour cent (20%). Le prix à pratiquer pour ces ventes est le prix de vente normal FOB obtenu par le Titulaire à l'occasion de ses ventes à l'exportation diminué de dix pour cent (10%).*'
>
> [Free translation into English: '[t]o cover Tunisia's domestic consumption needs, the Licensing Authority is entitled to purchase, as a priority, a share of the production of the Liquid Hydrocarbons extracted by the Holder, or on its behalf, of its Concessions in Tunisia. The quantities intended for the local market under this purchase are calculated pro rata to the quantities produced by each Concession up to twenty per cent (20%). The price used for these sales is the normal FOB price obtained by the Holder for its export sales less ten per cent (10%)'].

[9] On the obligation to prioritise supply of the local market, see paras 23.48 et seq below.

may have an obligation to notify the local authorities of the destination of the extracted resources.

If such operation is contrary to the political rights and interests of the host state, the private operator must comply with any restrictions notified to it regarding the destination.

b Transportation of the production

13.17 The transportation of the natural resources remains a very sensitive stage of exploitation operations. The lack of infrastructure constitutes a challenge both for the private operators and the State as their development is a key element of the development of the local economy.[10]

i *The right to transport production*

13.18 The issue of transport of the extracted resources by the private operator that holds mining rights is widely developed by mining and petroleum regulations.

13.19 The general rule is that the holder of mining rights over the extracted resources benefit from a right to transport the production inside and outside their perimeter. Besides, the local authorities will support the private operator when route authorisations need to be obtained and to resolve possible easement issues.

However, it should be stressed that the holder of mining rights may have to obtain special permission from the competent authorities in order to perform transport operations.[11]

13.20 National legislation sometimes anticipates the possibility of several holders joining forces to transport the extracted products of their operations.[12] This option to pool transport facilities can sometimes become an

---

[10] On the issue of the construction of transportation infrastructure see below para 24.14 et seq.

[11] See, in particular, to that effect, in Benin, Article 36 of Decree No. 2008-804 of 31 December 2008 Implementing the Mining Code and Mining Taxation in the Republic of Benin:

> *'[l]e droit de traiter, de transporter, de transformer, et de faire le commerce de substances minières ou de carrières est soumis à une autorisation particulière délivrée par arrêté du Ministre chargé des mines.'*
>
> [Free translation into English: '[t]he right to process, transport, transform, and trade mineral or quarry substances is subject to special permission granted by Order of the Minister in charge of mines'].

[12] See, in particular, in Cameroon, Article 48(1) of Law No. 99-013 of 22 December 1999 Enacting the Petroleum Code:

> *'[p]lusieurs Titulaires peuvent s'associer pour assurer le Transport des produits extraits de leurs Exploitations. Ils peuvent également s'associer avec des tiers qualifiés et l'État, soit directement, soit par l'intermédiaire d'un établissement ou organisme public dûment mandaté à cet effet, pour la réalisation et l'exploitation des canalisations et installations.'*
>
> [Free translation into English: '[s]everal Holders may join forces to transport the extracted products of their Operations. They may also join forces with qualified third parties and the State, either directly or through a public establishment or organisation duly appointed for this purpose, to construct and operate pipes and installations.'].

obligation when it comes to the construction by the joint holders of mining rights of a transport infrastructure.[13]

ii *The right to have production transported*

13.21 Some agreements, while acknowledging that the private operator has a right to transport its share of production using installations it owns, offer the possibility to call upon third parties to transport the products extracted. This principle is also reiterated in national legislation for the benefit of the holders of mining rights.[14]

13.22 The holder of mining rights may, in accordance with the application regulation, use the existing transport infrastructure up to the limit of the excess transport capacity. The company operating these infrastructures may be under an obligation to accept the transport of products from other operations.[15] Such use may give rise to payment by the operator of financial

---

See also, in Mauritania, Article 61 of Law No. 2010-033 of 20 July 2010 Enacting the Hydrocarbons Code amended by Law No. 2011-044 of 25 November 2011.

[13] See in particular, in Tunisia, Article 77 of Law No. 99-93 of 17 August 1999 Enacting the Hydrocarbons Code:

> '*[l]es Co-Titulaires d'une Concession d'Exploitation doivent s'associer entre eux pour assurer en commun le transport des Hydrocarbures extraits de la même Concession.*'
>
> [Free translation into English: '[t]he Joint Holders of an Exploitation Concession must join forces to transport the extracted Hydrocarbons of that Concession'].

[14] See, in particular, in Niger, Article 108 of the amended Ordinance No. 93-16 of 2 March 1993:

> '*[l]e titulaire d'un titre d'exploitation peut pendant la période de validité de son titre et dans les six mois qui suivent son expiration transporter ou faire transporter, les produits d'exploitation jusqu'au lieu de stockage, de traitement ou de chargement.*'
>
> [Free translation into English: '[t]he holder of an exploitation right can, during the period of validity of its right and within the six months following expiry thereof, transport the exploitation products or have them transported to the place of storage, processing or loading'].

See also, in Guinea, Article 137(1) of Law No. L/2013/No. 053/CNT of 8 April 2013 Amending Law No. 2011/006/CNT of 9 September 2011 Enacting the Mining Code:

> '*[l]e titulaire d'un Titre minier ou d'une Autorisation peut, pendant la durée de validité de ce Titre et les six mois qui suivent, transporter ou faire transporter les produits de l'exploitation qui lui appartiennent jusqu'aux lieux de stockage, de traitement et de chargement.*'
>
> [Free translation into English: '[t]he holder of mining rights or a licence can, during the period of validity of these rights and for six months afterwards, transport the extracted products belonging to it or have them transported to the place of storage, processing or loading'].

[15] See, in particular, in Benin, Article 46 of Law No. 2006-18 of 17 October 2006 Enacting the Petroleum Code:

> '*[l]'entreprise assurant l'exploitation d'une canalisation de transport peut, à défaut d'accord amiable, être tenue par décret pris en Conseil des ministres, d'accepter, dans la limite et pour la durée de sa capacité excédentaire, le passage des produits provenant d'autres exploitations productrices d'hydrocarbures. Les tarifs de transport sont soumis à homologation par le ministre chargé des hydrocarbures, ils ne peuvent en aucune manière être discriminatoires.*'
>
> [Free translation into English: '[t]he company operating a pipeline can, in the absence of an amicable agreement, be obligated by a Decree of the Council of Ministers to accept,

compensation to the owner of the infrastructure.[16]

3 PROCESSING OF THE EXTRACTED PRODUCTS

13.23 In principle, the holders of production mining rights may transform, in their own installations, the resources extracted or have them processed in other transformation installations in the national territory.[17] Besides, mining laws may provide that the holders of mining rights can establish packaging, processing and refining installations in the national territory.[18]

---

within the limit and for the duration of its surplus capacity, the transport of products from other hydrocarbon-producing exploitations. The transport prices are subject to approval by the Minister in charge of hydrocarbons and may not be discriminatory in any way'].

[16] See, in particular, in Algeria, Article 79 of Law No. 05-07 of 28 April 2005 On Hydrocarbons:

'[t]oute personne a le droit d'utiliser les infrastructures de transport par canalisation et de stockage des produits pétroliers sur la base du principe de libre accès des tiers moyennant le paiement d'un tarif non discriminatoire.'

[Free translation into English: '[a]ny person is entitled to use the oil pipeline and storage infrastructures based on the principle of free access by third parties in return for payment of a non-discriminatory price'].

[17] See, in particular, in Benin, Article 123 of Law No. 2006-18 of 17 October 2006 Enacting the Petroleum Code:

'[l]es titulaires des permis d'exploitation des hydrocarbures peuvent raffiner, dans leurs propres installations montées sur le territoire de la République du Bénin, les bruts extraits de leurs gisements ou les faire traiter dans d'autres raffineries existantes dans le pays.'

[Free translation into English: '[t]he holders of hydrocarbon exploitation permits may refine, in their own installations established in the territory of the Republic of Benin, the crudes extracted from their deposits, or have them processed in other refineries that exist in the country'].

[18] See, in particular, in Central African Republic, Article 38 of Law No. 09-005 of 29 April 2009 Enacting the Mining Code:

'[l]e permis d'exploitation industrielle comporte également l'autorisation d'établir en République Centrafricaine, des installations d'extraction et de conditionnement, de traitement, de raffinage, d'affinage et de transformation de substances minérales.'

[Free translation into English: '[t]he industrial exploitation permit also includes permission to set up installations in Central African Republic for the extraction and packaging, processing, refining and transformation of mineral substances'].

See also, in Niger, Article 109 of the amended Ordinance No. 93-16 of 2 March 1993 Enacting the Mining Law:

'[l]e titulaire d'un titre d'exploitation peut, conformément à la réglementation en vigueur, établir au Niger des installations de conditionnement, traitement, raffinage, et transformation de substances minières ou de carrière, y compris l'élaboration de métaux et alliages, de concentrés ou dérivés primaires de ces substances minières.'

[Free translation into English: '[t]he holder of exploitation rights may set up in Niger installations for packaging, processing, refining and transforming mineral or quarry substances, including the production of metals and alloys, and concentrates or primary derivatives of these mineral substances'].

3.24 The development of integrated projects in the territory of the host state represents a huge challenge for the latter. An integrated project, ranging from exploration to transformation, enables the State to maximise the profits generated by the exploitation of its natural resources.[19] Transformation *in situ* of the extracted natural resource can also represent a considerable advantage for the private operator, especially when production takes place in a landlocked country, which involves high transportation costs. However, transformation *in situ* of the extracted resources remains a challenge for many African states that do not have adequate infrastructure.

In order to promote the development of national economy and to encourage integrated projects, national legislation can sometimes limit the right to export the extracted natural resources by introducing the condition of being under an impossibility to process the natural resources in the national territory at an economically viable cost.[20]

13.25 When the production belongs to the private operator, the local authorities can define the place of transformation by only authorising its export in the form of raw materials. Similarly, the host state can urge the private operator to set up its transformation installations in the national territory.[21] Lastly, private operators may be obliged to supply the transformation units in the national territory as a priority.[22]

[19] For the benefits to local economy see below para 24.13 et seq.

[20] In this way, in the Democratic Republic of Congo, Article 85 of Law No. 2007-2002 of 11 July 2002 Enacting the Mining Code provides:

'*[s]ous réserve des dispositions de l'alinéa suivant, la commercialisation des produits miniers qui proviennent des Périmètres d'exploitation est libre. Le titulaire d'un Permis d'Exploitation peut vendre ses produits aux clients de son choix à des prix librement négociés. Toutefois, l'autorisation du Ministre est requise pour l'exportation des minerais à l'état brut pour traitement à l'extérieur du Territoire National. Cette autorisation ne sera accordée que si le titulaire qui la demande démontre à la fois: a) l'inexistence d'une possibilité de traitement dans le Territoire National à un coût économiquement rentable pour le projet minier; b) les avantages pour la République Démocratique du Congo au cas où l'autorisation d'exportation est accordée.*'

[Free translation into English: '[s]ubject to the provisions of the previous paragraph, mining products originating from the exploitation areas can be marketed freely. The holder of an Exploitation Permit can sell its products to the customers of its choice at freely negotiated prices. However, the Minister's authorisation is required for the export of minerals in their raw state for processing outside of the National Territory. This authorisation will only be granted if the requesting holder demonstrates that: a) there is no possibility for processing in the National Territory at a price that is economically viable for the mining project; b) the advantages for the Democratic Republic of Congo in the event that export authorisation is granted'].

[21] See, in particular, in Guinea, Article 139(1) of Law No. L/2013/N°053/CNT of 8 April 2013 Amending Law No. 2011/006/CNT of 9 September 2011 Enacting the Mining Code:

'*[l]e titulaire d'un Titre d'exploitation minière, ou tout autre investisseur guinéen ou étranger est exhorté à établir en République de Guinée des installations de conditionnement, de traitement, de raffinage et de transformation de Substances minières ou de carrières, y compris l'élaboration de métaux et alliages, de concentrés ou dérivés primaires de ces Substances minières, conformément à la réglementation en vigueur.*'

4 USE OF THE EXTRACTED PRODUCTS IN THE PRODUCTION PROCESS

13.26 The reasonable use by the private operator of the extracted products in the production process is permitted where the performance of the exploitation operations justifies this. Indeed, the performance of oil operations sometimes requires that part of the production be set aside for self-consumption, or even that the associated gases be re-injected for enhanced recovery of liquid hydrocarbons.[23]

Often, the private operator will have to provide accounts or measure the share of the production used within the production process[24] or lost, as long as this loss remains reasonable.

13.27 The associated gas in oil deposits has become a delicate issue.[25]

[Free translation into English: '[t]he holder of mining rights, or any other Guinean or foreign investor is urged to set up in the Republic of Guinea installations for packaging, processing, refining and transforming mineral or quarry substances, including the production of metals and alloys, and concentrates or primary derivatives of these mineral substances, in accordance with regulations in force'].

22 See, in particular, in Guinea, Article 139(2) of Law No. L/2013/N°053/CNT of 8 April 2013 Amending Law No. 2011/006/CNT of 9 September 2011 Enacting the Mining Code:

'*[l]es sociétés exploitant du minerai brut en République de Guinée sont tenues d'approvisionner en priorité les unités de transformation installées sur le territoire national. Les modalités de participation individuelle à cet approvisionnement font l'objet d'un arrêté du Ministre en charge des Mines sur avis conforme du Conseil des Ministres*'.

[Free translation into English: '[t]he companies exploiting the crude mineral in the Republic of Guinea must prioritise supply of the transformation units set up in the national territory. The conditions for individual participation in this supply are laid down in an Order of the Minister in charge of Mines acting on the advice of the Council of Ministers'].

23 In this way, in the Democratic Republic of Congo, Article 92(1) of Law No. 15/012 of 1 August 2015 Enacting the General Regime for Hydrocarbons provides:

'*[s]ans préjudice des dispositions de la loi relative à la protection de l'environnement, les gaz naturels associés produits des blocs d'exploitation peut, suivant les exigences de l'exploitation des gisements d'hydrocarbures liquides approuvées, par bloc d'exploitation, par le Ministre ayant les Hydrocarbures dans ses attributions, être affectés à l'autoconsommation liée aux opérations pétrolières, à la réinjection visant à améliorer la récupération des hydrocarbures liquides et à la consommation nationale ou à l'exportation*'.

[Free translation into English: '[w]ithout prejudice to the provisions of the Law on Environmental Protection, the associated natural gases produced by the exploitation blocks can, depending on the exploitation requirements of the liquid hydrocarbon deposits approved, for each exploitation block, by the Minister responsible for Hydrocarbons, be set aside for self-consumption in connection with the petroleum operations, for re-injection for enhanced recovery of liquid hydrocarbons and for national consumption or export'].

24 This concerns, in particular, enhanced recovery extraction scenarios.

25 Associated gas is the natural gas extracted at the same time as the crude oil. It is defined by some legislation as the '*gaz dissout dans le pétrole suivant les conditions de gisement et qui est séparé en surface dans les installations de traitement.*' [Free translation into English: 'gas dissolved in the oil depending on the deposit conditions and that is separated

Indeed, for a long time, it was often more profitable to flare associated gas rather than construct an infrastructure to recover it.

These questions have been a challenge for several decades. In April 1985, Gulf Oil was forced to pay a one million dollar fine by the Nigerian authorities, by virtue of a decree in force since 1 January 1985, for flaring the associated gas during extraction of crude oil.

In some agreements, the private operator must only exploit the associated gas if it considers it to be economically viable, and if it cannot be injected to perform enhanced recovery operations; otherwise, it is entitled to flare it. Nonetheless, the local State operator may then exploit the associated gas under sole risk, namely by bearing all of the costs itself.[26]

13.28 National petroleum laws may also include provisions relating to the use of the natural gas associated with hydrocarbon deposits. Under national legislation, gas flaring is in principle forbidden.[27] However, some legislation provides that, exceptionally, flaring authorisation may be granted by the petroleum authorities for a limited time and in return for payment of a special tax.[28]

Today, the development of technological capabilities has significantly reduced the problems associated with the storage and transport of the gas associated with hydrocarbon deposits.

---

on the surface in the processing installations'] (Article 5(31) of Law No. 011/2014 of 28 August 2014 Regulating the Hydrocarbons Sector in Gabon).

[26] On the sole-risk mechanism, see paras 19.110 et seq below.

[27] See, in particular, in Gabon, Article 129 of Law No. 011/2014 of 28 August 2014 Enacting the Hydrocarbons Code:

> '*[l]e torchage de gaz naturel est interdit en République Gabonaise. Toutefois, le Ministre chargé des Hydrocarbures fixe pour chaque champ, un seuil de gaz fatal, après examen des propositions dûment justifiées du contracteur et avis technique des services compétents. Ce seuil est révisé périodiquement. Toute quantité de gaz brûlée au-dessus du seuil du gaz fatal est passible de pénalités fixées par voie règlementaire.*'
> [Free translation into English: '[t]he flaring of natural gas is forbidden in the Republic of Gabon. However, the Minister in charge of Hydrocarbons sets a fatal gas threshold for each field after examining the duly reasoned proposals of the contractor and technical opinion of the competent departments. This threshold is reviewed regularly. Any quantity of gas burned above the fatal gas threshold is subject to penalties set by regulations'].

[28] See, in particular, in Algeria, Article 52 of Law No. 13-01 of 20 February 2013, Amending and Supplementing Law No. 05-07 of 28 April 2005 On Hydrocarbons:

> '*[l]e torchage de gaz naturel est interdit en République Gabonaise. Toutefois, le Ministre chargé des Hydrocarbures fixe pour chaque champ, un seuil de gaz fatal, après examen des propositions dûment justifiées du contracteur et avis technique des services compétents. Ce seuil est révisé périodiquement. Toute quantité de gaz brûlée au-dessus du seuil du gaz fatal est passible de pénalités fixées par voie règlementaire.*'
> [Free translation into English: '[g]as flaring is prohibited. However, exceptionally and for limited periods, the Minister of the Hydrocarbons Regulatory Authority can grant flaring authorisation upon request from the operator. The conditions for granting this exceptional authorisation and the permissible thresholds are laid down by regulations'].

14

TAXATION

D Financial issues

14.1 The amounts invested in and generated by extractive activities are significant, such that the financial challenges associated with mining and oil activities demand the utmost attention.

The profits generated by the extractive industries are correlated with the sums invested within the context of mining and oil activities. The size of the investments usually forces operators to seek external capital. Despite the profits generated, mining and oil projects can experience certain financing difficulties, in particular associated with the extremely volatile nature of raw material prices on international markets.

The host state, in return for the rights granted to private operators, intends to benefit from an increasing share of the profits generated by the mining and oil operations in its territory; in the mining sector essentially, besides taxation issues, the benefits for the host states also includes an equity participation that is generally free and non-contributory.

14.2 Within the context of extractive projects, the host state uses two methods for increasing its financial revenue: taxation and equity participation. In this way, it can get involved as a partner of the private operator through equity participation or intervenes as a public authority in tax matters.[1]

14.3 Initially, taxation within the host state was either non-existent or insignificant. This discrepancy, which was one of the reasons leading to nationalisations in the oil and mining industries in the 1950s and 1960s, has been reabsorbed by the constant development of local taxation.

Taxation provides the host state with a quick and flexible way to increase or decrease its revenue. This facility is all the more unfair as it can affect the private operator at any time during the project, and in particular before commercial quantities of a natural resource have even been discovered, therefore threatening the profitability of a project. Taxation is a tool for the host state that can hinder or enhance activity in the extractive sector.

States with reserves of natural resources have to weigh up the desire to appeal to foreign investors, by way of attractive taxation, against the need to obtain a right share of the mining income.[2] This delicate balance may be

---

[1] However, it should be noted that one of the main sources of revenue comes from the host state being awarded a share of the production.

[2] The concept of mining income means 'the difference between the revenue generated by the mining activity and the costs incurred by the activity, which include 'normal' remuneration of the capital factor. In terms of the micro-economy, there is therefore a surplus that can be taxed at 100% rate without affecting use of the resource, ie without

appreciated differently by each country depending on its mining history and the stability of its political regime.[3]

14.4 A general trend in terms of taxation, which is not specific to the extractive industry, is the great tax burden on investors on the African continent. The tax burden is, however, decreasing such that, in 2015, Africa no longer had the highest taxation rates[4] . Taxes still vary greatly from state to state, particularly within French-speaking Africa. For instance, in Gabon in 2015, the total tax rate was 40,6% compared with 73,3% in the Central African Republic.[5] In general, the tax burden is still not as high as in the English-speaking African countries.[6] An increase in the tax burden is particularly feared in the strategic sectors such as mining and oil activities. Frequent changes to mining and oil legislation on the African continent often give rise to fears of an increase in taxes in favour of the host state that may lead to a drop in investment in the extractive sector.[7]

14.5 The instability of the tax environment on the African continent should not be underestimated by private operators. Despite the many changes to tax rules, in particular those that apply to the extractive industry, foreign investors are often faced with a lack of access to, or even the non-existence of, administrative doctrine on the subject of taxation. Moreover, the lack of training of tax authorities on the African continent can lead to various difficulties, particularly within the context of tax inspections commissioned by the states' authorities which are not intended to reassure foreign investors, especially in the mining and oil sectors.[8] For greater security, investors can contemplate using tax rescripts.[9] Adequate training for the competent authorities is also expected to enhance legal security on the subject.

14.6 Reinforcement of inter-African cooperation, especially within the context of the *African Tax Administration Forum* (ATAF), as well as regional harmonisation of tax rules could also help to partially resolve these difficulties.

The application of regional rules is particularly noteworthy in terms of

affecting the investor's choice and with no economic distortions, hence the interest as a source of revenue for governments.' See Charlet, Laporte and Rota-Graziosi op cit 14.

[3] Id.

[4] For more information, see *World Bank Paying Taxes 2015* (2015) at 134 [online].

[5] For more information, see on this subject World Bank op cit 2.

[6] Id.

[7] In this way, the *Fédération des Entreprises du Congo* recently alerted the Government about the risk generated by the enactment of a new mining code that would result in an increase in taxes. See, to that effect, *Africa Mining Intelligence* 'Congo-K Fiscalité: la FEC cite la Zambie en contre-exemple' *African Mining Intelligence* (26 May 2015) at 1.

[8] In 2014, the tax adjustment of the Gabon subsidiary of the French oil company Total was worth approximately $805 million.

[9] The tax rescript enables an investor to obtain an interpretation regarding the meaning and the extent of a tax provision from the tax authority. This position is binding on the tax authority.

taxation. These unified tax rules, in the WAEMU[10] and CEMAC,[11] constitute taxation harmonisation tools intended to be applied alongside taxation specific to mining and oil operations.

14.7 The origin of the revenue that the states get from the exploitation of hydrocarbons and solid minerals in their territories has undergone a major structural change; to this end, we should mention the reduction in the share of royalties, the emergence of profit tax and the almost disappearance of customs duties, which is explained by the frequent exemption granted to operators. Insignificant during the first four decades of the 20th century, profit tax overtook royalties as the main source of oil and mining revenue for host states since the middle of the 1960s.

14.8 Besides taxation, and with a view to bringing in easy revenue, the host state often receives bonuses. This cash bonus principle was adopted in international contracts in the form of signature bonuses[12] . In the Middle East, this bonus first appeared on 29 May 1933 in the concession contract signed in Saudi Arabia with SOCAL, although it already existed beforehand unofficially. Despite the sums involved being very modest initially, bonuses became more substantial in mining and petroleum projects. This mechanism is primarily used in areas with demonstrated or highly probable production potential as a way to apportion conflicting demands by forcing the payment, at the time the

---

[10] The unified rules that apply in the WAEMU zone in terms of the ordinary rules of taxation include: Regulation No. 08/2008/CM/UEMOA of 26 September 2008 Enacting the Rules Introduced to Avoid Double Taxation Within WAEMU and the Rules Governing Tax Assistance; Directive No. 02/98/CM/UEMOA of 22 December 1998 On the Harmonisation of the Laws of the Member states On the Value Added Tax, amended by Directive No. 02/2009/CM/UEMOA of 27 March 2009; Directive No. 07/2001/CM/UEMOA of 26 November 2001 Establishing the Harmonised Regime for Advance Payments of Profit Tax Within WAEMU; Directive No. 01/2008/CM/UEMOA Harmonizing the Modalities for the Determination of the Taxable Result for Legal Entities within WAEMU; Directive 08/2008/CM/UEMOA of 26 September 2008 On the Harmonisation of the Rates of Taxes Levied on the Profits of Legal Entities in WAEMU Member states.

[11] The unified rules that apply in the CEMAC zone in terms of the ordinary rules of taxation include: the CEMAC Tax Treaty (Act No. 5/66 UDEAC-49 of 13 December 1966); Directive No. 1/99/-CEMAC-028-CM-03 of 17 December 1999 On Harmonising the Legislation of the Member states on Value Added Tax (VAT) and Excise Duty; Act No. 10/88-UDEAC-257 of 7 December 1988 Harmonising Registration, Stamp and Curatorship Duties; Appendix to Directive No. 02/01/UEAC-050-CM-06 Revising Act 3/72-153-UDEAC of 22 December 1972 Enacting the Corporation Tax; Regulation No. 17/99/CEMAC-20-CM-03 of 17 December 1999 Enacting the Investment Charter.

[12] See, in particular, in Côte d'Ivoire, Article 74 of Law No. 96.669 of 29 August 1996 Enacting the Petroleum Code, which provides:

'*[l]e contrat pétrolier peut prévoir une prime dénommée "bonus de signature" que son titulaire s'oblige à verser à l'État pour la conclusion du contrat, ainsi qu'une prime dénommée "bonus de production" que le titulaire a l'obligation de verser à l'État en fonction des quantités d'hydrocarbures produites.*'

[Free translation into English: '[t]he petroleum contract may provide for a "signature bonus" that the contractor must pay to the State in order to sign the contract, as well as a "production bonus" that the contractor must pay to the State based on the quantity of hydrocarbons produced'].

exploitation agreement enters into force, of a freely negotiable sum, which depends on the country and region in question.[13]

14.9 Alongside these signature bonuses, and in the same vein, mining and petroleum legislation requires the holder of mining rights to pay a fixed fee at the time mining rights are granted,[14] or even when submitting the application to obtain a mining title. As a general rule, a fixed fee will be paid each time an operation on the mining permit takes place (renewal, assignment, etc).

14.10 In principle, the financial results of exploitation are subject to taxation both in the host state and in the state of origin of the private operator. Indeed, almost all contracts and legislation impose mining and oil companies that want to perform exploitation operations to create a subsidiary under local law[15] at the exploitation stage. The host state taxes the revenues of the local subsidiary, and the net profits after taxation that will be transferred to the parent company are subject to taxation in the state of origin of the private operator. This taxation at the place of exploitation has been one of the major demands of producing states for decades.

14.11 Furthermore, the taxation applicable to oil and mining operations varies depending on the exploitation phase in question. As an incentive, the tax advantages granted to the private operator will be greater during the exploration and development phases.[16]

14.12 Besides the tax rules contained in national legislation and regional texts, the taxation issue is also addressed through contractual provisions contained in exploitation agreements or other instruments such as establishment agreements.

The superimposition of applicable rules can sometimes affect the intelligibility of the tax rules that apply to mining and oil operations.

1 THE TAX REGIME APPLICABLE TO MINING AND PETROLEUM OPERATIONS

14.13 The tax regime that applies to the private operator within the context of its operations encompasses the ordinary rules of taxation and the taxation specific to extraction activities.

14.14 In terms of the ordinary rules of taxation,[17] the private operator will be liable for various taxes including VAT, corporation tax, profit tax,

[13] The payment of a signature bonus is provided, in particular, in Algeria under the Law No. 05-07 of 28 April 2005 On Hydrocarbons.

[14] On the payment of fixed fees at the time of granting the mining rights, see para 9.39 above.

[15] With regard to the conditions relating to the nationality of the private operator, see paras 9.10 et seq above.

[16] See, to that effect, Charlet, Laporte & Rota-Graziosi op cit 18.

[17] The principle of applying the ordinary rules of taxation to the holder of mining rights, in particular, was reiterated in Rwanda by Article 41 of Law No. 13-2014 of 20 May 2014 Enacting Exploitation of Mines and Quarries:

> '*[l]e titulaire d'un permis minier ou de carrière est soumis à l'impôt sur le revenu et autres taxes conformément aux lois fiscales*'.
>
> [Free translation into English: '[t]he holder of a mining or quarry permit is liable to income tax and other taxes in accordance with taxation laws.']

business levy,[18] property tax or even business tax, unless stated otherwise in the exploitation agreement[19] or other incentives provided for in the petroleum and mining codes.

Furthermore, in principle, the private operator continues to be subject to reporting requirements.[20] However, the operator may, in certain situations, benefit from a simplified reporting system.[21]

See also, in Togo, Article 52 of Law No. 2003-012 of 4 October 2003 Amending and Supplementing Law No. 96-004 of 26 February 1996 Enacting the Mining Code:

> '*[l]e titulaire et ses prestataires de services et fournisseurs sont assujettis au régime fiscale de droit commun.*'
>
> [Free translation into English: '[t]he holder and its service providers and suppliers are subject to the ordinary rules of taxation'.]

In Liberia, Article 10(1) of the 2002 Petroleum Law: '[a]ll holders of petroleum contracts and their partners shall be subject to the payment of taxes and fees in accordance with the Revenue Code of Liberia of 2000 of General Application.']

18 The business levy is a tax that is still very common in Sub-Saharan Africa. '*[C]et impôt comporte généralement un droit fixe et un droit proportionnel sur la valeur des biens d'équipement*'. [Free translation into English: '[t]his tax generally includes a fixed duty and a duty proportional to the value of the capital goods']; see Charlet, Laporte & Rota-Graziosi, op cit 18. See, in Mauritania, Article 32 of Law No. 2012/012 Governing Mining Agreements and Approving the Standard Mining Agreement: '*[à] compter du 1er janvier qui suit la fin de la phase d'installation, la patente est due par le titulaire conformément au CGI, plafonnée toutefois à un maximum de 2 700 000 ouguiyas.*' [Free translation into English: '[f]rom 1 January following the end of the installation phase, the business levy is payable by the holder in accordance with the General Taxation Code (CGI), capped at 2,700,000 ouguiyas']. On the other hand, some legislation provides that the operator of licensable substances is exempt from the business levy (see, in Burundi, Article 160 of Order No. 1/138 of 17 July 1976 Enacting the Mining and Petroleum Code).

19 See, in particular, in Chad, Article 59 of the Law No. 011-PR-1995 of 20 June 1995 Enacting the Mining Code:

> '*[l]es titulaire de titres miniers ou bénéficiaires d'autorisations en vertu du Code Minier, ainsi que les entreprises travaillant pour leur compte ou en association avec eux, sont assujettis au Code Général de Impôts notamment, à la patente, à la taxe forfaitaire, à l'impôt sur les bénéficies industriels et commerciaux et à l'impôt sur les dividendes, sauf stipulation contraire dans la convention minière*'.
>
> [Free translation into English: '[t]he holders of mining rights or beneficiaries of permits by virtue of the Mining Code, as well as the companies working on their behalf or in partnership with them, are subject to the General Taxation Code in particular, the business levy, flat-rate tax, the tax on industrial and commercial profit and the tax on dividends, unless stated otherwise in the mining agreement'].

20 See, in particular, in Côte d'Ivoire, Law No. 2014-138 of 24 March 2014 Enacting the Mining Code, and, more specifically, Article 159:

> '*[l]e titulaire d'un permis de recherche reste soumis à l'obligation fiscale de souscription annuelle de la déclaration du compte d'exploitation et de résultats et des éléments de détermination de la patente*'.
>
> [Free translation into English: '[t]he holder of an exploration permit continues to be subject to the tax requirement for annual subscription of the statement of the operating account and the results and elements for determining the business levy'], and Article 160:
> '*[l]e titulaire d'un permis d'exploitation reste assujetti aux obligations déclaratives applicables aux sociétés soumises à l'impôt sur les bénéfices industriels et commerciaux et notamment, à l'obligation de souscription annuelle de la déclaration de son compte d'exploitation et de résultats*'.

14.15 Considering the tax regime specific to extraction activities, it is important to distinguish between general provisions on the one hand and the specific tax provisions pursuing a given objective on the other hand.

a The general tax regime: surface fees and royalties

14.16 The host state will tax the exploitation activity before it produces any financial results, ie either prior to the discovery of commercial quantities, or before the natural resources extracted are marketed (in their crude state or in the form of raw material). The private operator, to be subject to these payments, must hold mining rights.

The tax base is constituted by either the surface area defined by the contract or the mining title or the level of production belonging to it.

14.17 Any summary of property taxation is sensitive because of the diversity of the situations in the 54 African countries and the overlaying of many piecemeal solutions. Therefore, we will consider the two main tax measures that apply to the exploitation of natural resources even before the private operator receives any financial revenue from production: property taxation and royalties.

i *Taxation based on the perimeter covered by the mining right*

14.18 Property taxation is in principle related to the holding of mining rights.

The surface fees depend on the surface area allocated to the mining and petroleum operations. In this way, the surface fee is proportional to the perimeter area defined by the mining title.[22]

The principle of property taxation is widely accepted, especially in

---

[Free translation into English: '[t]he holder of an exploitation permit continues to be subject to the reporting requirements applicable to companies liable to tax on industrial and commercial profits and, in particular, the requirement for annual subscription of the statement of its operating account and results'].

[21] See, in particular, in Guinea, Article 163-I of Law No. L/2013/No. 053/CNT of 8 April 2013 Amending Law No. 2011/006/CNT of 9 September 2011Enacting the Mining Code:

'[l]es titulaires d'un Titre d'exploitation minière qui extraient des Substances minières en Guinée dans le but exclusif de les exporter à l'état brut, sans les revendre sur le marché intérieur, peuvent solliciter l'application d'un régime de déclaration simplifiée. Ce régime les autorise à déclarer la taxe à l'extraction sur les Substances minières visée à l'article 161 du présent Code et la taxe à l'exportation sur les Substances minières autres que les substances précieuses visée à l'article 163 du présent Code sur une seule et même déclaration.'

[Free translation into English: '[t]he holders of mining rights who extract mineral substances in Guinea with the sole aim of exporting them in their crude state, without selling them on the domestic market, can request application of a simplified reporting system. This system authorises them to declare the extraction tax for the Mineral Substances referred to in Article 161 of this Code and the export tax for Mineral Substances other than those precious substances referred to in Article 163 of this Code in a single declaration'].

[22] See, in particular, in Guinea, Article 160 of the Law No. L/2013/N°053/CNT of 8 April 2013 Amending Law No. 2011/006/CNT of 9 September 2011 Enacting the Mining Code.

Portuguese-speaking,[23] Spanish-speaking[24] and French-speaking Africa. To this end, the payment of a surface fee is asserted regionally within the context of the WAEMU by the Community mining code.[25]

14.19 Surface fees depends on many parameters, for instance the phase of operations in question or the substance covered by the mining right. Traditionally, at the production stage, surface fees are higher than those applied during prospecting and exploration.[26] In addition, within the same exploitation phase, regulations may provide for the payment of different amounts from one year to the next,[27] the principle being that the increase in surface fee depends on the time elapsed. Lastly, regulations may provide for an increase in the amount of the surface fee if the mining rights are renewed.[28]

The amounts payable are likely to vary depending on the substance involved. Thus, in the mining sector, surface fees may vary depending on whether the mining rights are granted for the exploitation of mineral substances, construction materials or precious metals and minerals.

---

[23] See, in Mozambique, Articles 16 et seq. of the Law No. 28/2014 of 23 September 2014; in Angola, Article 180 of the Mining Law No. 31/11 of 23 September 2011.

[24] See Article 46 of the Equatorial Guinea Law No. 9/2006 of 3 November 2006 Enacting the Mining Code.

[25] See Article 22 of Regulation No. 18/2003/CM/UEMOA of 23 December 2003 Enacting the Community Mining Code:

> '*[t]out titulaire d'un titre minier est soumis au paiement annuel d'une taxe superficiaire dont le montant et les modalités de règlement sont fixés, en l'absence de textes communautaires, par la législation minière nationale de chaque État membre*'. [Free translation into English: '[a]ny holder of mining rights is subject to the annual payment of a suface fee, the amount and payment terms of which are fixed, in the absence of Community regulations, by the national mining legislation of each Member State'].

In English-speaking African countries, the payment of a land tax is not however commonplace. As an example of application of land tax in English-speaking Africa see, in particular, in Libya, Article 13(1) of the Petroleum Law No. 25/55 of 11 May 2012.

[26] The discrepancies are often important. See, in particular, Article 46 of the Equatorial Guinea Mining Law No. 9/2006 of 3 November 2006.

[27] In the Central African Republic, Article 18(1) of Law No. 09-005 of 29 April 2009 Enacting the Mining Code: Land taxes for mining rights and permits are fixed as follows:

- exploration permit (first two years: 3,000 XOFXOF/km^2 pa; years 3 and 4: 6,000 XOFXOF/km^2 pa; year 5: 12,000 XOFXOF/km^2; from year 6: 24,000 XOFXOF/km^2 pa);
- small-scale exploitation permit: (5,000 XOFXOF/ha pa);
- semi-mechanised, small-scale exploitation permit (year 1: 10,000 XOFXOF/ha; subsequent years: 15,000 XOFXOF/ha pa;);
- industrial exploitation permit for small or large mines (60,000 XOFXOF/km^2 pa).

[28] See, in particular, in Chad, Appendix I to Law No. 011-PR-1995 of 20 June 1995 Enacting the Mining Code, which provides for the payment of a surface fee at a rate of 200 XOFXOF per km^2 per annum for the first period of validity of the exploration permit. This rate then increases to 400 at first renewal and 500 at the second. Similarly, for exploitation permits, the rate of 100,000 that applies for the first period of validity of the permit increases to 500,000 in the event of renewal.

Surface fees are generally payable on an annual basis.[29] This principle, widely accepted by mining and petroleum laws, is also asserted at regional level.[30]

The advance payment of surface fee is sometimes provided for in mining and petroleum laws.[31] In this case, the principle reiterated by certain texts is that, in the event of abandonment of land in the course of a year, the private operator will not be entitled to reimbursement of any surface fees already paid.[32]

14.20 Sometimes, mining and petroleum laws take care to lay down the conditions for distributing the profits generated by the surface fees. In this way, the mining code of Central African Republic provides that 20% of the sum received will benefit the local authorities in the territory where the mining rights are located,[33] illustrating once again the desire to integrate extraction projects into the region where they are based.

ii *Royalties*

14.21 Royalty is the tax concept most closely related to the exploitation of natural resources. One of the oldest contemporary examples we are aware of concerns the USA, where, in 1885, the Government had imposed a tax of one dollar per barrel of oil extracted.

As already mentioned, until the beginning of the 1960s, royalties were the main source of revenue generated from mining and oil activities. Until then, the tax on profits only provided secondary income; subsequently, the order of

[29] See, in particular, in Togo, Article 50 of Law No. 96-004 of 26 February 1996 Enacting the Mining Code; in Côte d'Ivoire, Article 150 of the Law No. 2014-138 of 24 March 2014 Enacting the Mining Code.

[30] See to that effect, Article 22 of the WAEMU Regulation No. 18/2003/CM/UEMOA of 23 December 2003 Enacting the Community Mining Code.

[31] See, in particular, in Togo, Article 50 of Law No. 96-004 as of 26 February 1996 Enacting the Mining Code:

> '*[t]out détenteur d'un titre minier paiera annuellement par anticipation une redevance superficiaire pour la superficie objet de son titre, dont le montant et les modalités de règlement sont précisés en annexe de la présente loi*'.
>
> [Free translation into English: '[a]ny holder of mining rights will pay in advance each year a surface fee for the surface area covered by its rights, the amount and payment terms of which are indicated in the Appendix to this Law'].

See also, in Central African Republic, Article 65 of the Central African Republic Ordinance No. 93.007 of 25 May 1993 Enacting the Petroleum Code, as amended by the 2002 and 2003 Finance laws.

[32] See, in particular, in the Central African Republic, Article 65 of Ordinance of the No. 93.007 of 25 May 1993 Enacting the Petroleum Code, as amended by the 2002 and 2003 Finance laws:

> '*[e]n cas d'abandon de surface au cours de l'année ou de force majeure, la société n'aura droit à aucun remboursement des redevances superficiaires déjà payees*'.
>
> [Free translation into English: '[i]n the event of abandonment of land in the course of the year or in a force majeure event, the company will not be entitled to reimbursement of any land taxes already paid'].

[33] See Article 18(1) of Law No. 09-005 of 29 April 2009 Enacting the Mining Code of the Central African Republic.

priority was reversed by the widespread development of a tax on profits. Nowadays, royalties have, as a result of their relative importance, become one of the tools states use to attract investors. After bonuses and surface fees, these are another financial effort required from private operators. Nonetheless, this disadvantage is in part corrected when the host state claims its royalty in kind rather than in cash. In the event that it opts for payment in kind, the private operator is often asked to carry out the transport, transformation and marketing operations on its behalf.

14.22 The royalty is now widely accepted in African mining and petroleum laws. We will look at both the material and functional aspects of the royalty.

(1) Functional aspects of the royalty

14.23 The royalty constitutes a way of regulating the importance of mining and petroleum activities and can be used for economic and political ends.

14.24 The host state can adjust the royalty in various ways. First of all, the host state can modify the royalty rate, even going as far as making it disappear. The host state can also modify the bases for calculating the royalty; in this case, the host state's action can affect both the rate and the basis of the royalty. Lastly, the State can modify the royalty payment frequency in order to receive the royalty over periods that are as short as possible.

14.25 Some legislation lays down conditions for distributing the royalty between the State and the regional authorities involved in the extraction projects.[34]

(2) The royalty system

14.26 The royalty compensates the owner of the natural resources, for the part that is extracted and that will diminish the available reserves and therefore its mining estate.

---

[34] See, in particular, Article 527 of the Decree No. 038/2003 of the Democratic Republic of Congo of 26 March 2003 Enacting Mining Regulations:

'*[a]vant le quinzième jour du mois suivant celui du paiement de la redevance minière par le Titulaire, le Ministre des Finances, Ordonnateur Général du budget de l'État, vérifie le paiement effectué et ordonne la répartition des recettes à raison de 60% pour le Trésor public, 25% pour la province où se trouve le projet et 15% pour la ville ou le Territoire dans le ressort duquel s'opère l'exploitation minière. En cas de chevauchement des carrés du périmètre d'exploitation sur deux ou plusieurs provinces, villes ou territoires, la répartition des quotités leur revenant se fait conformément aux dispositions de l'article 40 du présent Décret*'.

[Free translation into English: '[b]efore the fifteenth day of the month following the month of payment of the mining royalty by the Holder, the Minister of Finance, the Paymaster of the State budget, checks the payment made and orders distribution of the income with 60% for the government treasury, 25% to the project province and 15% to the city or the Territory within whose area of jurisdiction the mining operation is taking place. If the squares of the exploitation area span more than one province, city or territory, the portions are distributed in accordance with the provisions of Article 40 of this Decree'].

14.27 In the first petroleum exploitation agreements, the royalty (paid in cash) was constant and not related to changes in production volume. The connection between royalty and production was so minor in the minds of negotiators that it was even envisaged to pay a royalty in advance for a hypothetical production. The establishment of a royalty based on production was not immediate.

In the Middle East, for example, an agreement on this point was only reached in Iran on 23 September 1919 and in Iraq in 1923; in this region, Egypt was the first to adopt this mechanism in 1912. It involved a royalty of 2,5 Egyptian piastres per 100 gallons of oil produced (and not per tonne like in other Middle Eastern countries), supplemented by a payment of 5% of its market value. These payments applied to the activities of Anglo-Egyptian Oilfields. On the other hand, for solid mineral natural resources, the idea of a royalty, often paid in kind, was introduced much earlier, as reflected in certain Chinese concession contracts, dating from 1902, relating to the exploitation of copper ore.

14.28 The basis for a variable royalty (a royalty of which the rate changes with the production level) has changed. Initially, the operators set the value of the production themselves, arbitrarily based on the natural resources extracted. Then, greater internationalisation of the extractive industries led to the insertion of new provisions enabling the value of the production to be set depending on a range of international price references.

14.29 Operators gradually based calculation of the royalty for the natural resource extracted on the prices of the transformed raw materials. However, some provisions deviate from this trend and calculate the basis of the royalty using only the natural resource extracted. This measure is, in reality, intended to promote the processing of natural resource where it is extracted. Indeed, in this case, the basis used to calculate the royalty is lower in value than if the natural resource extracted was considered as being fictitiously transformed.

On average, the royalty rate varies between 10–20% for hydrocarbons.[35] Specific rates, which are generally lower, may apply to gaseous hydrocarbons.[36]

As a general rule, the royalty rates applicable to mineral substances are lower than those that apply in the oil sector. It would, however, be risky to generalise; indeed, the royalty rates applicable to precious metals and precious and semiprecious stones are often much higher than those that apply to other minerals or quarries substances.[37]

---

[35] As an example, in Libya, Article 13 of Petroleum Law No. 22/55 provides for a royalty rate of 12,5%.

[36] See to that effect, in Burundi, Article 210 of Decree No. 100/130 of 14 December 1982 Establishing Measures for Implementing Order No. 1/138 of 17 July 1976 Enacting the Mining and Petroleum Code of the Republic of Burundi, which provides for a royalty of 12,5% applied to liquid hydrocarbons compared with 5% for gaseous hydrocarbons.

[37] Thus, in Botswana, the royalty rate applicable to precious stones is 10%, compared with 5% for precious metals and just 3% for other mineral substances (Article 66 of the Mine and Minerals Act). Similar rates apply in Lesotho, namely 10% for precious stones and 3% for other substances (Article 59(2) of the Mines and Minerals Act).

14.30 The royalty rate may be set by mining and petroleum laws, which is the solution normally adopted, or determined by contractual provisions.[38] In this case, mining and petroleum laws may set a maximum royalty rate that cannot be exceeded.[39]

14.31 In some legislation, the royalty is considered to be an operating cost[40] and may therefore appear as a deductible expense.

14.32 Exploitation agreements have established several methods for calculating the royalty, generally based on the level of production achieved. Three main categories can be differentiated depending on whether the royalty is based on a single rate, on variable criteria or on a mixed calculation method.

(a) Setting of a single royalty rate

14.33 Based on the definition of a tax base constituted by the level of production, a fixed and single royalty is established that can be expressed in kind or, the majority of the time, in cash. In the latter scenario, the amount can be set in local currency or in another currency converted at the official exchange rate.

In practice, there seems to be a shift from fixed royalties to a variable royalty, which is now the main method for determining the royalty.

(b) Establishing a variable calculation method

14.34 The variable nature of the royalty is defined using two parameters, ie rate and base. In this way, several rates can apply to bases that are themselves variable.

The shift to a variable calculation method has resulted in the establishment of a very important principle, the results of which involve basing calculation of the royalty no longer on just the level of production achieved, but also on its commercial value.

14.35 Royalties based on the level of production. The royalty calculation

---

[38] See in particular, to that effect, in Eritrea, Article 22(2) of the Proclamation No. 108/2000, A Revised Proclamation to Govern Petroleum Operations; and, in Ethiopia, Article 22(2) of the Proclamation No. 295/1986 Regulating Petroleum Operations.

[39] See, in particular, Article 41 of the Djibouti Law No. 117/AN/05/5eL of 16 October 2005 Enacting the Petroleum Code: '*[l]es taux de redevance applicables sur les productions de pétrole brut ou de gaz naturel n'excèdent pas les 10% et sont fixés par la convention.Les taux de cette redevance ainsi que les règles d'assiette et de recouvrement sont précisés dans la convention.*' [Free translation into English: '[t]he royalty rates applicable to productions of crude oil or natural gas do not exceed 10% and are set by the agreement. The rates of this royalty, as well as the tax base and recovery rules, are stipulated in the agreement.']

[40] See, in particular, in Burundi, Article 147 of the Law No. 1/21 of 15 October 2013 Enacting the Mining Code: '*[l]a taxe* ad valorem *sur les titres miniers d'exploitation est comptabilisée comme une charge d'exploitation*'. [Free translation into English: '[t]he *ad valorem* tax on exploitation mining rights is counted as an operating cost'].

method can be based on the production level. In this case, a royalty rate will be applied to a certain quantity of extracted substances.[41]

National legislation will generally lay down the methods used to determine the level of production, on which the amount of the royalty is established. In this respect, the quantities of hydrocarbons consumed during exploration works, the hydrocarbons lost at the measurement point, and even the quantities re-injected into the deposit will be excluded from the royalty calculation.[42] Determination of the quantity of mineral substances extracted implies fewer difficulties, because of the absence of losses associated with the extraction methods compared with those used in the oil industry, particularly within the context of reinjection.[43]

14.36 Royalties based on the selling price of the natural resources. Regardless of how it is calculated, the royalty remains independent of the operating costs, and therefore of the profit generated by the extractive

---

[41] Thus, in Algeria, Article 160 of the Law No. 01-10 of 3 July 2001 Enacting the Mining Law: '*[p]our les métaux précieux et les pierres précieuses et semi-précieuses, l'assiette de la redevance d'extraction est constituée par la quantité de produits marchands obtenus.*' [Free translation into English: '[f]or precious metals and precious and semiprecious stones, the base for the extraction royalty is made up of the quantity of commodities obtained.']

[42] See, in Algeria, Article 26 of Law No. 13-01 of 20 February 2013 Amending and Supplementing Law No. 05-07 of 28 April 2005 On Hydrocarbons:

'*[l]a redevance est établie sur la base des quantités d'hydrocarbures produites et décomptées après les opérations de traitement au champ, au point de mesure. Sont exclues pour le calcul de cette redevance les quantités d'hydrocarbures qui sont: soit consommées pour les besoins directs de la production; soit perdues avant le point de mesure; soit réintroduites dans le ou les gisement(s), à condition que ces gisement(s) ai(en)t fait l'objet d'un seul et même contrat. Les quantités d'hydrocarbures consommées ou perdues qui sont exclues du calcul de la redevance, doivent être limitées à des seuils techniquement admissibles, prévus dans le plan de développement approuvé par l'agence nationale pour la valorisation des ressources en hydrocarbures (ALNAFT). Ces quantités exclues font l'objet de justifications auprès de l'agence nationale pour la valorisation des ressources en hydrocarbures (ALNAFT). Les modalités d'application des dispositions du présent article sont définies par voie règlementaire*'.

[Free translation into English: '[t]he royalty is established based on the quantities of hydrocarbons produced and counted after field processing operations, at the measurement point. Excluded from calculation of this royalty are the quantities of hydrocarbons: consumed for the direct needs of production; lost before the measurement point; reintroduced into the deposit(s), as long as these deposits are covered by one and the same agreement. The quantities of hydrocarbons consumed or lost, which are excluded from calculation of the royalty, must be limited to technically permissible thresholds, provided for in the development plan approved by the national agency in charge of promoting hydrocarbon resources (ALNAFT). These quantities are justified to ALNAFT. The conditions for application of the provisions of this Article are defined by regulations'].

See also, in Madagascar, Article 51 of Law No. 96-018 of 4 September 1996 Enacting the Petroleum Code.

[43] For an example of a method for determining the quantity of resources extracted in the mining sector, see, in particular, in Algeria, the Algeria Order of 26 July 2003 On the Method for Determining the Quantity of Extracted Mineral Substances.

activities. However, fluctuations in oil prices have led to new tax systems being introduced, according to which royalties are calculated in relation to profits and not just in relation to production.

14.37 Various systems are used for determining the value of production. First of all, the value of production can be determined by taking into consideration the price at extraction: for hydrocarbons, we talk about the price at the wellhead and for mines the price ex-mine.[44] This involves setting a price according to operating costs only.

14.38 The value of production can also be determined taking into account the price at the field gate. For hydrocarbons, we talk about the field price.[45] This involves setting a price based on the operating costs and the processing costs in preparation for transport.

---

[44] See, in particular, Article 160 of the Burundi Order No. 1/138 of 17 July 1976 Enacting the Mining and Petroleum Code:

> '*[l]es exploitants de substances concessibles sont dispensés de la patente. Ils sont soumis à une taxe ad valorem, dont le pourcentage est fixé en fonction de la valeur de la production sur le carreau de la mine, ou de la valeur départ champ des hydrocarbures liquides ou gazeux.*'
>
> [Free translation into English: '[o]perators of licensable substances are exempt from the business levy. They are subject to an *ad valorem* tax, the percentage of which is set based on the value of the production at the pithead, or the wellhead price for liquid or gaseous hydrocarbons']

In Benin, Article 63 of the Benin Law No. 2006-17 of 17 October 2006 Enacting the Mining Code and Mining Taxation:

> '*[l]a valeur de la redevance "ad valorem" est déterminée en fin d'année pour chaque exploitant sur la base de la valeur carreau mine des produits vendus au cours de l'exercice considéré*'.
>
> [Free translation into English: '[t]he value of the "*ad valorem*" royalty is determined at the end of the year for each operator based on the ex-mine price of the products sold during the financial year in question'].

[45] See, in particular, to that effect, in Benin, Article 103 of Law No. 2006-18 of 17 October 2006 Enacting the Petroleum Code: '*[l]es exploitants d'hydrocarbures sont dispensés de la patente. Ils sont soumis à une redevance "ad valorem" dont le pourcentage est fixé en fonction de la valeur départ champ des hydrocarbures*'.

> [Free translation into English: '[o]perators of hydrocarbons are exempt from the business levy. They are subject to an ad valorem royalty, the percentage of which is set based on the wellhead price of the hydrocarbons.'].

See, in particular, in Burundi, Article 160 of Decree-Law No. 1-138 of 17 July 1976 Enacting Mining and Petroleum Code:

> '*[les exploitants] sont soumis à une taxe ad valorem, dont le pourcentage est fixé en fonction de la valeur de la production sur le carreau de la mine, ou de la valeur départ champ des hydrocarbures liquides ou gazeux.*'
>
> [Free translation into English: '[operators] are subject to an *ad valorem* tax, the percentage of which is set based on the value of the production at the pithead, or the wellhead price for liquid or gaseous hydrocarbons']

In the Central African Republic, Article 66 of Ordinance No. 93.007 of 25 May 1993 Enacting Petroleum Code:

> '*[l]es titulaires de contrats de concession sont tenus d'acquitter une redevance à la production égale à 12,5% au plus de la valeur départ champ des hydrocarbures extraits sur le territoire de la République Centrafricaine. Le taux est réduit à 5% en ce qui concerne la production de gaz naturel. Les taux de la redevance sur la production*

14.39 Lastly, the value of production can be determined according to the exportation port price. In particular, this involves legislation that takes the FOB[46] value of the raw material[47] into consideration. If the royalty is determined in relation to the FOB price of the extracted materials, the royalty will be calculated to include all of the costs up to the time of loading onto the cargo ships.[48] The price is therefore set based on the operating costs, the processing costs in preparation for transport and the cost of transport from the deposit to the port of export. However, some provisions state that the royalty will be calculated using the price at the point of export less the transport costs between the point of export and the point of field departure.[49]

Calculation of the royalty based on the selling price of the natural resources is of real interest to private operators given the current instability of the prices of raw material. Indeed, a reduction in international prices would in this case be automatically reflected in the amount of the royalty, which would also be reduced.

14.40 Royalties based on the quality of the natural resource

---

ainsi que les modalités de détermination de la valeur départ champ sont fixés dans le contrat de concession'.

[Free translation into English: '[t]he holders of concession contracts must pay a production royalty equal to 12,5% above the wellhead price of the hydrocarbons extracted in the territory of Central African Republic. The rate is reduced to 5% for natural gas production. The production royalty rates and the conditions for determining the wellhead price are laid down in the concession contract'].

[46] The FOB (Free On Board) value is the selling price of a commodity once it has been loaded onto the vessel.

[47] See, in particular, Article 108(1) of Mauritania Law No. 2008-011 of 27 April 2008 Enacting the Mining Code:

'*[l]e titulaire d'un permis d'exploitation, d'un permis de petite exploitation minière, d'une autorisation d'exploitation de carrière industrielle est redevable du paiement d'une redevance d'exploitation calculée sur le prix de vente du produit résultant du dernier stade de transformation du minerai en Mauritanie ou de la valeur FOB du minerai si celui-ci est exporté avant d'être vendu. La combinaison de ce prix de vente et de cette valeur FOB vise à déterminer "la valeur assujettie" aux fins de l'application du présent article*.'

[Free translation into English: '[t]he holder of an exploitation permit, a small-scale mining permit or an industrial quarry exploitation permit is liable for payment of an exploitation royalty calculated using the selling price of the product resulting from the final stage of transformation of the ore in Mauritania or the FOB value of the ore if this is exported before being sold. The combination of this selling price and this FOB value aims to determine "the taxable value" for the purposes of application of this Article'].

[48] See, to that effect, Charlet, Laporte & Rota-Graziosi op cit note 36.

[49] See, in particular, in Madagascar, Article 49 of the Law No. 96-018 of 4 September 1996 Enacting the Petroleum Code:

'*[l]a redevance est assise sur le prix de référence fiscale qui sera égal au prix du marché international au point d'exportation diminué des coûts de transport entre ledit point d'exportation et le point de départ champ*'.

[Free translation into English: '[t]he royalty is based on the reference tax price, which will be equal to the price on the international market at the point of export less the transport costs between said point of export and the point of field departure.']

extracted. The royalty can in some cases depend on the quality of the natural resource extracted.

14.41 Royalties based on the place of processing of the natural resource. For the host state, the processing in its territory of the natural resource is of great political and economic interest; to this end, it can use the royalty as an incentive. In this way, a different royalty rate can be applied depending on whether or not production is processed in the national territory of the host state.

14.42 Thus, the applicable texts may in particular take into consideration the distance between the place of extraction and the port of export, or even establish a distinction depending on the onshore or offshore origin of the production.

Furthermore, a variable royalty rate can be applied to different areas of the national mining estate. For instance, in Algeria, the royalty rate applicable to the production of hydrocarbons varies depending on the quantity of resources extracted, but also according to the area where the petroleum operations took place.

(c) Setting a mixed calculation method

14.43 Calculation of the royalty is based on the two previous systems, namely application of a fixed rate and the establishment of a mixed calculation method; its scope mainly concerns hydrocarbons. It adopts two rate systems, one that is variable and another that is static, distinguishing several natural resources or not.

A first fixed-rate royalty applies; so, whatever the production's selling price, the local state freezes its revenue. In addition, a second royalty changes depending on the production's selling price; its rate is expressed as a percentage.

b Tax provisions with defined objectives

14.44 Tax provisions may have a defined purpose and be implemented by the host state as a tool to increase its revenue or reduce it (in the short term) to promote exploration of its subsoil.

i *Provisions intended to increase the host state's revenue*

14.45 Host states are interested in the profits that can be generated by the exploitation of natural resources and in particular hydrocarbons; such profits, although considerable, are more limited in the case of solid mineral natural resources.

Faced with these exceptional profits, the states sometimes intervened in order to favour a more equitable sharing of profits and capture a larger portion of the gain generated by mining and petroleum activities. This situation resulted in the creation of taxes on exceptional profits.

Faced with these exceptional profits, the states sometimes intervened in order to favour a more equitable sharing of profits and capture a larger portion of the gain generated by mining and petroleum activities. This situation resulted in the creation of taxes on exceptional profits.

The first major manifestation of the host state's right to unilaterally modify the financial equilibrium of an exploitation agreement using legislation dates back to 30 December 1950 when the Saudi Government imposed an income tax on ARAMCO.[50]

Recent legislative modifications have provided us with examples of the implementation of taxes on super profits, often declared as temporary or exceptional. In Algeria, the tax on exceptional profits is one example of this.[51] This tax applies to the share of production of liquid and gaseous hydrocarbons belonging to SONATRACH's foreign partners.[52] The tax on exceptional profits is only designed to be applied when oil price exceeds a certain amount.[53]

The establishment of this tax, the aim of which is to adjust the revenue of foreign companies associated with the national oil company Sonatrach, has sparked a great deal of controversy.[54] Indeed, the retroactive nature of such a tax, which is not conducive to the security and stability of the investment, has led to revision of various agreements. Such a consequence has in particular been illustrated by the reconsideration of production-sharing agreements between the national oil company and the two oil companies Anadarko and Maersk. Although these two companies resorted to arbitration to dispute the application of the tax on exceptional profits, the different parties were nonetheless able to reach an amicable settlement.

14.46 Alongside the taxation by the host state of the super profits generated by the exploitation of natural resources, some legislation has

---

[50] This involved the application of Royal Decrees No. 17/2/23/3321 and No. 17/2/7634. The increase in taxation enabled the Saudi Government to reject an equity participation that would have meant considerable financial costs for it.

[51] See on this issue, Article 101 of Ordinance No. 06-10 of 29 July 2006 Amending and Supplementing Law No. 05-07 of 28 April 2005 On Hydrocarbons and the Executive Decree No. 06-440 of 2 December 2006 Establishing the Procedure, the Conditions for Application and the Methodology for Calculating the Tax on Exceptional Exceptional Profits (TPE).

[52] See to that effect, Article 2 of the Executive Decree No. 06-440 of 2 December 2006 Establishing the Procedure, the Conditions for Application and the Methodology for Calculating the Tax on Exceptional Profits (TPE).

[53] See to that effect, Article 2 of Executive Decree No. 06-440 of 2 December 2006 Establishing the Procedure, the Conditions for Application and the Methodology for Calculating the Tax on Exceptional Profits (TPE):

> '*[l]a taxe sur les profits exceptionnels s'applique pour tout mois civil au cours duquel la moyenne arithmétique mensuelle des cotations moyennes entre la cotation basse 'low' et la cotation haute 'high' publiées par le PLATT'S CRUDE MARKETWIRE, sous la rubrique SPOT CRUDE ASSESSMENTS pour le pétrole 'BRENT (DTD)' est supérieure à 30 dollars des États-Unis d'Amérique par baril pour ledit mois civil.*'
>
> [Free translation into English: '[t]he tax on exceptional profits is calculated for the whole calendar month in which the monthly arithmetic mean of the average quotations of the low and high quotations published by PLATT'S CRUDE MARKETWIRE in the SPOT CRUDE ASSESSMENTS section for 'BRENT (DTD)' oil is higher than 30 US dollars per barrel for the said calendar month'].

[54] See the Elie Kleiman 'Les nationalisations' in *Le contentieux extractif* (ICC 2015) 88.

favoured the introduction of an additional tax based on the profitability of operations.[55]

Provisions intended to increase the host state's revenue often create difficulties, all the more so as this dissuasive taxation could discourage investment in the extraction sector.

However, consideration at fiscal level of the profitability of extraction projects significantly reduces the risk of subsequent modification of an exploitation agreement by the host state justified by the high profits generated by the extraction activity.

ii *Provisions intended to promote knowledge of national reserves of natural resources*

14.47 National legislators can use legal provisions to modify calculation of the tax base in order to promote exploration of licensable substances in the national territory.

14.48 With this in mind, petroleum and mining laws sometimes contain provisions that enable the private operator to establish a reserve for recovery of deposits. This reserve should be employed in prospecting, exploration and development works.[56]

---

[55] See, in particular, Article 46 of the Senegal Law No. 98-05 of 8 January 1998 Enacting the Petroleum Code:

'*[l]es titulaires de convention ou de contrat de services sont assujettis à un prélèvement pétrolier additionnel calculé sur un critère de rentabilité des opérations pétrolières, dont le taux, les modalités d'assiette, de déclaration, de liquidation et de recouvrement sont précisées dans la convention ou le contrat de services. Toutefois, lorsque la rémunération du titulaire de contrat de partage de production est déjà déterminée en fonction d'un critère de rentabilité des opérations pétrolières, ce mode de détermination tient lieu de prélèvement pétrolier additionnel. Le prélèvement pétrolier additionnel payé n'est pas une charge déductible pour la détermination du bénéfice soumis à l'impôt sur les sociétés.*'

[Free translation into English: '[t]he beneficiaries of agreements or service contracts are subject to an additional oil tax calculated based on a profitability of oil operations criterion, for which the rate and the tax base, declaration, liquidation and collection conditions are provided in the agreement or service contract. However, when the remuneration of the holder of a production-sharing agreement is already determined according to a profitability of oil operations criterion, this determination method is used instead of the additional petroleum tax. The additional petroleum tax paid is not a deductible expense for determining the profit subject to corporation tax'].

[56] See in particular, in the Democratic Republic of Congo, Article 78 of Ordinance-Law No. 81-013 of 2 April 1981 Enacting General Legislation on Mines and Hydrocarbons, amended and supplemented by Ordinance-Law No. 82-038 of 5 November 1982 and Law No. 86-008 of 27 December 1986:

'*[c]ette provision pour reconstitution de gisement devra être employée dans les deux ans de sa constitution en travaux de prospection, de recherches et de développement de gisements miniers, soit à l'intérieur, soit à l'extérieur des titres miniers ayant donné lieu à sa constitution*'.

[Free translation into English: '[t]his provision for recovery of deposits should be employed within two years of its establishment in prospecting, exploration and development works, either inside or outside the mining rights that gave rise to its establishment'].

These reserves for recovery of the deposit may in particular be established free of the tax on industrial and commercial profits, be accepted as expenses for determining the taxable profit[57] or even be deducted from the profit tax.[58]

Mining and petroleum laws can regulate the minimum and maximum amount of the sums included in this reserve.[59]

In order to control the use of the reserve for recovery of the deposit as effectively as possible, mining and petroleum laws can impose on the private operator an obligation to provide the tax authorities with evidence of the use of these reserves, otherwise the amount of the provision may have to be reinstated.[60]

---

[57] See, in Algeria, Article 169 of Law No. 01-10 of 3 July 2001 Enacting the Mining Code:
'*[s]ont admises pour la détermination du bénéfice imposable, les charges suivantes:* [. . .] *les provisions pour reconstitution de gisement constituées en franchises d'impôt sur les bénéfices industriels et commerciaux conformément à la réglementation en vigueur*'.
[Free translation into English: 'the following expenses are admitted in the calculation of the taxable profits: [. . .] the provisions for the reconstitution of the deposit which is tax-exempt of the tax on business profits as per the applicable regulation.'].
In Democratic Republic of Congo, Article 162 of Law No. 4-2005 of 11 April 2005 Enacting the Mining Code:
'*[l]es titulaires des titres miniers d'exploitation sont autorisés à constituer des provisions pour reconstitution du gisement. Ces provisions sont constituées déductibles de l'impôt sur les bénéfices.*'
[Free translation into English: '[t]he holders of exploitation mining rights are authorised to establish provisions for recovery of the deposit. These provisions are established as deductible from profit tax'].

[58] See in the Democratic Republic of Congo, Article 162 of the Law No. 4-2005 of 11 April 2005 Enacting the Mining Code.

[59] See, in particular, Article 78 of the Democratic Republic of Congo Ordinance-Law No. 81-013 of 2 April 1981 Enacting General Legislation on Mines and Hydrocarbons amended and supplemented by Ordinance-Law No. 82-039 of 5 November 1982 and Law No. 86-008 of 27 December 1986: '*[l]a provision pour reconstitution de gisement sera au minimum de 15% du montant brut des ventes, sans qu'elle puisse excéder 50% du bénéfice net comptable*'. [Free translation into English: '[t]he provision for recovery of deposits will be at least 15% of the gross sales, capped at 50% of the net accounting profit']. See also, Article 169 of the Algeria Law No. 01-10 of 3 July 2001 Enacting the Mining Code: '*[l]e taux maximum de cette provision est fixé à 1% du chiffre d'affaires annuel hors taxe*'. [Free translation into English: '[t]he maximum rate of this provision is set at 1% of the annual turnover before tax'].

[60] See, in particular, in Burundi, Article 222 of Decree No. 100/130 of 14 December 1982 Establishing Measures for Implementing Order No. 1/138 of 17 July 1976 Enacting the Mining and Petroleum Code:
'*[e]n cas de constitution d'une provision pour reconstitution de gisement, les exploitants doivent à peine de réintégration d'office fournir au vérificateur des impôts et au directeur des mines, toutes justifications sur l'emploi de ces provisions dans les délais impartis.*'
[Free translation into English: '[i]n the event of establishment of a provision for recovery of deposits, private operators must, on pain of compulsory reinstatement, provide the tax inspector and the Director of Mines with any evidence of the use of these provisions within the allotted time'].
See also, Article 169 of the Algeria Law No. 01-10 of 3 July 2001 Enacting the Mining Code:

If they are not used within the time allotted, the unused sums are reinstated in the taxable profit.[61]

c Tax relief applicable to mining and oil operations

14.49 National legislation and exploitation agreements often aim to improve in a way that is favourable to operators.

In practice, their action focuses, on the one hand, on reducing the risk of double taxation for operators and, on the other hand, on easing local tax rules.

i *Provisions intended to reduce the risk of double taxation of exploitation results*

14.50 The private operator is often faced with a non-negligible difficulty, running the risk of being taxed twice on the same base in both the host state and in its state of origin.

On the subject of double taxation, the basic problem is getting a sovereign state to recognise the tax legislation of another sovereign state and to take it into account when it comes to taxing the same taxpayer; this means that one of them has to waive some of the revenue to which it would be entitled according to its own legislation.

This dilemma often has the following backdrop: a multinational mining or oil company (whose decision-making body is usually based in an industrialised country) generates taxable results abroad, generally in industrialising countries. The company is taxed in the country where operations take place, then, with the authorisation of this country, the profits generated are transferred to their state of origin. On what base will it then be taxed? How is the tax already paid in the territory where the profits are generated taken into account?

The risk of the private operator's revenue being taxed twice has led to the emergence of solutions originating in and based on either the exploitation agreement itself, or elsewhere.

14.51 At the interstate level, the solution involves adding to the existing conventional framework on the subject of double taxation. Furthermore, the

---

'[c]ette provision doit être utilisée pour financer des travaux de recherche, dans un délai de (3) trois ans sous peine d'être obligatoirement réintégrée au résultat'.
[Free translation into English: '[t]his provision must be used to finance exploration work, within a period of three (3) years, otherwise it will be compulsorily reinstated in the results'].

[61] See, in particular, in Burundi, Article 167 of Order No. 1/138 of 17 July 1976 Enacting the Mining and Petroleum Code: *'[l]es provisions pour reconstitution de gisement dont l'emploi aux fins énumérées à l'article 166 ci-dessus, n'aura pas été justifié seront réincorporées aux résultats du premier exercice suivant; en cas de cessation d'activité avant un tel emploi, elles seront taxées au taux où elles l'auraient été lors de l'exercice de constitution.'* [Free translation into English: '[t]he provisions for recovery of deposits of which the use for the purposes listed in Article 166 above has not been justified will be reincorporated into the results for the first subsequent financial year; in the event of cessation of activity prior to such use, they will be taxed at the rate they would have been at in the financial year they were established'].

solution to a lack of coordination between two tax laws can be found by amending one of them.

14.52 At international level, the problems caused by double taxation are also taken into consideration within the context of supranational regulations. This is the case with the WAEMU, which since 2008 enacted a new regulation in this domain.[62]

14.53 When the issues caused by the risk of double taxation are not governed by national legislation, the exploitation agreements may fill this gap themselves by taking into account the taxation to which the private operator is subject in its country of origin. International practice will therefore extend the scope of influence of foreign legislation, not by applying it but by using it to implement the tax rate to be used in the state where exploitation is taking place, in order to preserve financial equity for the private operator. Practice therefore forces national tax legislation to internationalise its perception of situations. The different nationalities of the operators heavily influence exploitation agreements.

ii *Easing of the material scope of local tax rules*

14.54 After considering the existence and content of foreign tax legislation, contractual tax equilibrium needs to be established by acting on local tax legislation.

The insertion of such provisions into exploitation agreements goes back a long way. Two solutions enable easing of the material scope of local tax rules, namely their stability and their relaxation.

(1) Stabilisation of local tax legislation

14.55 Stabilisation clauses, widely applied in the extractive industry,[63] find in the tax domain a particularly adapted scope. Modification of the tax regime applicable to mining and petroleum operations can be fraught with consequences and greatly affect the profitability of an extraction project.[64]

In this way, very many exploitation agreements, taking into account the particularly long duration of the operations, contain stabilisation clauses that intend to fix the local tax legislation in force at the time they are signed, by declaring it applicable for the full term of the commitments.

The main objective of the stability guarantee in the tax domain is for the private operator to not be affected in the event of a change of the applicable legal framework that increases its tax burden.

Besides exploitation agreements, the possibility for the host state to sign an

---

[62] Regulation No. 08/2008/CM/UEMOA, adopting rules intended to prevent double taxation in the WAEMU and tax assistance rules.

[63] On stability clauses in general, see below para 19.6 et seq.

[64] See, in particular, to that effect, Sébastien Manciaux, 'Changement de législation fiscale et arbitrage international'2001 *Rev arb* 311.

agreement with the private operator that contains a stabilisation clause is widely accepted in mining and petroleum laws.[65]

14.56 On the subject of taxation, most of the legal or contractual provisions use the mechanism of unenforceability.[66] The mechanism of unenforceability assumes that new legal and regulatory provisions cannot be enforced on an operator benefiting from a stabilised regime.[67]

14.57 Generally speaking, the host state undertakes to ensure the stability of the tax burden in the broadest sense, ie tax, parafiscal and customs charges, as they result from the different taxes in force on the day the agreement is signed. Usually, the stability guarantee extends to the taxation and collection conditions, to the rates and even to the tax base rules.[68] In addition, the host state can guarantee not to subject the operating company to taxes that are created after stabilisation begins or to taxes that supplement those that already exist.

14.58 Sometimes, and to avoid disputes, a list is drawn up of the fees and taxes covered by stabilisation.[69] This assumes to provide for an update mechanism, especially in the case of disappearance, transformation or emergence of duties and taxes.

14.59 The issue of breach of a stabilisation clause has been brought before arbitral tribunals[70] on many occasions, as a result of the tightening of the tax provisions applicable to the investor. The arbitral tribunal in the case *Duke Energy v Peru*[71] went as far as to consider that stabilisation of the tax

---

[65] See, in particular, in Cameroon, Article 114 of Law No. 99-013 of 22 December 1999 Enacting the Petroleum Code and Article 99 of Law No. 001 of 16 April 2001 Enacting the Mining Code; in the Central African Republic, Article 134 of Law No. 09-005 of 29 April 2009 Enacting the Mining Code; in the Democratic Republic of Congo, Article 166 of Law No. 4-2005 as of 11 April 2005 Enacting the Mining Code; in Côte d'Ivoire, Article 164 of the Law No. 2014-138 of 24 March 2014 Enacting the Mining Code; in Guinea, Article 182 of Law No. L/2013/No. 053/CNT of 8 April 2013 Amending Law No. 2011/006/CNT of 9 September 2011 Enacting the Mining Code; in Madagascar, Article 154 of Law No. 99-022 of 19 August 1999 Enacting the Mining Code amended by Law No. 2005-021 of 17 October 2005; in Chad, Article 62 of Law No. 011-PR-1995 of 20 June 1995 Enacting the Mining Code and Articles 51 and 95 of Law No. 006-PR-2007 of 2 May 2007 On Hydrocarbons.

[66] See, in particular, to that effect, in Algeria, Article 12 of the Presidential Decree of 1 March 2003 Enacting the Standard Mining Agreement; in the Central African Republic, Article 134 of Law No. 09-005 of 29 April 2009 Enacting the Mining Code.

[67] On stabilisation, see below para 19.13 et seq.

[68] See, in particular in Benin, Article 74 of Law No. 2006-17 of 17 October 2006 Enacting the Mining and Mining Taxation Code; in Niger, Article 97 of amended Ordinance No. 93-16 of 2 March 1993 Enacting the Mining Law.

[69] See, in particular, in Guinea, Article 182 of Law No. L/2013/N°053/CNT of 8 April 2013 Amending Law No. 2011/006/CNT of 9 September 2011 Enacting the Mining Code.

[70] See, in particular, *Petrola Hellas SA v Greece* (Award of 22 April 1978) 1986 11 *YB Com Arb* 105; Award issued in the case *Company B (with its head office in State A) v State A* (1989) 7 (4) *ASA Bulletin* 382); Case *Power and Traction Finance Company Ltd v Greece* Award of 18 March 1930, in Sébastien Manciaux 'Changement de législation fiscale et arbitrage international' 2001 *Rev arb* 311.

[71] *Duke Energy International Peru Investments v the Republic of Peru* Award on the merits dated 18 August 2008 (ARB/03/28). The investor benefited from a Legal Stability Agreement (LSA), signed between the Peruvian Government and various investors, as well as a

regime concerned not only the changes made to the tax framework of the investment but also the possible change of interpretations by the administrative and legal authorities.[72]

14.60 Despite their widespread use in the tax domain, stability clauses have not been without criticism. The asymmetric nature of certain stability clauses, which freeze the applicable tax regime while acknowledging the right to opt for application of new and more favourable provisions, and the unreasonable duration of certain stability clauses have received criticism from some authors.[73]

(2) Relaxation of local legislation

14.61 In particular, relaxation of the regime applicable to the mining or petroleum project can result from an adaptation of the tax and customs regimes and of the accounting rules.

(a) Tax relief

14.62 Many countries use taxation as a tool for encouraging mining investment by granting relief on operator revenue.

14.63 International practice has long relied on the principle of local tax relief. These incentives have become uncommon in producing countries with a mining income but are often found in states whose national income is very low and whose potential in terms of natural resources is unproven. In particular, this may involve:

- exemption from taxes and duties collected in relation to company activities such as registration, an increase or decrease in share capital and assignment of shares;
- exemption from duties relating to filing of patents;
- exemption from certain local taxes;
- non-application of income tax on natural persons for expatriate employees of the private operator.

Like stabilisation of local legislation, the formal agreement of the host state is of course required for practical implementation of such provisions. Only a Decree from the local executive authority can effectively recognise the tax

stabilization clause contained in the Investment Code, under the terms of which: '[s]tability of the tax regime in force on the date of execution of the agreement. By virtue of the guaranteed stability of the tax regime, the foreign investor shall not be subject to a rate higher than the one established in the appropriate agreement for the purposes of the income tax payable by the company receiving the investment, and the tax assessing the profits attributed thereto and/or the dividends distribution in favour thereof; therefore, if the income tax payable by the company is increased, the rate applied to the foreign investor shall be reduced by the required amount so that the company's profits freely available thereto are at least equal to the guaranteed profits'. In this case, the investor suffered damages, not because of the change in the tax regulations applicable to the investment, but because of a change in how the tax regulations were interpreted by the host State.

[72] *Duke Energy International Peru Investments v the Republic of Peru* (Award on the merits dated 18 August 2008) (ARB/03/28), no 216.

[73] See Charlet, Laporte & Rota-Graziosi op cit 24 and 25.

exemption of the private operator. National State companies involved as local State operators do not have this authority. Nonetheless, they can agree to reimburse the private operator for any local taxes and customs duties it has had to pay.

14.64 Lastly, alongside tax relief, incentives granted to the private operator can also result from tax simplification measures.

(b) Relaxation of the accounting rules

14.65 The specific nature of natural resource exploitation costs and their accounting treatment no longer needs to be demonstrated.

Long ago, international practice adopted specific and more flexible solutions, in particular in terms of accounting procedures. National legislators have adopted some of them, primarily with regard to depreciation of the expenses incurred. A distinction is traditionally made between expenditures incurred before and after a commercial discovery is made.

14.66 The tax issue of the depreciation regime is very important. Indeed, rapid depreciation (sought by shrewd private operators) will reduce the host state's immediate tax income and may even, because of the low values of the initial productions, result in losses for a few years.

14.67 Since the beginning of the 1970s, accounting has been the subject of hard negotiations and many contractual amendments, especially when the level of complexity of the accounting regime is increased by the existence of a joint company between the private operator and the local State operator.

14.68 Mining and oil laws widely accept the principle of depreciation.[74]

---

[74] See, in particular, in Madagascar, Article 51 of the Law No. 2001-031 of 8 October 2002 Establishing a Special Regime for Major Investment in the Malagasy Mining Sector (LGIM) amended by Law No. 2005-022 of 17 October 2005, which provides:

> '*[f]ont partie des éléments amortissables, qui sont capitalisés en frais d'établissement et amortis au taux annuel maximum de un tiers (1/3), notamment: 1) les coûts de investissements effectués par le contribuable en recherche minière sur le périmètre qui fait l'objet de Permis miniers du Projet, en acquisition des droits miniers et/ou en Coûts de Développement préalables à la Date de Commencement de l'Exploitation Effective; 2) les contributions aux frais d'évaluation de l'étude d'impact environnemental du Projet; et 3) les dépenses sur investissement du Titulaire en recherche minière effectuée sur le périmètre qui fait l'objet des Permis miniers du Projet pendant la Phase d'Exploitation du Projet.*'
>
> [Free translation into English: '[t]he following in particular form part of the depreciable items, which are capitalised in start-up costs and amortised at the maximum annual rate of one third (1/3): 1) the costs of investment by the taxpayer in mining exploration in the area covered by mining permits for the Project, in acquisition of mineral interests and/or in Development Costs prior to the Effective Exploitation Start Date; 2) contributions to the evaluation costs of the Project's environmental impact study; and 3) the investment expenses of the Holder for mining exploration in the area covered by the mining permits for the Project during the Project Exploitation Phase'].

On the conditions for amortisation allocation, see also, in Tunisia, Article 98 of Law No. 2003-30 of 28 April 2003 Enacting the Mining Code:

> '*[p]our chaque exercice bénéficiaire, l'imputation des déficits et des amortissements sera effectuée dans l'ordre suivant 1- déficits antérieurs, 2- amortissements différés, 3- amortissements de l'exercice concerné.*'

Depreciation may also concern prospecting and exploitation costs,[75] fixed assets and even the costs of acquiring mining rights.

14.69 The principle of debit entry on the balance sheet of deductible expenses is also widely accepted by mining and petroleum regulations and by exploitation agreements.[76]

14.70 Similarly, the private operator may benefit from deductions for reinvestment in the different accounting years.

14.71 It is also worth mentioning that there is a movement of unification and harmonisation of accounting rules within the context of certain regional organisations.[77]

(c) Relaxation of the customs regime

14.72 The relaxation of local legislation may stem from customs incentive granted in favour of the private operator.

The principle of exemption of custom duties is widespread in terms of foreign investment and is covered by detailed provisions within investment codes. These incentives are also present in mining and petroleum codes, which grant the private operator, and sometimes its subcontractors[78] or contractors,[79] a customs derogation regime.

14.73 The customs regime that applies to mining and petroleum operations derives from mining and petroleum laws and exploitation agreements, although, exceptionally, investment codes must be considered.[80]

[Free translation into English: '[f]or each profitable year, deficits and amortisation will be allocated in the following order 1- previous deficits, 2- deferred amortisation, 3- amortisation for the year in question'].

[75] See in particular, in Algeria, Article 168 of the Law No. 01-10 of 3 July 2001 Enacting the Mining Code.

[76] See, in particular, in Guinea, Article 83 of the Law No. L/2014/N°034/AN of 23 December 2014 Enacting the Petroleum Code.
See also, in Chad, Article 77 of Law No. 006-PR-2007 of 2 May 2007 On Hydrocarbons.

[77] This involves, in particular, within the WAEMU area, the West African Accounting System (SYSCOA), which was enacted by Regulation No. 04/96/CM/UEMOA dated 20 December 1996. This unified accounting system was later amended by Regulation No. 05/2013/CM/UEMOA of 28 June 2013 Amending SYSCOA and Implementing Regulation No. 005/2014/COM/UEMOA of 31 May 2014, which lays down the new accounting rules and methods of SYSCOA. In the OHADA region (Organisation for the Harmonisation of Business Law in Africa), it is also worth mentioning the Uniform Act on the Organisation and Harmonisation of Company Accounting supplemented in January 2017 by the Uniform Act relating to Accounting and Financial Information.

[78] See, in particular, in Niger, Article 130 of Law No. 2007-01 of 31 January 2007 Enacting the Petroleum Code.

[79] See, in particular, in Morocco, Article 50 of Law No. 21-90 of 1 April 1992 On the Exploration and Exploitation of Hydrocarbon Deposits amended and supplemented by Law No. 27-99 of 15 February 2000; in Tunisia, Article 116(1) of Law No. 2002-21 of 14 February 2002 Supplementing Law No. 99-93 of 17 August 1999 Enacting the Hydrocarbons Code and Supplementing and Amending the Hydrocarbons Code.

[80] On the application of investment codes to mining and oil projects, see para 6.5 above.

Besides mining and petroleum law, national legislation can grant a more advantageous customs regime to certain extraction projects.[81]

Lastly, the customs regime that applies to mining and oil operations is also determined, at a regional level, by the Community Mining Code of the WAEMU.[82]

14.74 The two main advantages given to private operators within the context of their mining and oil operations consist of temporary admission and exemptions.

14.75 The temporary admission regime enables, under certain conditions, the import, completely or partially free from customs duties and taxes, of goods intended for re-export with a view to using them temporarily in the host state's territory[83] .

14.76 The customs incentives granted under exploitation agreements and national legislation tend to fluctuate depending on the exploitation phase considered.[84] While the exploration and development phases generally give rise to the granting of many benefits, the production phase sees a decrease or disappearance of the customs incentives. In this way, the customs exemptions granted to private operators may be limited to prospection and exploration works.[85] They may also extend to the exploitation phase, only with regard to the development phase[86] or until the date of first production.[87]

[81] See, in particular, in Niger, Law No. 2008-30 of 3 July 2008 Granting Derogation Benefits for Investments in Major Mining Projects.

[82] See Articles 25 et seq of Regulation No. 18/2003/CM/UEMOA of 23 December 2003 Adopting the Community Mining Code.

[83] As examples of admission under the temporary admission regime, see, in particular, in Togo, Article 53 of Law No. 96-004 of 26 February 1996 Enacting the Mining Code.

[84] This dichotomy between customs advantages granted during the exploration phase and the exploitation phase is also included in the WAEMU Mining Code (see Articles 26 and 28 of Regulation No. 18/2003/CM/UEMOA of 23 December 2003 Enacting the Community Mining Code).

[85] See in particular, in Morocco, Article 50 of Law No. 21-90 of 1 April 1992 On the Exploration and Exploitation of Hydrocarbon Deposits, amended and supplemented by Law No. 27-99 of 15 February 2000.

[86] See, in particular, in Senegal, Article 49 of Law No. 98-05 of 8 January 1998 Enacting the Petroleum Code:

'*[l]es matériels, matériaux, fournitures, machines et équipements, ainsi que les pièces de rechange, les produits et matériels consommables destinés directement et exclusivement aux opérations pétrolières sont exonérés pendant les phases de recherche et de développement de tous les droits et taxes, lors de leur importation en République du Sénégal par le ou les titulaires de convention ou de contrat de services ou par des entreprises travaillant pour leur compte et, dans la mesure où lesdits matériels, matériaux, fournitures, machines et équipements, pièces de rechange, produits et matières consommables ne sont pas disponibles en République du Sénégal dans des conditions équivalentes en termes de qualité, quantité, prix, délais de livraison et de paiement*'.

[Free translation into English: '[t]he equipment, materials, supplies and machinery, as well as the spare parts and consumables intended directly and exclusively of oil operations are exempt during the exploration and development phases from any duties and taxes, upon their import into the Republic of Senegal by the holders of agreements or service contracts or by companies working on their behalf and, insofar as the said

14.77 Furthermore, in order to promote the national market, mining and petroleum legislation can limit the granting of customs incentives to circumstances in which the imported goods are not available on the local market.[88]

The goods, materials and equipment benefiting from a preferential customs regime may be included in a list, which will generally be appended to the mining agreement[89] or attached to the mining rights.

14.78 In principle, the goods placed under the temporary admission regime or imported duty-free cannot be assigned, sold or transferred to another activity without the prior authorisation of the customs authorities.[90]

equipment, materials, supplies and machinery, spare parts and consumables are not available in the Republic of Senegal under equivalent conditions in terms of quality, quantity, price, lead times and payment terms'].

[87] See, in particular, in Togo, Article 53 of Law No. 96-004 of 26 February 1996 Enacting the Mining Code:

'*[l]e détenteur d'un permis d'exploitation, ses prestataires de services et fournisseurs bénéficieront de l'exonération de tous droit et de taxe de douane, de la taxe de statistique et de la TVA sur l'importation et l'exportation de tous les biens d'équipement, machines, véhicules utilitaires et outillages importés pour des activités minières relatives au titre minier jusqu'à la date de la première production commerciale, si le détenteur n'exporte pas sa production et, sans limite pour ses activités liées à l'exportation*'.

[Free translation into English: '[t]he holder of an exploitation permit, and its service providers and suppliers, are exempt from paying customs duties and taxes, the statistics tax and VAT on the import and export of any capital goods, machinery, commercial vehicles and tools imported for mining activities related to the mining rights until the date of first commercial production, if the holder does not export its production, and with no limit for its export-related activities'].

[88] See, in particular, in Morocco, Article 50 of Law No. 21-90 of 1 April 1992 On the Exploration and Exploitation of Hydrocarbon Deposits, amended and supplemented by Law No. 27-99 of 15 February 2000:

'*[t]outefois, les avantages ci-dessus ne seront pas accordés lorsque ces matériels, matériaux et produits consommables peuvent être fournis par le marché national dans la limite d'une majoration—prix CIF—de 10% à des conditions de qualités et de délais de livraisons équivalentes*'.

[Free translation into English: '[h]owever, the advantages mentioned above will not be granted if this equipment and these consumables can be supplied by the national market within the limit of an additional cost—CIF price—of 10% under equivalent quality and lead time conditions'].

[89] See, in particular, in Mauritania, Article 39 of Law No. 2012/012 Governing Mining Agreements and Approving the Standard Mining Agreement.

[90] See, in particular, in Mauritania, Article 45 of Law No. 2012/012 Governing Mining Agreements and Approving the Standard Mining Agreement:

'*[l]es biens mobiliers, matériels, équipements, véhicules et autres intrants importés dans le cadre du régime prévu à la présente convention ne peuvent être vendus sur le territoire national qu'après autorisation de l'Administration des Douanes et paiement des droits et taxes de douanes liquidés aux taux en vigueur et sur la base de la valeur résiduelle du bien à la date de la mise à la consommation. L'inobservation de cette disposition expose le titulaire aux pénalités édictées par la réglementation des douanes*'.

[Free translation into English: '[t]he movable goods, equipment, vehicles and other inputs imported under the regime provided for in this agreement can only be sold in the

14.79 The customs incentives granted to the private operator may also be applied to the export of the extracted substances. In this way, it is possible to provide that the raw materials extracted within the context of exploitation operations may be exported by the private operator free from any exit tax.[91]

14.80 In general, the personal effects of the expatriate personnel employed by the private operator are exempt from import and export duties and taxes.[92]

14.81 Lastly, the customs advantages granted to the private operator can be guaranteed by a stabilisation guarantee.

In principle, the private operator will not be exempted from following the customs clearance procedure and will have an obligation to complete the formalities provided for by regulations in force, particularly with regard to reporting requirements.[93]

2 THE INFLUENCE OF EXPLOITATION AGREEMENTS ON THE TAX REGIME APPLICABLE TO MINING AND OIL OPERATIONS

14.82 In addition to local tax legislation, exploitation agreements provide specific solutions. in terms of taxation. The interference of exploitation agreements in the tax domain results in particular from the inclusion, in the agreement itself, of tax obligations and in the interpretation by operators of local tax provisions.

a Creation of contractual tax obligations

14.83 The phenomenon associated with the creation of contractual tax obligations seems at first glance to be contradictory. How can a tax provision be created by an agreement that, as we have seen, brings together, on the one hand, a State or a public entity and, on the other hand, foreign or local private companies? This could only be possible with the express agreement of the legislative and executive authorities of the host state. These tax provisions will draw strength from the exploitation agreement itself. In any case, the tax

---

national territory following authorisation from Customs and payment of the customs duties and taxes at the rates in force and based on the residual value of the property on the date of its release for consumption. Failure to comply with this provision exposes the holder to the penalties laid down by customs regulations'].

[91] See, in particular, in Benin, Article 108 of Law No. 2006-18 of 17 October 2006 Enacting the Petroleum Code, which provides: '*[l]es substances minérales concessibles sont exemptes de droits de sortie à l'exportation*'. [Free translation into English: '[l]icensable mineral substances are exempt from exit fees upon export']. See also, in Tunisia, Article 106 of Law No. 2003-30 of 28 April 2003 Enacting the Mining Code; in Niger, Article 139 of Law No. 2007-01 of 31 January 2007 Enacting the Petroleum Code.

[92] See, in particular, in Mauritania, Article 95 of Law No. 2010-033 of 20 July 2010 Enacting the Crude Hydrocarbons Code amended by Law No. 2011-044 of 25 November 2011 and Article 44 of Law No. 2012/012 Governing Mining Agreements and Approving the Standard Mining Agreement; in Niger, Article 138 of Law No. 2007-01 of 31 January 2007 Enacting the Petroleum Code; in Chad, Article 89 of Law No. 006-PR-2007 of 2 May 2007 On Hydrocarbons.

[93] See in particular, in Mauritania, Article 50 of Law No. 2012/012 Governing Mining Agreements and Approving the Standard Mining Agreement.

provision established could not be applied generally. However, it can serve as a reference at a later date in terms of both contracts and regulations.

14.84 We can then talk about overlapping of specific 'contractual' taxation and general local 'legislative' taxation.[94]

14.85 Several elements can generate these financial obligations. In particular, they concern the recording of high profits for the private operator or low revenues intended for the local State operator.

14.86 The observation of low revenue to the detriment of the local State operator or the host state follows from the evaluation of the financial resulting from the application of local tax provisions or certain clauses of the exploitation agreement. Exploitation agreements can include provisions intended to take this situation into account.

14.87 Exploitation agreements can in particular provide that, in the event of operating results being considered too low, the private operator should pay a supplement to the local government.

b Contractual interpretation of tax provisions

14.88 In some agreements, operators assume supranational sovereign powers to assert the non-applicable nature of local tax legislation.

This involves contractual qualifications and assertions that should be ineffective and not binding on the host state. It is in fact impossible to accept that a contract between a public entity and a private entity can impose permanent solutions and decisions on the local tax authority.

This could only be conceived in the event that the agreement is elevated to a category of agreements directly sanctioned by the local public authority by granting it the force of law; again, the authority would have to confirm the contractual provisions in writing.

[94] This is what Pierre Legoux acknowledged as an 'ad hoc tax system' see Pierre Legoux 'Legal Aspects of Mining Development in Developing Countries' 1981 *NRF* 172 et seq.

15

FINANCING

15.1 During the first half of the century, the extractive industry—solid minerals and hydrocarbons—was virtually self-financing and external financing only accounted for a very small part of its needs. Gradually, the scale of exploitation projects affected to possibility of self-financing. The share of borrowing in the financing of mining and oil projects has continued to grow and new contractual forms of financing have emerged.

At present, the equity financing of mining and oil operations is an exception and the search by operators for outside capital is a major challenge. And yet, as things stand, mining and oil companies must deal with the difficulty, or even the impossibility, of raising the funds needed to finance their projects in Africa in particular.[1] Project selection is very stringent, favouring other parts of the world over Africa, in particular due to the lack of infrastructure.

The considerable financing needs push operators to call on third parties to the exploitation agreement, either for financing, or to guarantee proper fulfilment of the contractual obligations.

1 THIRD-PARTY INVOLVEMENT IN THE FINANCING OF EXPLOITATION OPERATIONS

15.2 A differentiation should be made depending on whether the project is financed on the local market or overseas. The exploitation agreement can in fact place restrictions on the third parties that operators can approach for financing and the conditions that they offer them, either by forcing them to look to the local market, or by forcing them to look to the external market. There may also be natural and political limitations requiring operators to look abroad. Foreign private operators sometimes prefer to rely on local financing so that they are less vulnerable to decisions involving a change in financial or tax conditions, or in very extreme cases nationalisation. As for the host state, it is the one most often opposed[2] , mainly because this weakens its financial market, which is generally already inadequate.

15.3 Often, the private operator, the host state or the local State operator spend a considerable amount of time looking for external financial support. This is often on top of the work that has already been performed and has revealed a commercial deposit of natural resources, the exploitation of which may require extensive and expensive infrastructures to be put in place. And yet, the financing of mining and oil projects is made difficult by the existence

[1] See on this subject, in the mining sector, Ilan Solomons op cit 42.

[2] Except when it wants to limit or even prevent any foreign industrial ownership in its territory.

of significant risks; many difficulties come to light within the context of obtaining financing, including the instability of the legal and fiscal conditions of exploitation. But the main obstacle standing in the way of investment is still the geological risk and the uncertainties regarding the quality and quantity of the natural resources contained in the deposit; this risk is never really eliminated.

15.4 Although the risks inherent to the exploitation of natural resources are distributed in the same way regardless of the sector, their coverage by third parties varies depending on whether or not hydrocarbons resources are concerned. Furthermore, natural gas and oil exploration often calls for much higher budgets than the search for solid mineral resources; this is especially the case since the exploration of hydrocarbons resources tends to concern offshore projects.

In the solid minerals sector, there is a tendency towards financial participation in the project by the buyers of the production. Indeed, before committing to any development of a site, operators seek outlets for the production and firm up these contacts through medium- or long-term agreements.

15.5 Because of the specifics of the subject, the international banking system has had to demonstrate innovation and the constantly evolving markets forces it to choose this path; indeed, in this field, traditional borrowing is generally unacceptable for operators.

15.6 Very few provisions in an exploitation agreement are dedicated to the organisation of financing, whereas it is the subject of fierce and long negotiation. A study of the provisions relating to financing contained in exploitation agreements would quickly prove inadequate. We therefore need to try and understand this question more broadly by studying both the third-party organisations (which, in practice) help to finance the extractive projects in Africa, and the contractual mechanisms that are used.

a Funding organisations

15.7 Funding organisations can be from a variety of sources, and their number has multiplied over time, in particular because of the expansion of the area of operation of international or national institutions and banks, but also as a result of the emergence of new players such as governments and investors from oil-producing countries. International oil companies have also maintained, perhaps even reinforced, their role.

i *International institutions*

15.8 Many international institutions were called upon to finance projects in Africa emerged during the second half of the 20th century. There are many international institutions, at the global and regional level. Using these institutions has numerous advantages, especially in that it helps to reduce the total financial cost of projects by obtaining generally preferential rates.

The loans awarded by these international institutions are based, in particular, on a desire to encourage large-scale projects on the African continent in order to provide for the development of the country and its local populations. However, the systemic weaknesses in the monitoring of projects funded by international backers have been greatly criticised by civil society.[3] Despite these criticisms, the existence of many control mechanisms helps to ensure that local development is taken into consideration within the context of mining and oil projects. These means include, first of all, the insertion into the financing agreement of provisions allowing for suspension of payments and even withdrawal of the loan in the event of the borrower not fulfilling its obligations in terms of local content. Moreover, it is worth mentioning the existence of inspection mechanisms aimed at taking better account of social requirements in projects financed by international backers.[4]

(1) The United Nations network

15.9 Since it was established in 1945, the United Nations has gradually focused its efforts on identifying a new international economic order. Its actions have mainly been channelled through the International Monetary Fund[5] and the World Bank Group.[6]

15.10 Historically, a very small part of the financial contribution of the World Bank Group went to the extractive industry. Indeed, the official guidelines of this institution have always been to leave private companies responsible for high-risk activities. However, these guidelines were revised in the wake of the 1973 oil crisis; imports of oil therefore broke the external equilibrium of the non-producing developing countries and the only way to re-establish it was to develop the national resources of these states to make these economies less dependent on foreign countries.

15.11 Nowadays, the United Nations' activities involving the financing of mining and oil projects are conducted by the World Bank and its various institutions, such as the International Bank for Reconstruction and Development (IBRD)[7] , the International Development Association

---

[3] See Valère Ndior 'Société civile et bailleurs de fonds internationaux: Un même combat?' in *Le contentieux extractif (ICC* 2015) 114. See, to that effect, Press Release No. 2015/332/ECR from the President of the World Bank Group, Jim Yong Kim, in which he recognises a lack of monitoring of projects financed by the institution.

[4] In this regard, see Ndior op cit 114.

[5] It was founded in 1945 in order to encourage smooth running of the international monetary system by aiding to eliminate exchange restrictions and establish a multilateral payment system. The Fund provides indirect technical support for the financing of projects.

[6] See in particular Efrain Friedmann 'Energy Activities of the World Bank' 1978 *NRF* 91; Hasan S Zakariya 'The Petroleum Lending Program of the World Bank' (1983) 17 (6) *Journal of World Trade* 471; Andrew Loewinger 'Multilateral Funding for Mineral Exploration in the Third World' (1980) 14 (6) *Journal of World Trade* 478 et seq; with respect to the World Bank's role in the financing of technology transfers, see in particular Georges R Delaume 'Le financement international des transferts de technologie et le développement: le rôle de la Banque mondiale' in *Transfert de technologie et développement* (Librairies Techniques 1977) 267.

[7] IBRD was founded in 1944. Originally set up to support the reconstruction of Europe after the Second World War, it now helps developing countries to reduce poverty and promote

(IDA)[8] and the International Finance Corporation (IFC).[9] The World Bank is heavily involved in mining and oil projects on the African continent as a backer or by way of equity investments.[10]

(2) *Financial institutions with a regional focus*

15.12 Many institutions have been set up at regional level; with regard to the African continent, the African Development Bank (AfDB)[11] occupies pride of place. Founded in 1964 by the Khartoum agreement, the AfDB has 78 member countries that have subscribed to its authorised capital, including 54 African countries.

With the goal of promoting private investment in the African Member states, the AfDB helped to launch the African Development Fund in 1972. This organisation is the concessional window of the AfDB and helps to promote economic and social development in the least developed African countries by providing funding.

In 1975, the African Development Bank played a key role in the creation of the Association of African Development Finance Institutions (AADFI), which groups together banks and financial institutions that are committed to financing development in Africa.

ii *Banks*

15.13 In the extractive industries sector, the inadequacy of public funding options makes private investment unavoidable.

Nonetheless, banks seem particularly reluctant when it comes to taking risks when financing mining and oil projects and prefer to provide short-term

---

economic growth. The projects financed by the IBRD cover all sectors and in particular the extractive industry. (http://www.banquemondiale.org/fr/about/what-we-do/brief/ibrd).

[8] Founded in 1960, IDA lends money to the poorest nations on concessional terms. This means that IDA credits have a zero or very low interest charge and repayments are stretched over 25–38 years, including a 5–10 year grace period. IDA also provides grants to countries at risk of debt distress. In addition to concessional loans and grants, IDA provides significant debt relief through the Heavily Indebted Poor Countries (HIPC) Initiative and the Multilateral Debt Relief Initiative (MDRI). In the past, this organisation approved a credit of $2,6 million to Liberia in April 1985 in order to promote oil exploration work and reinforce the resources of the Hydrocarbons Office of its Ministry of Mines and Energy. IDA is also heavily involved in technical support. In 1985, it selected two projects of this type in Kenya and Northern Yemen, financing two thirds of them. This project involved providing support in the form of consultants to help reinforce the new Ministry of Petroleum and Mineral Resources after discoveries were made in the north-east.

[9] Unlike IDA, the IFC (International Finance Corporation)'s activity is focused exclusively on the private sector. Founded in 1956, it is legally independent and its funds are separate from those of the World Bank.

[10] The IFC has been recently involved in the Ghana Jubilee offshore project in the form of a $100 million loan to Kosmos Energy and $165 million to Tullow Oil. In Guinea, the IFC acquired a 5% share in the project in 2006 and since then has invested $185 million. In Tanzania, the IFC loaned Petra Diamonds Limited $40 million. In Mozambique, the IFC got involved within the context of a mining project operated by Baobab Resources plc.

[11] The Agreement Establishing the AfDB was adopted and opened for signature at the Khartoum Conference on 4 August 1963.

finance;[12] in this way, they generally get involved at the development stage when commercial deposits have been discovered and even more easily at the end of a successful exploration campaign.

15.14 Often banks are not getting involved on their own because of the high risk and the amount of capital needed. They organise themselves into banking pools as long as financing of a project can be combined by involving several contracts.

iii *States*

15.15 Some states, primarily exporting states, with high financial revenue seek to make the most out of this revenue. There are two possible types of action depending on whether they want to act alone or in groups.

(1) Isolated action of one state

15.16 A single state can contact a multinational company, either to sign a global agreement or to take part in an exploitation project.

The State can in particular participate within the context of creation of a joint body with an international oil or mining company that has technological expertise and is therefore sufficiently credible.

15.17 A state can also get involved through equity investment in selected exploitation projects. In this case, the State appears as a third party to the exploitation agreement.

15.18 Moreover, the host state can grant subsidies, for research in particular.

15.19 Less commonly, a government can provide a grant to another state, specifically intended to directly finance exploitation of natural resources or develop the corresponding infrastructure.

(2) Joint action by several states

15.20 Joint financial action by several states is particularly developed in Arab countries. Within the context of this collaboration, several institutions have been set up; these include in particular the Islamic Development Bank, the Arab Fund for Economic and Social Development (AFESD) and the OPEC Fund.

15.21 The aim of the Islamic Development Bank, founded in 1975, is to promote the economic and social development of its Member states and the Muslim community as a whole in accordance with Sharia law. It acts as a backer and through equity investment.

15.22 The Arab Fund for Economic and Social Development helps to finance economic and social projects in Arab countries. Its actions involve in particular awarding loans for public projects and financing projects in the private sector.

---

[12] See in particular, to that effect, Thierry Tanoh 'Le financement de projet en Afrique subsaharienne' (2006) 235 *Banque & Stratégie* 14.

15.23 The OPEC Fund for International Development, created in 1976, has a clear external focus and assists all industrialising countries apart from the OPEC Member states. It aims to promote development and encourage South-South partnerships. Its actions involve in particular providing financial support in the form of development programme loans.

iv *Multinational companies*

15.24 Concerning multinational corporation, the parent company very rarely gets involved directly to find funds for its subsidiaries' activity abroad, but it will be present as a guarantor. Of course, it may still participate directly in financing out of its equity.

b Financing techniques

15.25 Financing techniques for mining and oil projects in Africa respond to issues related not only to the specific nature of the African continent,[13] but also the economic and financial significance of such projects.

15.26 Traditionally, mining and oil projects were financed by way of loans granted by financial institutions, plus the funds invested by the project's participants. Since the recent economic difficulties, national and international legislation has become stricter in terms of control of banking transactions. Besides these restrictions, there are other factors put forward by lenders to justify their reluctance to finance long-term projects on the African continent, in particular the political instability and the lack of confidence in the abilities of the public authorities.[14] However, the primary cause of a reduction in lender action is undoubtedly related to the current economic climate and the drop in raw material prices, which greatly affects the profitability of mining and oil projects. This reluctance by the banks has encouraged the development of alternative financing methods.

15.27 The financing needs of mining and oil projects vary over time, depending on the progress made with the project. Traditionally, it is possible to break performance of an oil project down into several phases: an initial exploration and investigation phase, a second development and production phase, and then a transformation and transport stage. Because of the risks inherent in the first phase, project operators tend to use in-house financing techniques during the exploration period. Then, when the resource has been identified, operators can call upon players outside of the project during the development and production phase.[15]

15.28 When looking for funding, operators should also take account of purely financial criteria in order to achieve the right balance between the different techniques and optimise the project in terms of taxation and finance. By way of an example, debt financing has a cost as the debt is

[13] As an example, the operator should take cultural, religious and environmental requirements into consideration. Some African countries demand compliance with Sharia law, which means that projects must be funded in accordance with the principles of Islamic finance.

[14] Tanoh op cit 14.

[15] Ernst & Young *Funding challenges in the oil and gas sector* (2014) at 2.

associated with an interest rate, so arbitration between equity financing (self-financing or increase in capital) and debt financing depends on the interest rate and the level of debt already completed. Tax-related factors are also taken into consideration because if it is associated with an interest rate, debt financing also has the advantage of appearing as an expense in the company's accounts, reducing its result and therefore its taxable base.

15.29 Generally speaking, bank loans are still the most common source of finance, followed by bond loans, then project finance. The evolving and case-based nature of the subject is not conducive to a full understanding of all of the financing techniques. Indeed, the financing of mining and oil projects is a field that is constantly changing, in which the pressing need for 'made-to-measure' financing forces those involved to be innovative.

15.30 When it comes to financing techniques, a distinction exists between equity financing on the one hand, and debt financing on the other.

i *Equity financing*

15.31 Financing techniques have changed over time, adapting to the needs of the economic operators in this sector. Although self-financing was possible before, it now seems completely obsolete to rely solely on this mechanism. Indeed, the considerable sums invested in mining and oil projects often make it difficult, or even impossible, to use equity financing for exploitation operations, especially for juniors involved in the extractive sector.

However, equity financing is still current and refers to several concepts, namely self-financing, private equity and financing by capital markets.

(1) Self-financing

15.32 Self-financing means a company funding its own activity, and its investments in particular, using its existing equity, such as its savings or its depreciations.

All financial investment arrangements must in part be self-financed. Self-financing, which is a source of financing, is also essential for obtaining a loan. Indeed, financial institutions will judge the seriousness of the operation, and the importance of the financial risk it represents, by taking heed of the level of investment in the project by the operating company.

(2) Private equity financing

15.33 Private equity financing concerns investment in companies not listed on the stock exchange. With this method of financing, the investor provides new funds to the company in the form of subscription of shares newly issued by it within the context of an increase in capital.

The aim of the operation is both to share the profits generated by the company and also—in the longer term—to realise capital gains when disposing of the company's securities.

15.34 One of the most recent trends in closed-end private equity is 'crowdfunding', or 'equity-based crowdfunding'. This method is used to appeal to a large number of people to finance a project without going through traditional financing players.[16] In practice, financing is done via dedicated online platforms. Although originally restricted to non-profit, artistic or leisure projects, we are now seeing the emergence of crowdfunding platforms dedicated to financing mining and oil projects.[17] This can become a real alternative to traditional financing mechanisms.

15.35 Another trend is the emergence, within the context of mining and oil projects, of '*business angels*'. A business angel is an individual who invests in a company that has potential and who not only provides capital, but also puts his expertise, experience and network at the company's disposal. His investment in the company represents high added value for this company and the investment is referred to as angel financing. This technique for financing mining and oil projects tends to be developing within the context of fluctuating raw material prices.[18]

(3) Capital market financing

15.36 With financing through the capital markets, a distinction exists between companies offering their securities on financial markets, on the one hand, and unlisted companies, on the other.

In the same way that a company raises capital through private equity, it can, if it is a listed company, get financing from the capital markets.

It is also worth mentioning Private Investment in Public Equity (PIPE) whose aim is to increase a listed company's capital. This does not result in a public offering but is reserved to certain private equity funds.

ii *Debt financing*

15.37 The considerable amount of the sums invested in mining and oil projects makes it difficult, or even impossible, to fund operations relying exclusively on the equity of the operators and often requires these operators to resort to other financing methods, including bank loans, bond financing, the use of national institutional credit, factoring or even project finance.

(1) Bank loans

15.38 Bank loans are the most conventional solution for financing mining and oil projects. For well-established companies, bank financing is a good solution, as the only cost of the financing is a relatively low interest rate.

Due to the tightening of banking legislation following the economic crises of recent years, the conditions for accessing this type of financing are increasingly tough. The companies to which the banks can award loans are those that do not need it, so this method is not easily accessible to juniors. This financing calls for guarantees and securities that are hard to come by.[19]

[16] To that effect, it is said to be 'disintermediated'.
[17] See, in particular, on this subject, the EnergyFunders crowdfunding platform.
[18] For further information on this financing method, see in particular the Oil and Gas Angel Investors website.
[19] Ernst & Young op cit 2.

In addition, the current context, which is marked by a drop in raw material prices, does not encourage the banks to get involved. They are still reluctant when it comes to investing in mining and oil projects, of which the profitability is threatened by the international economic climate.

(2) Bond financing

15.39 Given the difficulties encountered by the operators of exploitation projects to finance their activities through bank loans, bond financing has in recent years become more widespread.[20]

The mechanism involves the operator issuing negotiable debt securities to investors.

The advantage of this technique is that it can be easily customised by the issuer of the securities, in terms of maturity, return or opting between conversion and redemption as shares.

The disadvantage lies in its inaccessibility for medium-sized companies or lower-rated companies.

(3) The use of national institutional credits

15.40 This financing mechanism has been widely used in the oil industry and is becoming popular in mining. It involves financial contributions from the state of origin of the private operators with the intention of obtaining equipment and the provision of services in its territory or in the territory of the host state.

It is generally very difficult to put a figure on the extent of these commitments and to evaluate the actual impact on the economic development of the State in the territory where exploitation takes place.

It involves the government in the country where the private operator is encouraging its companies and banks to intervene in certain foreign states.

(4) Factoring

15.41 Factoring is a financing technique founded on the assignment of receivables to the project of a specialist establishment, the factor. This alternative technique can be used in particular to respond to a need for cash flow. The factoring operation presents no specific risks for the factor, especially if the transferred debtor has a good reputation for debt settlement.

In terms of natural resources, this technique enables operators to obtain cash flow for the needs of their activities. The private operator can use factoring for its accounts receivable or for invoices that have been raised but not yet paid.

(5) Project finance

15.42 Project finance[21] is a method of long-term financing of industrial or infrastructure projects, repayment of which is based not on the debtor's

[20] Id.

[21] See in particular on the subject, Fiona Scott and Claus-Peter Martens *International Project Finance* (Transnational Publishers 2000); Deborah Demasi & Kenneth B Weiner *Inter-*

assets but on the revenue that will be generated by the project.[22] This type of financing assumes the presence of various stakeholders, including those investing in the operating company's capital, the so-called sponsors, a banking pool, lending institutions and even insurers.

Project finance increased significantly in the 1960s under the impetus of the large US banks. It initially appeared for the exploitation of natural resources in the North Sea, and experienced significant growth in the 1970s. The project finance mechanism then expanded considerably in emerging countries in the 1990s.[23]

15.43 The most important feature of project finance is that lenders assume some of the risks inherent in the exploitation operations; they agree that repayment depends solely on the revenue generated by the project. Therefore, the lender's decision is based solely on a projected evaluation of profitability[24] . The identification and allocation of risks is therefore an essential component of this type of financing; indeed, the project may be subject to a number of technical, environmental, economic and political risks, especially in developing countries.

Project financing requires the constitution of an independent legal entity, the project company. With project finance, lenders only sign a contract with this project company. The guarantee given is therefore limited to its assets (non-recourse financing). In practice, lenders will demand guarantees directly from the operators to guard against the said structure's low levels of equity and to encourage operators not to withdraw from the project. This temptation would be even greater if failure were to occur during exploitation, whereas the capital invested has already been recovered and the current net value of future cash flow falls below the refinancing cost.

(6) Mezzanine debt

15.44 Mezzanine financing is a hybrid of debt and equity financing.

15.45 It is a loan, repayment of which depends on the repayment of 'senior' debt; the mezzanine lender will be the last one to be paid in the event of difficulties. In return, the lender will be able to convert its debt into equity at a given price if it is not repaid in full and on time.

As Mezzanine debt is by nature very risky for the lender, it tends to be more expensive than other types of financing. The mezzanine lender will only get involved if the operators have good cash flow, good earnings, growth history and a good reputation.

---

national Project Financing 4th ed (Juris Publishing 2011); Tanoh op cit 14; John D Finnerty *Project Financing: Asset-Based Financial Engineering* 2nd ed (John Wiley & Sons, Inc 2007).

[22] Yves Simon and Samir Mannaï *Techniques Financières Internationales* 7th ed (Economia 2002) 639 et seq.

[23] For a historical look at how the use of project financing has changed, see Michel Cremieux 'Le financement des infrastructures: le rôle des investisseurs financiers' (1999) 51 (51) *Revue d'économie financière* 217.

[24] See in particular, Arlète Tonye *Pratique juridique des financements structurés en Afrique* (L'Harmattan 2010) 65.

2 THIRD-PARTY INVOLVEMENT FOR GUARANTEEING FULFILMENT OF CONTRACTUAL OBLIGATIONS

15.46 The involvement of third parties may be required to guarantee proper fulfilment of the project. Such guarantee may cover the fulfillment of the exploitation agreement and the related obligations. It may also concern the risks associated with performing the exploitation operations.

a Third-party involvement to guarantee fulfilment of the exploitation agreement

15.47 The third parties involved to guarantee proper fulfilment of the exploitation agreement may be connected to the operating company. This would most likely be the parent company, which will intervene to guarantee fulfilment by its subsidiary of the exploitation agreement. The guarantee given within the context of fulfilment of the agreement may also result from the involvement of independent third parties.

i *Guarantee given by a parent company to its subsidiary*

15.48 Recourse to a guarantee given by a parent company to its subsidiary is commonly used within the context of mining and oil projects. This guarantee may take the form of a parent company guarantee[25] or be issued by the company's shareholders.

The parent company may act as guarantor for all of its subsidiary's obligations. In principle, the parent company's guarantee subsists until the start of production, ie until the financial institutions take over entirely.

The huge responsibility of the parent company in some cases justifies the direct involvement of its representatives, who may for example add their signature to the exploitation agreement.

In general, the guarantee given by the parent company will be appended to the exploitation agreement using a predefined template, negotiated and accepted prior to signature of the exploitation agreement.

15.49 The parent company's involvement in guaranteeing fulfilment of the obligations arising from the exploitation agreement is fully justified because of the links between the parent company and the operator. The guarantor and the operator benefiting from the guarantee are separated in appearance only. Of course, the independence between the private operator and the parent company exists in legal terms but is much more questionable from an economic point of view.

ii *Guarantee given by financial institutions: the performance guarantee*

15.50 The purpose of giving a performance guarantee is to guarantee completion of the whole project through the fulfilment by the private operator of all of its obligations. Through these clauses, an appointed third

[25] See in particular, in the Seychelles, the 2013 Model Petroleum Agreement (Second Schedule Part II).

party provides a financial guarantee of correct performance of the exploitation works as provided for in the agreement. Such provisions are long-standing and appeared in some of the first major concession contracts, although the amount of the guarantee has increased significantly since then. Through this mechanism, the guarantor undertakes not only to pay penalties, but also to assume responsibility itself for performance of the mining or oil project according to the terms of the exploitation agreement. The risks are therefore considerable, all the more so as the guarantor is not protected from discretionary decisions made by the host state that are harmful to the interests of the private operator that it guarantees, and that are therefore indirectly harmful to its own interests.

15.51 The performance guarantee provided by the private operator may take several forms, such as a letter of credit,[26] a security[27] or even a first-demand guarantee.[28]

The financial guarantee may cover all of the private operator's obligations or only concern one phase of the operations. In this way, if the guarantee is for the operation in its entirety, it guarantees all of the obligations contained in the exploitation agreement.[29] The guarantee may also concern a specific phase of the operations, such as the exploration period.[30] In this case, the amount of the guarantee will cover the minimum financial commitments to be spent during the exploration phase(s).[31] Lastly, the operator may have a duty to provide a specific guarantee that covers the decommissioning and rehabilitation of mining sites.

---

[26] See, in particular, in the Seychelles, the 2013 Model Petroleum Agreement (Second Schedule Part II).

[27] See in particular, in Chad, Article 10 of Law No. 006-PR-2007 On Hydrocarbons.

[28] See, in particular, in the Democratic Republic of Congo, Article 7 of the Production-Sharing Agreement Between the Democratic Republic of Congo and the Association: Dominion Petroleum Congo & Soco Exploration-Production RDC & La Congolaise des Hydrocarbures sur le Bloc V du Graben Albertine (November 2007).

[29] See, in particular, in South Africa, section 89 of the Mineral and Petroleum Resources Development Act 28 of 2002; in Algeria, Article 36 of the Presidential Decree of 1 March 2003 Enacting the Model Mining Agreement; in Angola, Article 32 of Petroleum Law No. 10/04 of 12 November 2004; in Libya, Article 9 of Petroleum Law No. 25/55 of 11 May 2012; in Togo, Article 56 of Law No. 99-003 of 18 February 1999 Enacting the Hydrocarbons Code.

[30] See, in particular, in Democratic Republic of Congo, Article 7 of the Production-Sharing Agreement Between the Democratic Republic of Congo and the Association: Dominion Petroleum Congo & Soco Exploration-Production RDC & La Congolaise des Hydrocarbures sur le Bloc V du Graben Albertine (November 2007); in Djibouti, Article 17 of Law No. 117-AN-05 of 16 October 2005 Enacting the Petroleum Code; in Guinea, Article 25 of Law No. L/2014/N°034/AN of 23 December 2014 Enacting the Petroleum Code; in Mauritania, Article 24 of Law No. 2010-033 of 20 July 2010 Enacting the Crude Hydrocarbons Code amended by Law No. 2011-044 of 25 November 2011; in Mozambique, Article 46 of Petroleum Law No. 21-2014 of 18 August 2014; in Sierra Leone, Article 47(1) of the Petroleum Exploration and Production Act, 2011.

[31] See, in particular, in Sierra Leone, Article 47(1) of the Petroleum Exploration and Production Act, 2011; in Madagascar, Article 39 of Law No. 96-018 of 4 September 1996 Enacting the Petroleum Code.

15.52 While certain legislation provides that the requirement to subscribe a performance guarantee is left to the discretion of the competent authority,[32] other texts make the provision of the guarantee a prerequisite for granting mining rights.[33]

In general, mining and petroleum laws require the guarantee to be contracted from a leading financial institution,[34] which should in certain cases be approved.[35]

Some texts make the obligation to provide a performance guarantee conditional on the nationality of the private operator. In terms of these provisions, only contracting companies of foreign origin will have to provide a guarantee.[36] At the same time, the host state can also require foreign operators to domicile in the national territory the funds required to cover all of its local needs.[37]

15.53 The fees for obtaining a performance guarantee may be considered as a deductible expense in the financial year in which the expenses are incurred.[38]

15.54 The performance guarantee sometimes represents a significant risk for the private operator, particularly within the context of first-demand guarantees.[39] This mechanism is also the subject of many disputes within the context of international agreements, in particular as a result of abuse or fraud. Indeed, the security can be called by the beneficiary upon claim on its part without it having to establish non-performance of one of the services

---

[32] See in particular, in Ethiopia, Article 71 of Proclamation No. 678/2010 Promoting Sustainable Development of Mineral Resources; in Malawi, Article 91 of the Mines and Minerals Act, 1981 and Article 46 of the Petroleum Act, 1983; in Togo, Article 48 of Law No. 96-004 of 26 February 1996 Enacting the Mining Code and Article 56 of Law No. 99-003 of 18 February 1999 Enacting the Hydrocarbons Code.

[33] See in particular, in Chad, Article 27(3) of Law No. 006-PR-2007 of 2 May 2007 On Hydrocarbons.

[34] See in particular, in Algeria, Article 43(2) of Law No. 13-01 of 20 February 2013, amending and supplementing Law No. 05-07 of 28 April 2005 On Hydrocarbons; in Democratic Republic of Congo, Article 7(1) of the Production-Sharing Agreement between Democratic Republic of Congo and the Association: Dominion Petroleum Congo & Soco Exploration-Production RDC & La Congolaise des Hydrocarbures sur le Bloc V du Graben Albertine (November 2007); in Guinea, Article 25 of Law L/2014/No. 034/AN of 23 December 2014 Enacting the Petroleum Code; in São Tomé and Principe, Article 7(11) of the Model Production Sharing Agreement.

[35] See in particular, in Algeria, Article 43(2) of Law No. 05-07 of 28 April 2005 On Hydrocarbons; in Tunisia, Article 7(2) of Law No. 99-93 of 17 August 1999 Enacting the Hydrocarbons Code.

[36] See in particular, in Algeria, Article 36 of the Presidential Decree of 1 March 2003 enacting the Model Mining Agreement.

[37] See in particular, in Madagascar, Article 40 of Law No. 96-018 of 4 September 1996 Enacting the Petroleum Code.

[38] See in particular, in Mauritania, Article 120 of Law No. 2008-011 of 27 April 2008 Enacting the Mining Code.

[39] The terminology may vary, so the OHADA law uses the term 'letter of guarantee' (Article 28 of the OHADA Uniform Act organising securities, adopted on 15 December 2010). See on this subject, Roger Nevry 'La garantie autonome à première demande: droit OHADA, droit français' (2007) 3 *Revue de droit des affaires internationales* 313.

provided for under the exploitation agreement.[40] As the guarantor is not released from its obligation in the event of manifest fraud or abuse, it is an independent guarantee of which the risk of wrongful calling must not be minimised.

However, the parties—which have complete flexibility when it comes to inserting conditions into the guarantee contract—can choose to use the so-called revolving guarantee mechanism. The amount of the cover will gradually be reduced as the obligations covered are fulfilled.[41] This condition has definite advantages in that it ensures cleaning up of the guarantee and reduces the risks associated with the automatic nature of demand guarantees, as the amount called corresponds to the amount of the non-performance.[42] .

b Third-party involvement in guarantees against the risks of the project

15.55 The considerable risks generated by the exploitation of natural resources make it essential to take out specific insurance policies.

i *Contractual insurance*

15.56 Operators are particularly at risk of suffering losses resulting from a variety of events and may be required to compensate third parties that have suffered damages as a result of the mining or oil operations. This is why it is often mandatory to take out insurance pursuant to the exploitation agreement,[43] or even mining and petroleum laws themselves.[44] Less commonly,

[40] The first-demand guarantee is independent of the agreement, such that no exception may be made with regard to its non-fulfilment, other than the case of manifest fraud. With respect to the concept of first demand guarantee in international trade, see in particular Jean-Pierre Mattout *Droit bancaire international* 4 ed (Revue Bancaire, 2009) 207 et seq. See also, the ICC Rules for First Demand Guarantees (on this subject, see in particular Stéphane Piedelièvre, 'Les nouvelles règles de la CCI relatives aux garanties à première demande' 2010 *RTD Com* 513).

[41] See in particular, in Angola, Article 32(5) of Petroleum Law No. 10/04 of 12 November 2004; in São Tomé and Principe, Article 7(12) of the Model Production Sharing Agreement.

[42] Stéphane Piedelièvre 'Garantie à première demande' *Rep com* (Dalloz, 2008) (updated September 2011) no 112 et seq.

[43] See in particular, in Algeria, Article 8(4) of the Presidential Decree of 1 March 2003 Enacting the Model Mining Agreement; in the Seychelles, Article 20 of the 2013 Model Petroleum Agreement.

To this end, some mining and petroleum laws make reference to the exploitation agreements when it comes to laying down the terms and conditions of the insurance that the operator must suscribe. For instance, in particular, in Cameroon, Article 62(3) of Law No. 99-013 of 22 December 1999 Enacting the Petroleum Code; in Côte d'Ivoire, Article 64 of Law No. 96-669 of 29 August 1996 Enacting the Petroleum Code; in Namibia, Article 49(2) of the Minerals Prospecting and Mining Act, 1992 and Article 13(2) of the Petroleum (Exploration and Production) Act, 1992; in Uganda, Article 18(2) of the Mining Act, 2003.

[44] This concerns in particular, in Algeria, Article 72 of Law No. 01-10 of 3 July 2001 Enacting the Mining Code; in Angola, Article 153 of Mining Law No. 31/11 of 23 September 2011; in Eritrea, Article 30 of Proclamation No. 108/2000, A Revised Proclamation to Govern Petroleum Operations; in Ethiopia, Article 19 of Proclamation No. 295/1986 to Regulate Petroleum Operations; in Guinea, Article 45 of Law No. L/2014/No. 034/AN of 23 Decem-

the applicable texts may refer to a decision by the responsible Minister to determine whether the insurance taken out by the operator within the context of its activity is mandatory.[45]

The main purpose of insurance taken out with third parties to the exploitation agreement is both to safeguard the operators and to guarantee continuation of the operations despite the occurrence of certain events that may endanger the mining or petroleum project. The risks covered by the insurer include, in particular, damages caused to mining installations, third parties, environment and even employees working within the context of mining and oil operations.[46] The damages caused to the environment have gradually taken on primary importance in the cover required.

It is the responsibility of the private operator to take out these policies and to bear the costs, which are generally deductible[47] as charges from the operating result.

The obligation to take out insurance can be extended to third parties to the exploitation agreement who are involved in the mining and oil operations, such as subcontractors.[48]

---

ber 2014 Enacting the Petroleum Code; in Malawi, Article 61 of the Petroleum Act, 1983; in Mali, Article 136 of Law No. 2012-015 of 27 February 2012 Enacting the Mining Code; in Mauritania, Article 32 of amended Law No. 2010-033 of 20 July 2010 enacting the Crude Hydrocarbons Code; in São Tomé and Principe, Article 75 of the Framework Law on Oil Operations; in South Sudan, Article 84 of the Petroleum Act, 2012; in Swaziland, Article 153(1) of the Mines and Mineral Act, 2011; in Tanzania, Article 111 of the Mining Act, 2010; in Chad, Article 57(1) of Law No. 006-PR-2007 of 2 May 2007 On Hydrocarbons; in Tunisia, Article 71 of the Mining Code and Article 89(2) of the Hydrocarbons Code; in Zambia, Article 64 of the Petroleum (Exploration and Production) Act, 2008.

[45] See in particular, in the Seychelles, Article 40(2) of the Petroleum Mining Act; in Zambia, Article 64(3) of the Petroleum (Exploration and Production) Act, 2008.

[46] A list of the risks covered by the insurance policy is generally contained in mining and petroleum laws, as well as in the exploitation agreements. See in particular, in Angola, Article 153 of Mining Law No. 31/11 of 23 September 2011; in Cameroon, Article 118 of Decree No. 2000-485 of 30 June 2000 Establishing the Conditions for Implementation of Law No. 99-013 of 22 December 1999 Enacting the Petroleum Code; in Eritrea, Article 30 of Proclamation No. 108/2000, A Revised Proclamation to Govern Petroleum Operations; in the Seychelles, Article 20(3) of the Model Petroleum Agreement, 2013; in Tanzania, Article 111(2) of the Mining Act, 2010; in Tunisia, Article 71 of Law No. 2003-30 of 28 April 2003 Enacting the Mining Code and Article 89(2) of Law No. 99-93 of 17 August 1999 Enacting the Hydrocarbons Code.

[47] Indeed, in many cases, the costs arising from insurance premiums are included in the overheads category, accepted as charges and therefore debited to the balance sheet, especially for calculating tax. See in particular, to that effect, in Benin, Article 73 of Law No. 2006-17 of 17 October 2006 Enacting the Mining Code and Mining Taxation in the Republic of Benin; in Democratic Republic of Congo, Article 256 of the Law No. 007-2002 of 11 July 2002 Enacting the Mining Code and Article 523 of Decree No. 038-2003 of 26 March 2003 Enacting Mining Regulations; in Mali, Article 11(14) of the Production-Sharing Agreement; in Chad, Article 60 of the Law No. 011-PR-1995 of 20 June 1995 Enacting the Mining Code and Article 77 of Law No. 006-PR-2007 of 2 May 2007 On Hydrocarbons.

[48] See in particular, in Eritrea, Article 30 of Proclamation No. 108/2000, A Revised Proclamation to Govern Petroleum Operations; in Mauritania, Article 32 of Amended Law No. 2010-033 of 20 July 2010 Establishing the Hydrocarbons Code; in the Seychelles, Article

15.57 The host state generally intends to establish a requirement, during negotiations, to use local insurance firms; and yet, their effectiveness is often relative considering the specific issues at stake in extractive projects and their limited international experience. This lack of knowledge and maturity often appears in the policy wording. Mining and petroleum laws echo this tendency by stating that the insurance policy should be taken out with a national insurance company, an approved company[49] or a company represented in the host state.[50]

The private operator is usually free to choose the insurance firm, even though some regulations require certain conditions to be met when exercising this discretion. The operator should in particular demonstrate the financial capabilities of the chosen institution,[51] or use an insurance firm that is internationally recognised or has a solid reputation.[52]

15.58 Furthermore, the operator may be required to comply with certain provisions whereby it must take out a policy for a high enough amount. The Minister in charge of mining or petroleum operations may, in some situations, set the amount of cover the operator gets.[53] However, texts tend to merely state that the insurance policy should meet international standards applied in the mining or oil industry.[54]

Some laws or exploitation agreements go as far as imposing the main clauses that must be inserted into the insurance policy. In particular, these clauses include provisions that mention the host state as an additional insured and provide for a subrogation in its favour.[55]

15.59 In order to ensure that the operator meets its insurance obligations, mining and petroleum laws and exploitation agreements may provide

---

20(8) of the Model Petroleum Agreement, 2013; in Chad, Article 61(1) of Law No. 006-PR-2007 of 2 May 2007 On Hydrocarbons.

[49] See in particular, in Guinea, Article 140 of Law No. 2011/006/CNT of 9 September 2011 Enacting the Mining Code and Article 45 of Law No. L/2014/No. 034/AN of 23 December 2014 Enacting the Petroleum Code; in Mauritania, Article 32 of Law No. 2010-033 of 20 July 2010 Enacting the Hydrocarbons Code amended by Law No. 2011-044 of 25 November 2011.

[50] See in particular, in Guinea, Article 45 of Law No. L/2014/N°034/AN of 23 December 2014 Enacting the Petroleum Code; in Mauritania, Article 32 of amended Law No. 2010-033 of 20 July 2010 Enacting the Hydrocarbons Code.

[51] See in particular, in Chad, Article 57 of the Law No. 006-PR-2007 of 2 May 2007 On Hydrocarbons amended by Law No. 2011-044 of 25 November 2011.

[52] See in particular, in the Seychelles, Article 20 of the Model Petroleum Agreement, 2013.

[53] See in particular, in Malawi, Article 61(1) of the Petroleum (Exploration and Production) Act, 1983.

[54] See in particular, in Cameroon, Article 117 of Decree No. 2000-485 of 30 June 2000 Implementing Law No. 99-013 of 22 December 1999 Enacting the Petroleum Code; in Eritrea, Article 30 of the Proclamation No. 108/2000, A Revised Proclamation to Govern Petroleum Operations; in the Seychelles, Article 20 of the Model Petroleum Agreement, 2013.

[55] See in particular, in Cameroon, Article 117 of the Decree No. 2000-485 of 30 June 2000 Establishing the Conditions for Application of Law No. 99-013 of 22 December 1999 Enacting the Petroleum Code; in Eritrea, Article 30 of Proclamation No. 108/2000, A Revised Proclamation to Govern Petroleum Operations.

that the private operator should provide the competent authority with certificates vouching for the fact that insurance has been taken out and maintained.[56]

15.60 Lastly, it is important to point out that some regional texts are to be applied in relation to insurance, in particular the CIMA (Inter-African Insurance Conference) insurance code.[57]

ii *Institutional insurance*

15.61 There are various mechanisms in place to stimulate the involvement of companies in the exploitation of natural resources. These include credit insurance, and various investment insurance systems.

(1) Credit insurance

15.62 The credit insurance mechanism guarantees the operator, in return for a premium, against payment defaults. The premium generally takes into account the duration of the risk, the financing structure and the country risk.

Export credit insurance has an additional function to that of credit institutions and involves a role as more than just a source of financing, also providing a guarantee service.

Some states have come up with export insurance systems in order to protect their national companies.

(2) Investment insurance systems

15.63 Investment insurance systems have now emerged. We need to distinguish between national insurance systems and international insurance systems.

15.64 It is at State level that the first investment insurance mechanisms appeared. The development of these systems at national level is justified by the desire of the states to support their investors overseas and to promote the investments of their nationals. In France, the State has put in place a public guarantee against political risks that may affect investments made overseas. This public guarantee system originally relied on the Banque Française du Commerce Extérieur (BFCE), privatised a few years later,[58] and Compagnie Française d'Assurance pour le Commerce Exterieur (COFACE). COFACE, which specialises in export credit insurance, was founded in 1946 and then

---

[56] See in particular, in Cameroon, Article 117 of Decree No. 2000-485 of 30 June 2000 Implementing Law No. 99-013 of 22 December 1999 Enacting the Petroleum Code; in the Seychelles, Article 20(2) of the Model Petroleum Agreement, 2013; in Swaziland, Article 153(2) of the Mines and Minerals Act, 2011; in Tanzania, Article 111(1) of the Mining Act, 2010.

[57] CIMA was developed from the Treaty Establishing an Integrated Organisation of the Insurance Industry in the African states signed on 10 July 1992 in Yaoundé by the Governments of the following 14 Member states: Benin, Burkina, Cameroon, Central African Republic, Comoros, Congo, Côte d'Ivoire, Gabon, Equatorial Guinea, Mali, Niger, Senegal, Chad, and Togo.

[58] The BFCE was founded in 1946 to promote financing of foreign trade operations. It went on to be privatised in 1996.

privatised in 1994. Listed on the stock exchange in 2014, COFACE's objective is to insure investors against the risk of financial failure. Outside France, national investment insurance systems have grown in many countries. To this end, it is worth mentioning in the USA, the *Overseas Private Investment Corporation* (OPIC), which provides loans and loan guarantees for investments in industrialising countries. This organisation was founded in 1971 and has been involved, on many occasions, in projects on the African continent.[59] In Germany, *Euler Hermes Aktiengesellschaft and PricewaterhouseCoopers Aktiengesellschaft Wirtschaftsprüfungsgesellschaft* are responsible for managing the national export credit insurance system.[60] Southern countries also have foreign trade insurance systems. For instance, Tunisia, through the Compagnie Tunisienne pour l'Assurance du Commerce Extérieur (COTUNACE), offers its national investors guarantee mechanisms to protect them within the context of their international operations. Although there may be many similarities, the protection mechanisms offered by these insurance systems vary quite a bit in practice.[61]

15.65 These insurance mechanisms are also developed at the regional level through the creation in 1974 of the *Compagnie Inter-arabe de Garantie des Investissements Internationaux*.[62] This organisation, founded under the Arab League, which offers investors insurance against non-commercial risks, paved the way for other international insurance mechanisms.

15.66 At international level, it is worth mentioning the involvement of the Multilateral Investment Guarantee Agency (MIGA), which guarantees investments that contribute to development.[63] This agency, founded on 11 October 1985 by the 1985 Seoul Convention creating the Multilateral Investment Guarantee Agency, has done a great deal.[64] The extractive industries are amongst the priorities of this organisation, which gets involved in complex projects, primarily in the extractive industries. In 2013, MIGA was involved in 14 projects in Sub-Saharan Africa.[65] In the same year, the amount

---

[59] For more information on OPIC projects on the African continent, see: https://www.opic.gov/opic-action/active-opic-projects.

[60] See on this subject, Scott & Martens op cit 88.
For other examples of national investment insurance systems, see in particular, in Luxembourg, Office Du Croire.

[61] See Arnaud de Nanteuil, *Droit International de l'Investissement* ed (A Pedone 2014) no 823 et seq.

[62] See de Nanteuil op cit no 822; Ibrahim Shihata 'Multilateral Investment Guarantee Agency' (1987) 33 *AFDI* 601.

[63] See de Nanteuil op cit no 822. See also, Ibrahim Shihata 'L'agence multilatérale de garantie des investissements' at 601.

[64] According to estimates provided by MIGA, since it was founded, it has issued more than $27 billion of guarantees (insurance against political risks), in support of more than 700 investment projects in around 100 developing countries. Its current portfolio is in excess of $10 billion.

[65] MIGA *Annual Report 2013* (2013).

of the guarantees provided within the context of mining and oil projects reached $652,1 million.[66]

15.67 Lastly, it should be noted that, in an indirect way, investment insurance also involves the introduction of contractual mechanisms at the international level, such as bilateral treaties to promote and protect investments through which the states intend to protect their nationals within the context of their international investments.[67]

[66] Id.

[67] See de Nanteuil op cit no 825.

16

Host state participation

16.1 The desire of producing states to benefit from the exploitation of natural resources resulted in the awarding in their favour of a share in the capital of the companies created at the development or production stage. This mechanism enables the host state to have access to some of the operating company's assets. This equity participation was initially designed to be free and non-contributory. Thereafter, this free participation was supplemented by a contributory participation of the host state.

Generally, natural resources are state properties and as such it seemed justified that the host states could participate in exploitation of their resources to compensate the reduction in their reserves.[1] This kind of participation has many advantages. First, it enables the State to increase its revenue, immediately in the case of a free participation, or after reimbursement of production costs in the case of a contributory participation. The host state will also benefit from greater involvement in mining and oil projects and can therefore increase its control over exploitation operations.[2]

[1] See, to that effect, Article 7(3) of Law No. 2014-138 of 24 March 2014 Enacting the Mining Code in Côte d'Ivoire, which provides:

> '*[l]'octroi par l'État des permis d'exploitation donne droit en contrepartie de la richesse distribuée et de l'appauvrissement du sous-sol, à l'attribution à l'État d'actions d'apport fixées à dix pour cent (10%) du capital de la société d'exploitation, pendant toute la durée de vie de la mine. Aucune contribution financière ne peut être exigée à l'État au titre de ces actions d'apport même en cas d'augmentation de capital. Dans tous les cas, la part de l'État reste au moins égale à dix pour cent (10%) du capital de la société d'exploitation*'.
>
> [Free translation into English: '[t]he granting by the State of exploitation permits entitles it, in return for the distributed wealth and the impoverishment of the subsoil, to the awarding to the State of shares fixed at ten per cent (10%) of the operating company's capital, for the lifetime of the mine. No financial contribution can be demanded of the State for these shares, even in the event of an increase in capital. In any case, the State's share remains at least equal to ten per cent (10%) of the operating company's capital'].

See also, to that effect, in Angola, Article 11(1) of Mining Law No. 31/11 of 23 September 2011, which provides:

> '*[o] Estado participa na apropriação do produto da mineração como contrapartida pela concessão dos direitos mineiros de exploração e comercialização, podendo usar uma das seguintes formas ou ambas conjugadas: a) participação, através de uma empresa do Estado, no capital social das sociedades comerciais a criar, não podendo essa participação ser inferior a 10%* [. . .]'
>
> [Free translation into English: '[t]he State participates in appropriation of the produce of the mineral products mines in return for a concession for exploration and marketing mining rights. It can use one or both of the following forms: a) participation, through a State company, in the share capital of the commercial companies created. This participation may not be less than 10% [. . .]']

[2] See, to that effect, Bret-Rouzaut & Favennec op cit 197.

16.2 The host state's equity participation in extraction projects can be awarded directly to the State, to the local State operator or to any other entity appointed by the host state.[3] Mining and petroleum laws may even provide for local and non-public equity participation.[4]

This financial participation, whether it is free or contributory, is becoming the norm, particularly in the mining sector. In the 1990s, because of the economic context, the host state's participation in exploitation projects tended to decline,[5] whereas the trend has now reversed.[6]

Even though the principle of the State's equity participation in the operating company is largely admitted in the relevant legislation, the implementation of this option can prove to be difficult. This is particularity true when it comes to the State's contributory participation, which requires the State to raise funds.

The conditions of the participation vary depending on the state in question and the natural resource involved. However, general observations can be made; while countries with unproven reserves tend to favour contributory host state participation, producing states generally favour free, non-contributory State participation. This situation results in states with few resources wanting to encourage exploitation operations and therefore discoveries of commercial deposits by assuming, alongside private operators, the risks associated with exploitation of natural resources.

---

[3] See, in particular, in Cameroon, Article 6 of Law No. 99-013 of 22 December 1999 Enacting the Petroleum Code; in Côte d'Ivoire, Article 6 of Law No. 96-669 of 29 August 1996 Enacting the Petroleum Code; in Guinea, Article 46 of Law L/2014/No. 034/AN of 23 December 2014 Enacting the Petroleum Code; in Niger, Article 8(2) of Law No. 2007-01 of 31 January 2007 Enacting the Petroleum Code; in São Tomé and Principe, Article 23(2) of the Framework Law on Oil Operations; in Senegal, Article 7 of Law No. 98-05 of 8 January 1998 Enacting the Petroleum Code.

[4] So, in Senegal, Article 31 of of Law No. 2016-32 of 8 November 2016 Enacting the Mining Code provides:

> '*[l]'État peut, en sus de sa part gratuite au capital, négocier pour lui et/ou le secteur privé national, à titre onéreux, une participation supplémentaire jusqu'à hateur de vingt-cinq pour cent (25%) au capital de la société d'exploitation, selon les modalités habituelles en vigueur en la matière.*'
>
> [Free translation into English: '[t]he State can, on top of its free carried interest, negotiate for itself and/or the national private sector, on a contributory basis, an additional participation up to twenty-five per cent (25%) to the share capital of the operating company and for the national private sector a participation in the operating company's capital, on the usual modalities applicable.]'

[5] See, to that effect, Bret-Rouzaut & Favennec op cit 197.

[6] Similarly, in Eritrea, the State's free participation in mining projects has increased from 5 to 10% following the 2011 reform (see Article 41 of Proclamation No. 165/2011, A Proclamation Amending the Mining Law No. 68/1995). In Mozambique, Article 34(2) of Mining Law No. 20-2014 of 18 August 2014 and Article 20(3) of Petroleum Law No. 21-2014 of 18 August 2014 provide that the State must gradually increase its participation in mining and oil projects.

1 NON-CONTRIBUTORY HOST STATE EQUITY PARTICIPATION

16.3 Within the context of signature of an exploitation agreement, host state equity participation in the capital of the operating company is generally provided for.[7] In principle, this participation is designed as the awarding, free of charge, in favour of the host state of shares in the operating company. No financial contribution can be required of the host state for operation of the operating company, including for its investments or even in certain cases for increases in capital.[8]

It is in the mining sector that the host state's non-contributory participation is most often contemplated, although some petroleum laws also provide for the host state's entitlement to free of charge State's equity participation.[9]

The free participation granted to the State is generally for a 10% share in the operating company,[10] although other provisions set a

[7] In Burundi, the granting of an industrial exploitation permit results in the awarding of a participation to the State (Article 63 of Law No. 1-21 of 15 October 2013 Enacting the Mining Code). In the Central African Republic, the participation is awarded to the State in the event that an industrial exploitation permit for large or small mines is granted (Article 36 of Law No. 09-005 of 29 April 2009 Enacting the Mining Code). In Côte d'Ivoire, the State's participation is awarded if an exploitation permit is granted (Article 7 of Law No. 2014-138 of 24 March 2014 Enacting the Mining Code). In Guinea, the Mining Code provides for the State's participation in the event that exploitation mining rights are granted (Article 150 of Law No. L/2013/No. 053/CNT of 8 April 2013 Amending Law No. 2011/006/CNT of 9 September 2011 Enacting the Mining Code). In Niger, the State's participation is subject to an exploitation permit being granted (Article 8 of amended Ordinance No. 93-16 of 2 March 1993 Enacting the Mining Code).

[8] To that effect, see in particular, Article 8 of Law No. 2006-17 of 17 October 2006 Enacting the Mining Code and Mining Taxation in the Republic of Benin.

[9] See, in particular, to that effect, in Liberia, Article 3(3) of the 2002 Petroleum Law, which provides:

> '[t]he National Oil Company, in addition to other rights, interests and benefits it is entitled to receive under any and all Production Sharing Agreements, it shall also receive free of charge, equity interest in all production operations and exploitation of hydrocarbon deposits in the Republic of Liberia. The value of such equity interest shall be twenty (20%) per cent of the authorized, issued and outstanding capital shares existing at any time, without dilution.'

See also, in Sierra Leone, Article 85 of the Petroleum Exploration and Production Act, 2011, which provides:

> '[w]here the State elects to participate in a petroleum licence, it shall have an initial carried interest of at least ten per cent and may acquire an additional paying interest up to a declared maximum within a specified period of time from the date a discovery is declared to be commercial discovery.'

[10] See, in particular, to that effect, in Benin, Article 8 of Law No. 2006-17 Enacting the Mining and Mining Taxation Code in the Republic of Benin; in Cameroon, Article 11 of Law No. 001 of 16 April 2001 Enacting the Mining Code as amended by Law No. 2010/011 of 29 July 2010; in Côte d'Ivoire, Article 7 of Law No. 2014-138 of 24 March 2014 Enacting the Mining Code; in Eritrea, Article 41 of Proclamation No. 165/2011, A Proclamation Amending the Mining Law No. 68/1995; in Mali, Article 65 of Law No. 2012-015 of 27 February 2012 Enacting the Mining Code; in Mauritania, Article 5 of Law No. 2012/012 Governing Mining Agreements and Approving the Standard Mining Agreement; in Niger, Article 8 of Ordinance No. 93-16 of 2 March 1993 Enacting the Mining Law; in Togo, Article 55 of Law No. 2003-012 of 04 October 2003 Amending and Supplementing Law No. 96-004 of 26 February

higher[11] or lower[12] threshold.

16.4 The participation the host state gets will generally be protected by legal provisions, particularly when, in the event of an increase in capital, the host state's participation is at risk of being diluted.[13]

16.5 In the majority of cases, the free participation granted to the host state does not prevent it from requesting an additional participation. Several mining codes provide for this possibility.[14] This additional participation will then be contributory, in which case the State will have the same rights and obligations as its partners. In practice, an agreement between the parties will lay down the conditions for the State's additional participation in the operating company's capital, which may be in cash or in kind.[15] This additional participation may however be limited.[16]

1996 Enacting the Mining Code. At a regional level, the principle of a free 10% State participation in mining projects is provided for in Article 12 of Regulation No. 18/2003/CM/UEMOA of 23 December 2003 Enacting the Community Mining Code.

[11] So, in the Central African Republic, Article 36 of Law No. 09-005 of 29 April 2009 Enacting the Mining Code provides for a State participation of at least 15% in the share capital of the operating company; in Guinea, Article 150 of Law No. L/2013/No. 053/CNT of 8 April 2013 Amending Law No. 2011/006/CNT of 9 September 2011 Enacting the Mining Code provides for a maximum free state participation of 15%.

[12] See, in particular, to that effect, in Ethiopia, where Article 70 of Proclamation No. 678/2010 Promoting Sustainable Development of Mineral Resources also provides for a free State participation of 5%.

[13] See in particular, in Burundi, Article 63 of Law No. 1-21 of 15 October 2013 Enacting the Mining Code; in Central African Republic, Article 36 of Law No. 09-005 of 29 April 2009 Enacting the Mining Code; in Mali, Article 65 of Law No. 2012-015 of 27 February 2012 Enacting the Mining Code; in Mauritania, Article 5 of Law No. 2012/012 Governing Mining Agreements and Approving the Standard Mining Agreement; in Niger, Article 8 of Amended Ordinance No. 93-16 of 2 March 1993 Enacting the Mining Law; within the WAEMU, Article 12 of Regulation No. 18/2003/CM/UEMOA of 23 December 2003 Enacting the Community Mining Code.

[14] See in particular, in Benin, Article 8 of Law No. 2006-17 of 17 October 2006 Enacting the Mining Code and Mining Taxation in the Republic of Benin; in Côte d'Ivoire, Article 7 of Law No. 2014-138 of 24 March 2014 Enacting the Mining Code; in Eritrea, Article 41 of Proclamation No. 165/2011, A Proclamation Amending the Mining Law Proclamation No. 68/1995; in Ethiopia, Article 70 of Proclamation No. 678/2010 Promoting Sustainable Development of Mineral Resources; in Guinea, Article 150 of Law L/2013/No. 053/CNT of 8 April 2013 Amending Law No. 2011/006/CNT of 9 September 2011 Enacting the Mining Code; in Mauritania, Article 5 of Law No. 2012/012 Governing Mining Agreements and Approving the Standard Mining Agreement; in Niger, Article 8 of amended Ordinance No. 93-16 of 2 March 1993 Enacting the Mining Law; in Togo, Article 55 of Law No. 2003-012 of 4 October 2003 Amending and Supplementing Law No. 96-004 of 26 February 1996 Enacting the Mining Code. At a regional level, the principle of an additional and contributory host State participation is provided for in Article 12 of Regulation No. 18/2003/CM/UEMOA of 23 December 2003 Enacting the Community Mining Code.

[15] In Niger, Article 8 of amended Ordinance No. 93-16 of 2 March 1993 Enacting the Mining Law provides for a State participation in cash or in kind.

[16] In Côte d'Ivoire, Article 7 of Law No. 2014-138 of 24 March 2014 Enacting the Mining Code provides that the State's additional participation may not exceed 15% (however, the State's contributory participation will not be limited in the event that the State has invested from the exploration phase). In Eritrea, Article 41 of Proclamation No. 165/2011 A Proclamation Amending Proclamation No. 68/1995 provides that the State's participation may not exceed 40% (the free 10% participation is included in this calculation). In Guinea, Article 150 of

2 CONTRIBUTORY HOST STATE EQUITY PARTICIPATION

16.6 Traditionally the contributory participation of the host state may or may not be awarded free of charge, ie with no contribution by the host state to the initial capital of the operating company.

Apart from cases where the State's free participation is supplemented by an additional, contributory participation, it is in the oil sector that the host state's contributory participation is usually contemplated.

Unlike free participation, contributory host state participation within the context of exploitation operations supposes that the host state assumes the rights and obligations arising from its participation. The State therefore becomes a real partner and, where appropriate, is expected to contribute to losses. The host state or the local state company should bear the costs associated with the exploitation operations pro rata to its participation.

16.7 The exploitation agreement generally lays down the terms and conditions for this participation.[17]

At the same time, the host state's equity participation may give rise to the signature of an association agreement (a mandatory requirement of certain legislation).[18] This agreement will generally define the respective rights and obligations of the parties and the rules of conduct and supervision of the joint

---

Law No. L/2013/N°053/CNT of 8 April 2013 Amending Law No. 2011/006/CNT of 9 September 2011 Enacting the Mining Code provides that the total State participation may not exceed 35% (the 15% granted as the State's free participation is included in this calculation). In Mauritania, Article 5 of Law No. 2012/012 Governing Mining Agreements and Approving the Standard Mining Agreement provides that the State's additional participation may not exceed 10%. In Niger, Article 8 of the amended Ordinance No. 93-16 of 2 March 1993 Enacting the Mining Code provides that the total State participation may not exceed 40% (this calculation includes the free 10% participation awarded to the State). In Togo, Article 55 of Law No. 2003-012 of 4 October 2003 Amending and Supplementing Law No. 96-004 of 26 February 1996 Enacting the Mining Code provides that the additional participation may be as much as 20% of the capital.

[17] See in particular, to that effect, in Benin, Article 39 of Law No. 2006-18 of 17 October 2006 Enacting the Petroleum Code in the Republic of Benin; in Cameroon, Article 6 of Law No. 99-013 of 22 December 1999 Enacting the Petroleum Code; in Côte d'Ivoire, Article 6 of Law No. 96-669 of 29 August 1996 Enacting the Petroleum Code; in Djibouti, Article 7 of Law No. 117-AN-05-5 of 16 October 2005 Enacting the Petroleum Code; in Guinea, Article 46 of Law L/2014/No. 034/AN of 23 December 2014 Enacting the Petroleum Code; in Morocco, Article 32 of Law No. 21-90 of 1 April 1992 On the Exploration and Exploitation of Hydrocarbon Deposits amended and supplemented by Law No. 27-99 of 15 February 2000; in Namibia, Article 13(2) of the Petroleum (Exploration and Production) Act, 1991; in Niger, Article 8(2) of Law No. 2007-01 of 31 January 2007 Enacting the Petroleum Code; in São Tomé and Principe, Article 23(3) of the Framework Law on Oil Operations; in Senegal, Articles 7, 17 and 34(2) of Law No. 98-05 of 8 January 1998 Enacting the Petroleum Code.

In the field of mining, see, in particular, in Equatorial Guinea, Article 55 of Mining Law No. 9/2006 of 3 November 2006; in Chad, Article 10 of Law No. 011-PR-1995 of 20 June 1995 Enacting the Mining Code.

[18] See, in particular, in Guinea, Article 48 of Law No. L/2014/N°034/AN of 23 December 2014 Enacting the Petroleum Code; in Mauritania, Article 44 of amended Law No. 2010-033 of 20 July 2010 Enacting the Crude Hydrocarbons Code amended by Law No. 2011-044 of 25 November 2011; in Togo, Article 55 of Law No. 2003-012 of 4 October 2003 Amending and Supplementing Law No. 96-004 of 26 February 1996 Enacting the Mining Code.

operations, such as the appointment of the operator or the establishment of an operating committee.

16.8 Usually, the host state's contributory participation in the exploitation operations is optional,[19] and the host state has the discretion to decide whether to exercise the option it gave, under the conditions laid down by the law or by the exploitation agreement. Regulations can therefore make provision for the competent entity to decide to exercise the option the host state has[20] and the timeframe for this decision. In this respect, regulations can provide for the exercise of the option following a commercial discovery,[21] after production mining rights have been granted[22] or even after the application for production mining rights has been submitted.[23] Usually, the regulations do not provide a time limit for the State to exercise its option.

16.9 In very few cases, the host state's participation may be implemented during any phase of the operations.[24] Generally, the State's equity participa-

[19] See, in particular, in Cameroon, Article 6(1) of Law No. 99-013 of 22 December 1999 Enacting the Petroleum Code; in Côte d'Ivoire, Article 6 of Law No. 96-669 of 29 August 1996 Enacting the Petroleum Code; in Djibouti, Article 7 of Law No. 117-AN-05 of 16 October 2005 Enacting the Petroleum Code; in Guinea, Article 46 of Law No. L/2014/N°034/AN of 23 December 2014 Enacting the Petroleum Code; in Mali, Article 22(1) of the Standard Production-Sharing Agreement; in Mauritania, Article 44 of Law No. 2010-033 of 20 July 2010 Enacting the Hydrocarbons Code amended by Law No. 2011-044 of 25 November 2011; in Mozambique, Article 20(1) of Petroleum Law No. 21-2014 of 18 August 2014; in Niger, Article 8(2) of Law No. 2007-01 of 31 January 2007 Enacting the Petroleum Code; in Uganda, Article 124(1) of the Petroleum (Exploration, Development and Production) Act, 2013; in Senegal, Article 7 of Law No. 98-05 of 8 January 1998 Enacting the Petroleum Code; in the Seychelles, Article 12(1) of the Petroleum Mining Act; in Sierra Leone, Article 162(1) of the Mines and Minerals Act, 2009 and Article 84(1) of the Petroleum Exploration and Production Act, 2011; in Tunisia, Article 94(1) of Law No. 99-93 of 17 August 1999 Enacting the Hydrocarbons Code.

It should be noted that in some cases the law refers to mining rights, which may, depending on the case, also contain a participation option for the host state. These include, in particular, in Malawi, Article 31 of the Mines and Minerals Act, 1981 and Article 17(2) of the Petroleum (Exploration and Production) Act, 1983.

[20] As an example, in São Tomé and Principe, Article 23(2) of the Framework Law on Oil Operations provides that the decision regarding equity participation will be made by the Government. In Chad, Article 13(6) of Law No. 006-PR-2007 of 2 May 2007 On Hydrocarbons provides that the option will be exercised by the National Company, which is also the case in Tunisia (Article 94(2) of Law No. 99-93 of 17 August 1999 Enacting the Hydrocarbons Code).

[21] See, in particular, in Sierra Leone, Article 85 of the Petroleum Exploration and Production Act, 2011; in Tunisia, Article 94(4) of Law No. 99-93 of 17 August 1999 Enacting the Hydrocarbons Code.

[22] See in particular, in Chad, Article 13(6) of Law No. 006-PR-2007 of 2 May 2007 On Hydrocarbons.

[23] See in particular, in Tunisia, Article 94(2) of Law No. 99-93 of 17 August 1999 Enacting the Hydrocarbons Code.

[24] See in particular, in Mozambique, Article 20(2) of Petroleum Law No. 21/-2014 of 18 August 2014; in São Tomé and Principe, Article 23(4) of the Framework Law on Oil Operations.

tion will be limited to the production phase, ie the phase in which revenue will be generated.[25]

Moreover, regulations may set a minimum[26] or maximum[27] participation rate for the host state. To this end, some countries, for instance Angola[28] and Algeria,[29] provide that State participation in oil projects should be in the majority and may not therefore be less than 51%.

16.10 Generally, the host state should, from the start of its equity participation, contribute to the cost of operations and reimburse those incurred previously.[30] However, conditions that are more advantageous may be provided for by the law or by the exploitation agreement. For example, the private operator who alone bears the costs of exploration may assume the exploration risk. These expenses are retroactively shared between the parties in the event of a commercial discovery. In this way, the expenses for which the host state is responsible will be paid in advance by the private operator and, where appropriate, reimbursed out of any production.[31]

---

[25] See in particular, in Niger, Article 14 of Decree No. 2007-082 of 28 March 2007 Enacting the Conditions for Application of Law No. 2007-01 of 31 January 2007 Enacting the Petroleum Code.

[26] As an example, in Guinea, Article 46 of Law L/2014/No. 034/AN of 23 December 2014 Enacting the Petroleum Code provides that the State's participation will be at least 10%. In Lesotho, Article 34(1) of the Mines and Minerals Act, 2005 provides that the State's participation may not be less than 20%.

[27] As an example, in Botswana, Article 40(1) of the Mines and Minerals Act provides for a maximum State participation of 15%. In Morocco, Law No. 21-90 of 1 April 1992 On the Exploration and Exploitation of Hydrocarbon Deposits amended and supplemented by Law No. 27-99 of 15 February 2000 provides that the State's participation may not be greater than 25%.

[28] In Angola, Article 15 of Petroleum Law No. 10/04 of 12 November 2004 provides: '1. In the associations referred to in Article 14, subparagraphs 2*(a)* and *(b)* where the National Concessionaire holds a participating interest, such interest shall, as a rule, exceed fifty per cent. 2. The Government may authorise the National Concessionaire, in duly grounded cases, to hold a participating interest lower than that established in the previous paragraph.'

[29] In Algeria, Article 32 of Law No. 13-01 of 20 February 2013, Amending and Supplementing Law No. 05-07 of 28 April 2005 On Hydrocarbons provides:

'*[l]es contrats de recherche et d'exploitation et les contrats d'exploitation contiennent obligatoirement une clause de participation de l'entreprise nationale Sonatrach SPA, dont le taux de participation est fixé à un minimum de 51%, préalablement à chaque appel à concurrence, dans lesdits contrats.*'

[Free translation into English: '[e]xploration and exploitation agreements and exploitation agreements must contain a clause on the participation of the national company Sonatrach SPA, the rate of which is set at a minimum of 51%, prior to each call for tenders, in the said agreements.']

[30] See, in particular, in Mauritania, Article 44 of Law No. 2010-033 of 20 July 2010 Enacting the Hydrocarbons Code amended by Law No. 2011-044 of 25 November 2011.

[31] See, in particular, in Chad, Article 13.4 of Law No. 006-PR-2007 of 2 May 2007 On Hydrocarbons, which provides:

'*[l]e remboursement de la quote-part de la Société Nationale dans les coûts et frais de Recherches s'effectue par le Contractant sur prélèvement en priorité, au point champ, sur la part de Production revenant à la Société Nationale. Le Contractant ne sera soumis à aucun impôt ou taxe de quelque nature que ce soit, à raison de tels remboursements. Lors de la phase de Développement d'un gisement d'Hydrocarbures,*

16.11 The host state also very often gives precedence to the expenses incurred previously in the mining area in order to make the most of its participation in the new project and therefore avoids it having to make any actual payments. This is likely to create tension, or even disputes with the previous operator, hence the importance of resolving this matter beforehand, either in the law itself, or through contractual means before or at the time of the operator's departure.

---

les coûts de la participation de la Société Nationale dans ledit Développement sont remboursables à hauteur d'un pourcentage (%) qui est défini dans le Contrat à même la quote-part de la Production revenant à la Société Nationale. Lors de la phase de Développement d'un gisement d'Hydrocarbures, les coûts de la participation de la Société Nationale dans ledit Développement sont remboursables à hauteur d'un pourcentage (%) qui est défini dans le Contrat à même la quote-part de la Production revenant à la Société Nationale.'

[Free translation into English: '[t]he National Company's share of the costs of Exploration is reimbursed by the Contractor on the taking as a priority, at the field point, out of the share of Production belonging to the National Company. The Contractor will not be subject to any taxes or duties of any kind whatsoever as a result of such reimbursements. During the Hydrocarbon deposit Development phase, the costs of the National Company's participation in the said Development are reimbursable up to a percentage (%) defined in the Agreement directly from the share of the Production belonging to the National Company.']

See, in particular, in Tunisia, Article 96.3 of Law No. 99-93 as 17 August 1999 Enacting the Hydrocarbons Code, which provides:

'*[l]es dépenses d'exploration et/ou de recherche et d'appréciation réalisées sur une Concession d'Exploitation, dans laquelle l'Entreprise Nationale a exercé son option de participation sont à la charge de son (ou ses) associé(s) et la quote-part de l'Entreprise Nationale desdites dépenses sera remboursée par elle si cette dernière participe au développement complémentaire de la Concession d'Exploitation concernée conformément aux conditions et modalités définies dans les Accords Particuliers. 94.4 L'Entreprise Nationale rembourse sa quote-part des dépenses susmentionnées par la contre-valeur d'un pourcentage de sa quote-part de production conformément aux modalités définies dans les Accords Particuliers.*'

[Free translation into English: '[t]he exploration and/or investigation and evaluation expenses for an Exploitation Concession, in which the National Company has exercised its participation option are to be borne by its partner(s) and the National Company's share of the said expenses will be reimbursed by it if it participates in further development of the Exploitation Concession in question, in accordance with the terms and conditions laid down in the Specific Agreements. 94.4 The National Company reimburses its share of the aforementioned expenses by the exchange value of a percentage of its share of production in accordance with the conditions laid down in the Specific Agreements.']

Part 3

OPERATIONAL ASPECTS OF EXPLOITATION

17

Disciplinary rules

17.1 The legal environment in which natural resource exploitation takes place includes a set of rules designed to monitor their progress; it includes disciplinary and control rules that both private and State operators must comply.

The operational legal aspects of exploration also refer to the implementation of a set of provisions promoting contractual management of risks. These provisions, which are mainly contractual, will enable the risks inherent to mining and oil operations to be managed, mainly for the benefit of private operators.

The operational legal aspects of exploration also assume that a legal framework which takes consideration of the host state's concerns is established and followed. Thanks to the transfer of technology, the host state should be able to exploit its natural resources itself. Similarly, the development from which the host state might benefit because of the exploitation of its resources is far from being insignificant. In practice, these concerns are clear in the setting up of contractual and legal provisions that promote development and technology transfer for the benefit of the host state.

E Modalities of performance of exploitation operations

17.2 The performance of exploitation operations involves several preliminary stages. First, a programme must be drawn up which brings together the various works to be undertaken and the schedule for these anticipated works; the performance of this programme should then be periodically verified and monitored.

It is these events which we shall examine by first considering the drawing up of the work programme, which constitutes the disciplinary rules, and secondly, the control of this performance, which constitutes the control rules.

17.3 The exploitation agreement may mention the several works programmes to be applied to the same contract:

- an exploration work programme fixed from the start of works and supplemented annually by specific programmes, then, according to discoveries;
- a development programme;
- a production programme which can also be supplemented annually.[1]

In practice, these work programmes are distinct from minimal work programmes establishing a minimum threshold of expenses and correspond-

[1] We shall use the term 'work programme' in the singular in a generic fashion.

ing works which must be incurred and performed which by the private operator, generally during the exploration phase.

The implementation of a work programme includes two successive phases: a drawing up phase and an implementation phase.

1 THE INITIAL DRAWING UP OF THE WORK PROGRAMME

17.4 The success or failure of a natural resource exploitation project very often depends on the proper implementation of the work programme. Therefore, a general framework for the performance of the project must be scrupulously designed, through the drawing up of a work programme.

Besides, it is based on the work programme that the operators will distribute the tasks between them as well as responsibilities.

The general rule requires the exploitation agreement to specify the work programme;[2] however, the work programme can be stipulated by the mining rights themselves. The most widely used rule is that where the work programme is specified in terms of its principles by the mining rights and detailed in the exploitation agreement.

a Drawing up a work programme

17.5 The elaboration of a work programme is a decisive stage in the exploitation of natural resources. By drawing this up, potential communication problems between operators can be detected as early as possible and thus highlight crisis situations.

Initially, the programmes to be carried out were defined very simply. These objectives, which most of the time were very ambitious, demonstrated a complete lack of knowledge about the exploitation of natural resources, particularly in terms of hydrocarbons. Over time, these work programmes have been the subject of ever more detailed provisions.

The conception of a work programme includes two milestones: the preparation of a proposal and its acceptance.

i Preparing a work programme

17.6 The constant insertion of a clause relating to the technical performance of the works is recent. It also results from the fact that all states impose a minimum amount of works to be carried out by operators, in particular during exploration phases.

---

[2] See, in particular, in the Central African Republic, Article 52 of Law No. 09-005 of 29 April 2009 Enacting the Mining Code: 'The following clauses relating to materials are included in the mining agreement: [. . .] the work programme, and the commitment of the Investor to start it and continue it with diligence, both in the preservation of the environment and archaeological wealth as well as by complying with the minimum amount of expenses per km^2 stipulated by mining regulations.' See also in the Côte d'Ivoire Article 18 of Law No. 16-669 of 29 August 1996 Enacting the Petroleum Code: '[t]he petroleum agreement must particularly lay down: [. . .] d) the conditions for establishing work programmes and budgets, the control of their execution, the provision to the Minister responsible for hydrocarbons of reports, data and information relating to oil operations'.

The preparation of a work programme proposal may be unilateral or performed as a group depending on the degree of collaboration that exists between the operators. In practice, the preparation of a work programme is left to the private operator.

Sometimes, a private operator will be under an obligation to present a work programme within an established timeframe.

In practice, work programmes are very complex and include numerous operations that must be organised and coordinated. This in particular means making an accurate determination of the works that must be carried out, drawing up a budget which corresponds to these works and finally of planning the timeframes in which these works should be completed in a precise order and sequence. Difficulties can arise, mainly in the use of subcontractors; knowledge of both the services that they are proposing and the timeframes which they currently need to execute the services should be correctly controlled.

17.7 Concerning the minimum work programmes to be performed by the operator during the exploration phase, mining and petroleum codes may provide a framework for the elaboration of the work programme, in particular by mentioning the amount of expenses to be incurred by square kilometres.

ii *Acceptance of a work programme*

17.8 Acceptance of a work programme takes on particular significance when one of the operators has prepared it unilaterally, which is generally the case, as the work programme is drawn up by the private operator most of the time.

17.9 Acceptance of the work programme concentrates most of the difficulties around the decision of the local State operator. Usually, the local State operator has to give its response within a mandatory time limit. Furthermore, objections raised by the local State operator must be genuine, logical and technically well founded.

17.10 The principle is that the local authorities control the work programme, once drawn up by the operators.[3] Indeed, without their authorisation, it is prohibited in most countries to build any kind of exploitation facilities or to perform drilling works for instance.

17.11 Finally, it should be underlined that the performance of the work programme is often delayed; this situation justifies the insertion of penalty clauses. These penalty clauses, which are a source of conflicts, must be

[3] See in particular, in Guinea, Article 28 of Law No. L/2014/N°034/AN of 23 December 2014 Enacting the Petroleum Code:

> '*[l]e Contractant doit soumettre à l'Administration Pétrolière pour approbation selon les modalités précisées dans le Contrat Pétrolier un programme annuel de travaux et un budget correspond*'.
>
> [Free translation into English: '[t]he Contracting Party must submit an annual work programme and corresponding budget to the Petroleum Administration for approval according to the conditions specified in the Petroleum Agreement'].

considered and applied with much discernment because of the difficulties inherent to the extractive sector.

b Performers of the chosen work programme: the roles of Operator and Contractor

17.12 The roles of Operator and Contractor have been defined and developed specifically for contracts relating to the exploitation of natural resources more especially in the hydrocarbon sector. Prior to examining the general features connected to these notions of Operator and Contractor, we will introduce the difficulties these notions represent.

i *Difficulties connected with the notions of Operator and Contractor*

17.13 It is very difficult to provide strict and exhaustive details on the notions of Operator and Contractor.[4] Indeed, they are used both in legislation and in agreements with various (sometimes even contradictory) meanings and definitions.

The distinction between these two concepts is dominated by the wish to allocate different responsibilities connected to the performance of the work programme and to organise the operators by appointing one of them at their head, ie the 'Operator', as opposed to the 'Contractors'.

ii *General features of the roles of Operator and Contractor*

(1) Appointment and functions of Operator and Contractor

17.14 We shall distinguish the question of the appointment of Operator and Contractor from the functions they imply.

(2) The choice of the Operator

17.15 The Operator is the entity in charge of the administration and conduct of the operations on the ground; it has a role of direct performer and manager. The Operator carries out its functions as manager of the partnership as part of the general directives given by the management board.

The New Guinean petroleum code defines this notion by specifying that the Operator is the company to which responsibility for conducting and performance petroleum operations in accordance with the stipulations of the petroleum agreement is entrusted.[5]

---

[4] A Mehani, for example, has defined 'operator' in terms that are less than specific as being 'a kind of entrepreneur who makes his technology available for a lead contractor to use.' A Mehani 'Les accords algéro-français du 29 juillet 1965 sur les hydrocarbures' 1966 *R Alg SJPE* 270.

[5] See Article 1 of Law No. L/2014/N°034/AN of 23 December 2014 Enacting the Petroleum Code in the Republic of Guinea. See also, in Mauritania, Article 2 of Law No. 2010-033 of 20 July 2010 Enacting the Hydrocarbon Code Amended by Law No. 2011-044 of 25 November 2011, which provides that:

> '*[o]pérateur: La personne morale responsable de la direction et de l'exécution des opérations pétrolières conformément aux dispositions du contrat d'exploration-production.*'

17.16 The appointment of an Operator is often imposed by the legislation when several private operators are involved in the project.[6]

During the performance of the exploitation agreement, the role of Operator may pass from one operator to another, for example, successively during the exploration and development phases and during the production phase.

A change in operator during exploitation operations is generally subject to the approval of the host state.[7] Furthermore, in certain exceptional circum-

---

[Free translation into English: '[o]perator: the legal entity responsible for managing and carrying out oil operations in accordance with the provisions of the exploration-production agreement.']

[6] See, in particular, in Mauritania, Article 46 of Law No. 2010-033 of 20 July 2010 Enacting the Hydrocarbon Code Amended by Law No. 2011-044 of 25 November 2011:

'[s]i le contractant est composé de plusieurs personnes morales, le contrat spécifie laquelle de ces personnes agira comme opérateur.'

[Free translation into English: '[i]f the contracting party is made up of several legal entities, the agreement shall specify which of these entities shall act as the operator.']

See also, in Algeria, Article 29 of Law No. 13-01 of 20 February 2013 Amending and Supplementing Law No. 05-07 of 28 April 2005 On Hydrocarbons:

'[p]endant la période de recherche, le contrat spécifie laquelle des personnes est l'opérateur. Pendant la période d'exploitation, le rôle d'opérateur pour la conduite des opérations pétrolières est assuré par toute personne composant le contractant ou par toute autre forme d'organe conjoint convenue par les personnes constituant le contractant et préalablement approuvée par l'agence nationale pour la valorisation des ressources en hydrocarbures (ALNAFT). Tout changement d'opérateur doit être soumis à l'accord préalable de l'agence nationale pour la valorisation des ressources en hydrocarbures (ALNAFT).'

[Free translation into English: '[d]uring the exploration period, the agreement shall specify which of the entities is the operator. During the exploitation period, the role of operator for conducting oil operations if provided by any entity making up the contracting party or by any other form of joint body agreed by the entities making up the contracting party and group beforehand by the Agence Nationale pour la Valorisation des Ressources en Hydrocarbures (ALNAFT—National Agency for the Promotion of the Hydrocarbon Resources of Algeria). Any change of operator must be subject to the prior agreement of the Agence Nationale pour la Valorisation des Ressources en Hydrocarbures (ALNAFT—National Agency for the Promotion of the Hydrocarbon Resources of Algeria).']

[7] See in particular, in Mauritania, Article 46 of Law No. 2010-033 of 20 July 2010 Enacting the Hydrocarbon Code Amended by Law No. 2011-044 of 25 November 2011:

'[s]i le contractant est composé de plusieurs personnes morales, le contrat spécifie laquelle de ces personnes agira comme opérateur. Tout changement d'opérateur doit être soumis à l'accord préalable du Ministre.'

[Free translation into English: '[I]f the contracting party is made up of several legal entities, the agreement shall specify which of these entities shall act as the operator. Any change in operator must be subject to the prior agreement of the Minister.']

See also in Algeria, article 29 of Law No. 13-01 of 20 February 2013 Amending and Supplementing Law No. 05-07 of 28 April 2005, On Hydrocarbons:

'[t]out changement d'opérateur doit être soumis à l'accord préalable de l'agence nationale pour la valorisation des ressources en hydrocarbures (ALNAFT).'

[Free translation into English: '[a]ny change of operator must be subject to the prior agreement of the Agence Nationale pour la Valorisation des Ressources en Hydrocarbures (ALNAFT-National Agency for the Promotion of the Hydrocarbon Resources of Algeria)'].

stances, the decision to carry out a change of operator may also result in a decision by the local authorities.[8]

(3) The roles of Operator and Contractor

17.17 The Operator will receive a mandate from the other operators, the content of which varies in very large proportion from one agreement to another.

The role of Operator primarily concerns information on the progress of the works.

In addition, the Operator generally negotiates all of the agreements needed in respect of third parties and represents the other operators to the extent that the latter have accepted the broad outlines of its action.

The Operator must act in order to guard against failures by subcontractors. It may therefore provide for a rapid substitution procedure, enabling the services entrusted to the subcontractors to be completed properly; in this case, the latter is obliged to send all documents, equipment and information to the company replacing them. It is therefore important that the Operator harmonises the technical guarantees between the exploitation agreement and the related contracts.

17.18 Generally, the Operator must keep very strict accounts of the operations that it carries out and invoice its costs to the other operators.[9]

The Contractor should only have a role of provider of funds or technical advisor. However, in the event of inadequacy on the part of the Operator, it may have the authority to become involved but always limiting the period or the tasks to be carried out.[10]

17.19 In practice, the situations are quite different, with the Contractor therefore sometimes being able to be responsible for the work programme.

The extent of the responsibility of the operators therefore varies according to the contractual framework. Most of the time, joint liability of the private operators is instituted.[11] The Operator is not bound by a performance

[8] See, in particular, in Uganda, Article 86(3) of the Petroleum (Exploration, Development and Production) Act 2013: '[n]otwithstanding subsection (2), the Minister may in exceptional circumstances prescribed in regulations, change an operator.'

[9] See, in particular, in the Seychelles, article 1.3.5 of the Seychelles Model Petroleum Agreement of 2013:

> '[t]he Operator, for and on behalf of all entities constituting the Company, shall maintain the accounts of the Petroleum Operations under this Agreement, prepare and submit to PETROSEYCHELLES and the Seychelles Revenue Commission the various Statements required under subsection 1.3.2, and shall do so in such a manner as to permit each such entity to fulfill its obligations under this Agreement.'

[10] Sometimes, the terms 'contractor/consultant' are found in agreements drawn up in English.

[11] See, in particular, in Guinea, Article 23 paragraph 3 of Law L/2014/no. 034/AN of 23 December 2014 Enacting the Petroleum Code:

> '*[s]i le Contractant est constitué de plusieurs sociétés, les sociétés en question sont conjointement et solidairement responsables pour toutes les obligations incombant au Contractant sous les dispositions du présent Code et du Contrat Pétrolier à l'exception des obligations en matière d'impôt sur les sociétés qui incombent à chaque société individuellement*'.

obligation; it can only be individually liable in the event of negligence or serious breach. Such a wording undeniably includes management faults that have resulted in potential loss of income for the operators. It does not appear that an intentional element needs to be present.

17.20 The Operator does not always receive special remuneration; this is more visible in the hydrocarbon sector than in the mining sector. However, the function of Operator is sought after because it allows a better control over the performance of the works and the acquisition of undeniable practical knowledge, which may subsequently be useful for other exploitation agreements, particularly those involving industrial cooperation.

2 THE ABILITY TO AMEND THE CONTENT OF THE WORK PROGRAMME DURING PERFORMANCE

17.21 The performance of the work programme is carried out, overall, within a framework of trust. However, parallel to liberal rules, the performance of the work programme is also governed by rigid rules, which give little room for the freedom of the private operator.

a The liberal rules of performance of the work programme

17.22 The principle is that the performance of the works should start within a certain timeframe following the signature of the agreement or the granting of mining rights. These periods are very variable, particularly with regard to the exploitation phase concerned.

However, the performance of the work programme does have a certain flexibility. This flexibility may result in subsequent authorised modifications of the work programme.

17.23 It is often very difficult to plan accurately what should be undertaken in order to obtain the optimum performance of exploitation operations; this is why the work programme is accompanied by a great flexibility, both in relation to the possibilities of exceeding the original budgets and in terms of carrying out works that were not planned initially.

i *Exceeding the initial budgets*

17.24 Exceeding the initial budgets is limited at two levels. The first limitation lives in the reason for the budgets being exceeded. It is usually necessary that this is duly justified by significant changes in the exploitation conditions. A second limitation concerns the amount by which the budgets are exceeded, which generally corresponds to a share of the budget initially accepted or, in some cases, to a threshold. It should be emphasised that some

---

[Free translation into English: '[i]f the Contracting Party is made up of several companies, the companies in question are jointly and severally liable for all the obligations incumbent upon the Contracting Party under the provisions of this Code and the Oil Agreement, except for tax obligations on companies which are incumbent upon each company individually'].

provisions require the private operator to prevent the execution of works that are not needed for it to fulfil its obligations.[12]

17.25 The failure to match up budget estimates with the works carried out may be marginalised in the exploitation agreement itself. The upward or downward readjustment of the initial budgets allows for the difficulty of time control in the execution of operations to be overcome.

17.26 The best operating rule remains that of flexibility. It is in effect very difficult to predict in advance what the operators are going to discover, the geological structures encountered, but also the changes in the political, social and economic environment throughout the term of the exploitation agreement, as well as, of course, the fluctuation of the international price of raw materials.

ii *Carrying out unplanned works*

17.27 Within the context of service agreements *stricto sensu*,[13] it is crucial to delimit the works to be carried out. If during the performance of the works, some services which were not initially planned, prove to be necessary, the private operator should in general carry them out; payments of these can then be fixed in a separately negotiated agreement.

17.28 If during the performance of an exploitation agreement the operators cannot find agreement on some unplanned works to be carried out, one of the operators can undertake them at its sole cost and risk.[14] These works will therefore be accounted for separately.

b Stringent rules of performance of the work programme

17.29 Alongside the provisions that we have just examined, the legal and contractual provisions may restrict certain direct or related aspects of the performance of exploitation operations, particularly with regard to the free choice of subcontractors.

17.30 Mining and petroleum codes may stipulate the conditions for using subcontractors. In principle, the operator is guaranteed free choice of its subcontractors.[15] However, the use of subcontractors may be limited, as the private operator is supposed to be able to carry out the services contained in the exploitation agreement itself, which in practice means that the coordination problems caused by the existence of multiple agreements can be avoided.

---

[12] Therefore, some provisions can force the private operator to prevent any expenditure which is not necessary.

[13] See above para 10.12 et seq.

[14] See below para 19.110 et seq.

[15] See in this regard, in Algeria, Article 18.1 of the Presidential Decree of 1 March 2003 Enacting a Standard Mining Agreement; Côte d'Ivoire Article 14 of Regulation No. 18/2003/CM/UEMOA of 23 December 2003 Enacting the Mining Code.

17.31 Despite the principle of free choice of subcontractor by the private operator, legislation may arrange a certain control in this respect, particularly through the transmission of subcontracts to the administration[16] or the necessity to obtain its prior approval.[17]

17.32 Furthermore, national legislation may impose some requirements regarding the choice of subcontractors, particularly by requiring that subcontracting agreements are concluded by technically qualified persons approved by local authorities.[18]

---

[16] See, to that effect, in Côte d'Ivoire, Article 51 of Law No. 96-669 Enacting the Petroleum Code and Article 131 of Law No. 2014-138 of 24 March 2014 Enacting the Mining Code; in Niger, Article 19 of Law No. 2007-01 of 31 January 2007 Enacting the Petroleum Code; and in Central African Republic, Article 58 of Ordinance No. 93.007 of 25 May 1993 Enacting the Petroleum Code. Some legislation specifies that only subcontracting agreements which exceed a certain amount will be sent to the local authorities. See, in particular, in Cameroon, Article 75(3) of Law No. 99-013 of 22 December 1999 Enacting the Petroleum Code.

[17] See in this regard, in Côte d'Ivoire, Article 131 of Law No. 2014-138 of 24 March 2014 Enacting the Mining Code:

> '*[l]e titulaire d'un titre ou le bénéficiaire d'une autorisation d'exploitation peut, sous sa responsabilité, sous-traiter à des entreprises qualifiées, des opérations minières dont il a la charge. Il doit accorder la préférence aux entreprises ivoiriennes, à conditions équivalentes de qualité, de prix et de quantités. Les contrats de sous-traitance doivent être communiqués à l'Administration des Mines. Les sous-traitants sont agréés dans les conditions fixées par décret.*'
>
> [Free translation into English: '[t]he holder of a right of the beneficiary of an exploitation authorisation may, under its liability, subcontract the mining operations for which it is responsible to qualified companies. It must give preference to Côte d'Ivoire companies, under equivalent quality, price and quantity conditions. Subcontracting agreements must be sent to the Mining Administration. Subcontractors are approved under the conditions laid down by decree.']

[18] See, in particular, in Gabon, Article 205 of Law No. 0112014 of 28 August 2014 Enacting the Regulations for the Hydrocarbon Sector in the Gabonese Republic:

> '*[s]ans préjudice des autres conditions exigées par les textes en vigueur, ne peuvent conclure un contrat de soustraitance pétrolière que les personnes physiques ou morales techniquement qualifiées dans les métiers des hydrocarbures et titulaires d'un agrément technique délivré par le Ministre chargé des Hydrocarbures. La durée de validité de l'agrément est de trois (3) ans. Les modalités de délivrance, de renouvellement et de retrait, de l'agrément technique sont fixées par voie réglementaire. Le ministère en charge des hydrocarbures tient un répertoire des sous-traitants agréés qu'il met à jour régulièrement et tient à la disposition des opérateurs*'.
>
> [Free translation into English: '[w]ithout prejudice to the other conditions required by the legislation in force, only natural persons or legal entities that are technically qualified in hydrocarbon professions and who hold a technical approval issued by the Minister responsible for Hydrocarbons may conclude an oil subcontracting agreement. The period of validity of the approval is three (3) years. The procedures for issuing, renewing and withdrawing the technical approval are laid down by regulations. The Minister responsible for hydrocarbons keeps a list of approved subcontractors which the Minister updates regularly and makes available to operators.']

17.33 In addition, private operators can formally subject themselves to embargo measures taken by the host state against another state in the exploitation agreement. This kind of provision will have an effect on the operator's ability to choose freely its subcontractors because of embargo measures, both unilateral and multilateral aimed at nationals of another state. These measures are intended to be applied to subcontractors under extractive contracts, but also downstream in transport or even distribution contracts.[19]

[19] Regarding the question of the impact of embargoes on the execution of exploitation operations, see Audit op cit 62 et seq.

18

Surveillance rules

18.1 It is essential that surveillance is carried out on exploitation operations in order to monitor the compliance with the works programme.

1 THE RULES CAUSING SURVEILLANCE

18.2 Monitoring that operators have fulfilled their obligations is accompanied beforehand by an exchange of information.

18.3 Originally, the communication of information was designed in order to protect the interests of the owner of the deposits under exploitation. Now, the exchange of information gathered during exploitation is rather more envisaged as being a dominant element of technology transfer.[1]

a The scope of surveillance through an exchange of information

18.4 The private operator has remained, for many years, the only holder of an obligation to transmit data and information relating to mining and oil operations.

18.5 However, the principle of reciprocity within the context of an exchange of information between private operators and the local State operator has, over time, been confirmed as part of exploitation agreements.

i *The scope of the exchange of information*

18.6 Ever more frequently, the obligation to transmit the information gathered during exploitation works has a legal or regulatory origin. Numerous mining and petroleum codes impose the obligation upon the private operator to communicate all the data, information and other details relating to mining and petroleum operations.[2]

18.7 In principle, the information concerned is essentially geological in nature.

---

[1] This concern appears clearly during negotiations.

[2] See in particular, in Central Africa, article 57 of the Ordinance No. 93.007 of 25 May 1993 Enacting the Petroleum Code. See also, in Algeria, Article 22 of Law No. 13-01 of 20 February 2013 Amending and Supplementing Law No. 05-07 of 28 April 2005, On Hydrocarbons:

> '*[t]outes données et résultats issus des travaux de prospection doivent être mis à la disposition de l'agence nationale pour la valorisation des ressources en hydrocarbures (ALNAFT) selon des procédures établies par voie réglementaire.*'
>
> [Free translation into English: '[a]ll data and results deriving from prospecting works must be made available to the Agence Nationale pour la Valorisation des Ressources en Hydrocarbures (ALNAFT- National Agency for the Promotion of the Hydrocarbon Resources of Algeria) according to the procedures established by regulations.'].

18.8 Parallel to the obligation for the operator to disclose the data gathered, some provisions may provide that the data becomes, once the host state receives it, the property of the latter.

18.9 The importance of information stems much more from the interpretations and analyses which result from it than from its raw form.

18.10 Therefore, the local State operator may receive complete files of seismic profiles from the private operator without being able to draw one single piece of concrete information from it.

This observation reinforces the importance of technology transfer from the private operator to the State operator.

ii *The reciprocity of the exchange of information*

18.11 In the first concession agreements, the principle of a flow of information between the operators and the host state was practically non-existent.

18.12 The parties involved in the exploitation of natural resources promptly realise the importance of information relating to the perimeter. This observation resulted in the institution of an obligation in exploration agreements to exchange the information obtained during the mining and petroleum operations, generally planned for the entire term of the agreements.

Parallel to contractual provisions which favour a reciprocal exchange of information between the private operator and the local State operator, it should be mentioned that there are some legislative provisions under the terms of which geological data and infrastructure documents are freely accessible to the public and, consequently, to the private operators who may benefit from them.[3]

b The limits of surveillance through an exchange of information: the issue of confidentiality

18.13 Provisions are inserted in all exploitation agreements, the purpose of which is to channel and control the disclosure of knowledge related to petroleum and mining operations. These clauses, even though they have different names, have comparable content and purpose. The aim of these clauses is to prevent the disclosure of know-how or information gathered by the operators.

18.14 Parallel to these contractual provisions, the mining and petroleum codes also contain provisions aimed at ensuring the confidentiality of information, data or documents relating to the operations.

---

[3] See, in particular, in Algeria, Article 27 of the Law No. 0 1-10 of 4 July 2001 Enacting the Mining Code which provides that:

> '*[l]es données et documents de l'infrastructure géologique, qui revêtent un caractère d'intérêt général, sont ouverts au public et utilisables par tous les secteurs de l'activité socio-économique et culturelle*.'
>
> [Free translation into English: '[g]eological data and documents which are of general interest are open to the public and can be used by all sectors of socio-economic and cultural activity.']

i *The agreed limits*

18.15 The principle of confidentiality, widely acknowledged by national legislation and exploitation agreements, has certain limitations. Therefore, the confidentiality guaranteed to the private operator is limited both in its scope and duration.

(1) The extent of confidentiality

18.16 The principle of confidentiality towards a third party is practically unanimously applied when it relates to protecting the information and data gathered during mining and oil operations. The application of this principle does, however, present more difficulties when it relates to asserting a principle of confidentiality of exploitation agreements.

(a) The confidentiality of exploitation agreements

18.17 The publication by the host state of confidential information contained in the exploitation agreement may present difficulties and create tension between the parties to the exploitation agreement.[4]

18.18 The issue of the publication of exploitation agreements or, on the other hand their confidentiality, highlights the dual nature of these agreements, which are both commercial agreements, requiring a certain confidentiality to be maintained, and State contracts, which would justify their publication based on transparency requirements.[5]

18.19 The main argument put forward in favour of maintaining the confidentiality of exploitation agreements is that of the protection of sensitive information. Private operators will consider that the contractual conditions under which mining and petroleum operations take place must remain confidential.

18.20 However, numerous arguments may be asserted in favour of the publication of extractive agreements. Firstly, an argument in the interest of private operators who may have an interest in the publication of exploitation agreements when the contractual provisions contain a most-favoured investor clause, which, for its application, supposes knowledge of the treatment granted to other investors. A second argument is that in practice, confidential information is rarely contained in the exploitation agreement itself, but sometimes in its appendices, or even in amendments signed subsequently. Finally, often the local authorities are tied to making extractive agreements

---

[4] One of the oldest examples concerns Saudi Arabia. The Saudi authorities have made almost all the clauses of the oil concession agreement public since its signature in May 1933. This publicity caused strong protests from the American private operator who maintained that these provisions should remain secret. The latter was, in effect, experiencing huge difficulties in the United States because in March 1933, right in the middle of the Depression, President Franklin D Roosevelt had ordered the closure of the country's banks and put an embargo on exports of gold. Such publishing could only delay certain payments planned for the King of Saudi Arabia and cause strong opposition.

[5] On this issue, see Isidore Costade 'Publication des contrats et confidentialité de l'industrie extractive' in *Le contentieux extractif* (ICC 2015) 159.

public, for political, commercial or constitutional reasons, particularly when the agreement must be approved by decree or by a vote of the local parliament.

18.21 Currently, the movement in favour of the publication of exploitation agreements is gaining ground. Against a background of the fight against corruption and the development of good governance and transparency in the extractive industries, more and more states are campaigning for public access to exploitation agreements.[6] From this perspective, a practice is emerging in favour of inserting a clause in the exploitation agreement that asserts the principle that it will be published. The wish to promote public access to exploitation agreements is especially sensitive in Guinea, where the mining code provides that any confidentiality clause in a mining agreement that prohibits the publication of the latter is considered null and void.[7]

18.22 This trend is confirmed with the development of standard mining and petroleum contracts. These models, used as a basis for negotiation, are today largely accepted, both in the mining and in the oil sector.[8]

---

[6] Therefore, some African countries, such as the Democratic Republic of Congo and Guinea, make oil and mining agreements freely available on the Internet. Similarly, some national legislation provides that exploitation agreements will be subject to publication in the Official Journal. This is the case of Burkina Faso, where Law No. 036-2015/CNT of 16 June 2015 Enacting the Mining Code of Burkina Faso provides in its article 1 that provides '*[l]es titres miniers et autorisations ainsi que les contrats ou conventions minières font l'objet de publication au Journal officiel du Faso.*'

[Free translation into English: 'mining rights and permits as well as mining contracts or agreements shall be subject to publication in Faso's Official Journal.']

See also in Democratic Republic of Congo, Decree No. 011/26 of 20 May 2011 On the Obligation to Publish All Contracts Related to Natural Ressources.

[7] See, in this regard, in Guinea, Article 217-II of Law No. L/2013/N°053/CNT of 8 April 2013 Amending Law No. 2011/06/CNT of 9 September 2011 Enacting the Mining Code:

'*[t]ous les Titres miniers, ainsi que toute Convention minière, sont publiés dans le Journal officiel et sur le site Internet officiel du Ministère en charge des Mines, ou tout autre site désigné par le Ministre. Toute clause de confidentialité présente dans une Convention minière interdisant la publication d'une Convention minière est nulle et non avenue.*'

[Free translation into English: '[a]ll Mining Rights, as well as any Mining Agreement, shall be published in the Official Journal and on the official website of the Ministry in charge of Mines, or any other website designated by the Minister. Any confidentiality clause in a Mining Agreement which prohibits the publication of a Mining Agreement shall be null and void.']

[8] Therefore, numerous mining codes make reference to typical mining agreements. See in particular, in Benin, Article 23 paragraph 4 of Law No. 2006-17 of 17 October 2006 Enacting the Mining and Mining Taxation Code in the Republic of Benin. See also, in Chad Article 40(6) of Law No. 011-PR-1995 Enacting the Mining Code:

'*[l]e modèle de convention minière adopté par Arrêté du Ministre chargé des Mines peut servir de base à la négociation entre le Gouvernement et le titulaire éventuel.*'

[Free translation into English: '[t]he mining agreement model adopted by the Order of the Minister responsible for Mines may be used as a basis for negotiation between the Government and the potential successful bidder.']

See also, in Chad, Article 40 paragraph 6 of Law No. 011-PR-1995 Enacting the Mining Code:

'*[l]e modèle de convention minière adopté par Arrêté du Ministre chargé des Mines peut servir de base à la négociation entre le Gouvernement et le titulaire éventuel.*'

18.23 In any event, the publication of the exploitation agreement does not hinder a certain level of confidentiality, particularly with regard to the information and data gathered during exploitation works.

(b) The confidentiality of data relating to mining and petroleum operations

18.24 Several pieces of mining and petroleum legislation guarantee the private operator confidentiality of the data, information and documents transmitted to the host state during exploitation operations.[9] The administration may therefore be bound by a positive obligation to protect the data and information covered by the principle of confidentiality.

The implementation of this confidentiality principle is accompanied by an obligation on the part of the mining and oil administration's personnel to respect professional secrecy when carrying out their tasks.[10]

[Free translation into English: '[t]he mining agreement model adopted by the Order of the Minister responsible for Mines may be used as a basis for negotiation between the Government and the potential successful bidder.']

Or in Madagascar, Article 8 of Law No. 2005-021 Establishing the Modification of Certain Provisions of Law No. 99-022 of 19 August 1999 Enacting the Mining Code:

'*[l]a recherche, l'exploitation, la transformation, le conditionnement, le transport et la commercialisation des minerais et substances radioactifs font l'objet de conventions particulières avec l'État selon un modèle de convention type fixé par voie réglementaire et précisant notamment les mesures de radioprotection et de gestion des déchets radioactifs applicables à ces activités, ainsi que les mesures de sécurité concernant l'emploi, le transfert et la commercialisation des substances radioactives conformément à la loi et aux engagements internationaux de l'État en la matière.*'

[Free translation into English: '[t]he exploration, exploitation, processing, packaging, transport and marketing of minerals and radioactive substances is subject to special agreements with the State according to a typical model agreement set forth through regulations and particularly specifying measures for radio protection and managing radioactive waste applicable to these activities, as well as safety measures concerning the use, transfer and marketing of radioactive substances in accordance with the law and the State's international commitments in this regard.']

[9] See, in particular, in Mauritania, Article 59(3) of Law No. 2008-011 of 27 April 2008 Enacting the Mining Code: '*[t]ous les renseignements, documents et toute information obtenus par le Ministère et ses agents dans le déroulement de leur mission ou dans l'exécution des obligations du titulaire d'un permis de recherche, seront considérés comme strictement confidentiels, sauf indication contraire du titulaire, pendant la durée du titre dans la mesure où tout ou partie de celui-ci n'a pas été transformé en permis d'exploitation. Au-delà de ce délai, les informations de caractère technique seront mises à la disposition du public.*' [Free translation into English: '[a]ll the information, documents and any details obtained by the Ministry and its officers when carrying out their task or in the execution of the obligations of the holder of an exploration permit will be considered to be strictly confidential, unless otherwise indicated by the holder, for the duration of the rights insofar as all or part of these have not been converted into an exploitation permit. Beyond this period, information of a technical nature will be made available to the public.']

[10] See in particular, in Benin, Article 77 of Law No. 2006-18 of 17 October 2006 Enacting the Petroleum Code in the Republic of Benin:

'*[l]es agents de cette direction sont tenus au respect du secret professionnel.*'

[Free translation into English: '[a]gents of this department are obliged to respect professional secrecy.']

18.25 However, the confidentiality can, in some cases, be restricted. Indeed, the host state may be permitted to use the information disclosed to it by the private operator with a view to publishing statistical data relating to the geology and natural resources of the host state.[11] Similarly, the host state may reserve the right to use confidential information internally or to publish the geological service maps for the purposes of incorporating these in the surface mapping,[12] particularly as part of a campaign to promote its extractive sector.

18.26 The definition of the confidentiality criteria, too often forgotten, is an essential provision that avoids significant disputes. It is necessary to identify with the greatest of accuracy which information must remain confidential by carefully distinguishing between the studies carried out, data and information, knowledge, industrial know-how and technology. These terms are not always well defined in exploitation agreements.

(2) The period of confidentiality

18.27 The confidentiality guaranteed to the private operator is generally limited in time.

The first solution consists of limiting the confidentiality obligation to the term of the exploitation agreement,[13] or to that of the mining rights.[14] In this

[11] See in this regard, in Cameroon, Article 41(3) of Law No. 216/017 of 14 December 2016 Enacting the Mining Code.

[12] See in particular, in Cameroon, Article 107 of Decree No. 2000-485 of 30 June 2000 Determining the Conditions for Implementing Law No. 99-013 of 22 December 1999 Enacting the Petroleum Code:

> '*[n]onobstant les dispositions des articles 105 et 106 ci-dessus: les cartes géologiques de surface et leurs interprétations peuvent être utilisées par l'État à tout moment aux fins d'incorporation dans la cartographie officielle*; [. . .] *l'État, les établissements et organismes publics peuvent utiliser les documents visés à l'article 105 ci-dessus aux fins d'usage exclusivement interne dès leur obtention sans aucune restriction.*'
>
> [Free translation into English: '[n]otwithstanding the provisions of Articles 105 and 106 above: the geological surface maps and their interpretations may be used by the State at any time for the purposes of incorporation in the official cartography; [. . .] the State and public establishments and bodies may use the documents specified in article 105 above for the purposes of their exclusively internal use once they have been obtained without any restriction.']

[13] See, in particular, in Guinea, Article 64 of Law No. L/2014/N°034/AN of 23 December 2014 Enacting the Petroleum Code: '*[l]e Contrat Pétrolier prévoit les conditions dans lesquelles le Contractant et l'État peuvent communiquer les informations et données à des tiers pour les besoins des Opérations Pétrolières pendant la durée du Contrat Pétrolier* [. . .]'. [Free translation into English: '[t]he Oil Contract stipulates the conditions under which the Contracting Party and the State may communicate the information and data to third parties for the needs of the Petroleum Operations during the term of the Petroleum Agreement [. . .]'.]

[14] See in particular, in Madagascar, Article 222 of Law No. 99-022 of 19 August 1999 Enacting the Mining Code amended by the Law No. 2005-021 of 17 October 2005:

> '*[l]es rapports, comptes-rendus et études fournis par les titulaires sont confidentiels pour la durée de validité des permis miniers. Passé ce délai,ils sont accessibles au public.*'

case, the termination of the agreement or the withdrawal of the mining permit will cause, *de facto*, the lifting of the confidentiality obligation.[15]

A second series of provisions extends the obligation after the termination of the agreement or the expiration of the mining rights, either for an indeterminate time, or for a specific period,[16] sometimes adjustable depending on the elements concerned.

Finally, legislation can provide that confidentiality will be guaranteed for a certain time from the date on which the information was obtained.

ii *Agreed adjustments to confidentiality*

18.28 The confidentiality may also be limited in the event that the private operator expressly consents that some information is published.[17]

iii *Violation of the confidentiality undertaking*

18.29 A final provision concerns the violation of the obligation of confidentiality. There is a significant deficiency with regard to sanctions applied in case of non-compliance with the obligation of confidentiality. Sometimes, mechanisms are provided for, such as the temporary suspension of activities connected to the performance of the agreement. The nomination of a third-party operator responsible for reporting the violation and determining its effects and the nature and amount of the loss suffered, or finally, make

---

[Free translation into English: '[t]he reports, minutes and studies supplied by the successful bidders shall be confidential during the period of validity of the mining permits. After this period, they are accessible to the public.']

[15] See in particular, in Senegal, Article 83 of Law No. 2016-32 of 8 November 2016 Enacting the Mining Code:

'*[l]es documents et renseignements à caractère géologique, minier, industriel, commercial et de propriété intellectuelle recueillis auprès de titulaires de titres miniers ne peuvent être communiqués au public ou aux tiers que sur autorisation écrite des titulaires, ou qu'en cas de retrait ou d'expiration du titre minier*'.

[Free translation into English: '[t]he documents and information with a geological, mining, industrial, commercial character or intellectual property gathered from the holders of mining rights may only be disclosed to the public or to third parties on the written authorisation of the said holders, or in the event of the withdrawal or expiry of the mining rights'.]

[16] See, in particular, in Algeria, Article 32(2) of the Presidential Decree of 1 March 2003 Enacting the Model Mining Agreement, which provides that confidential information may only be disclosed to the public or communicated after one year from the expiry of the mining agreement.

[17] See in particular, in Algeria, Article 32(2) of the Presidential Decree of 1 March 2003 Enacting the Model Mining Agreement:

'*[l]es renseignements, documents, études soumis demeureront confidentiels et ne pourront, sauf avec le consentement préalable de l'Investisseur, être rendus publics ou communiqués à des Tiers par l'Administration Minière qu'à l'issue d'un an à compter de l'expiration de la Concession Minière, selon le cas.*'

[Free translation into English: '[t]he information, documents and studies submitted shall remain confidential and may only, except for the prior consent from the Investor, be made public or disclosed to Third Parties by the Mining Administration after one year has passed from the expiry of the Mining Agreement, as appropriate.']

an explicit reference to the arbitration clause by giving the arbitrator all powers to intervene in the matter.

2 THE RULES OF WHICH THE CONTROL IS THE SUBJECT

18.30 The auditing of exploitation operations is done at not only the internal level but also results from external interventions.

a The internal surveillance of works: surveillance carried out by the operators

18.31 The internal surveillance carried out by the operators is continuously planned on very wide bases. It is organised, on the one hand, between the local State operator and the private operator and, on the other hand, between the private operators.

As part of this internal surveillance, keeping strict accounts is essential. Therefore, the exploitation agreement specifies the method for presenting the analytical exploitation accounts and the balance sheet forecasts.

The procedures for surveillance also concern the bank accounts opened both in the host state and abroad.

The surveillance must be carried out at reasonable times without disturbing or hindering the proper performance of the exploitation operations.

We shall distinguish between surveillance carried out by the local State operator from that carried out by the private operator.

i *Surveillance carried out by the local State operator*

18.32 Surveillance carried out by the local State operator is at the financial expense of the latter.

18.33 The local State operator is frequently prevented from effectively carrying out its surveillance power because of a lack of knowledge and skill on its part; therefore, foreign consultants can advise representatives of the host state. This allows the State entity to carry out an effective and efficient control.

18.34 An internal audit of the operating company can also be arranged, always at the expense of the private operator. This audit concerns different aspects: results obtained, training of local staff, employees' accommodation, safety of personnel and equipment, organisation of work and distribution of tasks, the state of the facilities, the conduct of operations as well as the maintenance of the exploitation equipment.

ii *Surveillance carried out by the private operator*

18.35 Very often, the private operator must prepare the documents needed for surveillance of the works and operations that it has carried out; this check is carried out directly by the other operators or by external consultants.

Its action is subject to close monitoring by the local authorities with regard to the techniques used and the valuation of production.

However, as part of this audit, the private operator also has an ongoing role as technical adviser, and in practice, it may be led to make a judgement on the operations that it has carried out. This role has a basis that is not contractual but practical. The private operator must draw up very full reports, the content of which is specified in detail in the exploitation agreement.

b The external surveillance of works: surveillance carried out by the local authorities

18.36 Throughout the period of the mining and petroleum operations, it is up to the host state to organise their administrative, technical and financial monitoring.

The monitoring of exploitation operations may require technical skills that the host state may not have. In order to overcome this lack of knowledge, the host state may use external consultants, either national or foreign, in order to assist it in its monitoring and audit operations.[18] However, administrative monitoring may mobilise significant sums, particularly when the host state has to call on external experts, particularly international experts, to carry out its audit assignments.

We shall successively consider the audit procedures, then the sanctions that may apply in the event that one of the operators is at fault.

i *The audit procedures*

18.37 The monitoring and surveillance by the host state of mining and oil operations essentially takes two forms: on the one hand, an administrative and technical check and, on the other hand, financial monitoring.

---

[18] See, in particular, in this regard, in Niger, Article 237 of Decree No. 2007-082/PRN/MME of 28 March 2007 Determining the Conditions for Application of Law No. 2007-01 of 31 January 2007 Enacting the Petroleum Code:

> '*[l]'État du Niger exercera son droit de surveillance administrative et technique et de contrôle financier, soit en faisant appel aux agents habilités et assermentés de l'administration nigérienne, soit en faisant appel à des consultants mandatés par le Ministre chargé des Hydrocarbures*.'
>
> [Free translation into English: '[t]he State of Niger shall exercise its administrative and technical monitoring and financial surveillance rights, either by calling on agents authorised and sworn in by the Niger administration, or by calling on consultants appointed by the Minister responsible for Hydrocarbons.']

See also, in Algeria, Article 33(5) of the Presidential Decree of 1 March 2003 Enacting the Model Mining Agreement:

> '*[l]'ÉTAT se réserve le droit de se faire assister, à ses frais et à tout moment, par un cabinet d'audit national ou international reconnu afin de vérifier, sans gêner les opérations de l'Investisseur, les renseignements et les pièces justificatives que l'Investisseur doit lui fournir en vertu de la Convention Minière*.'
>
> [Free translation into English: '[t]he STATE reserves the right to the assisted, at its expense and at any time, by a recognised national or international firm of consultants in order to check, without hindering the operations of the Investor, the information and written proof that the Investor must supply to it by virtue of the Mining Agreement.']

(1) The administrative and technical monitoring of exploitation operations

18.38 A right to monitor exploitation operations may be arranged for the benefit of the local authorities. It is within the context of inspections that the surveillance of exploitation operations by the host state will take place.

The right of the local authorities to carry out the inspection of mining and oil operations is widely recognised by mining and petroleum codes. As part of these inspections, numerous prerogatives are generally devolved to government officers from the mining or petroleum administration.

The latter particularly have the right to enter extraction sites in order to carry out audit operations.[19] Representatives of the administration may, during their inspections on site, take samples and access all of the documents, data or studies relating to the exploitation operations.[20]

18.39 In order to facilitate surveillance by the host state of mining and oil operations, numerous codes provide for an obligation according to which the private operation is responsible to keep a register of the operations or to provide technical reports, in particular on a quarterly or annual basis, which are subject to acceptance by the competent authority.[21]

---

[19] See in Botswana, Article 7 of the Mines and Minerals Act ('*Power of certain officers to enter upon prospecting areas, etc.*').

[20] See in particular, in the Republic of Congo, Article 29 of the Law No. 24-94 as 23 August 1994 Enacting the Hydrocarbon Code:

'*[a]fin d'assurer l'exécution de la présente loi, le Ministre chargé des Hydrocarbures peut nommer des inspecteurs chargés de l'exécution de sa mission. Les inspecteurs ont les pouvoirs suivants: ils ont accès à tous terrains, installations, bureaux et autres zones ou immeubles servant aux opérations pétrolières, ils ont accès à tous documents, données, interprétations, études, comptes, documents et échantillons relatifs aux opérations pétrolières, ils peuvent prélever des échantillons, ils peuvent vérifier ou installer tous instruments de mesure, ils exercent une surveillance de police pour la conservation des édifices et la sûreté du sol, et ils observent la manière dont l'exploitation est faite soit pour éclairer les exploitants sur ses inconvénients ou son amélioration, soit pour avertir l'autorité compétente des vices, abus ou dangers qui s'y révéleraient ou seraient susceptibles de se présenter.*'

[Free translation into English: '[i]n order to ensure the performance of this law, the Minister responsible for Hydrocarbons may appoint inspectors responsible for carrying out his or her task. The inspectors have the following powers: they shall have access to all land, facilities, offices and other zones or buildings used for oil operations, they shall have access to all documents, data, interpretations, studies, accounts, documents and samples relating to oil operations, they may take samples, they can check or install any measuring instruments, they shall carry out police surveillance for the conservation of the buildings and safety of the ground, and they shall observe the manner in which exploitation is carried out, either to enlighten the operators regarding its drawbacks or its improvement, or to warn the competent authority of defects revealed there or which would be likely to occur.']

See also, in Cameroon, Article 98 of the Decree No. 2000-485 of 30 June 2000 Determining the Conditions for Application of Law No. 99-013 of 22 December 1999 Enacting the Petroleum Code.

[21] See as an example, in Burundi, Article 46 of Law No. 1-21 of 15 October 2013 Enacting the Mining Code:

'*[l]e titulaire du permis de recherche est tenu de faire trimestriellement et annuellement un rapport technique et financier qui sera transmis au Ministre ainsi que, le cas*

(2) The financial monitoring of exploitation operations

18.40 The financial monitoring of mining and oil operations is the responsibility of the local authorities. Therefore, the mining and petroleum administrations have the obligation to take the necessary measures in order to ensure the economic and accounting monitoring of exploitation operations.[22]

18.41 Audits are the main tool that administrations have in order to ensure the financial monitoring of mining and petroleum operations. An audit (the function of which is to check the quantity of services provided, their quality as well as the validity and amount of expenses incurred), is very widely provided for by mining and petroleum codes.[23] The private operator may, as part of these audits, be under an obligation to facilitate the work of

---

échéant, une estimation des ressources et des réserves évaluées au cours de l'année. Le rapport annuel doit être analysé et validé par l'autorité compétente dans un délai ne dépassant pas deux (2) mois. Passé ce délai, il est considéré comme accepté.'

[Free translation into English: '[t]he holder of the exploration permit is under an obligation to draw up a technical and financial report quarterly and annually which will be sent to the Minister as well as, if appropriate, an estimate of resources and reserves assessed throughout the year. The annual report must be analysed and approved by the competent authority within a period not exceeding two (2) months. Beyond this period, it shall be considered to be accepted'.]

[22] See, in particular, in the Côte d'Ivoire Article 65 of the Law No. 96-669 as 29 August 1996 Enacting the Petroleum Code:

'[l]e gouvernement veille à l'application des dispositions de la présente loi et de ses textes d'application, ainsi qu'à l'exécution de leurs obligations par les titulaires de contrats pétroliers. Il prend toute mesure réglementaire et dispose à cet effet de tout droit de surveillance et d'inspection des opérations pétrolières. A ce titre, le gouvernement prend au niveau des administrations compétentes, toutes les dispositions appropriées pour assurer la surveillance administrative et technique, le suivi économique et comptable des activités visées par la présente loi.'

[Free translation into English: '[t]he government shall monitor the application of the provisions of this law and its enforcement, as well as the performance of obligations by the holders of petroleum contracts. It shall take all regulatory measures and has to this effect all rights to monitor and inspect petroleum operations. To this end the government shall, at the competent administration level, take all appropriate measures to ensure administrative and technical monitoring, and the economic and accounting monitoring of the activities specified by this Law.']

[23] The right of mining and petroleum administrations to conduct audits is provided by numerous mining and Petroleum Codes. See in particular, in Algeria, Article 111 of the Law No. 05-07 of 28 April 2005 On Hydrocarbons:

'[p]our l'ensemble des missions de l'agence nationale pour la valorisation des ressources en hydrocarbures (ALNAFT) et de l'autorité de régulation des hydrocarbures pour lesquelles est nécessaire un contrôle d'application et de conformité à des règles édictées, notamment l'audit des comptes des contractants ou des concessionnaires, ces agences peuvent faire appel à des cabinets professionnels, nationaux ou internationaux incontestables.'

[Free translation into English: '[f]or all of the tasks of the Agence Nationale pour la Valorisation des Ressources en Hydrocarbures (ALNAFT-National Agency for the Promotion of the Hydrocarbon Resources of Algeria) and the Hydrocarbons Regulation Authority for which a check of the application of and compliance with the enacted rules is necessary, in particular an audit of the accounts of the contracting parties or agents, these agencies may call on unquestionable professional firms, either national or international.']

the State's officers.[24] Parallel to this, national legislation may specify that the private operator will be obliged to send the financial documents relating to mining and oil operations to the competent authority.

18.42 The financial monitoring of operators also involves keeping accounts. These accounts should be held, as appropriate, in local or foreign currency,[25] most often in accordance with the national accounting plan[26] or

---

See also, in Madagascar, Article 125 of the Law No. 2001-031 of 8 October 2002 Enacting a Special System for Major Investments in the Madagascan Mining Sector, amended by the Law No. 2005-022 of 17 October 2005:

'*[t]ous les deux (2) ans, l'État malagasy peut, à ses propres frais, faire procéder à un audit des états financiers du Titulaire et de l'Entité de Transformation et exiger d'eux tout document nécessaire à cet égard. Le cas échéant, le Titulaire et/ou l'Entité de Transformation seront informés au préalable par le Ministre chargé des Mines.*'

[Free translation into English: 'Every two (2) years, the Madagascan State may, at its own expense, carry out an audit of the financial status of the Holder and of the Processing Entity and require any document necessary from them in this regard. If appropriate, the Holder and/or the Processing Entity will be informed in advance by the Minister in charge for Mines.']

[24] See in this regard, in Niger, Article 101 of the Ordinance No. 93-16 of 2 March 1993 Enacting the Amended Mining Law.

[25] For an example of legislation imposing the keeping of accounts in local currency, see in particular, in Burundi, Article 12 of the Decree No. 100-130 of 14 December 1982 Implementing Decree Law No. 1/138 of 17 July 1976 Enacting the Mining and Petroleum Code of the Republic of Burundi and Mining Taxation:

'*[l]es personnes morales désireuses d'obtenir des titres miniers d'exploitation doivent être constituées suivant les lois de la République du Burundi, y avoir leur siège social, y conserver leurs archives et y tenir leur comptabilité en monnaie burundaise.*'

[Free translation into English: '[l]egal entities wishing to obtain exploitation mining rights must be incorporated according to the laws of the Republic of Burundi, have their registered office there, keep their archives there and keep their accounts there in Burundian currency.'].

On the contrary, for laws permitting the holding of accounts in foreign currency, see in particular, in Cameroon, Article 96(3) of the Law No. 99-013 of 22 December 1999 Enacting the Petroleum Code:

'*[l]e Titulaire d'un Contrat Pétrolier qui effectue des Opérations Pétrolières sur le Territoire Camerounais est autorisé à tenir sa comptabilité en dollars américains et à libeller son capital social en cette monnaie. Les modalités de cette tenue sont précisées au Contrat Pétrolier.*'

[Free translation into English: '[t]he Holder of a Petroleum Agreement who carries out Petroleum Operations on Cameroon Territory is permitted to keep its accounts in American dollars and to word its share capital in this currency. The conditions for this holding are specified in the Petroleum Agreement.']

[26] See in particular, in Algeria, article 33(1) of the Presidential Decree of 1 March 2003 Enacting the Model Mining Agreement: '*[l]'Investisseur s'engage pour la durée de la Convention Minière: (a) à tenir une comptabilité véritable et détaillée des opérations, accompagnée de pièces justificatives permettant d'en vérifier l'exactitude; cette comptabilité sera tenue en conformité avec le Plan Comptable National algérien.*'

[Free translation: '[d]uring the term of the Mining Agreement, the Investor undertakes: (a) to keep proper and detailed accounts of operations, accompanied by supporting documents enabling the accuracy thereof to be checked; these accounts will be kept in compliance with the National Algerian Accounting Plan'].

See also, in Benin, Article 101 of the Law No. 2006-17 of 17 October 2006 Enacting the Mining and Mining Taxation Code.

the applicable accounting rules within a defined area, as for instance within the OHADA area.[27] The private operator may be restricted to holding separate accounts,[28] to organising the latter in such a way as to distinguish exploration expenses from other expenses[29] or to keep different accounts for each area of exploitation.[30] The operator will sometimes be obliged to support its accounts with supporting documentation allowing the veracity of the information provided to be certified.[31]

In order to ensure better financial control, mining and petroleum codes may provide that the private operator is under an obligation to have its accounts audited,[32] or that the latter will have the obligation to use a statutory

---

[27] See in this regard, in the Central African Republic, Article 123 of the Law No. 09-005 of 29 April 2009 Enacting the Mining Code.

[28] See, in particular, in Djibouti, Article 43 of the Law No. 117/AN/05/5eL of 16 October 2005 Enacting the Petroleum Code:

'[c]haque entreprise visée à l'article précédent, quel que soit le lieu de son siège, tient, par année civile, une comptabilité séparée des opérations pétrolières qui permet d'établir un compte de résultats et un bilan faisant ressortir tant les résultats desdites opérations, que les éléments d'actif et de passif qui y sont affectés ou s'y rattachent directement.'

[Free translation into English: '[e]ach company stipulated in the previous Article, regardless of the location of its registered office, shall keep, during the calendar year, accounts separate from the oil operations which enable a profit and loss statements and a balance sheet to be drawn up showing both the results of the said operations, and the asset and liability elements allocated to it or directly attached to it.']

[29] See, in particular, in Central African Republic, Article 17 of the Decree No. 09-126 of 30 April 2009 Determining the Conditions for Application of Law No. 09-005 of 29 April 2009 Enacting the Mining Code:

'[e]n vue de la vérification de ces dépenses, la comptabilité doit être organisée en vue de permettre une distinction entre les dépenses au titre des travaux de recherche et les autres dépenses.'

[Free translation into English: '[w]ith a view to checking this expenditure, the accounts must be organised in such a way as to allow a distinction to be made between expenditure for exploration work and the other expenditures.']

[30] See, in particular, in Algeria, Article 56 of the Law No. 05-07 as 28 April 2005 On Hydrocarbons:

'[l]e contractant doit tenir, conformément à la législation et à la réglementation en vigueur et par exercice, une comptabilité par périmètre d'exploitation, permettant d'établir des comptes "valeur ajoutée" et "résultats d'exploitation" et un bilan faisant ressortir les résultats desdites activités, les éléments d'actif et de passif qui y sont affectés ou s'y rattachant directement ainsi que le résultat brut afférent à ces activités.'

[Free translation into English: '[t]he contracting party must keep, in accordance with legislation and the regulations in force and per financial year, accounts per area of exploitation, enabling "added value" and "exploitation results" accounts and a balance sheet to be drawn up showing the results of the said activities, the elements of the assets and liabilities allocated to it or directly related to it as well as the gross result relating to these activities.']

[31] See, in particular, in Algeria, Article 33(1) of the Presidential Decree of 1 March 2003 Enacting the Model Mining Agreement.

[32] See, in particular, in Mauritania, Article 98 of the Law No. 2010-033 of 20 July 2010 Enacting the Hydrocarbon Code amended by the Law No. 2011-044 of 25 November 2011:

'[l]es contractants sont tenus de participer aux mécanismes de transparence des paiements qu'ils effectuent à l'État au titre de la présente loi et aux autres initiatives

auditor in order to check its financial records at its expense.[33]

ii *Sanctions*

18.43 A distinction should be made between, on the one hand, criminal sanctions resulting from infringements of mining and petroleum legislation and, on the other hand, sanctions manifesting themselves by a withdrawal of mining rights or the termination of the exploitation agreement.

(1) Criminal sanctions

18.44 Breaches of the provisions of the mining and petroleum codes may have serious consequences and result in the application of criminal sanctions.

---

relatives à la bonne gouvernance et à la transparence des industries extractives. Aux fins de faciliter les exercices annuels de collecte et de rapprochement des données relatives aux revenus provenant des industries extractives, ils doivent en particulier: 1° faire procéder à un audit annuel de leurs états financiers conformément aux règles d'audit internationales.'

[Free translation into English: '[t]he contracting parties are obliged to adhere to transparency mechanisms regarding the payments that they make to the State under this law and in other initiatives relating to good governance and the transparency of the extractive industries. For the purpose of facilitating the annual financial years of the collection and reconciliation of data relating to income from the extractive industries, they must in particular: 1° carry out an annual audit of their financial records in accordance with international auditing rules.']

[33] See, in particular, in Niger, Article 101 of the Ordinance No. 93-16 of 2 March 1993 Enacting the Amended Mining Law:

'[l]e *titulaire d'un titre minier ou d'autorisation d'ouverture et d'exploitation de carrière permanente doit tenir au Niger une comptabilité conformément au plan comptable en vigueur au Niger notamment à faire certifier pour chaque exercice par un Commissaire aux Comptes son bilan et ses comptes d'exploitation et à communiquer ses états financiers à chaque fin d'exercice au Ministre chargé des finances et à celui chargé des Mines. Il doit donner accès aux documents et pièces justificatives au personnel autorisé de l'État pour fins de vérifications ou d'audit.*'

[Free translation into English: '[t]he holder of mining rights or a permanent permit to open and exploit a quarry must carry out accounting in Niger in accordance with the accounting plan in force in Niger, particularly to have its balance sheet and operating accounts certified for each financial year by a Statutory Auditor and to send its financial statements at the end of each financial year to the Minister responsible for Finance and to the Minister responsible for Mines. It must give access to the documents and supporting documentation to the authorised personnel of the State for the purposes of checks or audit.']

See also, in Benin, Article 101 of Law No. 2006-17 of 17 October 2006 Enacting the Mining Code and Mining Taxation in the Republic of Benin:

'*[l]e titulaire d'un titre minier a l'obligation de tenir en République du Bénin une comptabilité conformément au plan comptable béninois, de faire certifier pour chaque exercice par un commissaire aux comptes agréé, son bilan et son compte d'exploitation et de communiquer dans les deux (02) mois de la fin de chaque exercice fiscal deux (02) exemplaires de ces états financiers au directeur chargé des impôts et un (01) exemplaire au directeur chargé des mines.*'

[Free translation into English: '[t]he holder of mining rights is obliged to carry out accounting in the Republic of Benin in accordance with the Benin accounting plan, to have its balance sheet and its operating accounts for each financial year certified by a statutory auditor and to send two (2) copies of these financial statements within two (2) months at the end of each financial tax year to the director responsible for taxes and one (1) copy to the director responsible for mines.']

The sanctions that may take the form of fines or even imprisonment are most often specified in mining and petroleum codes.

The recognition of these infringements falls under the remit of the mining police and the police[34] who have extensive powers and can carry out investigations, seizures or searches.[35]

The events qualified as criminal offences which may give rise to criminal sanctions are numerous. This may in particular be a question of non-compliance with the rules relating to public security, to the protection of the environment and to the safety of workers, the nonpayment of taxes and royalties, the recording of cases of corruption,[36] or providing inaccurate information during procedures for the granting of mining rights.[37]

(2) Sanctions resulting in the withdrawal of mining rights or the termination of the exploitation agreement

18.45 The sanctions may be drastic and may go as far as withdrawing the mining rights and terminating the exploitation agreement.

Substantively, the grounds for termination are generally provided for by exploration agreements, even though mining and petroleum codes often specify the outlines of these. Among the grounds most frequently used to justify the termination of an exploitation agreement, failure to incur investment expenses, non-execution of the work programme, refusal to submit to a decision by the arbitration Court, nonpayment of taxes and royalties or the bankruptcy of the private operating company should be

[34] See, in particular, in Benin, Article 87 of Law No. 2006-18 of 17 October 2006 Enacting the Petroleum Code of the Republic of Benin:

'*[l]es infractions aux dispositions du présent code sont constatées par des procès-verbaux établis par les fonctionnaires assermentés de la direction chargée des hydrocarbures et par les officiers et agents de la police judiciaire conformément aux dispositions du code de procédure pénale*'.

[Free translation into English: '[i]nfringements of the provisions of this Code will be recorded by the minutes drawn up by sworn officers of the department responsible for hydrocarbons and by the officers and agents of the judicial police in accordance with the provisions of the Code de procédure pénale [Criminal Procedure Code].']

See also, in Algeria, Article 178 of Law No. 01-10 of 4 July 2001 Enacting the Mining Code.

[35] See in particular, in Benin, Article 139 of Law No. 2006-17 of 17 October 2006 Enacting the Mining and Mining Taxation Code:

'*[l]es officiers de police judiciaire et les agents assermentés de la direction chargée des mines ont qualité pour procéder aux enquêtes, aux saisies et aux perquisitions, s'il y a lieu. La recherche des infractions entraîne le droit de visite corporelle. La visite corporelle d'une femme ne peut être faite que par un médecin ou par une femme.*'

[Free translation into English: '[o]fficers of the judicial police and sworn agents of the department in charge of mines are entitled to carry out investigations, seizures and searches, if necessary. The detection of infringements leads to the right to conduct personal searches. The personal search of a woman may only be conducted by a doctor or by a woman.']

[36] See, in particular, in the Democratic Republic of Congo, Article 307 of the Law No. 007/2002 of 11 July 2002 Enacting the Mining Code.

[37] See in particular, in Guinea, Article 123 of Law L/2014/no. 034/AN of 23 December 2014 Enacting the Petroleum Code.

mentioned.[38] Similar grounds are generally used to justify the withdrawal of mining rights.[39]

In form, the withdrawal of mining rights, just like termination of the agreement, must generally comply with being subject to certain formalities. Therefore, the applicable legislation may provide for the publication of the withdrawal decision. Furthermore, the withdrawal of mining rights should be pronounced in the same forms as their granting.[40]

With regard to the termination of an exploitation agreement, the latter must follow the formalities laid down by the agreement. The termination should be subject to prior formal notice.[41]

Other sanctions that have less extreme consequences, such as penalty clauses, are inserted in some agreements. These are provisions that the host states may apply at their discretion.

---

[38] For examples of grounds which may justify the termination of an exploitation agreement see in particular, in Senegal, Article 59 of Law No. 98-05 of 8 January 1998 Enacting the Petroleum Code; in Djibouti, Article 58 of the Law No. 117-AN-05-5 of 16 October 2005 Enacting the Petroleum Code; and in the Seychelles, Article 35(1) of the Model Petroleum Agreement of 2013.

[39] See in particular, in this regard, in Chad, Article 47 of Law No. 011-PR-1995 Enacting the Mining Code; in the Republic of Congo, Article 61 of the Law No. 24-94 of 23 August 1994 Enacting the Hydrocarbon Code; in Niger, Article 59 of the Ordinance No. 93-16 of 2 March 1993 Enacting the Amended Mining Law.

[40] See in particular, in the Central African Republic, Article 85 of Ordinance No. 93.007 of 25 May 1993 Enacting the Petroleum Code, which provides that:

> '*[l]'annulation du permis de recherche ou le retrait de la concession sont prononcés dans les mêmes formes que leur octroi, après mise en demeure du Ministre restée sans effet dans le délai stipulé au contrat pétrolier.*'
>
> [Free translation into English: '[t]he cancellation of an exploration permit or the withdrawal of the concession must be pronounced in the same forms as their granting, after the formal notice from the Minister remains without effect within the period stipulated in the petroleum agreement.']

[41] See in particular, in Mauritania, Article 48 of Law No. 2010-033 of 20 July 2010 Enacting the Hydrocarbon Code amended by Law No. 2011-044 of 25 November 2011:

> '*[l]orsque le contractant ne satisfait pas aux engagements souscrits ou lorsqu'il cesse de remplir les conditions et obligations résultant de la présente loi et des textes pris pour son application, le contrat d'exploration-production peut, après mise en demeure restée infructueuse, être résilié, sans préjudice des dispositions de l'article 49 ci-dessous.*'
>
> [Free translation into English: '[w]hen the contracting party does not satisfy the obligations to which it has subscribed or when it ceases to fulfill the conditions and obligations resulting from this law and the legislation taken for its enforcement, the exploration-production agreement may, after a formal notice remains unanswered, be terminated, without prejudice to the provisions of article 49 below.']

19

CONTRACTUAL ARRANGEMENTS TO MANAGE RISKS SPECIFIC TO EXPLOITATION AGREEMENTS

F Contractual arrangements to manage risks

19.1 Operations to exploit natural resources take place in an economic, social, political, legal and cultural environment that is far from stable. The duration of exploitation agreements, which last for decades, reinforces this risk of instability. Changes in the social, political and economic environment that may occur during the enforcement of the contract constitute risks that operators attempt to mitigate. These risks are the subject of very detailed analyses performed by the parties when they decide to invest and during negotiations. These in-depth studies allow operators to calculate, but not avoid, the probability of adverse events occurring.

Operators attempt to channel risks inherent in exploitation agreements by way of rules specific to mining and oil operations. In addition, a set of rules that are not specific to exploitation agreements in the extractive sector may also be used by the private operator to allow for better management of these risks.

19.2 As long as a private operator exploits natural resources, he assumes risks that are political, natural and financial. There are two types of risk that must be considered: formal assumption of political risks and the substantial assumption of natural and financial risks. In other words, a distinction must be made between the role of the host state, which acts as a 'minimal state' regarding political risks, and that of a host state that acts as a partner state with regard to natural and financial risks.

1 ASSUMPTION OF POLITICAL RISKS

19.3 Most of the capital invested within the host state cannot be recovered over the short term. Unlike other international trade operations, which are once off, investments made over time presuppose amortisation of the sums invested over the long term.[1] Throughout the investment period, the operator faces many risks that may affect the project's profitability.

Although risks and uncertainties are recognised when contractual terms are negotiated, they are forgotten—or seemingly forgotten—when a project is successful. Ongoing social and political tension in certain regions could

[1] Duration of the operation is one of the criteria taken into consideration by arbitral tribunal when defining the notion of foreign investment. See to that effect the *Salini Costruttori SpA and Italstradee SpA v Kingdom of Morocco* (ARB/00/4), (Decision on jurisdiction of 23 July 2001) *JDI* 2002 at 196, obs. E. Gaillard; *Fedax v The Republic of Venezuela* (ARB /96/3), (Decision on jurisdiction of 11 July 1997) *JDI* 1999 at 276, obs. Emmanuel Gaillard. On this issue see, Manciaux op cit 68 et seq.

reinforce this instability, as the risk of termination or unilateral changes to exploitation agreements remains a reality.[2]

In order to reduce these risks and ensure the sustainability of the contractual relations, operators attempt to mitigate these political risks by inserting special provisions in exploitation agreements, the scope of which has been the subject of widespread debate.

a Contractual arrangements to mitigate political instability

19.4 The protection of foreign investment has always been a major concern of international trade operators[3] . Natural resources are not an exception to this rule; throughout the period of the exploitation agreement, operators will attempt to mitigate the political instability that may affect the investment, in particular by including stabilisation and intangibility clauses.

The existence of a political risk results from the very nature of exploitation agreements. As is the case with any contract with the State,[4] exploitation agreements involve a private investor and a party endowed with sovereignty attributes, allowing it to cause a prejudice to the contract or change the legal environment,[5] thereby modifying the expectations of the private operator.

19.5 In light of their exceptional nature, and with a view to ensuring the full effectiveness of these clauses, operators must have these terms ratified by decree or by parliamentary approval,[6] in particular when the insertion of these clauses is not provided for in the applicable mining or petroleum law.

However, a number of scenarios may arise depending on whether the exploitation agreement is signed directly with the State or with a local State operator. When the host state signs the contract itself, the situation does not present any difficulties, although reservations may be expressed regarding the authority of the signatory to bind the State.[7] However, when the contract is

---

[2] With regard to the renegotiation of exploitation contracts, see above paras 3.36 et seq.

[3] See, to that effect, Seymour J. Rubin 'Private Foreign Investment Legal and Economic Realities' (Johns Hopkins Press, 1956) 73.

[4] With regard to the State contract concept, see, in particular, Charles Leben 'L'évolution de la notion de contrat d'État' 2003 *Rev arb* 629; Prosper Weil 'Problèmes relatifs aux contrats passés entre un État et un particulier' (1969) 128 *RCADI* 95; Dominique Berlin 'Les Contrats d'États et la protection des investissements internationaux' 1987 *DPCI* 187; Charles Leben, 'La théorie du contrat d'état et l'évolution du droit international des investissements' (2003) 302 *RCADI* 200; Pierre Mayer 'La neutralisation du pouvoir normatif de l'État en matière de contrats d'État' 1986 *JDI* 5.

[5] See, to that effect, Andrea Giardina 'Clauses de stabilisation et clauses d'arbitrage: vers l'assouplissement de leur effet obligatoire?—Les états dans le contentieux économique international, I. Le contentieux arbitral' 2003 *Rev arb* 647.

[6] In particular, see Article 48(2) of Ghana's Minerals and Mining Act, 2006, which provides: '[a] stability agreement entered into under subsection (1) shall be subject to ratification by Parliament'. Similarly, Article 7(4) of ECOWAS Directive C/DIR 3/05/09 as 27 May 2009 on the Harmonisation of Guiding Principles and Policies in the Mining Sector of ECOWAS provides that, '[t]he Stability Agreement shall be subject to ratification by the National Parliament or any other body approved by a Member State.'

[7] In this regard, we mention the Award rendered in ICC Case No. 3493 (1983), *Spp (Middle East) and Southern Pacific Properties Ltd v The Arab Republic of Egypt and Egyptian*

signed by the local State operator and the exploitation agreement has not been ratified by the host state, can it truly be considered binding? Two hypotheses must be distinguished. The first consists in equating the specialised company to the State, of which it is only one part.[8] However, in terms of the exploitation of natural resources, it seems preferable to adopt the second hypothesis according to which the host state has delegated its powers to its specialised company; this delegation could be defined in the articles of association of the local State operator or provided for in the applicable legislation.

We will now proceed successively with the study of stabilisation clauses and that of intangibility clauses.

i *Stabilisation of local legislation*

19.6 The State's exercise of its sovereign prerogatives can lead to a change in legislation, in particular regarding tax, customs, social and environmental matters.[9] All of these changes, even if they do not directly affect the contractual framework, could affect the situation of the investor who is subject to a considerable legislative hazard. These changes in the legal framework could even jeopardise the initial contractual and financial balance of the project. The aim of stabilisation clauses is to channel this risk and enable the private operator to obtain the expected return from its investments.[10]

General Company for Tourism and Hotels, 22 ILM 752 (1983), overturned by the Court of Appeal of Paris on 12 July 1984 (Court of Appeal of Paris (1 Ch. suppl.), 12 July 1984, *Arab Republic of Egypt v Southern Pacific Properties Ltd, Southern Pacific Properties (Middle East)* 1986 *Rev arb* 75, note Philippe Leboulanger). See Nicolas David 'Les clauses de stabilité dans les contrats pétroliers. Question d'un praticien' 1986 *JDI* 98 et seq.

[8] In the past, this solution had been adopted by the arbitral tribunal in the case of *Sapphire International Petroleum Ltd v National Iranian Oil Company* (decision of 15 March 1963) 41 ILR 136 (1967). See to that effect David op cit 98 et seq.

[9] With regard to stabilisation clauses, see, in particular: Piero Bernardini, "Stabilization and adaptation in oil and gas investments' (2008) 1 (1) *Journal of World Energy Law & Business* 98; David op cit 79; Abdullah Al Faruque 'Typologies, Efficacy and Political Economy of Stabilisation Clauses: A Critical Appraisal' (2007) 5 (4) *OGEL* 4; Jean-Marc Loncle and Damien Philibert-Pollez 'Les clauses de stabilisation dans les contrats d'investissement' (2009) 3 *RDAI* 267; Catharine Titi 'Les clauses de stabilisation dans les contrats d'investissement: une entrave au pouvoir normatif de l'État d'accueil' 2014 *JDI* No. 2, doctr. 6; Deloitte 'La reconnaissance et l'efficacité des clauses de stabilisation dans les contrats miniers' (2015) 3 *Revue juridique et fiscale d'Afrique francophone* 2.

[10] Article 23 of Law No. 2006-17 of 17 October 2006 Enacting the Mining and Mining Taxation Code in Benin provides in particular that a mining agreement '*peut garantir au titulaire d'un titre minier la stabilité des conditions lui permettant de prendre le risque et de rentabiliser les investissements nécessaires*'.

[Free translation into English: 'can guarantee to the holder of a mining stability conditions which enable it to take the risk and return on investment required'].

Also see, to this effect, Article 12 of Law No. 99-013 of 22 December 1999 Enacting the Petroleum Code in Cameroon, which provides:

'*[l]e Contrat Pétrolier fixe*: [. . .] *r) les clauses de stabilisation des conditions économiques et fiscales relatives à la rentabilité des investissements*'.

19.7 This type of clause is of great interest as it limits the State's exercise of its legislative authority with regard to the private operator,[11] thereby reducing the political risk that results from the State's exercise of its sovereignty. The inclusion of stabilisation clauses, according to which a sovereign state agrees that the exercise of its legislative or regulatory authority cannot affect or modify the contractual terms to which the parties agreed, has become widespread and constitutes a commonly accepted practice.[12]

19.8 The importance of stabilisation clauses is entrenched in mining and petroleum codes, which at times indicate that the stability of the applicable legal framework is a determining factor when exploitation agreements are entered into.

In the mining sector, the stability guarantee exceeds the national framework; for instance, the WAEMU Community Mining Code[13] and ECOWAS Directive on the Harmonisation of Guiding Principles and Policies in the Mining Sector[14] entrench this stability guarantee in most West African nations.

[Free translation into English: '[a] Petroleum Contract shall set forth: [. . .] r) the clauses relating to the stabilisation of economic and tax provisions which impact the profitability of investments.']

[11] See David op cit 81 and 82.

[12] See David op cit 80.

[13] Article 17 of the WAEMU Regulation No. 18/2003/CM/UEMOA of 23 December 2003 Enacting the Community Mining Code guarantees investors the stability of the tax and customs regime during the period of validity of the mining rights:

'*[l]a stabilité du régime fiscal et douanier prévu dans la réglementation en vigueur au sein de l'Union est garantie aux titulaires de titres miniers pendant la période de validité de leurs titres. Pendant la période de validité de ces titres miniers, les règles d'assiette et de liquidation des impôts, droits et taxes prévus par la réglementation en vigueur demeurent telles qu'elles existent à la date de délivrance desdits titres miniers et aucune nouvelle taxe ou imposition de quelque nature que ce soit n'est applicable au titulaire ou bénéficiaire pendant cette période. Cependant, en cas de diminution des charges fiscales et douanières ou de leur remplacement par un régime fiscal et douanier plus favorable, les titulaires de titres miniers pourront opter pour ce régime plus favorable à condition qu'ils l'adoptent dans sa totalité.*'

[Free translation into English: '[t]he stability of the tax and customs regime provided in the regulation applicable within the Union is guaranteed to holders of mining rights during the period of validity of the said authorisation. During the period of validity of these mining rights, the rules regarding the tax base and settlement of taxes and duties provided by the regulation in effect remain as they existed on the date of issuance of the said mining rights and no new tax or duty of any nature whatsoever can apply to the holder or beneficiary during this period. However, if the tax and customs charges are reduced or replaced by a more favourable tax and customs regime, the holders of mining rights may opt for this more favourable regime as long as they adopt this regime in its entirety.']

[14] Article 7 of the ECOWAS Directive C/DIR 3/05/09 of 27 May 2009 on the Harmonisation of Guiding Principles and Policies in the Mining Sector provides for the ability to enter into a stability agreement in the framework of negotiations with a mining investor: '1.The competent authority of Member States as part of the granting of a mining right may enter into a Stability Agreement based on negotiations with an investor in mining. 2. A Stability Agreement shall reflect the national interest of the Member State and that of the Investor. [. . .].']

19.9 However, certain difficulties arise from the growing number of stabilisation clauses. The inclusion of these clauses, along with a rapid change in the laws governing the natural resource sector, can lead to the simultaneous application of a number of mining and petroleum codes. Except in situations where a private operator benefits from a stability guarantee and formally requests the application of the new law, the former law may survive. The former law would then apply concomitantly with the new legislation, which could cause problems in terms of the intelligibility of the legal standard.

19.10 The stability guarantee may result from the legal framework or a contractual agreement. Indeed, it may be granted directly by lawmakers in a mining or petroleum code.[15] The stabilisation guarantee may also result from an agreement; for example, national laws may refer to the terms of an exploitation agreement to guarantee the stability of the applicable regime in favour of the operator.[16] In other cases, the mining and petroleum codes do not refer to the exploitation agreement but to related agreements, such as development agreements[17] or at times to a specific agreement exclusively related to the granting of a stabilisation guarantee.[18]

In certain cases, the application of the stability guarantee is optional and

---

[15] See, in particular, in Benin, Article 74 of Law No. 2006-17 of 17 October 2006 Enacting the Mining and Mineral Taxation Code of the Republic of Benin; in the Central African Republic, Article 134 of the Law No. 09-005 of 29 April 2009 Enacting the Mining Code; in the Democratic Republic of Congo, Article 276 of the Law No. 007-2002 of 11 July 2002 Enacting the Mining Code; in Côte d'Ivoire, Article 164 of Law No. 2014-138 of 24 March 2014 Enacting the Mining Code; in Guinea, Article 182 of Law L/2013/N°053/CNT of 8 April 2013 Amending Law No. 2011/006/CNT of 9 September 2011 Enacting the Mining Code; in Niger, Article 97 of Ordinance No. 93-16 of 2 March 1993 Enacting the Amended Mining Law; in Chad, Article 62 of Law No. 011-PR-1995 of 20 June 1995 Enacting the Mining Code.

[16] See, in Algeria, Article 84, para. 4 of Law No. 01-10 of 4 July 2001 Enacting the Mining Law; in Benin, Article 23, para. 3 of Law No. 2006-17 of 17 October 2006 Enacting the Mining and Mineral Taxation Code of the Republic of Benin; in Cameroon, Article 12 of Law No. 99-013 of 22 December 1999 Enacting the Petroleum Code; in the Central African Republic, Article 52 of Law No. 09-005 of 29 April 2009 Enacting the Mining Code and Articles 33 and 83 of Ordinance No. 93.007 of 25 May 1993 Enacting the Petroleum Code; in Côte d'Ivoire, Article 12 of Law No. 2014-138 of 24 March 2014 Enacting the Mining Code and Articles 18 and 84 of Law No. 117-AN-05 of 16 October 2005 Enacting the Petroleum Code; in Djibouti, Article 66, para. 2 of Law No. 117-AN-05-5as5 of 16 October 2005 Enacting the Petroleum Code; in Eritrea, Article 9 of Proclamation No. 108/2000, A Revised Proclamation to Govern Petroleum Operations; in Ethiopia, Article 9 of Proclamation No. 295/1986 Regulating Petroleum Operations; in Guinea, Article 18 of Law No. 2011/006/CNT of 9 September 2011 Enacting the Mining Code; in Mauritania, Article 51 of Law No. 2010-033 of 20 July 2010 Enacting the Hydrocarbons Code amended by Law No. 2011-044 of 25 November 2011; in Senegal, Article 67 of Law No. 98-05 of 8 January 1998 Enacting the Petroleum Code; in Chad, Articles 10 and 95 of Law No. 006-PR-2007 of 2 May 2007 On Hydrocarbons.

[17] See, to this effect, in Tanzania, Article 10(4) of the Mining Act, 2010.

[18] Certain laws also allow for a 'stability agreement' to be entered into. This is the case in Ghana, more precisely provided by Article 48 of the Minerals and Mining Act, 2006.

the private operator must inform the competent authorities whether or not he wishes to be subject to a stabilised regime.[19]

19.11 Although stabilisation clauses are quite common in the extractive sector, one can question the existence of a stability guarantee in the absence of related contractual or legal provisions. For example, do bilateral investment treaties, and more particularly the concept of fair and equitable treatment, imply the existence of an obligation to maintain a stable, foreseeable legal and economic framework that meets the investor's legitimate expectations?[20] It would appear that certain arbitral tribunals adopted this position by accepting the existence of a breach of fair and equitable treatment when the investor's legitimate expectations included a guarantee of stability of the legislative regime.[21] This position, however, is debated in doctrine and international arbitration. The decision rendered in the *Parkerings v Lithuania* case reiterates the right of the host state to proceed to change in the legislative and regulatory framework in the absence of stabilisation clauses. Therefore, the arbitral tribunal held that, in the absence of a stabilisation clause, a state has the right to enact, modify or cancel a law at its own discretion.[22]

The scope of this issue remains limited due to the broad use of stabilisation clauses in the field of natural resource exploitation.

However, the content of stabilisation clauses varies. These differences can be seen in relation to the legal mechanism used, but also the scope of application of the stabilisation.

(1) The legal stabilisation mechanism

19.12 Stabilisation may be organised according to three distinct methods: stabilisation per se, an adaptation of the contract or the non-aggravation of the content of the law applicable to the contract.

(a) Stabilisation per se

19.13 Stabilisation per se of the content of the applicable law requires the use of two different techniques, the first being referred to as incorporation and the second as unenforceability.

---

[19] See, to this effect, in Madagascar, Article 154 of Law No. 99-022 of 19 August 1999 Enacting the Mining Code amended by Law No. 2005-021 of 17 October 2005.

[20] For elements of a response, see Eduardo Silva Romero & Audrey Caminades 'Actualité des clauses de stabilisation' in *Le contentieux extractif* (ICC 2015) 57.

[21] On this point, see Silva Romero & Caminades op cit 57 and the decisions cited in this regard, ie *CMS Gas Transmission Company v The Republic of Argentina* (ARB/01/8) (Decision of 12 May 2005); or *Enron Corporation and Ponderosa Assets, LP v The Republic of Argentina* (ARB/01/3) (Decision of 22 May 2007).

[22] *Parkerings-Companiet AS v Lithuania* (ARB/05/8) (Decision of 11 September 2007) No. 332: '[a] State has the right to enact, modify or cancel a law at its own discretion. Save for the existence of an agreement, in the form of a stabilisation clause or otherwise, there is nothing objectionable about the amendment brought to the regulatory framework existing at the time an investor made its investment.'

19.14 The incorporation technique corresponds to a transformation of the legal terms into contractual terms by including them in the exploitation agreement itself. In this case, a list of the relevant legislation is included in the exploitation agreement. The stabilised legal and regulatory terms are therefore a full part of the contract. Another method consists in giving the contract the force of law so that future legislation changes cannot apply to the contract.[23]

19.15 The unenforceability technique presupposes that the new legal and regulatory terms cannot be enforced against an operator who benefits from a stabilised regime. The unenforceability technique implies that, at the time the exploitation agreement is signed, it is agreed that the terms of the contract and local legislation are compatible; however, any future change to local legislation could not be enforced against it.[24]

In certain situations, the unenforceability of the new law is not absolute. If the amended legislation is more favourable, the private operator can opt for this more favourable regime.[25] This right of option may include an obligation to adopt the more favourable regime in its entirety,[26] which, in practice, guarantees greater legibility. Other legislation simply provides for the application of the more favourable legal and regulatory terms, without making the application of the more favourable regime conditional on the private operator exercising an option.[27]

(b) Adaptation of the contract

19.16 This is a very specific situation that can be associated with change of circumstances clauses. Pursuant to these clauses, the State undertakes to launch negotiations with the investor in order to restore the contract's economic balance if there is a change in the legal or regulatory framework that could have a negative impact on the private operator's investment.[28] In this situation, the content of the applicable law is amended and applies immediately to the exploitation agreement. However, operators agree to amend their contract to cancel or at least considerably reduce the impact of the legislative changes. For example, Article 111 of the Guinean Petroleum Code of 2014 provides that:

---

[23] Bernardini op cit 100.

[24] See, in the Central African Republic, Article 134 of Law No. 09-005 of 29 April 2009 Enacting the Mining Code.

[25] See Article 74 of Law No. 2006-17 of 17 October 2006 Enacting the Mining and Mineral Taxation Code in the Republic of Benin; in the Central African Republic, Article 134 of Law No. 09-005 of 29 April 2009 Enacting the Mining Code; in Chad, Article 62 of Law No. 011-PR-1995 of 20 June 1995 Enacting the Mining Code.

[26] See, Article 74 of Law No. 2006-17 of 17 October 2006 Enacting the Mining and Mining Taxation Code of the Republic of Benin; in the Central African Republic, Article 134 of Law No. 09-005 of 29 April 2009 Enacting the Mining Code; Article 17 of the WAEMU Regulation No. 18/2003/CM/UEMOA of 23 December 2003 Enacting the Community Mining Code.

[27] See in particular, Article 166 of Law No. 4-2005 of 11 April 2005 Enacting the Petroleum Code in Cameroon.

[28] On this point, see Titi 'Les clauses de stabilisation dans les contrats d'investissement: une entrave au pouvoir normatif de l'État d'accueil' loc cit.

'[t]he Petroleum Contract may include a clause on the stabilisation of the legislative and regulatory framework on the effective date, allowing the Contracting Parties to require either the non-application of the financially prejudicial terms or an adjustment of the contractual terms to restore the initial economic balance if legislative or regulatory provisions enacted subsequent to the effective date of the Petroleum Contract disrupt the contract's economic balance.'[29]

This is often used as a remedy for a refusal to insert a stabilisation clause.

(c) Non-aggravation

19.17 Concerning clauses referred to as 'non-aggravation' clauses, the host state undertakes to render legislative or regulatory amendments that aggravate an operator's situation inapplicable.[30] However, the aggravation criterion is very rarely defined, which can lead in practice to different interpretations by the parties. In general, the aggravation criterion relates to an aggravation of the financial situation.

The aggravating nature of a legislative or regulatory provision is assessed as a whole with regard to the entire exploitation agreement.

(2) Scope of stabilisation

19.18 The issue of the scope of stabilisation requires a determination of the persons who benefit from this regime but also the areas covered by the stabilisation. It should be noted that the scope of stabilisation clauses is most often limited from both a material and temporal perspective.[31]

(a) Persons concerned by stabilisation

19.19 The stability of the legal and regulatory framework is intended to apply mainly to private operators, ie the holder of mining rights, whether these rights result from an exploitation agreement or mining permit.

19.20 However, certain mining and petroleum codes limit the stability guarantee and only grant it to specific categories of operators. Thus, certain African states have decided to limit the benefits of stabilisation to investments that exceed a certain amount.[32]

---

[29] This is Law No. L/2014/N°034/AN Enacting the Petroleum Code.

[30] See, to this effect, in Guinea, Article 111 of Law No. L/2014/N°034/AN of 23 December 2014 Enacting the Petroleum Code; in Mauritania, Article 51 of Law No. 2010-033 as of 20 July 2010 Enacting the Hydrocarbons Code amended by Law No. 2011-044 of 25 November 2011; Article 30 of the Production Sharing Agreement between the Democratic Republic of Congo and the Association 'Dominion Petroleum Congo & Soco Exploration-Production RDC & La Congolaise des Hydrocarbures' in Bloc V in the Graben Albertine (November 2007).

[31] See, to that effect, Al Faruque op cit 4.

[32] For example, in Madagascar, see Article 157 of Law No. 99-22 of 19 August 1999 Enacting the Mining Code amended by Law No. 2005-021 of 17 October 2005, which provides:

'*[p]euvent bénéficier de la garantie de stabilité les investissements dans la recherché et dans l'exploitation minière, d'un montant minimum de cinq cent millions d'ariary (Ar 500.000.000), que les promoteurs s'engagent à realise suivant le plan présenté au moment de l'option pour la garantie.*'

19.21 On the contrary, other texts widen the scope of the stabilisation by accepting that private operators and persons involved in the exploitation operations, in particular their subcontractors, may benefit from stability of the legal and regulatory framework. Depending on the situation, this may involve contractors working on behalf of or in partnership with the operator[33] or affiliated companies.

(b) Areas covered by stabilisation

19.22 Whether the stability guarantee results from the law or a contract, the stabilisation guarantee will generally apply to limited fields, in particular taxation, foreign exchange rules and the customs regime.

The texts that provide for the stability of economic conditions—taxation, customs and foreign exchange regulation—are by far the most common.[34]

Other clauses provide more broadly for the stability of legal and administrative conditions.[35] More rarely, the guarantee may also include the stability of social[36] or environmental rules.[37]

[Free translation into English: 'investments in mining exploration and exploitation in a minimum amount of five hundred million Ariary (500,000,000 Ar) that developers undertake to realise according to the plan presented at the time they opt for the guarantee may benefit from the stability guarantee.']

See also, Article 74 of Law No. 2006-17 of 17 October 2006 Enacting the Mining and Mining Taxation Code of the Republic of Benin, which guarantees fiscal stability to holders of mining exploitation permits accredited under regime of the Investment Code.

[33] See, to that effect, in Chad, Article 62 of Law No. 011-PR-1995 of 20 June 1995 Enacting the Mining Code, which provides:

'*[l]a stabilisation des impôts et taxes assimilées prévus au Code Miner est garantie aux titulaires de titre minier ou bénéficiaires d'autorisation en vertu du Code Minier, ainsi qu'aux entreprises travaillant pour leur compte ou en association avec eux, pendant toute la période de validité du titre minier ou de l'autorisation.*'

[Free translation into English: '[t]he stabilisation of taxes and similar taxes provided for in the Mining Code is guaranteed to the holders of mining rights or beneficiaries of authorisations under the Mining Code, and to companies working on their account or in partnership with them throughout the period of validity of the mining rights or authorisation.']

[34] See, in Cameroon, Article 114 of Law No. 99-013 of 22 December 1999 Enacting the Petroleum Code; in the Central African Republic, Article 52 of Law No. 09-005 of 29 April 2009 Enacting the Mining Code; in the Democratic Republic of Congo, Article 166 of Law No. 4-2005 of 11 April 2005 Enacting the Mining Code; in Côte d'Ivoire, Article 164 of Law No. 2014-138 of 24 March 2014 Enacting the Mining Code; in Guinea, Article 182 of Law No. L/2013/N°053/CNT of 8 April 2013 Amending Law No. 2011/006/CNT of 9 September 2011 Enacting the Mining Code; in Madagascar, Article 154 of Law No. 99-022 of 19 August 1999 Enacting the Mining Code amended by Law No. 2005-021 of 17 October 2005; in Chad, Article 62 of Law No. 011-PR-1995 of 20 June 1995 Enacting the Mining Code and Articles 51 and 95 of Law No. 006-PR-2007 of 2 May 2007 On Hydrocarbons.

[35] See, in particular, in Senegal, Article 27 of Law No. 2016-32 of 8 November 2003 Enacting the Mining Code.

[36] With regard to provisions that guarantee the stability of social conditions, see, in particular, in Algeria, Article 84, para. 4 of Law No. 01-10 of 3 July 2001 Enacting the Mining Law; in Benin, Article 23, para 3 of Law No. 2006-17 of 17 October 2006 Enacting the Mining and Mineral Taxation Code of the Republic of Benin.

[37] See, in Algeria, Article 84, para. 4 of Law No. 01-10 of 3 July 2001 Enacting the Mining Code.

(c) Areas not subject to stabilisation: material limits

19.23 In some cases, provisions related to the stability of the legal framework remove certain fields from the scope of application of stabilisation. These fields mainly relate to the safety of individuals, environmental protection, control over exploitation operations and labour law.[38]

19.24 In line with the movement toward subregional regulations, certain legal texts exclude from the scope of stabilisation new regulations that could be enacted within regional organisations.[39] The inclusion of this type of provision is justified by the obligation incumbent on member states of regional organisations, such as WAEMU and ECOWAS, to comply with community regulations.

(d) Stabilisation period: time limits

19.25 The stabilisation period generally runs from the date the mining right is granted or the date the exploitation agreement is signed[40] and the guarantee applies during the entire term of the exploitation agreement[41] or the validity period of the mining title.[42] Certain provisions provide for an end to the stabilisation period if the mining right is withdrawn or cancelled.[43]

---

[38] See in particular Article 111 of Law No. L/2014/N°034/AN of 23 December 2014 Enacting the Petroleum Code in Guinea, which provides:

'*[s]ont toutefois exclus de la clause de stabilisation ci-dessus les coûts additionnels occasionnés par une modification de la réglementation en matière de sécurité des personnes, de protection de l'environnement, de contrôle des operations pétrolières ou de droit du travail, à moins que ces modifications ne soient pas conformes aux pratiques internationales ou qu'elles soient appliqués à un Contractant de manière discriminatoire*.'

[Free translation into English '[t]he following is excluded from the stabilisation clause above: the additional costs of a change in the regulation on the safety of individuals, environmental protection, petroleum operations and labour law controls, unless the said changes do not comply with international practices or are applied to a contracting party in a discriminatory manner.']

[39] See in the Central African Republic, Article 134 of Law No. 09-005 of 29 April 2009 Enacting the Mining Code.

[40] See, to this effect, in Guinea, Article 182 of Law No. L/2013/N°053/CNT of 8 April 2013 Amending Law No. 2011/006/CNT of 9 September 2011 Enacting the Mining Code; in Ghana, Article 48 of the Minerals and Mining Act, 2006.

[41] See, in Algeria, Article 12 of the Presidential Decree of 1 March 2003 Enacting the Standard Mining Contract; in the Central African Republic, Article 52 of Law No. 09-005 of 29 April 2009 Enacting the Mining Code; in the Democratic Republic of Congo, Article 30 of the Production Sharing Agreement between the Democratic Republic of Congo and the Association 'Dominion Petroleum Congo & Soco Exploration-Production RDC & La Congolaise des Hydrocarbures" in Bloc V in the Graben Albertine (November 2007).

[42] See, to this effect, in Algeria, Article 84, para. 4 of Law No. 01-10 of 4 July 2001 Enacting the Mining Law; in the Central African Republic, Article 134 of Law No. 09-005 of 29 April 2009 Enacting the Mining Code; in the Democratic Republic of Congo, Article 166 of Law No. 4-2006 of 11 April 2005 Enacting the Mining Code.

[43] See, to that effect, in Mauritania, Article 58 of Law No. 2012/012 Regulating Mining Agreements and Approving the Standard Mining Agreement.

However, variants are found tending toward a time limitation of the stabilisation guarantee.[44]

19.26 Finally, certain clauses provide that the stabilisation period depends on the sums invested in the project and increases in line with the scope of these investments.[45] The greater the investment, the longer the amortisation of the sums invested and the profitability of the project will be delayed, which consequently requires a longer stability guarantee period.

(3) Consequences of a breach of a stabilisation clause

19.27 The breach of a stabilisation clause may have a specific impact if a treaty to promote and protect investments, that includes an umbrella clause, was entered into between the host state and the investor's state of origin. This clause provides that each state will comply with its obligations vis-à-vis investors from the other state.[46] The existence of an umbrella clause raises contractual obligations to a treaty level;[47] therefore, when an umbrella clause exists, the breach of a stabilisation clause can be considered a breach of the treaty itself.

---

[44] In the Democratic Republic of Congo, Article 276 of Law No. 007-2002 of 11 July 2002 Enacting the Mining Code creates a 10-year stability period in favour of the investor. The planned reform of the Mining Code provides for a reduction in the stability guarantee period to five years.

[45] See, to that effect, in Madagascar, Article 159 of Law No. 99-022 of 19 August 1999 Enacting the Mining Code, amended by Law No. 2005-021 of 17 October 2005.

[46] For examples of umbrella clauses included in bilateral investment treaties entered into with African States, see, in particular, Article 10(2) of the Agreement between the Government of Lebanon and the Government of the Republic of Senegal for the Reciprocal Promotion and Protection of Investments (promulgated on 15 May 2004), which provides:

> '*[c]hacune des Parties contractantes s'engage à assurer à tout moment le respect des engagements particuliers qu'elle aura contractés à l'égard d'un investisseur de l'autre Partie contractante.*'
>
> [Free translation into English: '[e]ach Contracting Party undertakes to ensure compliance at all times with the special undertakings it has made in relation to an investor of the other Contracting Party.'

Also see Article 8(2) of the Agreement between the Republic of Benin and the Belgo-Luxembourg Economic Union concerning the reciprocal promotion and protection of investments (effective of 30 August 2007)]).

[47] To that effect, the *Noble Ventures Inc v Romania* Decision established that: '[a]n umbrella clause is usually seen as transforming municipal law obligations into obligations directly cognizable in international law' (*Noble Ventures Inc v Romania* (ARB/01/11) Decision of 12 October 2005 No. 53).

This solution is the subject of debates in international case law. Certain arbitral tribunals hold that the umbrella clause does not transform the contractual obligation but transforms the breach of this obligation into a breach of the BIT (also see, to that effect, the Decision of the *ad hoc* committee in the *CMS Gas Transmission Company v Argentine Republic decision* (ARB/01/8): '[t]he effect of the umbrella clause is not to transform the obligation which is relied on into something else, the content of the obligation is unaffected' (*CMS Gas Transmission Company v Argentine Republic* (ARB/01/8), Decision of 25 September 2007, No. 95). On this distinction, see Sophie Lemaire, 'La mystérieuse "umbrella clause" (Interrogations sur l'impact de la clause de respect des engagements sur l'arbitrage en matière d'investissements)' 2009 *Rev arb* 479.

19.28 A breach by the State of its stability undertaking could also be considered an expropriation[48] and could therefore give rise to compensation. However, it is preferable for this situation to be limited to cases in which a breach of the stabilisation clause substantially affects the profitability of the investment.

19.29 Another more widely accepted solution is to consider that the breach of the stabilisation clause—when it prejudices compliance with the investor's legitimate expectations—constitutes a breach of the fair and equitable treatment guaranteed.[49] Certain arbitral tribunals have held that a change in the legal framework resulting from a specific undertaking from the host state constitutes a breach of fair and equitable treatment.[50]

19.30 With regard to the financial consequences of a breach of the stabilisation clause,[51] part of the doctrine holds that *restitutio in integrum* is an adequate remedy.[52] For others, the breach of a stabilisation clause must admittedly result in compensation of the prejudice suffered by the investor, but also compensation for lost profit, *lucrum cessans.*[53] International case law is not decisive on this point. While certain arbitration decisions have applied the *restitutio in integrum*[54] principle, others have denied the application of this principle.[55] Finally, another doctrinal trend argues that the compensation for the prejudice suffered due to the breach of a stabilisation clause is dependent on the legality or illegality of the measure. If the measure that breaches the stabilisation clause is not discriminatory but taken in the interest of the public, the financial compensation due for the breach of the

[48] This solution was adopted by the arbitral tribunal in the case of *Methanex Corporation v the United States*: '[a]s a matter of general international law, a non-discriminatory regulation for a public purpose, which is enacted in accordance with due process and, which affects, inter alios, a foreign investor or investment is not deemed expropriatory and compensable unless specific commitments had been given by the regulating government to the then putative foreign investor contemplating investment that the government would refrain from such regulation.' (*Methanex Corporation v United States*, Final decision on jurisdiction and merit of 3 August 2005, No 7). On this point, see Titi 'Les clauses de stabilisation dans les contrats d'investissement: une entrave au pouvoir normatif de l'État d'accueil' loc cit.

[49] On this matter, see Titi 'Les clauses de stabilisation dans les contrats d'investissement: une entrave au pouvoir normatif de l'État d'accueil' loc cit.

[50] In the *El Paso v Argentina* case, the Arbitral Tribunal held that: '[a] reasonable general regulation can be considered a violation of the FET standard if it violates a specific commitment towards the investor. The Tribunal considers that a special commitment by the State towards an investor provides the latter with a certain protection against changes in the legislation.' (*El Paso Energy International v The Argentine Republic* (ARB/03/15), Decision of 31 October 2011, No. 375).

[51] See Abdullah Al Faruque 'Validity and Efficacy of Stabilisation Clauses' (2006) 23 (4) *Journal of International Arbitration* 329 et seq.

[52] Id.

[53] Eduardo Jiménez de Arechaga 'International Law in the Past Third of a Century' (1978) 159 *RCADI* at 1.

[54] See, for example, the *Texaco v Libya* 53 ILR 454.

[55] See, for example, the *American Independent Oil Co. (AMINOIL) v Government of Kuwait* (Decision of 24 March 1982) 1982 *JDI* 869.

stabilisation clause is calculated based on fair market value. Inversely, if the measure taken by the State in breach of the stabilisation clause is illegal, compensation must cover the investor's current prejudice as well as lost profits.[56]

ii *Intangibility of the contract*

19.31 The host state is vested with exorbitant powers and in particular the authority to unilaterally amend the terms of an exploitation agreement or even exercise its right of expropriation or nationalisation.

Exploitation agreements may include an intangibility clause, which will protect the private operator from these prerogatives.

(1) The powers of the host state to unilaterally amend an exploitation agreement

19.32 International public law is stable today in that it recognises the right of the host state to unilaterally change an exploitation agreement.

(a) Contribution of the doctrine

19.33 In the past, three doctrinal trends opposed each other regarding the determination of the extent to which the State had the power to unilaterally modify a contract: the theories of the absolute intangibility, essential mutability and conditional mutability of the contract.

19.34 According to the theory on the absolute intangibility of the contract, international law prohibits any change to the contract to protect the equality of the parties, and therefore refuses to recognise the supremacy of a local State partner by restrictively interpreting its undertakings. The basis for this theory consists either in the affirmation that the contract is the only law that applies between the parties or by considering the contract to equate a treaty.[57] This theory has been widely criticised by some scholars in particular due to its inapplicability and lack of pragmatism.[58] Regarding its potential application, it seems likely that the application of this type of theory would discourage states not from entering into agreements but from potentially acting outside of the law. However, this theory has proven to be inapplicable in practice, because it is obvious that inequality exists between the parties to an exploitation agreement.

19.35 The theory of the essential mutability of the contract holds that the State cannot desist from unilaterally changing or terminating a contract, as

---

[56] See Al Faruque op cit 330.

[57] 'When a State, in the exercise of its sovereignty, has bound itself by a concession contract, the terms of the contract are binding on both the State and the Concessionaire, and the State is under an obligation to perform its contractual obligations and to protect the rights of the concessionaire in relation to the contract. Contractual obligations freely assumed by a State are as binding on the State as its treaty obligations.' Siksek op cit 80 et seq.

[58] Weil op cit 213.

this authority is an attribute of its sovereignty.[59] This theory on the essential mutability of the contract is in direct opposition to the preceding theory. It represents the sovereignty theory in the extreme (an affirmation of the inescapable inequality of the parties, which some authors describe as an '*absurdity with major consequences*').[60]

19.36 Finally, the theory that a contract is mutable under certain conditions is a position that finds itself between absolute intangibility and essential mutability. Thus, the principle of binding force and due regard for acquired rights of private investors is recognised; however, the host state may exercise its sovereign right to terminate or amend the contract.[61] That said, the measure thereby taken would only be lawful under certain conditions. This is the theory adopted by international law.

(b) The contribution of international law

19.37 Although arbitral tribunals have entrenched the '*pacta sunt servanda*'[62] principle and the principle of due regard for acquired rights,[63] the right of states to nationalise assets within their territory, held by their nationals or foreigners, is commonly admitted under international law.[64] In

---

[59] In particular, see Robert Delson 'Nationalization of the Suez Canal Company: Issues of Public and Private International Law' 1957 *Columbia Law Review* 9.

[60] Weil op cit 214.

[61] See Zouhair A Kronfol *Protection of Foreign Investment—A Study in International Law* (Leiden 1972) 82.

[62] For example, in the *Sapphire International Petroleum Limited v National Iranian Oil Company* (Decision of 15 March 1963) 1964 *Int'l & Comp LQ* 1011:

'*c'est un principe fondamental du droit, constamment proclamé par les juridictions internationales, que les engagements contractuels doivent être respectés.*'

[Free translation into English: '[i]t is a fundamental principle of law, which has constantly being proclaimed by international courts, that contractual undertakings must be respected.']

[63] *Saudi Arabia v Arabian American Oil Co. (ARAMCO* (Decision of 23 August 1958) 1963 *Rev crit DIP* 272:

'[b]y reason of its very sovereignty within its territorial domain, the State possesses the power to grant rights which it forbids itself to withdraw before the end of the Concession. Nothing can prevent a State, in the exercise of its sovereignty, from binding itself irrevocably by the provision of a concession and from granting to the concessionaire irrevocable rights. Such rights have the character of acquired rights.'

[64] See in that regard Kleiman 84 et seq. Nationalisation presupposes a transfer of private property authoritatively decided for public interest reasons in favour of the State (see Daillier, Forteau & Pellet op cit 1220 and 1221). With regard to the difference between expropriation and nationalisation, see, in particular, Dominique Carreau and Patrick Julliard, *Droit international économique* 4ed (Dalloz 2010) 489:

'*[e]xpropriation comme nationalisation sont des actes de puissance publique entraînant transfert de propriété du secteur privé au secteur public mais des différences significatives peuvent être relevées entre l'une et l'autre. L'expropriation résulte d'une mesure administrative, alors que la nationalisation résulte d'une mesure législative.* [. . .] *L'expropriation vise des biens particuliers et individualisés; la nationalisation vise une collection de biens, moins individualisés qu'individualisables.*'

[Free translation into English: '[e]xpropriation, like nationalisation, is an act of the public authority that leads to a transfer of ownership from the private sector to the public sector,

the field of natural resources, the right to nationalise is an adjunct to the principle of the permanent sovereignty of states over their natural resources, solemnly affirmed by the United Nations.[65]

19.38 Whether expropriation is direct or indirect,[66] under international law, the validity of the expropriation is subject to compliance with a number of requirements.[67] These legality conditions imposed by international law are reiterated in World Bank Guidelines on the Treatment of Foreign Direct Investment, which provide that a state can only expropriate from an investor 'where this is done in accordance with applicable legal procedures, in pursuance in good faith of a public purpose, without discrimination on the basis of nationality and against the payment of appropriate compensation.'[68]

19.39 First, expropriation must be based on the grounds of public utility.[69] The absence of a public interest is rarely upheld by international case law, as only the State can, in principle, be the judge of the public interest of

---

but there are major differences. Expropriation results from an administrative measure while nationalisation results from a legislative measure. [. . .] Expropriation relates to specific, individual assets; nationalisation covers a group of assets, which are less individualised and individualisable. The objective of expropriation is to satisfy a general interest, but its scope is local; the objective of nationalisation is to satisfy a general interest with a national scope.'].

[65] In the area of natural resources, the principle was upheld by the United Nations' General Assembly Resolution 1803 (XVII) of 14 December 1962. The right to nationalise was also established by the United Nations' General Assembly in its Charter of Economic Rights and Duties of States (Resolution 3281 (XXIX)), which provides that each State is entitled to 'nationalise, expropriate or transfer ownership of foreign property, in which case appropriate compensation should be paid by the State adopting such measures, taking into account its relevant laws and regulations and all circumstances that the State considers pertinent.' With regard to the principle of the sovereignty of States over their natural resources, see above paras 4.1 et seq.

[66] While 'direct' expropriation leads to a direct transfer of the property of the investor to the host State, expropriation can also result from one or more measures that have the same effect (referred to as 'indirect' or 'creeping' expropriation). For a definition of indirect expropriation, see, in particular: *Parkerings-Companiet AS v Lithuania* (ARB/05/8), Decision of 11 September 2007, No. 437. A distinction should be made between direct expropriation and creeping expropriation, which includes a series of measures, each of which, taken individually, could not be considered a prejudice to the investment. For a definition of indirect expropriation, see, in particular: *El Paso Energy International v The Argentine Republic* (ARB/03/15), Decision of 31 October 2011, No. 518. For an analysis of the indirect and creeping expropriation concepts, see Yves Nouvel 'Les mesures équivalant à une expropriation dans la pratique récente des tribunaux arbitraux' 2002 *RGDIP* 79; Yves Nouvel 'L'indemnisation d'une expropriation indirecte' (2003) 5 (3) *International Law Forum* 198. Hamid G Gharavi & Marie-Lame Bizeau 'L'expropriation indirecte dans les conflits entre contractants extractifs' in *Le contentieux extractif* (ICC 2015) 167.

[67] On this point, see Daillier, Forte & Pellet op cit 1222 et seq.

[68] Article IV. 1 of the World Bank Guidelines on the Treatment of Foreign Direct Investment of 21 September 1992. On this issue, see Sébastien Manciaux op cit No. 625.

[69] Resolution 1803 (XVII) of the General Assembly states: '[n]ationalization, expropriation or requisitioning shall be based on grounds or reasons of public utility, security or the national interest which are recognized as overriding purely individual or private interests, both domestic and foreign.' On this point, see Manciaux op cit 486 et seq.

the measure it applies.[70] This consideration is therefore limited to an error regarding the legal qualification of the facts and misuse of power.

19.40 The nationalisation or expropriation measure must also comply with current procedures.[71] The domestic laws of the host state must therefore be examined to determine whether the procedure enforced by the State complies with the laws and regulations.

19.41 Furthermore, the measure must not result from any discrimination on the part of the host state,[72] in particular based on nationality. Discrimination consists in a differentiated treatment that aims at refusing rights and advantages to individuals, groups and states that are recognised for others.[73] In investment law, the principle of non-discrimination bars states from granting, recognising or treating certain investors differently or more favourably than others or from granting them advantages that are not granted to others.

19.42 To be lawful, the State's measure must include adequate compensation. In the field of natural resources, this principle was affirmed in the United Nations General Assembly Resolution 1803 (XVII), which provides that, in the event of nationalisation, '[t]he owner shall be paid appropriate compensation, in accordance with the rules in force in the State taking such measures in the exercise of its sovereignty and in accordance with international law'.[74] Therefore, the compensation granted to the investor must be adequate, effective and prompt.[75]

Arbitral tribunals endeavour to guarantee that investors dispossessed of their property receive full compensation.[76]

Nevertheless, it is generally accepted that a damage that results from a legitimate measure taken in response to the misconduct of an investor does not result in any compensation being awarded. A number of arbitral tribunals have held that a State measure that punishes an investor's failure to comply with his legal or contractual obligations cannot be qualified as expropria-

[70] This position has been confirmedby many arbitration decisions (the Decision rendered in ICC Case No. 1526 (1968), *Mojzez Lubelski v Burundi* 1974 *JDI* 915, note Y Derains; *AGIP SpA v People's Republic of Congo* (ARB/77/1) (Decision of 30 November 1970) 1982 *Rev crit* DIP 92; *Antoine Goetz and others v Burundi* (ARB/95/3) (Decision of 10 February 1999) (2000) 15 *ICSID Rev* 457.

[71] See in that regard, see Manciaux op cit 493 et seq.

[72] See in that regard Manciaux *Investissements étrangers et arbitrage entre États et ressortissants d'autres États* at 496 et seq; Jean Charpentier 'De la non-discrimination dans les investissements' 1963 *AFDI* 35.

[73] For a definition of the discrimination concept in investment law, see *Antoine Goetz and others v Republic of Burundi* (ARB/95/3) (Decision of 10 February 1999) No. 121:

'*Une discrimination suppose un traitement différencié appliqué à des personnes se trouvant dans des situations semblables*'. [Free translation: '[d]iscrimination presupposes that different treatment is applied to individuals in similar situations'.]

[74] United Nations' General Assembly Resolution 1803 (XVII), No. 4.

[75] Wording used by the US Secretary of State Hull in 1938.

[76] Daillier, Forteau & Pellet op cit 1124. For an example of arbitration decisions see, in particular, *Benvenuti and Bonfant v People's Republic of Congo* (ARB/77/2) (Decision of 8 August 1980) 21 ILM 1982 at 740.

tion.[77] The tribunal conditions the expropriation qualification on the merit of the government's allegations against the investor.

19.43 With regard to the amount of compensation, a distinction must be made as to the legality or illegality of the expropriation. Regarding illegal expropriations, the Permanent Court of International Justice in the case concerning the Factory at Chorzow established the principle. In its decision, the Court stated that reparation must, as far as possible, wipe out all the consequences of the illegal act and re-establish the situation that would, in all probability, have existed if that act had not been committed. Thus, in the case of an illegal expropriation, the investor will be entitled to compensation and restitution.[78]

A contrario, in the case of a legal expropriation (ie that complies with international law), the injured investor is entitled to compensation that corresponds to the *damnum emergens* or losses suffered on the date of expropriation.[79] The World Bank Guidelines of 1992 provide for compensation being calculated based on the fair market value of the asset expropriated, as determined just prior to the expropriation or the publication of the expropriation decision.[80]

19.44 The right of the State to expropriate, which is widely accepted in international law, is in fact reiterated in certain mining and petroleum codes.[81]

19.45 Although the host state is entitled to change the exploitation

---

[77] *Alex Genin and others v The Republic of Estonia* (ARB/99/2) (Decision of 25 June 2001) 200217 *ICSID Rev* 395.

[78] *Chorzow Factory case (Germany v Poland)* (Decision of 13 September 1928) Series A No. 17 at 47.

[79] See in that regard, Suzy H Nikiema 'Indemnisation de l'Expropriation' Institut international du développement durable, 2013, Best Practices Series.

[80] Daillier, Forteau & Pellet op cit 1124.

[81] See, in particular, in Mauritania, Article 102 of Law No. 2010-033 of 20 July 2010 Enacting the Hydrocarbons Code, amended by Law No. 2011-044 of 25 November 2011:

> *'[l]'État garantit aux contractants qu'aucune mesure de nationalisation, de réquisition ou d'expropriation ne sera prise à leur encontre, à moins qu'elle ne soit édictée par la loi, qu'elle ne soit prise pour des motifs d'utilité publique et sur une base non-discriminatoire, et qu'elle ne donne lieu à une juste et préalable indemnité.'*
>
> [Free translation: '[t]he State guarantees to the contractors that no act of nationalisation, requisition or expropriation will be taken against them, unless such actions are dictated by law, or taken on account of general interest on a non-discriminatory basis, and have given rise to fair compensation paid in advance.'

In Chad, Article 94 of Law No. 006-PR-2007 On Hydrocarbons:

> *'[l]es biens appartenant au Titulaire du Permis bénéficient de la protection de la loi. Ils ne peuvent être réquisitionnés ou expropriés que pour les motifs d'intérêt général moyennant une indemnisation juste et équitable. L'indemnité est réglée en devises convertibles.'*
>
> [Free translation into English: '[t]he assets of the Permit Holder benefit from the protection of the law. They can only be requisitioned or expropriated for reasons of general interest and in return for fair and equitable compensation. The compensation is paid in convertible currencies.']

agreement under certain conditions, the question is whether it can validly waive these prerogatives. This waiver takes the form of intangibility clauses.

(2) *Intangibility clauses*

19.46 Although the State's right to amend (unilaterally) an exploitation agreement is subject to certain limits, operators may protect themselves from exorbitant State measures of the host state by way of intangibility clauses. Intangibility clauses are intended to prevent any change to a contract by the host state.[82] They often combine a twofold undertaking by the public authority: an undertaking not to unilaterally modify the exploitation agreement and an undertaking to waive its right to nationalise.

19.47 Today, clauses that include the State's waiver of its nationalisation right are rarely included in exploitation agreements.

b Assessment of the effectiveness of clauses to mitigate political instability

19.48 In the past, the effectiveness of clauses that mitigate political instability was the subject of debate in terms of both doctrine and international case law.[83]

---

[82] Mustafa Erkan *International Energy Investment Law Stability through Contractual Clauses* (Kluwer Law International 2011) 105.

[83] The number of arbitration decisions rendered in the oil sector alone is significant. With regard to major decisions, see, in particular:

- *Texaco Overseas Petroleum Co and California Asiatic Oil Co v The Government of the Libyan Arab Republic* (Decision of 19 January 1977) 1977 *JDI* 350. See, in particular, Gérard Cohen-Jonathan 'L'arbitrage Texaco-Calasiatic c. Gouvernement Libyen' 1977 *AFDI* 452; Jean-Flavien Lalive 'Un grand arbitrage pétrolier entre un gouvernement et deux sociétés privées étrangères' 1977 *JDI* 319;
- *British Petroleum v The Government of the Libyan Arab Republic* (Decision of 10 October 1973) 1980 *Rev arb* 117;
- *Libyan American Oil Company (LIAMCO) v The Government of the Libyan Arab Republic* (Decision of 19 January 1977) 1977 *JDI* 350. See, in particular, to that effect, Patrick Rambaud 'Un arbitrage pétrolier: la sentence LIAMCO' 1980 *AFDI* 274; Brigitte Stern 'Trois arbitrages un même problème, trois solutions: les nationales pétrolières libyennes devant l'arbitrage international' 1980 *Rev arb* 3;
- *Sapphire International Petroleum Limited v National Iranian Oil Company* (Decision of 15 March 1963) 1964 *Int'l & Comp LQ* 1011.
- *Saudi Arabia v Arabian American Oil Co (ARAMCO)* (Decision of 23 August 1958) 1963 *Rev crit DIP* 272. See, in particular, Suzanne Bastid, 'Le Droit international public dans la sentence arbitrale de l'ARAMCO' 1961 *AFDI* 300; Henri Batiffol 'La sentence ARAMCO et le Droit international privé' 1964 *Rev Crit DIP* 647;
- *American Independent Oil Co (AMINOIL) v Government of Kuwait* (Decision of 24 March 1982) 1982 *JDI* 869. In particular, see P Kahn 'Contrats d'État et nationalisations: les apports de la sentence du 24 mars 1982' 1982 *JDI* 844; Geneviève Burdeau 'Droit international et contrats d'États. La sentence Aminoil/Koweït du 24 mars 1982' 1982 *AFDI* 454;
- *Petroleum Development Limited v Sheikh of Abu Dhabi* (Decision of 28 August 1951) (1951) 18 *ILR* 144;
- *AGIP SpA v People's Republic of Congo* (ARB/77/1) (Decision of 30 November 1970) 1982 *Rev crit DIP* 92;

The issue of the effectiveness of stabilisation and intangibility clauses has been debated, whether the operator's contract is governed by a national legal system or an international legal system.

i *The contract is governed by a national legal system*

19.49 As long as governments were not very demanding, political considerations related to foreign investment were not a major issue. Until the end of World War II, the contract fell exclusively under the domestic system and international law was not involved in contractual issues. Positive law recognised that relations between parties in no case fell under an international system; it was up to national law of the host state, to govern contractual relations, and determine appeal procedures and the bodies that would settle any disputes. This position was based on the idea that the foreign partner was in general a person under private law and could not be subject to international law. The famous decision of the Permanent Court of International Justice in 1929 concerning the Serbian loans, illustrates this concept: '[a]ny contract which is not a contract between States in their capacity as subjects of international law is based on the municipal law of some country.'[84]

19.50 Despite a vast 'de-nationalisation' and 'internationalisation' movement, we note that natural resource exploitation contracts tend to fall under an internal legal system; this is what appears from all of the *professio juris* clauses and those related to operator undertakings.[85]

Thus, the issue of the effectiveness of protection clauses in the case of an exploitation agreement governed by a national legal system may be debated; we will study this issue in two successive sections, indicating in the first the general framework for the application of a national legal system and, in the second, the effectiveness of this application.

(1) General framework for the application of a national legal system

19.51 The application of a national legal system to a contract raises two issues: first, should an exploitation agreement governed by a national legal

- *Elf Aquitaine Iran v National Iranian Oil Co* (Preliminary Decision of 14 January 1982) 1984 *Rev arb* 401. See, in particular, to that effect, P Fouchard 'L'arbitrage Elf Aquitaine Iran c. National Iranian Oil Company, une nouvelle contribution au droit international de l'arbitrage' 1984 *Rev arb* 333.

The validity of stabilisation clauses before international arbitrators does not seem to raise any difficulties. In particular, see Y Derains in the Decision rendered in ICC case No. 3093/3100 (1979) *JDI* 1981 951.

[84] *Case concerning the payment of various Serbian loans* (Decision of 12 July 1929) Series A No. 20 at 41. The *dictum* of the Court of Justice in 1929 was used again by the same Court on 22 July 1952 with regard to the dispute between the United Kingdom and Iran on the refusal to internationalise the exploitation agreement entered into by the government of Iran and the *AIOC (Anglo-Iranian Oil Co case (United Kingdom v Iran)* Decision of 22 July 1952 Recueil 93).

[85] On the application of *professio juris* clauses in mining and petroleum projects, see below para 20.31 et seq.

system be considered specific under private international law and, second, is there any presumption that the law of the host state applies?

(a) Should a distinction be made between contracts entered into by states based on whether or not they are acting as subject to international law?

19.52 To answer this question, we must no doubt differentiate undertakings that relate to sovereignty prerogatives and those that do not.

In the hydrocarbon sector, Henry Cattan established this distinction by separating elements of the exploitation agreement that fall under private law from those that fall under public law,[86] and found that these contracts are *sui generis* in nature. He also wrote:

> 'The elements of public law which undoubtedly exist in an oil concession do not, therefore, efface its basically contractual and commercial character. Accordingly, a concession can be said to combine elements of private and public law which are knitted together to create a relationship intended to exploit a State resource for the mutual benefit of the parties. It is a State contract which is *sui generis*.'[87]

19.53 The concept of an action exercised by a subject of international law is opposed by a certain doctrine according to which a state always acts as a subject of international law; thus '[a]lthough contractual relations and undertakings require the existence of a balance between the parties, States always remain States even when they engage in trade-like operations. They are always subject to public law, which grants them certain privileges [. . .].'[88]

To follow this reasoning would lead us inexorably to an aberration that, by legalising a *de facto* inequality, it would justify a legal inequality within the context of an exploitation agreement.

(b) Does a presumption exist as to the application of the host state's law?

19.54 In the Serbian loan case, the Permanent Court of International Justice affirmed that, in substance, as the loans had been contracted by the Serbian State, they could not be subject to any law other than the law of the borrowing state:

> '[a] sovereign State [. . .] cannot be presumed to have made the substance of its debt and the validity of the obligations accepted by it in respect thereof, subject to any law other than its own.'[89]

We are fully aware of the risk that this type of interpretation could create in the framework of exploitation agreements. In reality, we must ask the question as to whether these contracts present a specificity, not due to the presence of at least one public person as a party but due to the subject of the contract relating to the material foundation of the sovereignty of a state.

---

[86] Henry Cattan *The Law of Oil Concessions in the Middle East and North Africa* (Oceana Publications 1967) 20 et seq.

[87] Cattan op cit 30.

[88] Khalil Khalilian *Investissements privés étrangers et souveraineté économique* (Doctoral thesis Paris Panthéon-Assas 1978).

[89] Case concerning the payment of various Serbian loans (Decision of 12 July 1929) Series A No. 20.

(2) The ineffectiveness of the application of the national legal system

19.55 This ineffectiveness creates a twofold risk: the risk of non-effectiveness of the provisions freely chosen by the operators due to their autonomy of will and the lack of realism regarding an enforcement of the host state's liability.

(a) Ineffectiveness of the law chosen by the parties

19.56 The ineffectiveness of the law chosen by the parties results both from the non-application of the law defined as applicable to the contract by the operators as well as the quasi-ineffectiveness of the protection clauses in *inter partes* relations.

19.57 The non-application of the contract's law may benefit the law of the host state. Often, the host state will attempt—directly, if it is a party to the exploitation agreement or indirectly if it has only been called upon to guarantee the investment—to impose the application of its law on all contracts enforced on its territory regarding the exploitation of its natural resources.

19.58 Operators must consider the host state's laws, in particular when it can be considered public order legislation. The general trend that appears is a decline in the application of the contract's law. Protection clauses could remedy this situation; in a stabilisation clause, one specifies the application of the host state's law at the time the exploitation agreement is signed, thereby setting aside the terms of future laws.

19.59 However, protection clauses will not have much chance of success if the exploitation agreement is governed by a national legal system.

The effectiveness of protection clauses therefore depends on the applicable national law:

> '[j]ust as national law may authorise the application of the new law and allow the State to modify or terminate the contractual relationship, it is incumbent on it to determine whether the stabilisation clauses freezes or crystallises this relationship or if, on the contrary, it is deemed not to exist'.[90]

19.60 To have the application of protection clauses depend on the applicable national law runs the risk of a certain ineffectiveness of the protection clauses in *inter partes* relations when—if the exploitation agreement is governed by a national legal system—their effectiveness depends on the scope they are given under the applicable national law.

Certain authors have sought a remedy by attempting to invoke the international responsibility of the host state that is directly or indirectly a party to the exploitation agreement.

(b) The State's international responsibility: a remedy?

19.61 Note that we are still discussing an exploitation agreement governed by a national legal system. We will adopt the assumption that there

[90] See Prosper Weil 'Les clauses de stabilisation ou d'intangibilité insérées dans les accords de développement économique' in Charles Rousseau *Mélanges offerts* (Pédone 1974) 317.

is no international 'umbrella' treaty between the state in which the operations are located and the national State of the private operator pursuant to which the obligation to enforce the contract is an international obligation incumbent on the former in favour of the latter.[91]

In this scenario, can we consider that the non-performance of the contract intrinsically constitutes an illegal act from an international perspective in light of the international obligation incumbent on states to comply with this obligation? In other words, although the contract is governed by a national legal system, can the international responsibility of the host state be invoked? The doctrine is divided: some scholars only accept the international responsibility of the State in cases where an inter-state undertaking has been transgressed (this is the minimalist theory), which is not necessary for others.

19.62 According to the minimalist theory, as the contract falls under a domestic legal system, its non-performance cannot breach an international rule of law; the existence of an international violation is required to engage the international responsibility of the State in the field of contracts. It is therefore up to the internal legal system to determine whether a measure taken by a beneficiary of the investment regarding the private operator is legal. The State, as mentioned by Albert de La Pradelle and Nicolas Politis, is 'held responsible not because of the non-performance of the contract, but for bad faith, proven unfairness or flagrant injustice it has committed.'[92] This idea calls for a comment, as it allows a state that is party to the contract to be granted excessive powers: it simply needs to modify its domestic legislation to justify its intervention. This is a considerable objection as -it is worth

[91] See to that effect, Prosper Weil:

> '*[i]l n'y a en effet, pas de difficultés particulières* [en ce qui concerne la mise en jeu de la responsabilité contractuelle de l'État] *lorsqu'il existe entre l'État contractant et l'État national du cocontractant un traité de 'couverture' qui fait de l'obligation d'exécuter le contrat une obligation internationale à la charge de l'État contractant envers l'État national du cocontractant. L'intervention du traité de couverture transforme les obligations contractuelles en obligations international es et assure ainsi, comme on l'a dit, 'l'intangibilité du contrat sous peine de violer le traité'; toute inexécution du contrat, serait-elle même régulière au regard du droit interne de l'État contractant, engage dès lors la responsabilité internationale de ce dernier envers l'État national du cocontractant.*'
>
> [Free translation into English: '[t]here is, in fact, no particular difficulty [regarding the contractual responsibility of the State being invoked] when there is an 'umbrella treaty' between the contracting State and the State of the other contracting party, which turns the obligation to perform the contract into an international obligation of the contracting State vis-à-vis the State of the other contracting party. The intervention of the umbrella treaty transforms contractual obligations into international obligations thereby ensuring, as it has already been stated, the 'intangibility of a contract under the threat of violating the treaty'; any non-performance of the contract, even if it is legal under the national law of the contracting State, gives rise to the international responsibility of the latter vis-à-vis the national State of the other contracting party.']

Prosper Weil 'Problèmes relatifs aux contrats passés entre un État et un particulier' (1969) 128 *RCADI* 95.

[92] See Albert de la Pradelle & Nicolas Politis *Recueil des arbitrages internationaux* (Éditions internationale 1957) 548.

repeating- this situation involves a contract that is not 'de-nationalised' and where its validity is governed by a domestic legal system.

It is worth mentioning a few more 'liberal' theories, which are less demanding in terms of the acceptance of the international responsibility of states, ie those that accept State responsibility in the absence of any transgression of an inter-state undertaking; this constitutes the maximalist theory represented by Professor Prosper Weil.

19.63 According to the maximalist theory,[93] any breach of a contract between a State and a foreigner is contrary to international law and therefore directly engages the State's international responsibility. The '*pacta sunt servanda*' principle, in light of its universality, must apply to international contracts. The maximalist view has been subject to numerous criticisms, in particular in terms of its principles; 'International public law cannot govern contractual relations between parties when only one of the parties is subject to international law.'[94]

For his part, Prosper Weil examined the situation where the international responsibility of the host state is invoked by the state of the private operator. He proposed the idea that, if the applicable national law does not grant any value to protection clauses, the host state's disregard for its undertakings could not constitute an internationally illegal act; in fact, he argues this to be an injury to good faith and misconduct based on its co-contracting party being misled.[95]

There are many precedents according to which the non-performance of a contract by a host state is per se internationally illegal and engages its international responsibility, with no need for any offence to be identified. This is Commissioner Findlay's opinion as stated in the *Beals, Nobles and Garrison*[96] and the *Venezuelan loan*[97] cases.

19.64 However, it seems impossible to identify a single principle; we can only be certain that protection clauses lack effectiveness if an exploitation agreement is governed by a national legal system.

ii *Application of a new legal system*

19.65 Faced with the risk of the ineffectiveness of the protection clauses when the contract is governed by a national legal system, in particular the system of the host state, an attempt must be made to release the contractual

[93] For a list of its principal advocates, see, in particular, Hartley W Shawcross 'Le problème des investissements à l'étranger en droit international' (1961) 102 *RCADI* 369; Maurice Bourquin 'Arbitration and Economic Development Agreement' 1960 *Bus Lawyer* 860.

[94] Huseyin Pazarci 'La responsabilité internationale des États à l'occasion des contrats conclus entre États et personnes privées étrangères' 1975 *RGDIP* 375.

[95] '*La violation constitue une faute engageant sa responsabilité internationale.*' [Free translation into English: '[t]he violation constitutes misconduct that engages its international responsibility.'] Prosper Weil '*Les clauses de stabilisation ou d'intangibilité insérées dans les accords de développement économique' at 318.*

[96] John Bassett Moore *History and Digest of International Arbitration* (Government Printing Office 1898) 3555.

[97] Moore op cit 3616.

relations from the national law of the State to transfer them to a new system that would allow for the protection clauses to be effective.

19.66 The arbitration decision rendered in the *Sapphire*[98] case summarises the considerations that may be invoked to justify a 'de-localisation' of a contract:

> '[t]he foreign company was bringing financial and technical assistance to Iran, which involved it in investments, responsibilities, and considerable risks. It therefore seems natural that they should be protected against any legislative changes which might alter the character of the contract, and that they should be ensured of some legal security. This could not be guaranteed to them by the outright application of Iranian law, which is within the power of the Iranian State to change [. . .] It is in the interest of the two parties to these agreements that disputes that may divide them be exempted from the specificities of national laws [. . .]'.[99]

19.67 Furthermore, very often, a private operator encounters significant procedural difficulties before the courts of the host state. A certain number of elements must be extracted from the contract, illustrating the contract's shift outside of the State's realm. These elements, treated objectively, constitute a multitude of factors which, when taken into account, will allow the national legal system to be set aside and will justify recourse to a new legal system.[100]

(1) In which legal system should the contract be established?

19.68 Delocalisation allows a contract to forego the national legal system, but a new legal system must apply, as noted in the decision rendered in the ARAMCO case:

> '[i]t is obvious that no contract can exist in vacuo, i.e., without being based on a legal system. The conclusion of a contract is not left to the unfettered discretion of the Parties. Human will can only create a contractual relationship if the applicable system of law has first recognised its power to do so.'[101]

19.69 The complexity of this issue is illustrated by the many related theories. In this regard, two options are conceivable: either the contract falls under a specific legal system qualified as a third-party legal system or it falls under the international legal system.

(a) Application of an autonomous legal system

19.70 This theory is divided into two trends. According to some authors, the contract contains its own legal system; according to others, it belongs to a transnational legal system.

---

[98] *Sapphire International Petroleum Limited v National Iranian Oil Company* (Decision of 15 March 1963) 1964 *Int'l & Comp LQ* 1011.

[99] *Sapphire International Petroleum Limited v National Iranian Oil Company* (Decision mentioned by Weil op cit 150).

[100] These factors are mainly composed of the *professio juris* and arbitration clauses. See the Texaco Decision, which held that the presence of an arbitration clause in an exploitation agreement constituted a process of the contract internationalisation (*Texaco Overseas Petroleum Co and California Asiatic Oil Co v Government of the Libyan Arab Republic* (Decision of 19 January 1977) 1977 *JDI* 350). Regarding contractual provisions common in State contracts, also see Leben op cit 248.

[101] *Saudi Arabia v Arabian American Oil Co (ARAMCO)* (Decision as of 23 August 1958) 1963 *Rev crit DIP* 272.

19.71 According to some scholars, an autonomous legal system created by the joint will of the parties exists. This system, which is different from national law and international law, is composed of all of the terms of the contract, which form an indivisible whole and are self-sufficient. The contract creates its own *lex contractus*, which will exclusively govern the rights and obligations of the parties. These are referred to as 'quasi-international contracts' to underline their *sui generis*[102] nature.

This theory is subject to a number of criticisms.[103] First, it restores the idea of a law-less contract, which is expressly rejected, in particular by the ARAMCO decision.[104] Furthermore, the *pacta sunt servanda* principle and the autonomy of will principle do not operate in a vacuum; a legal system is required to give these principles legal force. In addition, the *lex contractus* theory demonstrates its own insufficiency to the extent that it provides for the need to refer to rules of law from an external legal system in the event of a deficiency, which also leaves significant latitude to arbitrators for interpreting the contract. With regard to the application of another system, more specifically public international law, this referral implies that this system, as a subsidiary foundation of the contract, is conditional on the will of the parties. However, this is still contested and raises the problem of the scope of the parties' choice.

19.72 These criticisms led to the formulation of another theory, that of a transnational legal system. This is an international legal system that is neither a State system nor an international system, the content of which derives from *lex mercatoria*, general legal principles, international trade practices and arbitral decisions.

This theory, as with the preceding theory, also draws criticism. Uncertainty remains in relation to the determination of its substance due to three overlapping factors. The absence of distinct sources and a criterion to determine which rules would be taken from each of the other legal systems to constitute the synthesis that the transnational system operates, the absence of specific sanctions and, finally, the significant latitude granted to courts and arbitral tribunals to define the content thereof. Part of the doctrine summarises the criticisms expressed against the transnational legal system as

---

[102] See, in particular, Jean Honorat 'Rôle effectif et rôle concevable des quasi-contrats en droit actuel' 1969 *RTD Civ* 653; Malak El-Shishini 'Les contrats quasi-internationaux—Base juridique et droit applicable' 1968 *Rev Egypt* DI 81.

[103] With regard to criticisms expressed by the doctrine, see, in particular, Francis A Mann 'The Proper Law of Contracts Concluded by International Persons' (1959) 35 *BYIL* 49 et seq; Rouhollah K Ramazani 'Choice of Law Problems and International Oil Contracts: A Case Study' (1962) 11 *Int'l & Comp LQ* 1962 503; Prosper Weil 'Contrats entre États et particuliers' 1969 *RCADI* 117 et seq; Wilhelm Wengler 'Immunité législative des contrats multinationaux' 1971 *Rev Crit DIP* 637.

[104] *Saudi Arabia v Arabian American Oil Co (ARAMCO)* (Decision of 23 August 1958) 1963 *Rev crit DIP* 272.

being a system 'that is mysterious by nature and poorly defined in terms of its scope of application.'[105]

More recently, part of the doctrine has been favourable to the appearance of a transnational legal system specific to the petroleum sector and the emergence of a *lex petrolea,*[106] and potentially a *lex mineria*.

(b) Application of an international legal system

19.73 The argument in favour of the application of an international legal system is based on two understandings. The first applies classic international law to the contract and the second applies international contract law.

19.74 Regarding the exploitation of natural resources, the theory relative to the application of classic international law attempts to answer the following question: can international law, in strict terms, be applied to an exploitation agreement? One must note that this theory presupposes that a private operator is equated to a party subject to international law and the contract is considered equivalent to a treaty. However, these considerations were rejected both by the Permanent Court of International Justice in the Serbian loans case and by the arbitration decision in the ARAMCO case. It is based on these same considerations that the application of the international law *clausula rebus sic stantibus* principle must be rejected, which means that a contract between parties subject to international law remains valid as long as the circumstances that existed at the time it was signed have not changed. A transposition of this principle to exploitation agreements would also mean equating the contracts to treaties, which would be completely erroneous.

19.75 The second option subjects the exploitation agreement to international contract law.[107] This theory subjects exploitation agreements to a series of international rules of law, which would constitute a specific branch of the public international law referred to as international contract law. This proposition avoids the pitfalls of the preceding theory to a certain extent. By turning the investment contract into an international legal deed of a new type, the comparison to a treaty is avoided. By creating a specific branch of international law based on international rules of law related to contracts, these rules can be differentiated from the rules that govern treaties and cannot be transposed to contracts. In this regard, Professor Weil argues that the principles borrowed from general international law (*pacta sunt ser-*

[105] Georges R Delaume 'Des stipulations de droit applicable dans les accords de prêt et de développement économique et de leur rôle' 1968 *RBDI* 363.

[106] Alfredo de Jesus 'The prodigious story of the lex petroleum and the rhinoceros philosophical aspects of the transnational legal order of the petroleum society' (2012) 1 *TPLI Series on Transnational Petroleum Law* No. 1.

[107] This theory is defended in particular by Professors Prosper Weil (see his work cited in our bibliographic appendices) and Clive Schmitthoff 'Nature and Evolution of the Transnational Law of Commercial Transactions' in *The transnational law of international commercial transactions* Studies in Transnational Economic Law vol 2 (Kluwer 1982)19. Against the existence of an international contract law, see, in particular, Muthucumaraswamy Sornarajah 'The Myth of International Contract Law' (1981) 15 *JWTL* 187.

vanda, respect for acquired rights, *rebus sic stantibus*) will have a specific connotation in contractual matters. Professor Weil distinguishes international law as a legal system on which the contract is based from the *lex contractus* intended to govern the substantive aspects of the contract. International law as the foundation for the contract would therefore allow for the contract to be turned into an international legal deed. Therefore, the state's disregard for a contractual provision that international law considers as mandatory would suffice for the state to be held internationally responsible.

19.76 Having examined the various doctrinal studies that have attempted over recent decades to subject the contract to a legal system, it appears to us that the most compelling and applicable to practices in the extractive sector is the theory proposed by Professor Weil. The autonomous legal system theory is in fact unrealistic; the transnational legal system theory is vague in terms of the substance of this law, and the theory regarding the application of classic international law to the contract creates confusion and targets the applicable substantive law and not the legal system.

19.77 We previously reached the conclusion that the protection clauses were ineffective in a situation where the contract was subject to a national legal system. We now find ourselves in a situation wherein the contract is subject to an international system, and we must repeat our study on the effectiveness of the protection clauses while changing one variable—the applicable legal order considered herein being international.

(2) Assessment of the effectiveness of protection clauses in the case of a contract subject to international law

19.78 As a starting assumption, we will consider a contract entrenched in the international legal system whether or not it refers exclusively to the national law of the host state. This does not override internationalisation, as it results from objective factors in the contract or the contractual context. In this context, what is the value of the protection clauses? We will distinguish stabilisation clauses from intangibility clauses.

(a) Stabilisation clauses

19.79 In reality, there is no antagonism between the legislative competence of the host state and the stabilisation clauses. The full powers of the State remain, as, by definition, the contract eludes the national legal system. The effect of the reference to national law simply allows the parties to integrate into their contract rules borrowed from national legislation.

19.80 Thus, if no stabilisation clause exists, there is a presumption that the rules chosen will be those that provide for an evolution, and legislative modifications will be taken into account. However, the presence of a stabilisation clause inverses this presumption.

Therefore, the stabilisation clause does not prejudice the legislative competence of the host state.

(b) Intangibility clauses

19.81 Having solved the issue of whether international law allows the State to exercise its exorbitant prerogatives to unilaterally modify a contract, we must now examine whether, under international law, the State can legally waive its recognised prerogatives under the same international law, as this would be the impact of the inclusion of an intangibility clause in the body of the contract itself.

19.82 The doctrine is not unanimous but divided into two major trends. For certain authors,[108] the intangibility clause has no value and is considered non-existent because the State cannot waive prerogatives intended to protect the general interest.

For other authors,[109] the inclusion of protection clauses does not constitute a waiver of sovereignty but on the contrary, the exercise thereof.

19.83 In addition, when operators have included an intangibility clause in the exploitation agreement, it is because they believed it was valid and effective. A private operator could reasonably make undertakings on this assumption. If, thereafter, the State thwarts this assumption, it breaches the good faith principle. The host state must therefore effectively waive the exercise of part of its sovereignty prerogatives if it agrees to an intangibility clause; this clause must be fully effective throughout the entire exploitation agreement period.

19.84 A number of historic arbitration decisions uphold the effectiveness of intangibility clauses; the most important relate to natural resources—the *Revere* decision for the solid minerals sector, and the *Texaco* decision rendered by Professor Dupuy on 19 January 1977 in the hydrocarbon sector.[110] In this case, the problem regarding the intangibility of international contracts was settled for the first time by way of international arbitration. The exploitation agreement included a stabilisation clause and an intangibility clause; the Libyan government decided, however, to take nationalisation measures. The private operators asked the arbitral tribunal to answer the question of whether a completely sovereign government could be entitled to limit the exercise of its sovereignty by signing an international contract. The arbitral tribunal ruled that the exercise of the State's right to nationalise is considered an expression of its territorial sovereignty, but a nationalisation cannot prevail against an internationalised contract entered into by a state and a foreign private company which includes stabilisation provisions, as the sovereign nationalising state cannot disregard it.[111] The decision was rendered with no reservation in favour of the validity and effectiveness of the intangibility clauses.

19.85 In current arbitral case law, a private operator benefits from a guarantee of security when protection clauses are included in the exploitation agreement as long as the contract is subject to the international legal system.

---

[108] See, in particular, Gérard Fouilloux *La nationalisation et le Droit international public* (LGDJ 1962) 303; Arghyrios A Fatouros *Government Guarantees to Foreign Investors* (Columbia University Press 1962) 342 et seq.

[109] See, in particular, various publications of Professor Weil in our bibliographical appendices.

[110] *Texaco Overseas Petroleum Co and California Asiatic Oil Co v Libyan Arab Republic* (Decision of 19 January 1977) 1977 *JDI* 350.

[111] Id.

19.86 However, is this application acceptable relative to exploitation agreements? Furthermore, can a private operator's obligations truly be subject to international law? Throughout our study, we noted all of the antinomy created by contracts relative to natural resource exploitation and the related inability to reach a consensus. The private operator has every interest in having the international system apply to the exploitation agreement, although the host state is often hostile to this application as it loses absolute control that it increasingly can and wants to exercise on the petroleum or mining project. Should certain contractual clauses prevail over others, then, in order to find elements of a response and justify the application of one system rather than another? These questions reopen the debate and demonstrate that the difficulties have not been resolved, which fully confirms our initial hesitation on the effectiveness of clauses to mitigate political instability.

19.87 In reality, the element that is troubling from a legal perspective is factual: it is undeniable that a state has full authority over its territory and that private operators are trapped due to the immoveable assets they own on this territory and the investments they have made. The host state can unilaterally modify or terminate an exploitation agreement in breach of all of the legal principles we have mentioned. It knows that no political, economic, commercial, and legal constraint can be bind it to any significant degree. This means that, in practice, a state can do whatever it wants with the contractual framework related to the exploitation of natural resources on its territory.[112] Among lawyers, it is this feeling of a lack of legal power and sufficiency that prevails.

As stated by Jean-Flavien Lalive:

> 'Among the obstacles to overcome, one must observe—a lesson in humility for lawyers—that in international economic relations, the law occupies a space that, although not negligible, is limited. Many decisions regarding international contracts are taken in ignorance, disdain and even disregard for the law for reasons quite foreign to the law. This is regrettable. In fact, considerations of expediency are short-lived, while the law takes time into account and, as such, serves the interests of everyone.'[113]

19.88 Thus, the search for a solution to provide true stability to undertakings made in exploitation agreements requires not internationalisation thereof but depoliticisation. This has resulted in an extreme entanglement. It demonstrates the recognition of an idiosyncrasy—if not a sort of malaise—in State contracts, at the head of which are those related to the exploitation of natural resources. The solutions the doctrine proposes and the responses provided by positive law reinforce the idea that they are highly specific and provide us with a new illustration thereof.

---

[112] Werner Goldschmidt supports the idea according to which asymmetrical relations are in fact created among the parties resulting from one of the said parties being a public authority. Werner Goldschmidt 'Transactions between States and Public Firms and Foreign Private Firms' (1972) 136 *RCADI* 203.

[113] Lalive op cit 231.

The problems encountered and their solutions differ with regard to natural and financial risks.

2 ASSUMPTION OF A SIGNIFICANT SHARE OF THE NATURAL AND FINANCIAL RISKS

19.89 Unlike political risks, natural and financial risks correspond to significant immediate and concrete undertakings in terms of investment and liability.

The assumption that the producing states are unable to manage the exploitation of their natural resources on their own is no longer accurate. Indeed, whether alone or with the help of service providers, producing states—often via dedicated public companies—have the technical and financial capabilities required to exploit their natural resources.

However, contracting with a private operator is still the most common practice due to the unchanged scope of natural and financial risks related to exploitation.

As such, private operators generally undertake the financial and geological risks.

a Risks related to the content of contractual obligations

19.90 The exploitation of mining and petroleum resources requires considerable investments. These investments are especially important during the exploration phase where the financial and geological risk undertaken by the private operator will in particular result from the obligation to carry out a minimum work programme and to spend minimum expenditures in the project.

i *Minimum investment to be committed: minimum work programmes*

19.91 The minimum amount of investment required to exploit natural resources is generally provided in a work programme.

Clauses related to minimum work programmes present in most exploitation contracts provide for a minimum financial commitment on the part of the operator. These minimum work programmes define the expenditure amounts that must be incurred by the operator, while indicating the details on the work to be performed as well as the related schedule and budget.

19.92 The provisions related to minimum work programmes and related expenditure are usually defined during the exploration stage, as the objective is to ensure that the private operator will take every measure to establish the existence of natural resources in quantities that are commercially viable.

Mining and petroleum laws attach considerable importance to the private operator's compliance with the minimum work programme and related financial commitments. In many mining and petroleum laws, a minimum work programme is considered a prerequisite for mining rights as the application for an exploration permit will be supported by a work programme and related budget. Similarly, the renewal of mining rights is often conditional upon compliance with the minimum undertakings the private operator has undertaken.[114] The provisions related to the withdrawal of mining rights[115] or the termination of the exploitation agreement also illustrate the intent to ensure that the private operator complies with his undertakings.

19.93 Mining and petroleum laws grant private operators some flexibility by allowing changes to the minimum work programme, generally after having received the opinion or agreement of the public operator or licensing authority. Similarly, some flexibility is granted to private operators who, in certain cases, may benefit from a carry-over of sums not spent to the following period[116] or, *a contrario*, a carry-over of sums invested in excess of

---

[114] See, in particular, in Tunisia, Article 32 of Law No. 2003-30 of 28 April 2003 Enacting the Mining Code:

> '*[l]e Titulaire d'un Permis de Recherche a droit au renouvellement de son Permis pour deux périodes successives d'une durée maximale de trois ans chacune et sous réserve qu'il ait: a—rempli les obligations auxquelles il est tenu, notamment celles relatives aux minima de dépenses et de travaux à réaliser dans le périmètre couvert par le Permis, au cours de la période de validité arrivée à échéance.*'
>
> [Free translation into English: '[t]he Holder of the Title to an Exploration Permit has the right to renew his permit for two (2) successive periods, each with a maximum period of three (3) years, provided that it has: a—fulfilled all obligations, notably those relative to minimum required expenditures and works, within the area of the Permit and prior to the end of the validity period.']

Also see, in The Comoros, Article 9 Order No. 54-1110 of 13 November 1954 Enacting the Mineral Substances Regime.

[115] See, in particular, in Mozambique, Article 64 of Mining Law No. 20-2014 of 18 August 2014; in Rwanda, Article 25 of Law No. 13-2014 of 20 May 2014 On Mining and Quarry Exploitation.

[116] See, in particular, in Chad, Article 23 of Law No. 011-PR-1995 of 20 June 1995 Enacting the Mining Code:

> '*[l]e titulaire d'un permis de recherches a l'obligation, pendant la période initiale du permis et pendant chaque période de renouvellement, d'exécuter le programme de travaux de recherches soumis et de dépenser en recherches le montant minimum établi dans la réglementation. Toutefois, si le titulaire du permis de recherches ne dépense pas pour une année quelconque une partie du montant qu'il est tenu de dépenser selon la réglementation, il peut maintenir son permis de recherches en vigueur en dépensant la somme non dépensés l'année suivante, à la condition toutefois d'avoir dépensé au moins cinquante pour cent (50%) de la somme qu'il était tenu de dépenser pour l'année déficitaire. Le report du déficit ne peut se faire que pour une année; la somme ainsi reportée s'ajoute à celle devant être dépensée pendant l'année du report. Par contre, si le titulaire dupermis de recherches dépense pendant une année une somme supérieure à celle qu'il est tenu de dépenser en recherches selon la réglementation, l'excédent des sommes dépensées peut être reporté sur l'année suivante en réduction de l'obligation de dépenses de cette anné-là*'.
>
> [Free translation into English: '[t]he holder of an exploration licence has an obligation, during the initial period and during each licence renewal period, to run the exploration works presented and spend the minimum amount established in the regulations. However, if within one year the licencee does not spend any part of the amount he is required to spend under the regulations, he can keep his licence effective by spending the amount not spent the following year, provided, however, he has spent at least fifty per cent (50%) of the amount he was required to spend during the year in deficit. The carry-over of the deficit can only apply to one year; the amount so deferred must be

the work programme in the form of an advance on expenditure for the following period.[117]

19.94 If an operator relinquishes its mining rights during the period of validity, the holder is usually not released from his obligations under the minimum work programme and may be obliged to pay the state compensation for the financial commitments that were not performed.[118]

19.95 Generally, the area in which the investments must be made is left to the discretion of the operators. Nevertheless, in order to allocate the investments equitably, certain contracts may provide for a breakdown of the contractual areas into different sections, with each one being assigned a portion of the investments that is proportional to its surface.

19.96 The minimum investment is generally provided in the exploitation agreements although they may also be defined in the mining rights themselves[119] or directly in the mining and petroleum laws. The provisions

---

added to that spent during the year of deferral. However, if within one year the exploration licensee spends an amount greater than he is required to spend on exploration according to the regulations, the amount of the sums spent can be carried forward to the next year as a reduction in the expenditure obligation for that year.']

[117] See, in particular, in Libya, Article 11(2) of Petroleum Law No. 25/55 of 11 May 2012: '[a]ny sum spent during any of the working periods specified above in excess of the minimum sum prescribed for that period shall be carried forward as a credit to the Company against the expenditure requirements for the following period or periods'.

[118] See, in Tunisia, Article 36(2) of Law No. 99-93 of 17 August 1999 Enacting the Hydrocarbons Code:

> '*[d]ans le cas où le Titulaire n'a pas accompli ses engagements minima de travaux et/ou de dépenses, il peut renoncer au Permis de Recherche après avoir versé à l'Autorité Concédante, une indemnité compensatrice égale à la différence entre le montant minimum de dépenses à réaliser et le montant de dépenses réalisées ou le montant nécessaire à l'achèvement des travaux fixés pour la période de validité du Permis de Recherche durant laquelle la renonciation a eu lieu.*'
>
> [Free translation into English: '[i]n case the commitments for the minimum work programme and/or expenditures are not accomplished, the Holder may relinquish his permit by paying to the Licensing Authority, a compensating indemnity equal to the difference between the minimum expenditures agreed upon and the expenditures incurred or the amount necessary to conclude the minimum work programme relative to the considered validity period of the Exploration Permit.'].

[119] See, to that effect, in Zambia, Article 17(1) of the Petroleum (Exploration and Production) Act, 2008:

> '[a] holder of a petroleum exploration licence shall [. . .] expend on exploration operations not less than the amount prescribed or required by the terms and conditions of the exploration licence to be so expended.'

Also see, in Guinea, Article 26 of Law No. 2011/006/CNT of 9 September 2011 Enacting the Mining Code:

> '*[l]'arrêté institutif du Permis de recherche fixe le programme minimum de travaux qui devra être exécuté par le titulaire pendant la durée de validité du Permis ainsi que l'effort financier minimum qu'il devra consacrer chaque année à ses recherches pendant la durée de validité du Permis et de ses renouvellements éventuels. A cet effet, il est institué, pour les Permis de recherche, des dépenses minimales par km² dont les montants sont fixés par la Réglementation minière.*'
>
> [Free translation into English: '[t]he order granting the Exploration Permit defines a minimum work programme which the holder must carry out while the Permit is in effect,

related to minimum work undertakings are divided into two broad categories according to whether the planned investments are based on objective or subjective evaluation criteria. The drafting of certain clauses may be hybrid and cover both objective and subjective elements.

19.97 In certain situations, the contract (or the law) allows the private operator to determinc the sums to be invested in exploration operations.

19.98 At times, the private operator's free determination of its minimum investment programme will be limited, in particular in light of minimum legal thresholds that investors must abide by. Similarly, in certain cases, a private operator must take various factors imposed by the law into consideration, such as the natural resources sought or the potential of the zone under exploration.

(1) Minimum investment commitments based on objective criteria

19.99 Although they are of less interest to the private operator than the preceding clauses, the goal of these clauses is to keep the project active regardless of the geological and technical results obtained.

These clauses are structured based on two principal elements, ie time and surface.

(a) Determination of a minimum investment programme in accordance with time spent

19.100 According to this concept, the minimum investment programme will vary in accordance with the periods under consideration.[120]

19.101 These expenditures are established as of the date the exploitation agreement is signed, or the mining rights are granted. Only time will determine the scope of the investments that were discussed during the initial project negotiations. The breakdown of these sums varies and every situation is possible: a linear framework, progressive evolution,[121] digressive evolution or a combination of the latter two.

19.102 At times, the contractual zone is divided into two parts and therefore two different investment programmes are planned. This difference

---

and the minimum financial outlay it is expected to devote to exploration each year during the term of the Permit and any renewals thereof. To this end, minimum expenditures per km^2 are instituted for Exploration Permits, the amounts of which are set out in the Mining Regulations.']

[120] See, in the Central African Republic, Article 15 of Decree No. 09-126 of 30 April 2009 on the Application Conditions for Law No. 09-005 of 29 April 2009 Enacting the Mining Code:
'*[p]endant la validité du permis de recherche, les dépenses doivent augmenter chaque année de cinquante pour cent (50%) au moins.*'
[Free translation into English: '[d]uring the period of validity of the exploration permit, expenditures must increase by at least fifty per cent (50%) every year.']

[121] See, in the Central African Republic, Article 15 of Decree No. 09-126 of 30 April 2009 Implementing Law No. 09-005 of 29 April 2009 Enacting the Mining Code, which provides that, during the period of validity of the exploration permit, expenditures must increase by at least fifty per cent every year.

can be explained by the host state's intent to proceed with the development of zones referred to as 'priority' zones.

(b) Determination of a minimum investment programme in accordance with the surface of the perimeter

19.103 A minimum investment programme may also be determined in accordance with the perimeter allocated to mining and petroleum works. Certain legal or contractual provisions may provide an obligation incumbent on the private operator to invest a precise amount per hectare or km^2,[122] at times with no details on the work to be performed. This method is largely applied in practice when it comes to determine the minimum investments to be performed by the private operator during the exploration phase.

(c) Determination of an overall investment

19.104 An overall investment amount may also be established. This amount covers either the maximum period of the exploitation agreement, which is utopian as, at the time the contract is signed, the nature and scope of potential discoveries or the exploration phase alone, is unknown.

The determination of a minimum investment plan and the overall amount may include penalty clauses to guarantee the satisfactory execution of the operations.

ii *Financing costs incurred by an operator*

19.105 The private operator's search for financing is most often part of complex financial arrangements.[123] A very large portion of the sums required to finance exploitation operations is covered by self-financing, in particular for the phases that precede the production launch, ie the prospection, exploration and part of the development phase.

The cost of financing incurred by an operator—most often a private operator—may result from the exploitation agreement itself, in which case the financing of the exploitation operations will be the subject of negotiations. It may also result from a decision taken by the operator during the exploitation operations.

---

[122] See, in the Central African Republic, Article 14 of Decree No. 09-126 of 30 April 2009 on the Application Conditions of Law No. 09-005 of 29 April 2009 Enacting the Mining Code: [Free translation into English: 'In accordance with Article 23 of the Mining Law, holders of exploration permits must commit to minimum annual expenditure per square kilometre in the amount of five hundred thousand (500,000) F CFA on work undertaken under their permit.'] Also see, in Côte d'Ivoire, Article 21 of Decree No. 2014-148 of 26 March 2014 Implementing Law No. 2014-138 of 24 March 2014 Enacting the Mining Code:

> '*[l]e budget de recherche des quatre années présenté par le demandeur du permis de recherche ne doit en aucun cas être inférieur à un million six cent mille francs par kilomètre carré.*'
>
> [Free translation into English: '[t]he four-year exploration budget presented by the exploration permit applicant must in no case be inferior to one million six hundred thousand francs per square kilometre.']

[123] With regard to financing techniques, see above para 15.25 et seq.

(1) Negotiated financing cost

19.106 Financing is an essential part of the negotiations and constitutes a prerequisite, as it is one of the key elements in selecting a private operator along with technological aspects. Two types of clauses should be distinguished: those that limit the unilateral financing obligation in time and those that, on the contrary, extend it and maintain the obligation throughout the term of the exploitation agreement.

(a) A negotiated, limited cost

19.107 In general, the unilateral financing obligation incumbent on the private operator is limited to the exploration phase during which the private operator runs the risk of non-discovery alone. It may be presented as consideration for services provided by the local State operator, such as the initial demarcation of the zones that present a geological interest in which the exploration area is located, or the State company's provision of the exploration zone.

When viable commercial quantities are discovered, the exploration investments incurred by a single operator are, in principle, reimbursed either in whole or according to his holdings in the share equity of the operating company. However, in zones with known natural resource potential, there are examples to the contrary, in which the exploration expenditures are paid by the operator who undertook to make these investments regardless of the result.

(b) A negotiated, extended cost

19.108 In certain cases, an operator who contributes funds must continue to play this role throughout the entire term of the exploitation agreement. The local State operator may waive this obligation; the private operator then finances all of the exploration and development costs until the local State operator can take over thanks to the revenue from the initial exploitation results.

The financial obligations of the private operator may be subject to a security in favour of the host state. This is meant to guarantee the financial commitments of the private operator. The funds provided as security may be released progressively as the works progress. The amount of the financial guarantee is becoming essential in relation to the investments guaranteed.

(2) Choice of financing

19.109 The exploitation of natural resources often requires different operators to partner in an attempt to share the costs and risks related to mining and petroleum operations. However, this type of partnership may at times encounter difficulties, in particular when it consists of operators with unequal financial resources. Furthermore, in certain cases, exploration results may be subject to varied interpretations and differences may arise among operators regarding the continuance or discontinuance of exploitation operations.

19.110 To settle disputes between operators, certain contracts initially define the concept of 'commercial wells' and a 'commercial deposit'. These terminological definitions, although they are key, may seem insufficient thereafter. To overcome deadlocks that may arise in the framework of a joint venture, operators may resort to clauses called 'sole risk' clauses.[124]

19.111 In accordance with the 'sole risk' mechanism, if operators disagree on whether or not to continue operations and on the scope of the proven reserves of natural resources, one of the operators may decide to continue with the deposit development and production launch alone. In this case, the operator who implements the sole risk clause may continue with the operations alone at its own risk and cost, and benefit from any future revenue from this exploitation.

19.112 One of the very first times the 'sole risk' mechanism was put into practice involved the discovery of natural gas in Abu Qir, Egypt. The private operator, Phillips, did not believe the reserves present were sufficient to justify bringing the deposit into production, which was not the opinion of the Egyptian Government, who was essentially targeting local energy independence. Production was nevertheless launched, but at the exclusive cost of the Egyptian public entity. The existence of an imbalance in the implementation of this clause should be noted as only the public operator has a right to reconsider. Thus, if the private operator takes the risk, either he was mistaken and loses all of his investments or he was not mistaken, and the local State operator can at any time reconsider and share the profits. Today, these clauses are essentially found in contracts related to hydrocarbon exploitation.

19.113 Widely used in *joint operation agreements*, this mechanism allows an operator who wishes to continue operations to override a decision of the 'operating committee'. The implementation of the sole risk clause requires compliance with a procedure described in the exploitation agreement or one of the resulting contractual documents. The operator, who specifies the planned work and expenditures, and the vote of the operators, usually bases this procedure on a proposal.

In principle, a refusal by one of the operators to participate in the operations does not preclude his participation *a posteriori*.

b Risks related to penalties

19.114 Along with the major financial commitments, the exploitation contract establishes considerable liability. This liability comes into play both among the operators themselves and in relation to third parties.

---

[124] For example, the Petroleum Code of Madagascar provides in Article 23:

> '*[c]haque signataire d'un contrat pétrolier qui n'a pas accepté de financer des travaux supplémentaires décidés par ses partenaires devra se conformer aux termes et conditions énoncés dans le contrat en vertu du principe du "Sole risk".*'
>
> [Free translation into English: '[e]ach signatory of an oil contract that has not agreed to finance supplementary work decided upon by its partners must comply with the terms and conditions expressed in the contract under the "sole risk" principle'.] (Law No. 96-018 Enacting the Petroleum Code of 20 August 1996).

i *Content of the risks related to penalties*

19.115 When an exploitation contract is negotiated, the parties will attempt to obtain mutual guarantees, the objective of which is fair and rapid compensation of a prejudice that one of the parties may suffer. These provisions are generally very rigid and most often to the detriment of the private operator.

(1) Rigid inter-operator liability mechanisms

19.116 Joint liability between a private operator and a public operator is generally not accepted in exploitation agreements, unlike the joint liability of private operators.

This position leads to very burdensome financial liability clauses, which may involve the obligation to provide a financial performance guarantee or the inclusion of a penalty clause. The goal of these mechanisms—whether contractual or legal—is to ensure that the private operator complies with his undertakings, in particular those related to the work and expenditure.

19.117 If the private operator fails to comply with its financial undertakings and when the contract or law so provides, it may be sanctioned by a penalty clause, ie a clause pursuant to which an operator undertakes to pay a lump-sum compensation if he fails to fulfil his obligation.

19.118 Certain mining and petroleum laws include provisions to this effect. Therefore, if the private operator fails to perform all or part of the work programme, the holder of the exploration permit may be required to pay penalties,[125] whether or not the failure in fact relates to the operator's non-compliance with the annual work programme or with the minimum investment programme.

19.119 The legal texts usually provide the amount of these penalties, which, in principle, total the amount of the non-performed obligations.[126]

---

[125] See in the Central African Republic, Article 14, para. 2 of Law No. 93.007 of 25 May 1993 Enacting the Petroleum Code; in Côte d'Ivoire, Article 23 of Law No. 96-669 of 29 August 1996 Enacting the Petroleum Code; in Liberia, Article 6.5.1 of the Petroleum Law, 2002; in Chad, Article 27(8) of Law No. 006-PR-2007 of 2 May 2007 On Hydrocarbons.

[126] See, in the Central African Republic, Article 14, para. 2 of Ordinance No. 93.007 of 25 May 1993 Enacting the Petroleum Code:

> '*[s]i le titulaire ne satisfait pas à ses obligations de travaux dans les délais impartis, il doit verser à l'État, dans les conditions fixées dans le contrat pétrolier, une indemnité égale au montant des obligations non exécutées.*'
>
> [Free translation into English: '[i]f the holder does not satisfy its work obligations by the defined deadlines, it must pay the State an indemnity equal to the amount of the non-fulfilled obligations under the conditions defined in the petroleum agreement.']

In Chad, Article 27(8) of Law No. 006-PR-2007 of 2 May 2007 On Hydrocarbons:

> '*[s]ans préjudice du programme de travaux, si, au terme de chaque période de validité du Permis de Recherches, les sommes dépensées n'atteignent pas le montant de l'engagement financier minimal souscrit pour ladite période, le Titulaire du Permis est tenu de verser à l'État une indemnité égale à la différence entre le montant souscrit pour ladite période et les sommes dépensées au titre de ladite période.*'
>
> [Free translation into English: '[w]ithout prejudice to the work programme, if, at the end of each validity period of the Exploration Permit, the sums spent do not equal the amount

However, the implementation of certain provisions may result in a dispute regarding the calculation of the compensation owed by the private operator. This is the case of Liberia's Petroleum Law, which simply refers to 'reasonable compensation';[127] this term may obviously lead to different interpretations. One must therefore refer to the terms of the exploitation agreement for details on the manner in which the compensation is assessed. In fact, certain laws refer to the terms of the exploitation agreement in order to determine the amount of the compensation to be paid by the private operator.[128] Although the amount of the penalty may be indicated in the legal and regulatory texts, the methods to pay this penalty are most often set out in the exploitation agreement.[129]

19.120 These rigid liability mechanisms can result in the termination of an exploitation agreement. The rigidity of liability mechanisms was recently illustrated in the framework of an arbitration between Arcelor Mittal and the State of Senegal in an environment of rising commodity prices.[130]

As mentioned earlier, mining and petroleum projects are particularly sensitive to market fluctuations, which can affect the overall financial balance of a project and lead to the suspension of operations. Faced with this type of change in circumstances, operators might be tempted to suspend the performance and delay the completion of projects whose profitability have come into question.[131] However, in light of such changes, can a private operator unilaterally decide to suspend compliance with his contractual obligations? In 2007, *Arcelor Mittal* and the State of Senegal entered into a series of agreements related to the exploitation of iron deposits in La Falémé located in southeast Senegal. As compensation for the exploitation of the deposits, *Arcelor Mittal* undertook to design, finance, build and operate an integrated mining project, which included the construction of a number of infrastructures. After the preliminary studies performed in the framework of this project, *Arcelor Mittal* announced that it could not pursue the project in an environment in which iron ore prices were dropping, in particular due to the cost of the infrastructures and the technical quality of the mineral resource. Further to this decision, Senegal initiated arbitration proceedings under the aegis of the International Chamber of Commerce and requested a

of the minimum financial undertaking for the said period, the Holder of the Permit is bound to pay the State an indemnity equal to the difference between the amount committed for the said period and the sums spent for the said period.']

[127] Article 6.5.1. of the Petroleum Law of Liberia, 2002.

[128] See, in Côte d'Ivoire, Article 23 of Law No. 96-669 of 29 August 1996 Enacting the Petroleum Code; in Niger, Article 73 of Law No. 2007-01 of 31 January 2007 Enacting the Petroleum Code.

[129] See, in the Central African Republic, Article 14, para. 2 of Ordinance No. 93.007 of 25 May 1993 Enacting the Petroleum Code.

[130] See, to that effect, Thierry Lauriol and Emilie Raynaud 'La rupture unilatérale des contrats d'exploitation' in *Le contentieux extractif* (ICC 2015) 141.

[131] See, to that effect, in Mauritania, the suspension of all of the iron ore exploitation projects, in particular the Askaf mining project, in Mauritania by *Sphere Minerals*, a subsidiary of the Glencore group, or Rio Tinto's hesitations in the framework of the Simandou project in Guinea presented as a 'mega project', and the related impact on the host State.

termination of the various agreements entered into in the framework of this project, claiming *Arcelor Mittal* was liable. In a partial decision rendered on 3 September 2013, the arbitral tribunal found that *Arcelor Mittal* had breached its undertakings and ordered the termination of the agreements. Prior to the issuance of the final decision on liability and the awarding of potential damages, the parties settled.[132]

This decision illustrates how little room to manoeuvre an operator has in a case where a change in circumstances calls the project's profitability into question. The operator's failure to comply with his obligations, even when the project's profitability is no longer certain, does not justify a cessation of the operations and may therefore lead to a termination of the exploitation agreement.

(2) Flexible mechanisms for the adjustment of obligations

19.121 The clauses described below provide for a readjustment of certain obligations while they are being performed and operations are not halted. They are therefore based on a will to achieve contractual fairness and maintain a legitimate balance between operators.

19.122 A contractual adjustment in favour of a private operator may take the form of the inclusion of clauses referred to as 'most favoured enterprise' and 'non-discrimination clauses'.

19.123 The most favoured enterprise clause is often an essential provision in light of multiple changes in legislation, mining policies and state policies which operators must face throughout the lengthy term of their contractual relations. The most favoured enterprise clause can in fact be hazardous as it binds the host state significantly and implies that it must inform the private operator of all of the contracts it has signed in the relevant sector. Furthermore, the extension obligation can only be determined based on the existence of an analogous activity, with no consideration of geological, geographic or other specificities. Prior contractual balances may therefore be suddenly overturned.[133]

19.124 We understand the difficulty inherent in drafting a most favoured enterprise clause: should a specific provision of the general contractual framework be the focus, or should the overall balance of the contract be taken into consideration? The second option seems more logical to us with

[132] Settlement Agreement between Senegal and Arcelor Mittal dated 30 May 2014.

[133] However, the clause may have a limited spatial field of application and may only cover one part of the host State's territory. This is the case, in particular, of the service agreement between PETROPERU, on the one hand, and Occidental Peruana Inc., Bridas Exploraciones y Produccion S.A. and Canadian Occidental Limited, on the other, dated 23 July 1980, which provides in its Article 18-1 that: '[i]n the unlikely case that PETROPERU grants in the future a set of more favourable conditions to other contractors of Risk Petroleum Services, in operations similar to ones established in the Contract, in the area of the Northwest, PETROPERU shall, within 90 days following Contractors' request, grant to them the said set of more favourable conditions, which will be in force of the date on which the said set of conditions was granted to the other contractor of Risk Petroleum Services.'

regard to the extreme diversity of geological situations—hence the risks taken—and the amount of the investments required.

19.125 In international law, a general non-discrimination principle exists regarding the origin of investments.[134] However, contracts related to the exploitation of natural resources often defy this principle to the extent that they derogate from the general tax and legal regime.

19.126 With regard to the contractual application of the non-discrimination principle, the host state, by seeking to drive local economic activity towards the objectives it defines, subjects investments to individualised requirements that include a significant share of discretionary power. When an exploitation agreement is signed, it may go even further than the requirements of international law by undertaking to forego some of its discriminatory practices during the performance of the contract either in general or in a limited manner. Furthermore, the host state's waiver of its use of discriminatory practices can cover certain specific areas only but that are very important in practice, such as the customs regime.

19.127 The adjustments to the contractual framework can also benefit jointly to the operators. Indeed, two types of clauses allow for an adjustment of the contractual framework in favour of parties to an operating agreement without the said agreement particularly benefiting any specific operator (the private operator or the local State operator). These are the cooperation clause and the review clause.

19.128 The cooperation clause provides that, if the initial contractual balance changes, the host state or the operator who benefits from this change must grant the same advantages that were granted when the contract was signed but only by modifying certain of its clauses. This clause is distinct from the changed circumstance clause in that it does not provide for renegotiations but a simple readjustment of conditions that have already been negotiated.

19.129 A variant of the cooperation clause may provide for consultation between operators to take into consideration significant changes in the conditions under which the exploitation agreement is performed; however, this consultation may not take place before the end of a given period, which is generally long.

19.130 Unlike the cooperation clause, the review clause is independent from any extraneous event. It provides operators with an opportunity to

[134] International law has adopted the principle of non-discrimination between foreign operators and national operators. The discrimination that is prohibited is based on national origin. Not all unequal treatment between nationals and foreigners is discriminatory; discrimination only relates to unfavourable treatment of a foreigner because he is a foreigner. Furthermore, an issue to be resolved is whether international law imposes an obligation on the host State to treat foreign operators the same, without discriminating against any of them. There is no obligation incumbent on the State to equalise the privileges it grants to foreigners of different nationalities. The principle is therefore as follows: States are free to limit the benefits they wish to grant to the nationalities of their choice. However, the most favoured nation clause diminishes the scope of this principle. The host State is not bound to treat foreign operators equally but must avoid penalising them on the grounds of their nationality.

request the partial renegotiation of the exploitation agreement. The danger of this type of clause is that it may slow down or even bring a halt to exploitation operations. However, it is justified by an intent to improve the effectiveness of the means in place for the exploitation work in light of information obtained during the operations.

ii *Consequential changes of the risks related to penalties*

19.131 Mining and petroleum legislation, as well as exploitation agreements, detail operators' liability vis-à-vis third parties. These liability regimes are, in principle, governed by national law. However, exploitation agreements complete these regimes and adapt them to the factual circumstances of each project. With regard to these provisions, a distinction should be made between provisions establishing preventive measures and those establishing rules relating to compensation for losses suffered.

(1) Provisions related to the prevention of damages

19.132 It is in the interest of an operator in charge of operations to take measures to prevent incidents for which he may be held liable. Therefore, the operator must comply with a certain number of basic obligations that vary from one project to another in order to minimise the impact of mining and petroleum operations, in particular environmental and social consequences, or in order to protect cultural heritage.

19.133 The very essence of the exploitation of a natural resource consists in extracting a component that is intrinsic to a geographic zone, so it implies a constant modification thereof. In fact, the extraction of mining and petroleum resources presupposes the destruction of the original geological and natural balance; this may involve the disappearance of hills or mountains, the diversion of rivers or groundwater, land subsidence, deforestation, etc. Environmental challenges are therefore significant,[135] which, in practice, means that provisions to prevent environmental risks are included in exploitation agreements and mining and petroleum laws. Certain states have created special national institutions to oversee the protection of the environment and take action in the mining and petroleum sectors.[136]

19.134 Prevention consists of an obligation incumbent on operators to conduct mining and petroleum operations according to the rules to minimise

[135] See the preamble of ECOWAS Directive C/DIR 3/05/09 of 27 May 2009 on the Harmonisation of Guiding Principles and Policies in the Mining Sector, which provides: 'CONSCIOUS of the fact that mining affects the ecosystems and is not limited to the immediate area of the mine site and also impacts on communities in the environmental, social and economic zone surrounding the mine; and that certain "communities of interest" require special consideration by mining companies, governments, and investors; and such groups include local peoples, artisanal miners, mine workers and people within the communities who are marginalized.']

[136] For example, in Ghana and in Sierra Leone, the Environmental Protection Agency; in Guinea, the Direction Nationale de l'Envrionnement; and in Kenya, the National Environmental Management Authority.

the impact of the exploitation on the environment and protect the latter.[137] The exploration and extraction methods used by operators must therefore be rational in order to ensure that the exploitation is as profitable as possible while remaining compatible with environmental protection requirements.[138] The management of the environmental impact of mining and petroleum operations requires, in particular, the conduct of an environmental impact study, which mining and petroleum codes consider of utmost importance.[139]

[137] See, in particular, in Angola, Article 7(2) of Petroleum Law No. 10/04 of 12 November 2004; in Botswana, Articles 58(3) and 65 of the Mines and Minerals Act; in Benin, Article 98 of Law No. 2006-17 of 17 October 2006 Enacting the Mining and MiningTaxation Code; in Cameroon, Article 82 of Law No. 99-013 of 22 December 1999 Enacting the Petroleum Code; in the Central African Republic, Article 56 of Ordinance No. 93.007 of 25 May 1993 Enacting the Petroleum Code and Article 104 of Law No. 09-005 of 29 April 2009 Enacting the Mining Code; in the Republic of Congo, Article 132 of Law No. 4-2005 of 11 April 2005 Enacting the Mining Code; in Côte d'Ivoire, Article 49 of Law No. 96-669 of 29 August 1996 Enacting the Petroleum Code and Article 140 of Law No. 2014-138 of 24 March 2014 Enacting the Mining Code; in Djibouti, Article 50 of Law No. 117-AN-05-5 of 16 October 2005 Enacting the Petroleum Code; in Guinea, Article 143 of Law No. 2011/006/CNT of 9 September 2011 Enacting the Mining Code; in Lesotho, Article 58(1) of the Mines and Minerals Act, 2005; in Madagascar, Article 99 of Law No. 99-022 of 19 August 1999 Enacting the Mining Code; in Nigeria, Article 118 of the Minerals and Mining Act, 2007; in Senegal, Article 51 of Law No. 98-05 of 8 January 1998 Enacting the Petroleum Code; in Chad, Article 66 of Law No. 011-PR-1995 of 20 June 1995 Enacting the Mining Code.

[138] See, to that effect, in Algeria, Article 3, para. 2 of Law No. 05-07 of 20 April 2005 On Hydrocarbons:

> '*[les substances et les ressources en hydrocarbures] doivent être exploitées en utilisant des moyens efficaces et rationnels afin d'assurer une conservation optimale, tout en respectant les règles de protection de l'environnement.*'
>
> [Free translation into English: '[hydrocarbons substances and resources] must be extracted using efficient and rational means in order to ensure optimum preservation while at the same time observing environmental protection regulations.']

[139] In South Africa, sections 5(4) and 39 of the Mineral and Petroleum Resources Development Act 28 of 2002 refer to an '*environmental management programme*' or an '*environmental management plan*'; in Algeria, Law No. 01-10 of 3 July 2001 Enacting the Mining Law refers to an '*étude d'impact*' (Articles 149 and 150) supplemented by a '*plan de gestion environnementale*' (Article 151), Law No. 05-07 of 20 April 2005 On Hydrocarbons refers to an '*étude d'impact environmental* and a '*plan de gestion de l'environnement*' (Article 18); in Angola, the conduct of an '*étude d'impact*' is provided in Article 64 of Law No. 31/11 of 23 September 2011; in Benin, Law No. 2006-17 of 17 October 2006 Enacting the Mining and Mineral Taxation Code of the Republic of Benin provides for the conduct of an '*étude d'impact sur l'environnement*' and a '*programme de protection de l'environnement*' [Free translation into English: 'programme for the protection of the environment' (Article 35), Law No. 2006-18 of 17 October 2006 Enacting the Petroleum Code of the Republic of Benin provides for the conduct of an '*étude d'impact environmental*' (Article 64); in Botswana, the Mines and Minerals Act refers to an '*environmental impact assessment*' (Article 65(2)); in Burundi, Law No. 1-21 of 15 October 2013 Enacting the Mining Code refers to an '*étude d'impact environnemental* ' (Article 138); in Cameroon, Law No. 99-013 of 22 December 1999 Enacting the Petroleum Code provides for the conduct of an '*étude d'impact environnemental*' (Article 83, completed by Article 67 of Decree No. 2000-485 of 30 June 2000 Defining the Application Peculiarities of Law No. 99-013 of 22 December 1999); in the Central African Republic, Law No. 09-005 of 29 April 2009 Enacting the Mining Code refers to an '*étude d'impact environnemental et social*' and a '*plan de gestion environnementale et sociale du projet*' (Articles 34 and 105); in the Republic of Congo, Law No. 4-2005 of

19.135 The objective of these studies is, in particular, to assess the

---

11 April 2005 Enacting the Mining Code refers to an '*étude d'impact sur l'environnement*' and a '*programme de protection de l'environnement*' (Articles 50 and 101); in the Democratic Republic of Congo, Law No. 007-2002 of 11 July 2002 Enacting the Mining Code provides for the conduct of an '*étude d'impact environnemental*' and a '*plan de gestion environnementale du projet*' (Article 204); in Côte d'Ivoire, Law No. 2014-138 of 24 March 2014 Enacting the Mining Code refers to an '*étude d'impact environnemental et social*' and a '*plan de gestion environnementale et sociale*' (Article 141); in Djibouti, Law No. 117-AN-05-5 of 16 October 2005 Enacting the Petroleum Code refers to an '*étude d'impact sur l'environnement*' (Article 28); in Ethiopia, Proclamation No. 678/2010 Promoting Sustainable Development of Mineral Resources provides for the conduct of an '*environmental impact assessment*' (Article 60); in Guinea, Law No. 2011/006/CNT of 9 September 2011 Enacting the Mining Code provides for the conduct of an '*étude d'impact environnemental et social*' and preparation of a '*plan de gestion environnementale et sociale*' (Article 142), Law No. L/2014/N°034/AN Enacting the Mining Code refers to an '*environmental and social impact study*' and an '*environmental and social management plan*' (Article 60); in Equatorial Guinea, Law No. 9/2006 of 3 November 2006 Enacting the Mining Code refers to a '*plan detallado para la proteccion y saneamiento del medio ambiente*' (Article 41); in Lesotho, the Mines and Minerals Act 2005 refers to an '*environmental impact assessment*' (Article 58(3)); in Madagascar, Law No. 99-022 of 19 August 1999 Enacting the Mining Code amended by Law No. 2005-021 of 17 October 2005 refers to a '*plan d'engagement environnemental*' an '*étude d'impact environnemental*' (Articles 33 and 100); in Mali, Law No. 04-037 of 2 August 2004 on the Organisation of Exploration, Exploitation, Transportation and Refining of Hydrocarbons refers to as an '*étude d'impact sur l'environnement*' (Article 83); in Mauritania, Law No. 2010-033 of 20 July 2010 Enacting the Hydrocarbons Code, amended by Law No. 2011-044 of 25 November 2011 provides for the conduct of an '*étude d'impact environnemental*' (Article 36); in Mozambique, Mining Law No. 20-2014 of 18 August 2014 (Articles 36 and 70) and Petroleum Law No. 21-2014 of 18 August 2014 Enacting the Petroleum Law (Article 15) provide for the conduct of an '*environmental impact study*'; in Namibia, the Minerals Prospecting and Mining Act, 1992 (Article 50) provides for the conduct of an '*environmental impact assessment*' and the preparation of an '*environmental management plan*'; in Niger, Law No. 2007-01 of 31 January 2007 Enacting the Petroleum Code provides for the conduct of an '*étude d'impact sur l'environnement*' (Article 26); in Nigeria, the Minerals and Mining Act, 2007 refers to an '*environmental impact assessment statement*' and an '*environmental protection and rehabilitation program*' (Article 119); in Uganda, the Mining Act, 2003 (Article 108) and the Petroleum (Exploration, Development and Production) Act, 2013 (Article 76) refer to the conduct of an '*environmental impact assessment*'; in Rwanda, Law No. 13-2014 of 20 May 2014 On Mining and Quarry Operations refers to a '*programme de gestion de l'environnement*' and a '*plan de gestion de l'environnement*' (Articles 20 and 22); in Senegal, Law No. 98-05 of 8 January 1998 Enacting the Petroleum Code provides for the conduct of an '*étude d'impact sur l'environnement*' (Article 28); in the Seychelles, the Model Petroleum Agreement of 2013 provides for the conduct of an '*environmental impact assessment study*' (Article 21); in Sierra Leone, the Mines and Minerals Act, 2009 provides for the conduct of an '*environmental impact assessment*' (Article 133); in South Sudan, the Mining Act, 2012 provides for the conduct of an '*environmental and social impact assessment*' (Article 121(2)), the Petroleum Act, 2012 provides for the preparation of an '*environmental management plan*' (Article 60), a '*strategic environmental assessment*' and an '*environmental and social impact assessment*' (Article 59); in Chad, Law No. 011-PR-1995 of 20 June 1995 Enacting the Mining Code provides for the implementation of a '*programme de protection et de gestion de l'environnement*' (Article 30); Law No. 006-PR-2007 of 2 May 2007 On Hydrocarbons refers to an '*étude d'impact sur l'environnement*' (Articles 26 and 29).

impact of exploitation operations on the environmental balance, to define measures to mitigate this impact and to establish implementation and follow-up methods for the said measures.

19.136 Furthermore, site rehabilitation is taken into consideration in almost all of the mining and petroleum laws,[140] in particular by requiring the implementation of a rehabilitation plan. The rehabilitation of mining sites is an essential component of the prevention of environmental damage. It includes, for example, removing all of the installations, and levelling and reforesting the area.[141] In particular, site rehabilitation should allow for the recovery of stable conditions regarding security and agricultural productivity.[142] The rehabilitation obligation therefore requires the operator to bear the financial burden of the restoration of mining and petroleum sites.

19.137 To this end, a private operator may be under an obligation to

---

[140] For examples of national laws providing for the rehabilitation of sites, see, in particular, in South Africa, section 38 of the Mineral and Petroleum Resources Development Act 26 of 2002; in Algeria, Article 81, para. 4 of Law No. 05-07 of 20 April 2005 On Hydrocarbons; in Angola, Article 24(2) of Petroleum Activities Law No. 10/04 of 12 November 2004; in Benin, Article 35 of Law No. 2006-17 of 17 October 2006 Enacting the Mining and Mining Taxation Code; in Burundi, Articles 137 and 140 of Law No. 1-21 of 15 October 2013 Enacting the Mining Code; in the Central African Republic, Article 104 of Law No. 09-005 of 29 April 2009 Enacting the Mining Code; in the Republic of Congo, Articles 128 et seq. of Law No. 4-2005 of 11 April 2005 Enacting the Mining Code; in the Democratic Republic of Congo, Article 61 of Law No. 007-2002 of 11 July 2002 Enacting the Mining Code; in Côte d'Ivoire, Articles 144 et seq. of Law No. 2014-138 of 24 March 2014 Enacting the Mining Code; in Ghana, Article 28 of the Minerals and Mining Act, 2006; in Guinea, Article 64 of Law No. 2011/006/CNT of 9 September 2011 Enacting the Mining Code; in Lesotho, Article 58(4) of the Mines and Minerals Act, 2005; in Liberia, Article 8(2) of the Petroleum Law, 2002; in Madagascar, Article 102 of Law No. 99-022 of 19 August 1999 Enacting the Mining Code amended by Law No. 2005-021 of 17 October 2005; in Mali, Article 110 of Law No. 2012-015 of 27 February 2012 Enacting the Mining Code; in Mozambique, Article 71 of Mining Law No. 20-2014 of 18 August 2014; in Namibia, Article 54 of the Minerals Prospecting and Mining Act, 1992 and Article 54 of the Petroleum (Exploration and Production) Act, 1991; in Niger, Article 46 of Law No. 2007-001 of 31 January 2007 Enacting the Petroleum Code; in Nigeria, Article 114 of the Minerals and Mining Act, 2007; in Uganda, Article 66 of the Mining Regulation, 2004 and Article 110 of the Mining Act, 2003; in Rwanda, Article 31 of Law No. 13-2014 of 20 May 2014 on the Exploration of Mines and Quarries; in Sao Tome and Principe, Article 55(2) of the Framework Law of Petroleum Operations; in South Sudan, Article 122 of the Mining Act, 2012; in Swaziland, Article 127 of the Mines and Minerals Act, 2011; in Chad, Article 52.5 of Law No. 006-PR-2007 On Hydrocarbons; in Zambia, Article 69 of the Petroleum (Exploration and Production) Act, 2008.

With regard to the rehabilitation of mining and petroleum sites, in particular, see Hery Ranjeva 'L'obligation de réhabilitation d'un site d'extraction minière à la charge de l'investisseur' Liber Amicorum Raymond Ranjeva *L'Afrique et le droit international: variations ou l'organisation internationale* (Pédone 2013) 549.

[141] With regard to the content of rehabilitation obligations in the mining sector, in particular see Article 31 of Law No. 13-2014 on the Exploitation of Mines and Quarries in Rwanda.

[142] See, to this effect, in Guinea, Article 144 of Law No. 2011/006/CNT of 9 September 2011 Enacting the Mining Code.

provide a security deposit or pay a provision,[143] which may be considered, in certain circumstances, as a deductible expense.[144] If applicable, a rehabilitation fund may be created for this purpose.[145]

19.138 If the operator fails to comply with his obligations relative to rehabilitation, the host state may use the sums deposited in the rehabilitation fund. It should be noted that certain legal texts extend the obligation to pay a provision to guarantee all of the operator's environmental obligations.[146]

19.139 In general, holders of mining rights must comply with the laws related to environmental protection in effect on the territory of the host state, which supplement the special terms of the mining and petroleum laws.[147]

19.140 In addition to environmental consequences, mining and petroleum projects can have a considerable impact on the local populations and, in certain circumstances cause their displacement. The laws tend to prevent the negative social impacts of mining and petroleum projects, in particular by providing for mechanisms to prevent any prejudice to populations. The operator's obligation to minimise the social impact of its operations is one aspect of these laws.[148] The conduct of impact assessment studies also allows for the prevention of prejudice being caused to local populations as they often include not only an environmental aspect but also a social one.[149]

---

[143] See, in South Africa, section 41 of the Mineral and Petroleum Resources Development Act 28 of 2002; in Algeria, Article 82 of Law No. 05-07 of 20 April 2005 On Hydrocarbons; in the Democratic Republic of Congo, Article 204 of Law No. 007-2002 of 11 July 2002 Enacting the Mining Code; in Guinea, Article 144 of Law No. 2011/006/CNT of 9 September 2011 Enacting the Mining Code; in Mozambique, Article 71(2) of Mining Law No. 20-2014 of 18 August 2014.

[144] See, in Algeria, Article 82 of Law No. 05-07 of 20 April 2005 On Hydrocarbons.

[145] The creation of an environmental rehabilitation fund constitutes an obligation incumbent on Members States of the WAEMU (Article 6(7) of Regulation No. 18/2003/CM/UEMOA of 23 December 2003 Enacting the Community Mining Code).

[146] See, in Uganda, Article 112 of the Mining Act, 2003; in Rwanda, Article 32 of Law No. 13-2014 of 20 May 2014 On the Exploitation of Mines and Quarries; in Swaziland, Article 127 of the Mines and Minerals Act, 2011.

[147] In this regard, mining and petroleum laws generally provide for the operator's compliance with the environmental protection laws in effect. For example, see, in Guinea, Article 59 of Law No. L/2014/N°034/AN of 23 December 2014 Enacting the Petroleum Code; in Uganda, Article 3 of the Petroleum (Exploration, Development and Production) Act, 2013; in Sierra Leone, Article 65(1) of the Mines and Minerals Act, 2009; in Swaziland, Article 122 of the Mines and Minerals Act, 2011.

[148] See, in Benin, Article 98 of Law No. 2006-17 of 17 October 2006 Enacting the Mining and Mining Taxation Code.

[149] See, in the Central African Republic, Law No. 09-005 of 29 April 2009 Enacting the Mining Code, which provides for a '*plan de gestion environnementale et sociale*' and an '*étude d'impact environnemental et social*' in Côte d'Ivoire, Law No. 2014-138 of 24 March 2014 Enacting the Mining Code provides for an '*étude d'impact environnemental et social*' as well as a '*plan de gestion environnementale et sociale*'; in Guinea, Law No. 2011/006/CNT of 9 September 2011 Enacting the Mining Code provides for an '*étude d'impact environnemental et social*' and a '*plan de gestion environnemental et social*'; Law No. L/2014/N°034/AN Enacting the Petroleum Code refers to an '*étude d'impact environnemental et social*' and a '*plan de gestion environnementale et sociale*'; in Mali, Law No. 2012-015 of 27 February 2012 Enacting the Mining Code provides for an '*étude d'impact*

19.141 The legal texts may also require the private operator to implement regular inspections of the sites to protect the health and well-being of populations in proximity to mining and petroleum sites.[150]

19.142 Also, the prevention of damage to cultural heritage is included in certain mining and petroleum laws.[151] The operator must, in particular, inform the competent authorities of any archaeological finds,[152] take all the measures required to protect historical sites[153] and minimise the impact of mining and petroleum activities on local uses and ancestral traditions.[154]

19.143 With regard to the content of these preventive measures, it must be noted that the operator is not bound by an obligation of result, which would consist in eliminating risks, but only by an obligation of means with the sole intent of minimising risks.

The prevention of risks is structured around the implementation of insurance mechanisms that cover the risks generated by mining and petroleum operations.[155] These insurance mechanisms, which protect the private operator, also guarantee that potential damage may be compensated.

(2) Compensation for damage

19.144 In practice, the private operator is bound to compensate personal injury and material damage related to the exploitation operations; this principle is enacted in many mining and petroleum laws.[156]

---

environnemental et social'; in Sierra Leone, the Petroleum Exploration and Production Act, 2011 provides for the conduct of a '*social and environmental impact assessment*'; in South Sudan, the Mining Act, 2012 and the Petroleum Act, 2012 provide for the conduct of an '*environmental and social impact assessment*'.

150 See, in Côte d'Ivoire, Article 141 of Law No. 2014-138 of 24 March 2014 Enacting the Mining Code.

151 The protection of cultural heritage is also provided by legal provisions that ban exploitation activities in certain specific zones, such as burial grounds, sacred places and religious buildings.

152 See, in Sudan, Article 17 of the Mineral Resources and Mining Development Act, 2007; in Ethiopia, Article 17 of Proclamation No. 295/1986 Regulating Petroleum Operations.

153 See, in Ethiopia, Article 17 of Proclamation No. 295/1986 Regulating Petroleum Operations.

154 See, in Benin, Article 98 of Law No. 2006-17 of 17 October 2006 Enacting the Mining and Mining Taxation.

155 See above paras 15.55 et seq.

156 See, in Algeria, Article 152 of Law No. 01-10 of 3 July 2001 Enacting the Mining Law; in Angola, Article 25 of Petroleum Law No. 10/04 of 12 November 2004; in Benin, Article 100 of Law No. 2006-17 of 17 October 2006 Enacting the Mining and Mining Taxation Code; in Cameroon, Article 62 of Law No. 99-013 of 22 December 1999 Enacting the Petroleum Code; in Côte d'Ivoire, Article 148 of Law No. 2014-138 of 24 March 2014 Enacting the Mining Code; in Djibouti, Article 51 of Law No. 117-AN-05-5 of 16 October 2005 Enacting the Petroleum Code; in Eritrea, Article 19(2) of Proclamation No. 108/2000, A Revised proclamation Governing Petroleum Operations; in Ethiopia, Article 19(2) of Proclamation No. 295/1986 Regulating Petroleum Operations; in Guinea, Article 106 of Law No. 2011/006/CNT of 9 September 2011 Enacting the Mining Code and Article 45 of Law No. L/2014/N°34/AN of 23 December 2014 Enacting the Petroleum Code; in Equatorial Guinea, Article 51 of Law No. 9/2006 of 3 November 2006 Enacting the Mining Code; in Mauritania Article 32 of Law No. 2010-033 of 20 July 2010 Enacting the Hydrocarbons Code amended by Law No. 2011-044 of 25 November 2011; in Mozambique, Article 36 of Mining Law No.

19.145 In the framework of joint operations, the holders of mining rights are in principle jointly and severally liable.[157]

The private operator will have to compensate damage it has caused but also damage caused by its subcontractors[158] or, more generally, entities working on its behalf.[159]

In certain cases, the liability regime applicable to the private operator is described. At times, and under certain conditions, the private operator may benefit from a liability exemption, in particular when no fault is found. A distinction must therefore be made between the texts that provide for liability without fault on the part of the operator[160] and the others, which are much more infrequent.

19.146 Certain texts go beyond the general principle of the operator's liability for damage caused in the framework of mining and petroleum operations by providing specific liability regimes to compensate damage caused to the environment, surface land and the operators of neighbouring mines.

(a) Compensation for environmental damage

19.147 Operations to exploit natural resources have a major impact on the environment; for example, open-pit mines destroy the landscape, at times definitively. As we have seen, an obligation is increasingly imposed on operators to rehabilitate exploitation sites to the state in which they were prior to operations.[161]

19.148 In addition to the preventive aspects referred to above, the laws and regulations provide for the operator's liability in the event of environmental pollution or damage,[162] in particular marine pollution.[163]

20-2014 of 18 August 2014 and Articles 15 and 40 of Petroleum Law No. 21-2014 of 18 August 2014; in Senegal, Article 52 of Law No. 98-05 of 8 January 1998 Enacting the Mining Code; in Sudan, Article 18 of the Mineral Resources and Mining Development Act, 2007; in Tunisia, Article 86 of Law No. 2003-30 of 28 April 2003 Enacting the Mining Code.

[157] See, in South Sudan, Article 93 of the Petroleum Act, 2012.

[158] See, in the Central African Republic, Article 50 of Ordinance No. 93.007 as of 25 May 1993 Enacting the Petroleum Code; in Côte d'Ivoire, Article 64 of Law No. 96-669 of 29 August 1996 Enacting the Petroleum Code; in Guinea, Article 45 of Law No. L/2014/N°034/AN of 23 December 2014 Enacting the Petroleum Code; in Mauritania, Article 32 of Law No. 2010-033 of 20 July 2010 Enacting the Hydrocarbons Code amended by Law No. 2011-044 of 25 November 2011.

[159] See, in Djibouti, Article 51 of Law No. 117-AN-05-5 of 16 October 2005 Enacting the Petroleum Code; in Senegal, Article 52 of Law No. 98-05 of 8 January 1998 Enacting the Petroleum Code; in Angola, Article 25 of Law No. 10/04 of 12 November 2004; in Guinea, Article 106 of Law No. 2011/006/CNT of 9 September 2011 Enacting the Mining Code.

[160] In Cameroon, Article 62 of Law No. 99-013 of 22 December 1999 Enacting the Petroleum Code; in Uganda, Article 129 of the Petroleum (Exploration, Development and Production) Act, 2013 (in terms of environmental damage).

[161] With regard to rules on the rehabilitation of mining and petroleum sites, see above para 882.

[162] See, in particular, in Guinea, Article 61 of Law No. L/2014/N°034/AN of 23 December 2014 Enacting the Petroleum Code; in Madagascar, Article 129 of Law No. 99-022 of 19 August 1999 Enacting the Mining Code amended by Law No. 2005-021 of 17 October 2005; in Mozambique, Articles 54 and 56 of Petroleum Law No. 21-2014 of 18 August 2014; in

Certain provisions provide that the host state can be held liable by third parties, in which case the private operator must compensate the host state,[164] which can generally not be held liable vis-à-vis third parties in any way for damage that results from mining or petroleum operations when it is not involved therein as an operator.[165]

(b) Compensation for damage caused to surface land

19.149 To access subsoil resources, the holder of mining rights must first be able to access the surface area. The holder of mining rights will, in principle, be authorised to occupy the land required for exploration and exploitation works to be carried out.[166]

During this occupancy, exploitation may cause damage to the surface land; mining and petroleum laws have considered this risk by setting out methods to compensate related damage.[167]

Compensation will, in principle, be granted to the landowner. However, the legal texts extend the benefits of these terms to holders of customary land rights[168] and holders of land tenure rights.[169] Finally, in the framework of the

---

Namibia, Article 71 of the Petroleum (Exploration and Production Act), 1991 and Article 130 of the Minerals (Prospecting and Mining) Act, 1992; in Uganda, Article 129 of the Petroleum (Exploration, Development and Production) Act, 2013; in South Sudan, Article 61 of the Petroleum Act, 2012; in South Africa, section 38(1) of the Mineral and Petroleum Resources Development Act 28 of 2002; in Malawi, Article 55 of the Petroleum (Exploration and Production) Act, 1983; in Sierra Leone, Article 93 of the Petroleum Exploration and Production Act, 2011; in South Sudan, Article 60 of the Petroleum Act, 2012.

[163] See, in particular, in Algeria, Article 214 of Law No. 01-10 of 3 July 2001 Enacting the Mining Code.

[164] See, in particular, in Sierra Leone, Article 146 of the Petroleum Exploration and Production Act, 2011.

[165] See, in Cameroon, Article 62(2) of Law No. 99-013 of 22 December 1999 Enacting the Petroleum Code.

[166] With regard to relations between operators and land owners, see above para 8.68 et seq.

[167] See in the Democratic Republic of Congo, Article 280 of Law No. 007-2002 of 11 July 2002 Enacting the Mining Code; in Ghana, Article 74 of the Minerals and Mining Act, 2006; in Guinea, Article 126 of Law No. 2011/006/CNT of 9 September 2011 Enacting the Mining Code; in Madagascar, Article 129 of Law No. 99-022 of 19 August 1999 Enacting the Mining Code amended by Law No. 2005-021 of 17 October 2005; in Morocco, Article 35 of Law No. 21-90 of 1 April 1992 On the Exploration and Exploitation of Hydrocarbon Deposits amended and completed by Law No. 27-99 of 15 February 2000; in Niger, Article 117 of Ordinance No. 93-16 of 2 March 1993 Enacting the amended Mining Law; in Rwanda, Article 29 of Law No. 13-2014 of 20 May 2014 On the Exploitation of Mines and Quarries; in the Seychelles, Article 7 of the Minerals Act and Article 15 of the Petroleum Mining Act; in Sierra Leone, Article 35 of the Mines and Minerals Act, 2009; in Chad, Article 78 of Law No. 011-PR-1995 of 20 June 1995 Enacting the Mining Code; in Tunisia, Article 80 of Law No. 2003-30 of 28 April 2003 Enacting the Mining Code; in Togo, Article 44 of Law No. 99-003 of 18 February 1999 Enacting the Hydrocarbons Code.

[168] See, in Chad, Article 78 of Law No. 011-PR-1995 of 20 June 1995 Enacting the Mining Code.

[169] See, in Guinea, Article 126 of Law No. 2011/006/CNT of 9 September 2011 Enacting the Mining Code; in Kenya, Article 10(2) of the Petroleum Act; in the Seychelles, Article 15 of the Petroleum Mining Act; in Sierra Leone, Article 35 of the Mines and Minerals Act, 2009; in Chad, Article 78 of Law No. 011-PR-1995 of 20 June 1995 Enacting the Mining Code.

exploitation of artisanal mines, the texts also provide for the compensation of damage suffered by farmers.[170]

19.150 In principle, the compensation awarded corresponds to the value of the prejudice caused,[171] although certain texts provide that the damage caused to assets will be compensated based on their actual replacement value, increased by 50%, unless they are restored to their state of origin.[172] In certain cases, the texts may provide that the claim for compensation must be filed within a certain time frame as of the date the events that have led to the damage cease.[173]

(c) Compensation for damage caused to a neighbouring mine

19.151 In the framework of its activities, an operator may cause damage to a neighbouring mine. In practice, the damages are often caused by abnormal quantities of water or polluted water penetrating into the neighbouring mine.

Many mining and petroleum laws that govern relations between holders of mining rights include provisions regarding the compensation of damage to a neighbouring mine.[174]

19.152 The texts may describe the methods to determine the compensation due for this prejudice or provide for an expert appraisal for this purpose.[175]

---

[170] See, in the Democratic Republic of Congo, Article 112 of Law No. 007-2002 of 11 July 2002 Enacting the Mining Code.

[171] See, in Benin, Article 68 of Law No. 2006-17 of 17 October 2006 Enacting the Mining and Mining Taxation Code; in Cameroon, Article 61 of Law No. 99-013 of 22 December 1999 Enacting the Petroleum Code.

[172] See, in the Democratic Republic of Congo, Article 280 of Law No. 007-2002 of 11 July 2002 Enacting the Mining Code.

[173] See, in Cameroon, Article 61 of Law No. 99-013 of 22 December 1999 Enacting the Petroleum Code; in Sierra Leone, Article 35(7) of the Mines and Minerals Act, 2009.

[174] See, to this effect, in Burundi, Article 123 of Law No. 1-21 of 15 October 2013 Enacting the Mining Code; in The Comoros, Article 36 of Order No. 54-1110 of 13 November 1954 Reforming the Mineral Substance Regime; in the Republic of Congo, Article 116 of Law No. 4-2005 of 11 April 2005 Enacting the Mining Code; in Guinea, Article 116 of Law No. 2011/006/CNT of 9 September 2011 Enacting the Mining Code; in Madagascar, Article 131 of Law No. 99-022 of 19 August 1999 Enacting the Mining Code, amended by Law No. 2005-021 of 17 October 2005; in Niger, Article 119 of Ordinance No. 93-16 of 2 March 1993 Enacting the Amended Mining Law; in Chad, Article 83 of Law No. 011-PR-1995 of 20 June 1995 Enacting the Mining Code; in Tunisia, Article 90 of Law No. 2003-30 of 28 April 2003 Enacting the Mining Code.

[175] See, in the Republic of Congo, Article 116 of Law No. 4-2005 of 11 April 2005 Enacting the Mining Code.

20

Rules not specific to exploitation agreements

20.1 The sharing of risks inherent in the performance of an exploitation agreement also requires the application of rules that are not specific to the mining and petroleum sectors. Exploitation agreements and mining and petroleum laws will refer to mechanisms that international trade operators in all sectors are familiar with. In particular, this involves the inclusion of provisions intended to avoid litigation, such as force majeure and renegotiation clauses, but also clauses that govern potential litigation, such as the choice of governing law and dispute resolution clauses.

1 CLAUSES INTENDED TO PREVENT LITIGATION

20.2 In principle, a contract is signed with a view to compliance therewith until its purpose has been achieved.[1] However, in light of the long-term nature of exploitation agreements, it is reasonable to believe that the legal, political and economic environment that existed at the time the exploitation agreement was signed may change during the enforcement thereof.[2] This is not surprising, as contracts related to the exploitation of natural resources generate a series of successive operations spread out over lengthy periods.

20.3 New events that occur over the lifespan of the contract may have two very distinct consequences: to prevent the performance of contractual obligations, or to make compliance with these obligations completely inequitable for one party. The inclusion of clauses to mitigate, or at least diminish, the consequences of these events and adapt them to the contract is a key concern.[3]

20.4 The significant economic value of an exploitation agreement explains the measures enacted to ensure that it remains in effect when various changes could disrupt its performance. Two types of clauses have

[1] However, it is not unusual to see certain African countries adopt renegotiation policies regarding mining agreements that have already been signed. With regard to the renegotiation of exploitation agreements in Africa, see above para 3.36 et seq.

[2] 'The long duration of a petroleum contract reveals the reality that it can never be perfectly drafted because the parties' foreseeability to predict all the contingencies is limited. (Due to) the fact that petroleum contracts are negotiated under conditions of considerable complexity and uncertainty, it is not possible for the parties to specify in advance every conceivable contingency in (the) contract.', Abdullah Al Faruque 'Adaptation of Petroleum Contract by Third Party: An Overview'(2006) 3 (4) *TDM* 3 *in* Erkan op cit 242.

[3] With regard to the management of these concerns when contracts are drafted, see, in particular, Philippe Fouchard 'L'adaptation des contrats à la conjoncture économique' 1979 *Rev arb* 67.

been adopted in practice: force majeure clauses and hardship (changed circumstance) clauses.[4] The objective of force majeure clauses is to provide an excuse for the operators' failure to comply with their obligations due to events beyond their control and expectations. The aim is to obtain an exemption of liability. Hardship clauses allow the contract to be renegotiated when its outcome will be fundamentally different from the outcome expected when it was signed; hardship clauses are intended to apply when compliance with contractual obligations creates a situation that is particularly severe and unfair for one of the contractors.

20.5 Terminological differences have appeared regarding hardship clauses. For some scholars, the clause should correspond to the French concept of unpredictability (ie 'unpredictability' clause),[5] while for others, the appropriate term would be a 'revision' clause or a 'safeguard' clause.[6] In reality, safeguarding the contract is a concern all operators share when drafting these clauses. The 'change of circumstance clause' wording presents the advantage of immediately indicating its essential meaning. Finally, major differences in the force majeure and hardship concepts in the different national laws must be noted.[7]

20.6 Contracts related to the exploitation of natural resources have adopted these clauses with some adaptations required by the specificities of mining and petroleum projects. These clauses appeared rather late in petroleum contracts. The first concession contracts did not include this type of clause, which, however, did not prevent them from being modified during the performance of the contract.

Today, some African mining and petroleum laws take the necessary adaptation of contracts in the event of change of circumstances into consideration and include provisions in this regard; but more often they provide for the insertion of a force majeure clause rather than a hardship clause.

a Force majeure clauses

20.7 The force majeure concept[8] is interpreted differently in accordance with the legal tradition of the relevant state. Laws derived from French law

[4] The term 'hardship' stems from common law.

[5] The English term 'hardship' has arisen from practice; in French, it could be considered an 'unpredictability clause'. Pierre Van Ommeslaghe 'Les clauses de force majeure et d'imprévision (hardship) dans les contrats internationaux' 1980 *RDIDC* 10 note 3. Also see Ole Lando 'Renegotiation and Revision of International Contracts' 1980 *GYIL* 49.

[6] As an example, see Jean-Marie Deleuze *Le contrat de processus de transfert de technologie, know-how* (Masson 1982) 97.

[7] See, in particular, Henry Lesguillons 'Frustration, Force Majeure, Imprévision, Wegfall der Geschaftsgrundlage' 1979 *DPCI* 507.

[8] We consider force majeure to be a fortuitous event. See in particular Georges R Delaume 'Excuse for Non-Performance and Force Majeure in Economic Development Agreements' (1971) 10 *Columbia Journal of Transnational Law* 242; Marcel Fontaine 'Les clauses de force majeure dans les contrats internationaux' 1979 *DPCI* 469; Philippe Kahn 'Force majeure et contrats internationaux de longue durée' 1975 *JDI* 467. These scholars argue

usually adopt criteria such as unavoidability, unpredictability and exteriority to define a force majeure event.[9] However, other laws (most often based on the common-law system) embrace a more flexible concept and provide that a force majeure event may be one over which the parties have no control and that reasonable efforts have not been able to remedy. Note that in certain cases, the force majeure concept is extended to events in which one party is totally or partially prevented from complying with his obligations. In general, force majeure may cover events such as war, embargo, acts of terrorism, natural disasters, acts of government, sabotage or civil unrests, to name a few examples.

20.8 The structure of the force majeure clause may vary significantly from one contract to another, but also from one law to another. First, the provision may only include an abstract definition of force majeure based on its impact on the operators' obligations.[10] The definition may be replaced by an exhaustive (or non-exhaustive) list, which is a drafting technique inspired

---

that a fortuitous event is an internal event related to the activities of the parties, such as fires or defective equipment; for its part, force majeure is an external event such as lightning or war.

[9] See in Algeria, Article 31(2) of the Presidential Decree as 1 March 2003 Enacting the Model Mining Agreement, which provides:

> '*[p]our les fins de la Convention Minière, la définition de force majeure comprend tout acte ou événement imprévisible, irrésistible et indépendant de la volonté des Parties et les empêchant provisoirement ou définitivement d'exécuter leurs obligations.*'
>
> [Free translation into English: '[f]or the purposes of the Mining Agreement, the definition of force majeure includes any unpredictable, unavoidable event outside the control of the Parties and that temporarily or definitively prevents them from fulfilling their obligations.']

In Burundi, Article 2(12) of Law No. 1-21 as 15 October 2013 Enacting the Petroleum Code provides:

> '*[f]orce majeure: s'entend de tout acte ou évènement imprévisibles, insurmontable et indépendant de la volonté des parties.*'
>
> [Free translation into English: '[f]orce majeure: means any unpredictable, insurmountable act or event outside the will of the parties.']

In the Democratic Republic of Congo, Article 297 of Law No. 007-2002 of 11 July 2002 Enacting the Mining Code provides: '*[c]onstitue un cas de force majeure, tout événement imprévisible, irrésistible, insurmontable et indépendant de la volonté du titulaire.*' [Free translation into English: '[a]ny event which is unforeseeable, unavoidable, insurmountable and outside the control of the holder constitutes a case of force majeure.']

[10] With regard to laws and agreements that include an abstract definition of the force majeure concept, the following should be mentioned: in Angola, Article 41 of Petroleum Law No. 10/04 of 12 November 2004; in the Republic of Congo, Article 188 of Law No. 4-2005 of 11 April 2005 Enacting the Mining Code; in the Democratic Republic of Congo, Article 297 of Law No. 007-2002 as 11 July 2002 Enacting the Mining Code and Article 24 of the Production Sharing Agreement Between the Democratic Republic of Congo and the Association Dominion Petroleum Congo & Soco Exploration-Production RDC & La Congolaise des Hydrocarbures in Bloc V in the Graben Albertine (November 2007); in Guinea, Article 1 of Law No. L/2013/No. 053/CNT of 8 April 2013 On the Amendment of Certain Terms of Law No. 2011/006/CNT of 9 September 2011 Enacting the Mining Code; in Madagascar, Article 180 of Law No. 99-022 of 19 August 1999 Enacting the Mining Code, amended by Law No. 2005-021 of 17 October 2005; in the Seychelles, Article 1 of the Model Petroleum Agreement of 2013.

by US legal practices and is found essentially in English-speaking African countries; these are called descriptive clauses.[11] These descriptive clauses introduce some instability and lack of clarity as each element thereof may give rise to interpretation; in order to avoid any uncertainty, each of these elements should be defined, which is not often the case. A force majeure clause may also consist in a combination, ie an abstract definition illustrated by a few non-exhaustive examples.[12] In certain cases, the definition of force majeure may be supplemented by a negative definition, which lists scenarios that do not constitute force majeure.[13] Finally, mining and petroleum laws may refer to a definition included in external instruments, in particular the definition and implementation methods provided in the petroleum or mining agreement.[14]

20.9 Most clauses manage the force majeure issue from a domestic perspective, in other words, force majeure events that occur in the host state. However, a force majeure clause may take events originating outside of the host state into consideration.

i *Recognition of force majeure and implementation of the clause*

20.10 The implementation of the force majeure clause presupposes, first, that the parties involved immediately notify of the force majeure event. In most cases, it is incumbent on the party that cannot comply with one of its contractual obligations and alleges a force majeure event to inform the other party[15] or the competent authority[16] thereof by the time limits set in the

[11] The following legal provisions provide an exhaustive or non-exhaustive list of force majeure events: in Malawi, Article 42 of the Petroleum (Exploration and Production) Act, 1983; in the Seychelles, Article 56(2) of the Petroleum Mining Act; in Uganda, Article 188(1) of the Petroleum (Exploration, Development and Production) Act, 2013.

[12] The following legal provisions include an abstract definition of force majeure illustrated by a few examples: in Algeria, Article 31(2) of the Presidential Decree of 1 March 2003 establishing a model mining agreement; in Burundi, Article 2 of Law No. 1-21 of 15 October 2013 Enacting the Mining Code.

[13] See, to this effect, in Guinea, Article 87 of Law No. 2001/006/CNT of 9 September 2011 Enacting the Mining Code.

[14] With regard to mining and petroleum laws that refer to the definition included in the petroleum contract, the following should be mentioned: in Benin, Article 41 of Law No. 2006-18 of 17 October 2006 Enacting the Petroleum Code; in Cameroon, Articles 12 and 114 of Law No. 99-013 of 22 December 1999 Enacting the Petroleum Code; in Côte d'Ivoire, Articles 18 and 84 of Law No. 96-669 of 29 August 1996 Enacting the Petroleum Code; in Niger, Article 100 of Law No. 2007-01 of 31 January 2007 Enacting the Petroleum Code; in Sao Tome and Principe, Article 33 of the Framework Law on Petroleum Operations; in Chad, Article 10 of Law No. 006-PR-2007 of 2 May 2007 On hydrocarbons; in Togo, Article 27(6) of Law No. 99-003 of 18 February 1999 Enacting the Petroleum Code.

[15] See, to this effect, in Algeria, Article 31 of the Presidential Decree of 1 March 2003 Enacting the Model Mining Agreement; in Guinea, Article 87 of Law No. 2011/006/CNT of 9 September 2011 Enacting the Mining Code.

[16] See, to this effect, in the Democratic Republic of Congo, Article 298 of Law No. 007-2002 of 11 July 2002 establishing the Mining Code, which provides for an obligation to inform the Mining Registry; in Malawi, Article 42(2) of the Petroleum (Exploration and Production) Act, 1983, which provides for an obligation to immediately inform the Minister; in Uganda,

applicable regulation. Certain legal or contractual provisions provide methods for the communication of this information, in particular in terms of time-limits[17] and the means to serve notice of the force majeure event.[18]

20.11 Notification of a force majeure event is at times followed by an investigation during which the competent authority must issue a decision as to whether or not force majeure exists.[19]

20.12 At times, the reliance on a force majeure event may be challenged. This generally occurs when a difference arises as to the interpretation of the facts presented to justify the existence of force majeure. Certain mining and petroleum laws provide appeal procedures to allow a party to contest a decision denying the existence of force majeure; this may involve an administrative appeal initiated against a decision of the national competent authority[20] or arbitration proceedings to settle a dispute regarding the existence of a force majeure event or remedial measures to be taken.[21]

20.13 Mining and petroleum laws, as well as exploitation agreements, may provide that the implementation of the force majeure clause will include precautionary measures to ensure that the parties make every effort to prevent the worsening of the event and comply with their contractual obligations to the greatest extent possible. To this end, the force majeure clause may, in particular, provide an obligation incumbent on the parties to minimise the impact of the force majeure on the smooth performance of the mining or petroleum operations[22] and take all the measures required to normally fulfil their contractual obligations.[23]

ii *Legal consequences of a force majeure event*

20.14 The principal impact of a force majeure event is to excuse the failure of one of the parties to comply with all or part of his contractual

---

Article 188(2) of the Petroleum (Exploration, Development and Production) Act, 2013, which also provides for an obligation to inform the Minister of the existence of a force majeure event.

[17] See, to that effect, in Algeria, Article 31 of the Presidential Decree of 1 March 2003 Enacting the Model Mining Agreement.

[18] For example, certain texts provide that notification must be served in writing, in particular: in the Democratic Republic of Congo, Article 83 of Decree No. 038-2003 of 26 March 2003 Enacting the Mining Code; in Guinea, Article 87 of Law No. 2011/006/CNT of 9 September 2011 Enacting the Mining Code.

Finally, certain texts require the transmission of reasoned notice, in particular, in Algeria, Article 31 of the Presidential Decree of 1 March 2003 Enacting the Model Mining Agreement.

[19] See, to that effect, Article 84 of Decree No. 038-2003 of 26 March 2003 Enacting the Mining Code in the Democratic Republic of Congo.

[20] See, to that effect, Article 89 of Decree No. 038-2003 of 26 March 2003 Enacting the Mining Code in the Democratic Republic of Congo.

[21] See, to that effect, in Guinea, Article 87 of Law No. 2011/006/CNT of 9 September 2011 Enacting the Mining Code.

[22] See, to that effect, in Guinea, Article 87(3) of Law No. 2011/006/CNT of 9 September 2011 Enacting the Mining Code.

[23] See, to that effect, Article 38(2) of the Seychelles Model Petroleum Agreement of 2013; in Sierra Leone, Article 149(3) of the Petroleum Exploration and Production Act, 2011.

obligations;[24] in this case, the failure of a co-contracting party to fulfil one of its obligations will not be considered a breach of contract. Throughout the period of the force majeure event, compliance with the obligations affected by the said event will be suspended,[25] it being understood that the normal performance of the contract will resume once the force majeure event has ended.

20.15 However, the occurrence of a force majeure event does not result in the parties being exempted from complying with all of their obligations. The parties must continue to fulfil their obligations with the exception of those affected by the force majeure event.[26] Certain provisions also provide that force majeure will not release the operator from its obligations related to the payment of a sum of money.[27]

20.16 The consequences related to the occurrence of a force majeure event will generally be progressive over time as the force majeure continues. In general, the exploitation agreement period will automatically be extended for a period that corresponds to that of the interruption caused by the force majeure event.[28] In practice, this often raises a difficulty relating to the obligations that are not affected by the force majeure in terms of whether they should or must also be extended. This principle also applies to mining

[24] See in Algeria, Article 31 of the Presidential Decree of 1 March 2003 Enacting the Model Mining Agreement; in Guinea, Article 109 of Law No. L/2014/N°034/AN Enacting the Petroleum Code; in Madagascar, Article 180 of Law No. 99-022 of 19 August 1999 Enacting the Mining Code, amended by Law No. 2005-021 of 17 October 2005; in Malawi, Article 42 of the Petroleum (Exploration and Production) Act, 1983; in Uganda, Article 188 of the Petroleum (Exploration and Production) Act, 2013; in the Seychelles, Article 38 of the Model Petroleum Agreement, 2013; in Sierra Leone, Article 149 of the Petroleum Exploration and Production Act, 2011.

[25] See, to that effect, in the Democratic Republic of Congo, Article 298 of Law No. 007-2002 of 11 July 2002 Enacting the Mining Code; in Côte d'Ivoire, Article 35 of Law No. 2014-138 of 24 March 2014 Enacting the Mining Code; in Tunisia, Article 62(1) of Law No. 99-93 of 17 August 1999 On the Promulgation of the Hydrocarbons Code.

[26] See, to that effect, in the Democratic Republic of Congo, Article 297 of Law No. 007-2002 of 11 July 2002 Enacting the Mining Code, Article 24(4) of the Production Sharing Agreement between the Democratic Republic of Congo and the Association Dominion Petroleum Congo & Soco Exploration-Production RDC & La Congolaise des Hydrocarbures in Bloc V in the Graben Albertine (November 2007); in the Seychelles, Article 38(2) of the Model Petroleum Agreement of 2013; in Sierra Leone, Article 149(4) of the Petroleum Exploration and Production Act, 2011.

[27] See, to that effect, in Malawi, Article 42(4) of the Petroleum (Exploration and Production) Act, 1983.

[28] See, to that effect, in Algeria, Article 31(1) of the Presidential Decree of 1 March 2003 Enacting the Model Mining Agreement; in the Democratic Republic of Congo, Article 298 of Law No. 007-2002 of 11 July 2002 Enacting the Mining Code, Article 86 of Decree No. 038-2003 of 26 March 2003 Enacting the Mining Regulation and Article 24(2) of the Production Sharing Agreement between the Democratic Republic of Congo and the Association Dominion Petroleum Congo & Soco Exploration-Production RDC & La Congolaise des Hydrocarbures in Bloc V in the Graben Albertine (November 2007); in Malawi, Article 42(3) of the Petroleum (Exploration and Production) Act, 1983; in the Seychelles, Article 38(1) of the Model Petroleum Agreement of 2013; in Sierra Leone, Article 149 of the Petroleum Exploration and Production Act, 2011.

and petroleum rights that may be extended for a period equal to that of the force majeure event.[29] If the force majeure lasts longer than a certain period, some legal and contractual provisions provide for the automatic termination of the contract. However, the question of whether or not the termination must be retroactive is unresolved; compensation may potentially be provided for if a single operator renders a sufficient number of services. These terms must imperatively be provided or detailed in the exploitation agreement.

With regard to the financial consequences of the force majeure event borne by a private operator, certain national laws provide that a force majeure event does not create a right to the reimbursement of taxes and royalties the said operator has paid.[30]

b Change of circumstances clause

20.17 As of the very early stage of the operations, aware of the lengthy time the projects require and the difficulty of foreseeing their outcome, the parties more and more frequently negotiate terms to implement a balanced legal framework to reconcile the various interests to the greatest extent possible. This balance, which results from the negotiations, may, during the performance of the mining and petroleum operations, change due to extraneous circumstances.[31] It is not uncommon to see the performance of exploitation agreements be affected by a disruption in the political, social or economic environment that radically alters the expectations of the parties to the agreement.[32] If the balance of the contract is materially and significantly affected due to an unpredictable change in circumstances, it may be essential to amend the contractual framework to allow the parties to effectively comply with their obligations and, at times, to ensure the very survival of the contract. With this in mind, in an attempt to protect the balance of their undertakings, the parties will ensure that they can renegotiate the terms of the contract by way of renegotiation clauses.[33]

---

[29] See, in particular, in the Democratic Republic of Congo, Article 342 of Law No. 007-2002 of 11 July 2002 Enacting the Mining Code and Article 88 of Decree No. 038-2003 of 26 March 2003 Enacting the Mining Regulation; in Togo, Article 27 of Law No. 96-004 of 26 February 1996 Enacting the Mining Code.

[30] See, in particular, to that effect, in the Central African Republic, Article 65 of Ordinance No. 93.007 of 25 May 1993 Enacting the Petroleum Code.

[31] To quote Jeswald W Salacuse 'Experienced lawyers and executives know that the challenge of international project negotiations is not just "getting to yes", but also staying there.' Jeswald W Salacuse 'Renegotiating International Projects Agreements' (2000) 24 (4) *Fordham International Law Journal* 1319.

[32] See, to that effect, John Gotanda 'Renegotiation and Adaptation Clauses in Investment Contracts, Revisited' (2003) 36 *Vand J Transnat'l L* 1462; also see, to this effect, Klaus Peter Berger 'Renegotiation and Adaptation of International Investment Contracts: the Role of Contract Drafters and Arbitrators' (2004) 1 (4) *TDM* 1 and 2.

[33] For a definition of renegotiation clauses, see, in particular, John Gotanda: '[r]enegotiation clauses are provisions in contracts that, upon the happening of a certain event or events, require all parties to return to the bargaining table and renegociate the terms of their agreements.' Gotanda op cit 1462.

20.18 Although the stabilisation and renegotiation clauses target different objectives—respectively, retaining the original terms of the contract in application of the *pacta sunt servanda* principle, and the adaptation thereof in accordance with the *rebus sic stantibus* principle[34]—these clauses have a single final objective: the stability of the contractual balance that existed when the agreement was signed. All of the stakeholders (both operators and the host state) benefit from this stability as it will allow for the projects to continue.

20.19 Some scholars have debated the expediency of including these types of clauses due to the unpredictability they create in terms of costs resulting from a potential renegotiation of the contract.[35] These provisions are no longer subject to controversy as evidenced by their frequent use.[36] Mining and petroleum laws as well as exploitation agreements include provisions related to renegotiations in the case of an event that changes the balance that existed when the agreement was signed.[37]

i *Recognition of a disruption of the contractual balance*

20.20 A change in the contractual balance on which the parties found an agreement results from a fundamental change in circumstances. However, it may be difficult, in particular for lawyers, to determine the initial contractual balance that prevailed when the exploitation agreement was signed. The initial balance of an exploitation agreement is based on complex economic assumptions and, in the extractive sector, these assumptions require consideration of additional factors related to the many uncertainties (in particular technical and financial) that apply to the exploitation of the resource. These different factors create major uncertainty and make it difficult to determine the initial balance that prevailed in terms of economic profitability.[38]

20.21 A change in the initial contractual balance must also be extraneous to the operators as none of them must be able to ward off or predict this change. Whether the cause is national or international in origin, it is the exteriority notion that prevails. In practice, especially when regulatory changes are applied by the host state, one can question the existence of an

---

[34] See, to this effect, Bernardini op cit 98.

[35] See, to that effect, Gotanda op cit 1463.

[36] With regard to the frequent use of renegotiation clauses in long-term investment contracts, see Stefan Kröll 'The Renegotiation and Adaptation of Investment Contracts' (2004) 1 (3) *TDM*.

[37] With regard to mining and petroleum laws that include renegotiation provisions, the following should be mentioned: in Sierra Leone, Article 120 of the Petroleum Exploration and Production Act, 2011, which provides : '[t]he terms and conditions of a petroleum licence or a permit for the laying and operation of a pipeline may be reviewed at any time that a significant change occurs under the circumstances prevailing at the time of the issue of the licence or permit.'

[38] With regard to the determination of the initial balance in the extraction sector, see Honoré Le Leuch 'Analyse économique et renégociation des contrats extractifs' in *Le contentieux extractif* 2015 ICC 31.

exteriority factor. This factor is in fact debatable in view of the ties that bind the executive branch and the local State operator. We note the antinomy of positions related to the interpretation of the regulatory intervention of the host state in terms of whether or not it is truly outside of the control of all operators.

20.22 Setting aside the case of regulatory intervention, a change in the initial contractual balance can, for example, result from an upheaval in the international monetary system or a significant drop in the prices of commodities. More generally, a change in circumstances that disrupts the contractual balance must result in a reasonable impossibility to continue to enforce the exploitation agreement.

National case law adopted distinct solutions; for certain courts, in particular English and American ones, a finding of a change in circumstance must result in the termination of the contract;[39] this is called the application of the 'frustration' theory.

20.23 Unlike force majeure clauses, an automatic implementation of renegotiation clauses must be excluded; in fact, the objective of these clauses is to trigger renegotiations or, more simply, a revision of the contract when circumstances have changed that make the enforcement of the said contract truly unfavourable to an operator. The recognition of a disruption in the contractual balance must be notified by an operator who wishes to avail himself of the renegotiation clause to the other parties to the contract.

20.24 The difficulties encountered by an operator who is the victim of a change of circumstances create a major risk of forcing the operator to abandon or waive his presence in the ongoing project if no solution is found. This generally leads to extensive litigation, causing major project delays, in which no party truly wins.

ii *Implementation of renegotiations*

20.25 Most contracts are silent in terms of the implementation of the renegotiation/hardship clause and refer to the underlying spirit at the time the exploitation agreement was signed, which, in practice, leads to many disputes and interpretations. In general, one of the operators will seek to reform and the other will want to maintain the contractual framework. A number of solutions are provided in exploitation agreements, the most common of which consists in the operator who believes all of the conditions for a 're-adaptation' exist notifying his partner and submitting a draft contract revision, with the other then notifying his position within a short period.

20.26 To ensure that renegotiation clauses are fully effective, the parties must therefore ensure that they indicate the conditions under which the renegotiation must take place as well as the parties' obligations in this respect.[40] However, even if the renegotiation clause does not include

[39] See, as an example, John J Gorman 'Commercial Hardship and the Discharge of Contractual Obligations under American and British Law' (1980) 13 *Vand J Transnat'l L* 107.

[40] Jeswald W Salacuse 'Renegotiating International Project Agreements' (2000) 24 (4) *Fordham International Law Journal*. 1325.

instructions in this regard, common principles govern the conduct of the renegotiations.

20.27 First, it must be noted that the duty to renegotiate does not include an obligation incumbent on the parties to reach an agreement.[41] That said, the absence of an obligation of result does not deprive the clauses of their effects, as the parties must undertake renegotiations in good faith.

20.28 In any event, it is not always easy to renegotiate a contract, and it may prove impossible if one of the parties fails to cooperate or if an agreed solution is not reached; it is therefore crucial that the contract or the law provide for the participation of a mediator or arbitrator to settle the fundamental issues surrounding the renegotiations.[42] Although there is a growing trend in favour of the existence of a renegotiation obligation,[43] it is all the more important to provide for the involvement of a third party in renegotiations as international case law does not always recognise the existence of an obligation to renegotiate and requires the participation of a third party be provided by law or the agreement,[44] or accepted by the parties themselves.

---

[41] See, to this effect, Berger 'Renegotiation and Adaptation of International Investment Contracts: the Role of Contract Drafters and Arbitrators' at 20 and 21: '[a]ccording to unanimous international opinion, renegotiation clauses therefore only contain an obligation on the parties to make the best possible effort to reach an agreement [. . .] they do not, however, require the parties to actually reach an agreement.' Also see, to this effect, *American Independent Oil Co (AMINOIL) v Government of Kuwait* (Award of 24 March 1982) 1982 *JDI* 869.

[42] Jelswald W Salacuse 'Renegotiating International Project Agreements' (2000) 24 (4) *Fordham International Law Journal* 1319 spec 1325; also see, to this effect, Berger op cit 15 et seq.

[43] It is sometimes argued that, even in the absence of contractual provisions that provide for contract renegotiations, parties who face a change in circumstances must renegotiate in good faith (see, to that effect, Berger op cit 9 et seq: '[i]t is often argued that even without a specific contractual clause a duty to renegotiate in light of changed circumstances is inherent to international investments contracts'.

[44] This point was referred to in the *AMINOIL* award: '[t]here can be no doubt that, speaking generally, a tribunal cannot substitute itself for the parties in order to [. . .] modify a contract unless that right is conferred upon it by law or by the express consent of the parties [. . .] arbitral tribunals cannot allow themselves to forget that their powers are restricted. It is not open to doubt that an arbitral tribunal—constituted on the basis of a "compromissory" clause contained in relevant agreements between the parties to the case [. . .] could not, by way of modifying or completing a contract, prescribe how a provision (for the determination of the economic equilibrium) must be applied. For that the consent of both parties would be necessary.' *American Independent Oil Co (AMINOIL) v Government of Kuwait* (Award of 24 March 1982) 1982 *JDI* 869. Award referred to by AL FARUQUE op cit 3. Also see the award rendered in UNCITRAL *Himpurna California Energy Ltd v PT (Persero) Perusahaan Listruik Negara* (Award as 4 May 1999) 2000 *YB Com Arb* 13 spec 61 and 62, §§. 200 et seq: '[i]t is not for the Arbitral Tribunal to question the motives or judgement of the Parties, but to assess their rights and obligations in light of their legally significant acts or omissions. That is all; that is enough. To go beyond this role would be to betray the legitimate expectations reflected in the Parties' agreement to arbitrate and indeed to impair the international usefulness of the arbitral mechanism. [. . .] The arbitrators cannot usurp the role of government officials or business leaders. They have no political authority, and no right to presume to impose their personal view of what might be an

iii *Implementation of renegotiations in the absence of a clause*

20.29 Although it has been widely accepted when a renegotiation clause exists, the authority of an international arbitrator or a court to adapt a contract due to changed circumstances is the subject of controversy.[45] That said, certain arbitral awards in which the arbitral tribunals allowed the good faith adaption of a contract should be mentioned.[46]

Finally, one may rightfully wonder whether renegotiation clauses allow for the affirmation of the existence of a principle of mutability of exploitation agreements. If so, they may have a negative impact on the proper enforcement of international contracts. In any event, it would then be necessary to recognise that this principle applies in favour of all operators, whether public or private. However, the mutability of the exploitation agreement should remain limited to very precise scenarios defined by the operators themselves when they sign the agreement. This therefore involves a limited internal mutability only, which is crucial in light of the period over which the operators must fulfil their obligations, as these obligations may face frequent economic, technological and even political changes. Far from acknowledging a mutability principle, we believe it is more appropriate to refer to an exception to the immutability principle.

2 CLAUSES DRAFTED IN ANTICIPATION OF LITIGATION

20.30 Two fundamental elements required for the proper performance of a project are, on the one hand, a clear choice of the law that governs relations between the parties and, on the other, the ability to resort to dispute resolution mechanisms that are suited to the parties' expectations. In exploitation agreements, it is therefore essential to establish provisions related to the choice of governing law, the *professio juris* clauses, and clauses related to the settlement of disputes, ie arbitration clauses.

i *Change of law provisions*

20.31 The objective of *professio juris* or change of law clauses is to determine the law that governs the contract.[47] This type of clause is

appropriate negotiated solution. Whatever the purity of their intent, arbitrators who acted in such fashion would be derelict in their duties, and would create more mischief than good.'

[45] See, to this effect, Mamadou Konate 'Le cadre juridique de la renégociation' in *Le contentieux extractif* 2015 ICC 41.

[46] See, in particular, the awards mentionned by Konate op cit 41, ie the Award rendered in ICC Case No. 4761(1987), 1987 1012, obs. S Jarvin, in which the tribunal affirms that the *lex mercatoria* 'gives effect to the unpredictability theory, based on the principle that the *pacta sunt servanda* rule is limited by the overriding good faith principle.' See also the award rendered in ICC Case No. 5953(1989), Recueil des sentences arbitrales de la CCI 1986-1990, comm. Yves Derains.

[47] In common law, the applicable law concept is known as the '*proper law of the contract*' or the '*governing law*' (see, to this effect, Abdul F Munir Maniruzzaman 'Choice of Law in International Contracts—Some Fundamental Conflict of Laws Issues' (1999) 16 (4) *Journal of International Arbitration* 141).

particularly important as the law that governs a contract determines the conditions of its enforceability, the obligations of the parties and the consequences of their failure to comply therewith. The law chosen as the law that applies to the contract may result from the parties' express or tacit choice, in which case reference must be made to the rules of law with which the contract has the closest ties.[48]

20.32 The principle of freedom of choice, ie the concept of subjectivism in private international law, allows the parties to have every freedom in choosing the law that applies to the contract[49] by referring to the law of a state, general principles of law, or by applying the technique referred to as splitting ('dépeçage')[50] . Furthermore, and although this solution is highly controversial, the parties may opt for a *sui generis* legal system that is established by and included in the contract itself; this is the contract without a proper law theory.[51] This theory remains difficult to apply to exploitation agreements as they can never be self-sufficient in light of the subjects they cover.

Although the parties' freedom to select the governing law remains the basic principle, the laws of the state where the mining and petroleum operations

---

[48] See, to that effect, Article 1 of the Institute of International Law's Resolution on The Proper Law of the Contract in Agreements between a State and a Foreign Private Person (Athens—1979), which provides: '[c]ontracts between a State and a foreign private person shall be subject to the rules of law chosen by the parties or, failing such a choice, to the rules of law with which the contract has the closest link.']

[49] The principle related to the parties' freedom in choosing the law that governs the contract is universally accepted (see, to that effect, Maniruzzaman op cit 161-162). In ICC Case No. 1512/24, the Tribunal stated as follows: '[t]here are few principles more universally admitted in private international law than that referred to by the standard terms of the "proper law of the contract", according to which the law governing the contract is that which has been chosen by the parties, whether expressly or (with certain difference or variations according to the various systems) tacitly. The differences that may be observed here between different national systems relate only to the possible limits of the parties' power to choose the applicable law or to certain special questions or to modalities, but not to the principle itself, which is universally accepted.' (ICC Case No. 1512/24(1971) 1975 YB Com Arb 130). With regard to the subjectivism concept, also see Henri Battifol 'Subjectivisme et objectivisme dans le droit international privé des contrats' in *Mélanges offerts à Jacques Maury* (Dalloz 1960) 39; Willis L M Reese 'Power of Parties to Choose Law Governing their Contract' 1960 *Proc ASIL* 49; M V A Naidu 'On the Choice of Law in Contracts: The Doctrine of Autonomy of Parties' (1973) 13 (4) *IJIL* 594; Lars O Lagerman 'Choice of Forum Clauses in International Contracts: What Is Unjust and Unreasonable?' 1978 *Int Lawyer* 779; Morley Friedman 'The Validity and Efficacy of Choice of Law Clauses in International Contracts' 1969 *LLJ* 71.

[50] The 'splitting' concept consists of the parties being able to choose that multiple applicable laws which govern different aspects of the contract (see in particular, to this effect, Maniruzzaman op cit 153 et seq).

[51] Under the 'contrat sans loi' or 'contract without a proper law' theory, the contract takes its compulsory effect solely from the will of the parties and bypasses any state law. With respect to the contract without a proper law concept, see in particular Lena Gannage 'Le contrat sans loi en droit international privé' (2007) 11 *Electronic Journal of Comparative Law* 3; Jean-Paul Beraudo 'Faut-il avoir peur du contrat sans loi ? ' in *Mélanges en l'honneur de Paul Lagarde* (Dalloz 2005) 93.

are performed will apply. As such, mining and petroleum laws may impose the law that will govern the contractual relations between the private operator and the host state.

20.33 Whether the governing law is chosen freely by the parties or imposed on the said parties by mining and petroleum laws, two types of *professio juris* clauses must be distinguished: on the one hand, those that provide for the application of a single law and, on the other hand, providing for the application of multiple laws.

ii *Clauses that designate a single law*

20.34 As it is most often the case, *professio juris* clauses may name a single law as the law that governs the contract, in which case a number of options must be taken into consideration. The first consists in giving priority to the application of a neutral law.[52] This choice allows the parties to maintain a certain balance by avoiding the application of the legal system of the producing state or that of the investor on the contract. The second possibility and the one most often preferred consists in choosing the law of the state that hosts the investment as the applicable law.[53] *A contrario*, the law of the state in which the oil company is registered might also apply.[54] Yet neither is acceptable for the two operators as '*[t]he presence in contractual relations of the State, the defender of national interests that has sovereignty prerogatives, has an influence on the determination of the applicable rules of law*'.[55] In the first case, the private operator is subject to uncertainties relative to a co-contracting party who is also a lawmaker; in the second case, the host state would fall under the authority of another state and would therefore be bound by all of the amendments to the said state's laws. When operators choose the local law, they are entitled to only adopt certain provisions thereof, either by mentioning them or by listing those that will not apply.

20.35 Other solutions are also available to the parties, in particular the application of the *lex mercatoria* or UNIDROIT principles.[56]

20.36 In the extractive sector, as most *professio juris* clauses recognise the application of the law of the state on whose territory the exploitation operations take place, the governance of a contract by national law is highly

[52] According to a survey conducted among the random 112 practising lawyers selected from organisations active in the oil sector and entities responsible for the settlement of oil disputes, the choice of a neutral law is a preference in 42% of cases (Erkan *International Energy Investment Law Stability through Contractual Clauses* at 230).

[53] This solution is preferred in approximately 34% of lawyer cases (Erkan *International Energy Investment Law Stability through Contractual Clauses* at 230).

[54] This solution is chosen in approximately 13% of cases by lawyers (ERKAN *International Energy Investment Law Stability through Contractual Clauses* at 230).

[55] Prosper Weil 'Principes généraux du droit et contrats d'État' in Berthold Goldman (ed) *Études offertes* (Litec 1982) 407.

[56] However, this choice still attracts a minority—only 5% of lawyers decide to adhere to it (Mustafa Erkan, *International Energy Investment Law Stability through Contractual Clauses* 230).

imbued with territorialism. It is often the law itself that imposes the application of the local law to an exploitation agreement,[57] although other mining and petroleum laws refer to the contractual terms to determine the applicable law.[58] At times, states impose their national law as applicable to exploitation operations via their model petroleum and mining agreements.[59]

20.37 The governing law may have a major impact on the interpretation of contractual clauses, the settlement of disputes and even on the very validity of the contract. For example, the choice of certain laws can be detrimental to effective arbitration proceedings as certain national laws may ban arbitration relative to State contracts.[60] Similarly, the choice of a law that applies the principles of Islamic law could contradict certain contractual provisions, in particular related to the payment of interest in respect of overdue payments.[61]

20.38 The parties must therefore ensure that they choose a law that is appropriate to govern their relations in a specific field in which the governing law is necessarily highly technical in nature. The law of the host state may at times be unable to govern legal relations between the parties in light of the extensive technicity of mining and petroleum operations. This was noted by

---

[57] With regard to mining and petroleum laws providing for application of the local law to the exploitation agreement, the following should be mentioned: in Angola, Article 94 of Petroleum Law No. 10/04 of 12 November 2004; in the Central African Republic, Article 82 of Ordinance No. 93.007 of 25 May 1993 Enacting the Petroleum Code; in Djibouti, Article 66 of Law No. 117-AN-05-5 of 16 October 2005 Enacting the Petroleum Code; in Eritrea, Article 26 of Proclamation No. 108/2000, A Revised Proclamation Governing Petroleum Operations; in Ethiopia, Article 26 of Proclamation No. 295/1986 Regulating Petroleum Operations; in Senegal, Article 67 of Law No. 98-05 of 8 January 1998 Enacting the Petroleum Code.

[58] With regard to mining and petroleum laws that refer to the terms of the contract to determine the law that applies to exploitation agreement, the following should be mentioned.

[59] The following should be mentioned in particular: in the Democratic Republic of Congo, Article 25 of the Production Sharing Agreement between the Democratic Republic of Congo and the Association Dominion Petroleum Congo & Soco Exploration-Production RDC & La Congolaise des Hydrocarbures in Bloc V in the Graben Albertine (November 2007); in Equatorial Guinea, Article 25(1) of the Model Production Sharing Contract; in the Seychelles, Article 40(1) of the Model Petroleum Agreement of 2013.

[60] See, to this effect, Erkan *International Energy Investment Law Stability through Contractual Clauses* at 229.

[61] For example, this is the case of Mauritanian law, providing for the dominance of the principle of equivalence of services, which bans the application of interest (*Riba*) (Article 276 of the Mauritanian Code of Obligations and Contracts). The scope of this ban is illustrated by an analysis of Article 1179 of the same code, which provides that the deficiencies in Mauritanian law must be remedied by referring to the *Maliki* rite (Article 1179 of the Mauritanian Code of Obligations and Contracts: '*[p]our combler les lacunes de cette ordonnance, il est fait référence au rite malékite*'. [Free translation into English: '[t]o remedy the deficiencies of this ordinance, reference is made to the Miliki rite.']. Islamic law, as interpreted by Imam Malek, refers to the Sunni Islam denomination's traditions. This denomination bans the increase granted by a debtor to a creditor in return for a postponement of the payment deadlines: the *Riba Al-Faddish*, which corresponds to the default interest concept. See Walid Ben hamida 'L'incidence des intérêts moratoires sur l'exécution des sentences arbitrales dans les pays arabes' (2012) *Rev arb* 548 and 575.

the arbitral tribunal in the *Petroleum Development v Sheikh of Abu Dhabi* case, which noted that the law of Abu Dhabi did not allow appropriate legal solutions to be found in the framework of the dispute in question.[62] In Africa, this issue tends to disappear with the constant adoption of more and more complete regulations, often at the avant-garde of mining and petroleum law.

20.39 The application of a national law to an international contract is not free of difficulties. The application of general principles of law may also be chosen directly or implicitly in order to supplement the provisions of the national law.

iii *Clauses that designate multiple laws*

20.40 There is no doubt that operators may choose multiple legal systems in the framework of a single contract. There is a broad diversity of clauses of this type, which can be classified in numerous categories.

20.41 First, the choice of law clause could provide for the application of rules that are common to two or more domestic legal systems and, as an ancillary source, resort to general principles of law.

Other clauses may provide for the application of the host state's law, supplemented by potential recourse to general principles of law. For example, Article 110 of Guinea's new Petroleum Code provides:

> '[a]ll Petroleum Agreements are governed and interpreted in accordance with Guinea's law and the general principles of international law that apply to these matters. Unless an exemption is provided in the present Code, the Contracting Parties are subject to all of the common law legal and regulatory provisions in effect in the Republic of Guinea.'[63]

An interesting alternative may provide for the initial application of general principles of law and, as an ancillary source, the national law of the host state. It should be noted that the application of general principles of law to govern international contracts guarantees predictability.

These solutions are summarised in the Resolution of the Institute of International Law's on the Proper Law of the Contract in Agreements between a State and a Foreign Private Person, which provides that '[t]he parties may in particular choose as the proper law of the contract either one or several domestic legal systems or the principles common to such systems, or the general principles of law, or the principles applied in international economic relations, or international law, or a combination of these sources of law.'[64]

20.42 Lastly, reference could be made to the general principles of petroleum or mining laws. This is of interest, as it tends to attest to the

---

[62] *Petroleum Development Limited v Sheikh of Abu Dhabi*, Award of 28 August 1951, (1951) 18 *ILR* 144.

[63] Article 110 of Law No. L/2014/No.° 034/AN establishing Enacting the Petroleum Code.

[64] Article 2 of the Institute of International Law's Resolution on Proper Law of the Contract in Agreements Between a State and a Foreign Private Person (Athens—1979).

existence of these general principles, and therefore to a '*lex petrolea*' and a '*lex mineraria*'.[65]

In addition to clauses determining the applicable law, arbitration clauses should also be taken into consideration in mining and petroleum laws as well as in exploitation agreements.

b Arbitration clauses

20.43 Although arbitration is not uncontroversial,[66] today, this dispute resolution mechanism is a constant in contracts related to the exploitation of natural resources.

The inclusion of arbitration clauses in these contracts results from the many advantages attributed to this dispute resolution mechanism.[67] Some of these advantages are particularly important in the mining and petroleum sectors (technical matters that often require the assistance of experts to understand and even settle the disputes).[68] Furthermore, projects related to the exploitation of natural resources take time and arbitration, unlike proceedings before national courts, is better able to preserve contractual relations and thereby allows the parties to more easily ensure that the contract continues to be performed when the dispute arises.[69] In this regard, the survival of contractual relations is taken into consideration in mining and petroleum contracts, as they require the parties to continue to fulfil their

[65] With regard to the emergence of a *lex petrolea*, see de Jesus 'The prodigious story of the lex petroleum and the rhinoceros. Philosophical aspects of the transnational legal order of the petroleum society' loc cit.

[66] The controversies were initiated in particular by states who, on a number of occasions, were critical of the position taken by the arbitrators, considered as beeing more favourable to the interests of investors. This trend resulted, in particular, in the denunciation by many South American countries of the Washington Convention. Bolivia was the first state to initiate this trend by withdrawing further to its notification to the World Bank in May 2007. This withdrawal was followed by Ecuador in July 2009 and more recently by Venezuela, who officially announced its withdrawal on 24 January 2012, definitively convinced that international arbitrators were partial to foreign investors after the International Chamber of Commerce ordered the national company (Petroléos de Venezuela) to pay $907 million in damages to Exxon for the prejudice caused by the 2007 nationalisations.

[67] Although some of these advantages are sometimes debatable, it is worth noting, in particular, among the advantages of arbitration as compared to local proceedings before State courts: the efficiently, the costs (although arbitration is increasingly seen as 'elitist' and costly), the flexibility, the confidentiality (although this is not a general rule, in particular in the field of investment arbitration, which, as a principle, is not confidential), the right to choose the arbitrator, the neutrality, the facilities granted in the execution of the award, etc. In particular, see, to that effect, Jérôme Ortscheidt & Christophe Seraglini *Droit de l'arbitrage interne et international* (Montchrestien 2013) 52; Richard Garnett, Henry Gabriel, Jeff Waincymer & Judd Epstein *A Practical Guide to International Commercial Arbitration* (Oceana Publications 2000) 11 et seq.

[68] In particular, see to that effect, Lazare Kopelmanas 'Le rôle de l'expertise dans l'arbitrage commercial international' 1979) *Rev arb* 205; Jean Robert 'Contrats et arbitrage international dans le domaine des matières premières' in *Les marchés internationaux des matières premières* (Economica 1982) 73.

[69] See, in particular, Ucheora Onwuamaegbu 'Resolution of oil and gas disputes at ICSID' (2004) 21 (1) *News From ICSID* 14.

contractual obligations during the arbitration proceedings[70] in order to avoid damage or greater damage to the project. However, in a field as strategic as the exploitation of natural resources, the main advantage of arbitration is undoubtedly its neutrality. This method to settle disputes allows a private operator, depending on the subject of the dispute, to bypass the host state's court system.

20.44 Arbitration clauses therefore constitute a fundamental provision in exploitation agreements. This type of clause first appeared in the petroleum field, more precisely in concession agreements, in the early part of the 20th century. These clauses—which were not very detailed at first—have seen their content grow to include more complete, and at times highly sophisticated, wording.

20.45 However, the parties to an agreement too often draft arbitration clauses at the last minute, despite the importance of such clauses. In general, these 'midnight clauses' are not granted sufficient consideration by the parties to a contract who—often after very lengthy negotiations—do not want to dwell on the dispute settlement issue.[71] Furthermore, the positive relations that may exist between new contractual partners, who target fruitful, long-standing business relations, result in the settlement of potential disputes being left to the side when the 'active' contractual clauses are drafted. That said, specific attention should be paid to arbitration clauses as careful wording thereof can avoid many difficulties if a dispute arises, in particular in order to limit damage or avoid the suspension of exploitation operations.[72]

20.46 The practice, which tends to almost systematically include arbitration clauses in contracts relative to the exploitation of natural resources, finds significant support in arbitral case law, which, historically, has ensured that they are fully effective despite the different issues their inclusion raises. In the past, awards on questions regarding the validity of arbitration clauses defined a series of principles to protect the initial choice of the parties to bring any dispute related to the interpretation or enforcement of a contract before an arbitral tribunal. For this reason, principles such as the validity and autonomy of arbitration clauses have been enshrined by arbitral awards, seeking to ensure that arbitration clauses are fully effective, in particular in the framework of mining and petroleum disputes. The legality of clauses relative to dispute settlement had been challenged by states that feared they would lose their sovereignty if these clauses were implemented. A solution to this

[70] To that effect, it is worth mentioning, in the Seychelles, Article 24(5) of the Model Petroleum Agreement of 2013.

[71] Dana H Freyer 'Practical Considerations in Drafting Dispute Resolution Provisions in International Commercial Contracts—A US Perspective' (1998) 15 (4) *Journal of International Arbitration* 7.

[72] Therefore, the following should be taken into consideration when the arbitration clause is drafted: the scope of the arbitration clause, the choice between *ad hoc* arbitration or institutional arbitration, the language of the arbitration, the choice of law governing the dispute, the confidentiality of the proceedings, the issue of interim measures, the allocation of the arbitration costs and the rules applicable to the production of evidence.

controversy was provided (in the field of petroleum contracts) in the *LIAMCO* award,[73] in which the arbitral tribunal found that the undertaking made by the State to waive some of its sovereignty by signing an arbitration agreement was valid. A state that is a party to the contract can therefore validly agree to be brought before arbitral tribunals without the legality of its waiver being called into question.

In addition to the argument based on the unlawfulness of arbitration clauses, states called upon to defend themselves before arbitral tribunals were able to challenge the jurisdiction of the arbitral tribunal in light of the absence of an arbitration clause resulting from the rescission of the contract that contained the said clause. 'What would be the use of providing that any difficulty arising from the enforcement of the contract will be subject to arbitration if, as soon as the first difficulty appears, the Contracting State can bypass the arbitration clause and hide behind its contractual prerogatives?'[74] This principle, extensively accepted in arbitration law, was entrenched in the petroleum sector in the case of *TOPCO v Libya.*[75] The issue was whether the rescission of the contract would result in the disappearance of the dispute settlement clause contained therein. The answer to this question was 'no', as the tribunal found that the arbitration agreement was valid despite the rescission of the concession agreement.[76] This arbitral case law—intending to protect the effectiveness of arbitration clauses—has contributed to the development of arbitration proceedings, in particular in the petroleum and mining sectors.

20.47 The inclusion of arbitration clauses in contracts related to the exploitation of natural resources has resulted in African laws considering this phenomenon and most of these countries now provide for arbitration as a dispute resolution mechanism. However, it should be noted that many of these laws require the parties to first seek an amicable settlement of the dispute as a prerequisite to seizing the arbitral tribunal. Furthermore, it is interesting to note that some of these laws do not simply set out a principle on arbitration but also provide terms to structure the conduct of the arbitration proceedings.

i *Arbitration as a preferred means to settle disputes in the mining and petroleum sectors*

20.48 The quasi-systematic inclusion of arbitration clauses in contracts related to the exploitation of natural resources has led lawmakers to insert this alternative method to settle disputes into mining and petroleum laws. Arbitration has thus become the preferred dispute settlement method in the

[73] *Libyan American Oil Company (LIAMCO) v The Government of the Libyan Arab Republic* (Award as 19 January 1977) 1977 *JDI* 350.

[74] See Prosper Weil 'Les clauses de stabilisation ou d'intangibilité insérées dans les accords de développement économique' at 301.

[75] *Texaco Overseas Petroleum Co. and California Asiatic Oil Co v The Government of the Libyan Arab Republic* (Award of 19 January 1977) 1977 *JDI* 350.

[76] See also the Award of the sole arbitrator, Dr S Mahmassani in the case of *LIAMCO v The Libyan Arab Republic (Libyan American Oil Company (LIAMCO) v The Government of the Libyan Arab Republic* (Award of 19 January 1977) 1977 *JDI* 350).

framework of mining and petroleum laws. In this regard, many laws provide for arbitration, whether directly or indirectly, by referring to the contractual terms. While certain texts explicitly provide for arbitration[77] , others are less explicit and simply require the inclusion of an arbitration clause in the exploitation agreement, investment agreement or establishment agreement to grant jurisdiction to arbitral tribunals[78] . Arbitration is then subject to the consent of the parties to the contract, which, at the end of the day, is one of the basic principles of arbitration.

20.49 However, resorting to arbitration in the framework of exploitation agreements for natural resources may, in the absence of arbitration clauses, result from the host state's dissociated consent to arbitration ('arbitration without privity') granted in the framework of an international treaty or law. Arbitration without privity therefore results from a lapse in time between the host state's consent to arbitration, which results from the *erga omnes* offer contained in the law or a treaty, and the acceptance of this offer by the foreign investor, usually when the arbitration proceedings are initiated[79] .

[77] With regard to laws that provide for arbitration, the following texts could be mentioned: in Algeria, Article 58 of Law No. 13-01 of 20 February 2013 Amending and Supplementing Law No. 05-07 of 28 April 2005 On Hydrocarbons; in Angola, Article 89 of Petroleum Law No. 10/04 of 12 November 2004; in Benin, Article 134 of Law No. 2006-17 of 17 October 2006 Enacting the Mining and Mining Taxation Code; in Botswana, Article 2(2) of the Mines and Mineral Act; in Cameroon, Article 115 of Law No. 99-013 of 22 December 1999 Enacting the Petroleum Code; in the Democratic Republic of Congo, Articles 317 et seq. of Law No. 007-2002 of 11 July 2002 Enacting the Mining Code; in Djibouti, Article 65 of Law No. 117-AN-05-5 of 16 October 2005 Enacting the Petroleum Code; in Eritrea, Article 48(5) of Proclamation No. 68/1995 Promotingthe Development of Mineral Resources and Article 25 of Proclamation No. 108/2000, A Revised Proclamation Governing Petroleum Operationspe-troleum operations; in Ethiopia, Article 76 of Proclamation No. 678/2010 Promoting Sustainable Development of Mineral Resources and Article 25 of Proclamation No. 295/1986 To Regulate Petroleum Operations; in Ghana, Article 27 of the Minerals and Mining Act, 2006; in Guinea, Article 219 of Law No. 2011/006/CNT of 9 September 2011 Enacting the Mining Code and Article 112 of Law No. 2014/034/AN Enacting Petroleum Code; in Libya, Article 20 of Petroleum Law No. 25/55 of 11 May 2012; in Mauritania, Article 49 of Law No. 2010-033 of 20 July 2010 Enacting the Hydrocarbons Code amended by Law No. 2011-044 of 25 November 2011; in Nigeria, Article 11 of the Petroleum Act; in Sao Tome and Principe, Article 74 of the Framework Law On Petroleum Operations; in Senegal, Article 66 of Law No. 98-05 of 8 January 1998 Enacting the Petroleum Code; in Chad, Article 104 of Law No. 006-PR-2007 of 2 May 2007 On Hydrocarbons.

[78] In this regard, the following should be mentioned: in the Central African Republic, Article 52 of Law No. 09-005 of 29 April 2009 Enacting the Mining Code and Article 84 of Ordinance No. 93.007 of 25 May 1993 Enacting the Petroleum Code; in the Republic of Congo, Article 99 of Law No. 4-2005 of 11 April 2005 Enacting the Mining Code; in Côte d'Ivoire, Article 85 of Law No. 96-669 of 29 August 1996 Enacting the Petroleum Code; in Morocco, Article 33 of Law No. 21-90 of 1 April 1992 On the Exploration and the Exploitation of Hydrocarbon Deposits amended and supplemented by Law No. 27-99 of 15 February 2000; in Niger, Article 158 of Law No. 2007-01 of 31 January 2007 Enacting the Petroleum Code.

[79] See, in particular, Sébastien Manciaux *Investissements étrangers et arbitrage entre États et ressortissants d'autres États* op cit 193 et seq; Michael Nolan & Frédéric G Caivano 'Limits of Consent—Arbitration without Privity and Beyond' in *Liber Amicorum Bernardo Cremades* (La Ley 2010) 873.

This mechanism will therefore allow private operators—when no arbitration clause exists—to refer to the mining and petroleum laws, investment laws (when they apply to mining and petroleum operations)[80] and international treaties signed by the host state.

20.50 Often, arbitration is intended to co-exist with other methods to settle disputes, whether a technical expertise[81] or the jurisdiction of state courts.[82]

20.51 With regard to the technical expertise, the existence of alternative dispute settlement methods should be mentioned, such as 'dispute boards' which, although extensively used in the construction sector, are rare in the field of natural resource exploitation.[83]

20.52 In limited circumstances, petroleum and mining laws set out restrictions on the jurisdiction of arbitral tribunals; some laws include the condition that at least one of the parties must be a foreign entity,[84] while other texts state that arbitration may only take place if law allows it.[85] Finally, certain provisions prohibit arbitration for national companies operating in the mining or petroleum sectors.[86]

---

[80] Regarding the application of investment laws to mining and petroleum operations, see above para 6.3 et seq.

[81] Due to the technicity of certain petroleum and mining disputes, technical appraisals are extensively used as a dispute settlement method in these sectors. Many African mining and petroleum laws directly or indirectly provide for a technical appraisal to settle disputes by referring to the contract. Among these laws, the following should be mentioned: in Algeria, Article 26(2) of the Presidential Decree of 1 March 2003 Enacting the Model Mining Agreement; in Cameroon, Article 119 (1) of Decree No. 2000-485 of 30 June 2000 Enacting the Implementation Rules for Law No. 99-013 of 22 December 1999 Enacting the Petroleum Code; in Mauritania, Article 49 of Law No. 2010-033 of 20 July 2010 Enacting the Crude Hydrocarbons Code, amended by Law No. 2011-044 of 25 November 2011; in Togo, Article 60 of Law No. 96-004 of 26 February 1996 Enacting the Mining Code.

[82] Among the African mining and petroleum laws that provide both for arbitration and the jurisdiction of national courts, the following should be mentioned: in Burundi, Article 169 of Law No. 1-21 of 15 October 2013 Enacting the Mining Code; in the Democratic Republic of Congo, Article 315 of Law No. 007-2002 of 11 July 2002 Enacting the Mining Code.

[83] On this point, see Florent Lager 'Les Dispute Board dans le secteur extractif' in *Le contentieux extractif* (ICC 2015) 213.

[84] See, in particular, in Morocco, Article 33 of Law No. 21-90 of 1 April 1992 On Exploration and the Exploitation of Hydrocarbon Deposits amended and supplemented by Law No. 27-99 of 15 February 2000: '[p]etroleum agreements may provide for arbitration when one of the parties is a foreign entity'.

[85] In this regard, Article 65 of Law No. 117-AN-05-5 of 16 October 2005 Enacting the Petroleum Code in Djibouti provides: '*[t]outes les infractions aux lois et règlements applicables relèvent des cours et tribunaux de la République de Djibouti. Toutefois, les différends qui pourraient naître de l'interprétation ou de l'application d'une convention ou d'un contrat de services sont, le cas échéant, soumis à une procédure d'arbitrage, lorsque celle-ci est autorisée par la loi.*' [Free translation into English: '[a]ll violations of the applicable laws and regulations are subject to the courts and tribunals of the Republic of Djibouti. However, disputes that may arise from the interpretation or application of a contract or service agreement are, where appropriate, subject to arbitration proceedings when the said proceedings are authorised by law.']

[86] This is the case in Algeria where Law No. 13-01 of 20 February 2013 Amending and Supplementing Law No. 05-07 of 28 April 2005 On Hydrocarbons provides in Article 58:

20.53 In addition, not all laws accept arbitration as an alternative dispute settlement method. Certain texts only provide for the jurisdiction of national courts,[87] leaving the private operator at times exposed to the lack of neutrality of local courts in a field where considerations related to the strategic importance of natural resources may weigh heavily on the national court's decision. For their part, other legal systems establish the principle of the national courts' jurisdiction, subject to the existence of an arbitration clause that grants jurisdiction to arbitral tribunals.[88]

20.54 Many African mining and petroleum laws provide for arbitration directly or indirectly by referring to the terms of the contract. However, arbitration is often subject to compliance with a prior conciliation process.

ii *The preliminary conciliation phase*

20.55 Recourse to arbitration provided in mining and petroleum laws is often preceded by an amicable settlement phase during which the parties are bound or are entitled to attempt to amicably settle their dispute.[89] There are many advantages to these amicable settlement clauses, also known as 'escalation clauses' or 'tiered dispute resolution clauses'. The search for a compromise avoids a decision being taken that is at times perceived as an injustice by one of the parties. The amicable search for a compromise also favours continued contractual relations between the parties.[90] The use of mediation or conciliation is increasing and attracts more and more parties to exploitation agreements, in particular for minor issues[91] for which arbitration is very costly not only in terms of time spent (due to a possible postponement in the performance of the project) but also in terms of arbitration costs. It is therefore to all of the parties' advantage to agree to a preliminary conciliation attempt[92] .

'*[l]e recours à l'arbitrage international susvisé concerne exclusivement les personnes constituant le contractant, autres que l'entreprise nationale Sonatrach Spa.*' [Free translation into English: '[r]ecourse to the international arbitration referred to above exclusively relates to the persons of the contractor other than the national company Sonatrach-Spa.']

[87] With regard to laws that provide for the exclusive jurisdiction of national courts, the following should be mentioned: in Chad, Article 98 of Law No. 011-PR-1995 of 20 June 1995 Enacting the Mining Code.

[88] See, in particular, in Ivory Coast, Article 85 of Law No 96-669 of 29 August 1996 Enacting the Petroleum Code; in the Central African Republic, Article 84 of Ordinance No. 93.007 of 25 May 1993 Enacting the Petroleum Code; in Mozambique, Article 69 of Petroleum Law No. 21-2014 of 18 August 2014; in Niger, Article 158 of Law No. 2007-01 of 31 January 2007 Enacting the Petroleum Code; in Togo, Article 60 of Law No. 96-004 of 26 February 1996 Enacting the Mining Code.

[89] With regard to the mandatory or optional nature of the preliminary conciliation phase, see below para 20.59.

[90] See, in particular, to this effect, Ortscheidt & Seraglini op cit 9 and 10.

[91] However, arbitration remains the only method to settle disputes in the extractive sector for major issues, obviously due to their complexity.

[92] Furthermore, an amicable settlement of disputes allows parties to confidentially reach an amicable solution. It is unanimously accepted that the amicable settlement phase must be confidential. The confidentiality of the amicable procedure allows the parties to investigate

20.56 This phase, during which the parties will attempt to settle their dispute, covers a number of concepts such as conciliation[93] and mediation[94] . Although these terms refer to two distinct mechanisms, the underlying idea is the same: to find a negotiated settlement for a dispute, if necessary with the help of a third party[95] .

20.57 With regard to the interests referred to above, many laws have adopted provisions to implement a phase that allows an amicable settlement to be sought.[96] Certain mining and petroleum laws have established conciliation systems and indicate that, in the event the conciliation fails, the most diligent petitioner may initiate arbitration proceedings. This general trend is highly apparent in the petroleum and mining sectors.[97] Certain laws, for their part, refer to the contractual terms and provide the parties to a petroleum or mining agreement with an opportunity to include provisions related to the implementation of a preliminary conciliation phase.[98]

20.58 Finally, in certain cases, the mining and petroleum laws are silent in terms of an obligation to try to reach an amicable settlement before initiating proceedings, in which case one must refer to the model contracts or

---

all potential compromises without their statements being used in future litigation (Xavier Lafarge 'L'efficacité des clauses de conciliation ou de médiation' 2000 *Rev arb* 377).

[93] Conciliation, unlike mediation, is a procedure that involves a third party where the third party offers recommendations that the parties are free to accept or not accept. While the mediator facilitates communication to allow the parties to find a solution to a dispute themselves, a conciliator takes a position on the dispute and offers a solution (Isabelle Vaugon 'Expertise et médiation dans le contexte du contentieux extractif' in *Le contentieux extractif* (ICC 2015) 202.

[94] A method to settle disputes that consists—for the person selected by the stakeholders (often in light of his personal authority)—in proposing a potential solution, without limiting himself to attempting to bring them together but without having the authority to impose a jurisdictional decision, unlike arbitration and state courts (Gérard Cornu *Vocabulaire juridique* (Universitaires de France 1987)). Unlike conciliation, mediation is a negotiation method assisted by a neutral, independent and impartial third party who helps the parties reach an agreement when they have failed to do so by direct negotiation (Vaugon, 'Expertise et médiation dans le contexte du contentieux extractif' at 200 and 201).

[95] Pascal Ancel 'L'efficacité procédurale des clauses de conciliation ou de médiation' 2003 *D* 1386. Also see, to this effect, Xavier Lafarge 'L'efficacité des clauses de conciliation ou de médiation' at 377.

[96] See, to that effect, Ortscheidt & Seraglini *Droit de l'arbitrage interne et international* at 11.

[97] The following provisions entrench the principle relative to the amicable settlement of disputes between a state and the operator: in Algeria, Article 58 of Law No. 13-01 of 20 February 2013 Amending and Supplementing Law No. 05-07 of 28 April 2005 On Hydrocarbons; in Benin, Article 134 of Law No. 2006-17 of 17 October 2006 Enacting the Mining and Mineral Taxation Code of the Republic of Benin; in Ghana, Article 27 of the Minerals and Mining Act, 2006; in Guinea, Article 219 of Law No. 2011/006/CNT of 9 September 2011 Enacting the Mining Code; in Sierra Leone, Article 113 of the Petroleum Exploration and Production Act, 2011; in Chad, Article 104(1) of Law No. 006-PR-2007 of 2 May 2007 On Hydrocarbons.

[98] This is the case in the Central African Republic of Article 52 of Law No. 09-005 as 29 April 2009 Enacting the Mining Code; in Cameroon, Article 115(2) of Law No. 99-013 of 22 December 1999 Enacting the Petroleum Code; in Côte d'Ivoire, Article 85 of Law No. 96-669 of 29 August 1996 Enacting the Petroleum Code.

agreements enacted by the State, which may provide for a preliminary conciliation stage.[99]

20.59 It is important to emphasise that the scope of the amicable settlement clause may vary: it may be either mandatory or optional. Although the mandatory nature of certain clauses does not raise any difficulties (for instance an amicable settlement clause that provide terms on the appointment of mediators or conciliators), other provisions may raise some difficulties when determining whether the amicable proceedings are mandatory or optional.[100] Among these clauses, Article 69 of Mozambique Petroleum Code is worth mentioning in that it provides that any dispute related to the contracts or concession agreements must be settled 'preferably' by negotiation.[101] The use of the term 'preferably' seems to indicate that the attempt to amicably settle a dispute is only optional. Therefore, when a doubt persists as to the mandatory nature of the amicable settlement clauses, the parties to an exploitation agreement will have every interest in offsetting the deficiencies of the applicable texts by indicating, in particular, the mandatory or optional nature of the mediation as well as the procedure that must be implemented to reach an agreement.

20.60 When a preliminary conciliation stage exists and is mandatory, the parties are bound by an obligation of result when they implement the clause. The parties are therefore under an obligation to initiate the conciliation process before seizing the arbitrator or the competent court. Once the amicable procedure has been implemented, the parties are only bound by an obligation of means.[102] Therefore, at the conciliation stage, the only burden on the parties is a good faith or contractual loyalty obligation, which presupposes that the parties make 'the efforts required to fairly reach a negotiated settlement.'[103]

20.61 In order to ensure that the implementation of the clause providing for an amicable conciliation stage is as effective as possible, it is important that the relevant laws (or the petroleum or mining agreements) include provisions that govern the conduct of the amicable settlement procedure. These provisions often prevent obstacles that may affect the progress of the conciliation stage. For instance, mining and petroleum may indicate a time limit after which the conciliation procedure will be considered to have failed, thereby allowing the parties to seize the arbitral tribunal. In practice, we can find examples where mediation period last between 30 days and three

---

[99] This is the case in Algeria, Article 26 of the Presidential Decree of 1 March 2003 Enacting the Model Mining Agreement; in Seychelles, Article 39 of the Model Petroleum Agreement of 2013.

[100] See, in particular Ancel 'L'efficacité procédurale des clauses de conciliation ou de médiation' at 1386.

[101] '*As disputas emergentes dos contractos e contractos de concessão devem ser solucionados, de preferencia, por negociação*' [Free translation into English: '[a]ny disputes arising from the contracts and concession agreements shall be resolved preferably by negotiation'].

[102] See Pascal Ancel 'L'efficacité procédurale des clauses de conciliation ou de médiation'at 1388.

[103] Id.

months.[104] Besides these indications on the duration of the conciliation stage, it is rare to find legal provisions that provide details on the procedure that applies to the conciliation stage. In Africa, Madagascar's Petroleum Code is the only law that provides many details on the means to be used by the parties to appoint conciliators and, if applicable, the appointment thereof by a third party if a problem arises.[105] Finally, it should be noted that no law provides for institutional mediation or conciliation, and that all of the legal texts refer to an *ad hoc* mediation or conciliation implemented by the parties.

20.62 Regardless of the conciliation or mediation procedure (selected by the parties in an exploitation agreement or provided by the texts that govern the mining and petroleum sectors), one question remains: how is the breach of a mandatory amicable settlement clause punished? To answer this question, one must refer to the domestic laws and, in particular, the terms of the procedural laws in effect in the host state. In any event, different penalties may be contemplated. They may be of two types: contractual or procedural. The refusal by one of the parties to participate in a conciliation or mediation proceeding may be considered a breach of contract. This breach may lead to two contractual penalties: first, the forced enforcement of the clause -although this penalty is difficult to imagine[106] as the parties cannot be compelled to meet to find an amicable settlement- and, on the other, the payment of compensation to the party that sought conciliation or mediation and that could not avoid resorting directly to arbitration.[107] In addition to these contractual penalties, a breach of a mandatory mediation or conciliation clause may also lead to procedural penalties, such as the inadmissibility of the notice of arbitration brought before an arbitral tribunal in breach of the mediation clause.[108] In this case, an arbitral tribunal will find itself having jurisdiction. This solution faces certain difficulties as it ends the mandate of the arbitral tribunal. Therefore, ending an arbitration proceeding would result

[104] In general, the laws provide for a 30-day period: in Ghana, Article 27(3) of the Minerals and Mining Act, 2006; in Madagascar, Article 67 of Law 96-018 of 4 September 1996 Enacting the Petroleum Code.

[105] Articles 62 et seq of Law 96-018 of 4 September 1996 Enacting Madagascar's Petroleum Code.

[106] See, in particular, Pascal Ancel 'L'efficacité procédurale des clauses de conciliation ou de médiation'at 1386.

[107] In the framework of ICC arbitration, see the Partial Award of 1985 in the ICC Case (1985) mentionned by Isabelle Vaugon 'Expertise et médiation dans le contexte du contentieux extractif' at 203. According to this Award, the Arbitration Tribunal held that: '[w]hen the parties have agreed that mediation constituted a mandatory prerequisite for any litigation, non-compliance with this agreement causes the breaching party to be held civilly liable.' The solution that holds that the breach of the mandatory amicable settlement clause, as a substantive clause, must be sanctioned in the contractual arena has been adopted, in particular, by the British courts. See, in particular, Christopher Boog 'How to Deal with Multi-tiered Dispute Resolution Clauses—Note—6 June 2007—Swiss Federal Supreme Court' 2008 *ASA Bulletin* 106.

[108] This solution was adopted by the French Court of Cassation in a decision of the *Chambre Mixte* of 14 February 2003 (for a commentary, see, in particular, Pascal Ancel 'L'efficacité procédurale des clauses de conciliation ou de médiation'at 1386.

in a waste of time and increase in procedural costs. Another solution may be available: to suspend the arbitration proceedings initiated in breach of the mandatory mediation or conciliation clause so that the parties can remedy this omission, in which case the sanction would consist of a stay of proceedings.[109]

20.63 African mining and petroleum laws therefore often simply provide the principle of a preliminary conciliation stage. It is up to the parties to ensure that the preliminary conciliation or mediation clauses are effective, in particular by inserting provisions in the mining and petroleum agreements that describe the conduct of these proceedings. In contractual practice, these provisions are very rare and, when difficulties arise, defer issues that have become sensitive and complex; this results in a delay in proceedings and may even prevent the search for an amicable settlement.

iii *Arbitration proceedings*

20.64 Although certain mining and petroleum laws simply refer to the terms of the contract to determine the rules related to the conduct of an arbitration proceeding[110], other legal texts provide details on certain essential points of the process, thereby limiting the parties' freedom in the conduct of the arbitration proceedings but also avoiding a delay in proceedings if a dispute arises. Certain texts that govern mining and petroleum operations include, in particular, provisions related to the distinction between domestic and international arbitration, the choice of *ad hoc* or institutional arbitration, the law that applies to the merits of the dispute, the determination of the seat of arbitration, the constitution of the tribunal, the language of the proceedings and the enforcement of the awards.

The laws in force also include provisions on precautionary measures and arbitration costs.[111]

[109] Part of the doctrine favours the solution which consists in sanctioning a breach of the mandatory mediation clause by a stay of proceedings, ie a procedural exception and not a demurrer (the French 'fin de non recevoir') (on this point, see, in particular, Xavier Lagarde 'L'efficacité des clauses de conciliation ou de mediation' 2000 *Rev arb* 337, esp at 390 et seq).

[110] In this regard, the following provisions must be mentioned: in Eritrea, Article 25(2) of Proclamation No. 108/2000, A revised Proclamation to Govern Petroleum Operations; in Mauritania, Article 49 of Law No. 2010-033 of 20 July 2010 Enacting the Hydrocarbons Code amended by Law No. 2011-044 of 25 November 2011; in Senegal, Article 66 of Law No. 98-05 of 8 January 1998 Enacting the Petroleum Code; in Sierra Leone, Article 113 of the Petroleum Exploration and Production Act, 2011; in Tunisia, Article 21 of Law No. 99-93 of 17 August 1999 Enacting the Hydrocarbons Code.

[111] The costs incurred by the parties during the arbitration proceedings, generally include the administrative costs of arbitral institutions, the fees and expenses of the arbitrators, the fees and expenses of experts appointed by the arbitral tribunal, the costs related to the hearings, etc. (see Ortscheidt & Seraglini op cit 772). In order to avoid any uncertainty regarding the distribution of these costs, certain model agreements include terms related to the breakdown of arbitration expenses, in particular: Article 39(2) of the Seychelles Model Petroleum Agreement of 2013: '[e]ach Party to the dispute shall bear the cost of its own arbitrator and its representation in the arbitral proceedings. The cost of the Chairperson in

(1) The difference between domestic and international arbitration

20.65 Some mining and petroleum laws differentiate between domestic and international arbitration. Most often, this distinction is based on legal or economic criteria, ie the existence of extraneous elements or the interests of international trade. The consequence of this duality is that domestic and international arbitration are subject to different legal regimes. Most often, petroleum and mining laws only refer to international arbitration[112] . However, certain texts mention the existence of two different types of arbitration: domestic arbitration and international arbitration[113] .

(2) The choice between ad hoc and institutional arbitration

20.66 In the mining and petroleum laws that set out details on the arbitration procedure, a difference should be noted between legal systems that provide for the application of institutional arbitration and those that provide for *ad hoc* arbitration.

20.67 Unlike institutional arbitration, the organisation of *ad hoc* arbitration is not entrusted to an arbitral institution.[114] In the framework of institutional arbitration, the designated arbitral institution will organise the arbitration under its aegis and ensure that it is conducted properly and efficiently. Institutional arbitration presupposes, in particular, that the conduct of the arbitration proceedings will be governed by procedural rules established by the institution. Although institutional arbitration is often criticised in that it leads to an institutionalisation of arbitration and increases procedural costs, it has the major advantage of ensuring the proceedings are conducted appropriately.[115] *A contrario*, *ad hoc* arbitration allows the parties to benefit from greater flexibility in the conduct of the arbitration, although,

discharging his or her duties in relation to the arbitration and the remaining costs of the arbitration shall be borne equally by the Parties to the dispute. The arbitral tribunal may, however, direct in its award that a higher proportion of costs shall be borne by one or other of the Parties to the dispute, and this award shall be binding on the Parties to the dispute.'

[112] With regard to texts that refer exclusively to international arbitration, the following should be mentioned: in Algeria, Article 58 of Law No. 13-01 of 20 February 2013 Amending and Supplementing Law No. 05-07 of 28 April 2005 On Hydrocarbons; in Benin, Article 134 of Law No. 2006-17 of 17 October 2006 Enacting the Mining and Mining Taxation Code; in Burundi, Article 169 of Law No. 1-21 as 15 October 2013 Enacting the Mining Code; in the Central African Republic, Article 84 of Ordinance No. 93.007 of 25 May 1993 Enacting the Petroleum Code; in Mauritania, Article 49 of Law No. 2010-033 of 20 July 2010 Enacting the Hydrocarbons Code amended by Law No. 2011-044 of 25 November 2011.

[113] With regard to texts that provide for the existence of two distinct arbitration systems: in the Central African Republic, Article 52 of Law No. 009-005 of 29 April 2009 Enacting the Mining Code; in the Democratic Republic of Congo, Articles 318 (related to domestic arbitration) and 319 (related to international arbitration) of Law No. 007-2002 of 11 July 2002 Enacting the Mining Code; in Guinea, Article 219 of Law No. 2011/006/CNT of 9 September 2011 Enacting the Mining Code.

[114] For a definition of *ad hoc* and institutional arbitration, see Ortscheidt & Seraglini op cit 23 et seq.

[115] For an overview of the advantages and disadvantages of institutional arbitration, see Ortscheidt & Seraglini op cit 24 et seq.

most often, procedural rules are applied, such as the Arbitration Rules of the United Nations Commission on International Trade Law.[116]

20.68 With regard to the laws that provide for institutional arbitration,[117] a number of arbitration forums are selected, in particular the International Centre for Settlement of Investment Disputes (ICSID) and the International Chamber of Commerce (ICC). Regarding the African continent, it is also worth mentioning arbitration under the aegis of the Common Court of Justice

[116] More commonly known as the UNCITRAL Arbitration Rules.

[117] With regard to laws that refer to institutional arbitration, the following should be mentioned in particular: in Botswana, Article 2(2) of the Mines and Mineral Act: '[w]here in this Act provision is made for a matter to be referred to arbitration, the matter shall be referred to a single arbitrator appointed by the parties and finally resolved under the rules of the International Centre for the Settlement of Investment Disputes in force from time to time or in accordance with such other rules or procedures as the parties may agree';

In the Democratic Republic of Congo, Article 319 of Law No. 007-2002 of 11 July 2002 Enacting the Mining Code:

> '*[n]onobstant les dispositions de l'article 318 du présent Code, les litiges pouvant survenir à l'occasion de l'interprétation ou de l'application des dispositions du présent Code, peuvent être réglés, à la requête de la partie la plus diligente, par voie d'arbitrage conformément à la Convention sur les Règlements des Différends Relatifs aux Investissements entre État et Ressortissants d'autres États, à la condition que le titulaire soit Ressortissant d'un autre État contractant aux termes de l'article 25 de ladite convention.*'
>
> [Free translation into English: '[n]otwithstanding the provisions of Article 318 of the present Code, disputes which might result from the interpretation or application of the provisions of the present Code may be settled, at the request of the party who proceeds first, by arbitration in accordance with the Convention on the Settlement of Disputes Relating to Investments between the State and Nationals of other States, provided that the holder is a "National" of another contracting state according to the terms of Article 25 of the said convention.'];

In Ghana, Article 27 of the Minerals and Mining Act, 2006:

> '[i]f the parties do not reach an agreement under paragraph (a) within thirty days, or a longer period agreed between the parties, of the matter being submitted to arbitration, in accordance with (i) firstly, the framework of a bilateral or multilateral agreement on investment protection to which the Republic and the country of which the holder is a national, are parties, or (ii) secondly, if no agreement contemplated by subparagraph (i) exists, the rules of procedure for arbitration of the United Nations Commission on International Trade Law, (UNCITRAL Rules)';

In the Seychelles, Article 39 of the Model Petroleum Agreement of 2013:

> 'where the Company is duly incorporated, constituted, set up, or otherwise duly organized under the law of a State Party to the 1965 Convention on the Settlement of Investment Disputes between States and Nationals of other States ('the Convention'), refer the dispute to the International Centre for Settlement of Investment Disputes ('ICSID') for arbitration pursuant to Article 36 of the Convention';

In Chad, Article 104(1) of Law No. 006-PR-2007 of 2 May 2007 Enacting the Petroleum Code:

> '*[l]es litiges opposant les parties au Contrat Pétrolier et relatifs à l'interprétation ou l'exécution dudit contrat qui n'ont pu être réglés à l'amiable seront soumis à l'arbitrage de la Chambre de Commerce International (C.C.I.) de Paris*'.
>
> [Free translation into English: 'Disputes between parties to the Petroleum Agreement and relative to the interpretation or enforcement of the agreement that have not been settled amicably will be subject to arbitration at the International Chamber of Commerce of Paris (I.C.C.)'].

and Arbitration (CCJA) governed by the CCJA Arbitration Rules.[118] In addition to its jurisdictional role, the CCJA thus ensures the administration of proceedings conducted under its aegis as an arbitration centre.

(3) The law applicable to the merits of the dispute

20.69 The choice of the law applicable to the merits is essential as it will be the one the arbitrators will apply to settle the dispute.[119] It is unanimously accepted that the law that applies to the merits of the dispute is not necessarily the law that applies to the proceedings or the *lex arbitri* (the law of the territory in which the arbitration takes place).[120]

In this regard, the principle is that of the freedom of the parties to choose the law that applies to the merits of the dispute.[121] However, certain provisions of mining and petroleum laws are such that they limit the parties' freedom by indicating the law that will govern the substance of the dispute.

20.70 Whether it is chosen by the parties or the arbitral tribunal or is imposed by law, the law that applies to the merits of the dispute can be a national law, the law of the host state or a neutral law that is unrelated to the parties, a series of laws[122] or transnational rules.[123] African mining and petroleum laws that include provisions on the law that applies to the merits of the dispute will provide for the application of the law of the host state.[124] However, certain legal provisions prefer the application of the host state's law, supplemented, where appropriate, by the principles and practices recognised in the relevant industry,[125] which illustrates the existence of an international practice in this area.

---

[118] The CCJA Arbitration Rules adopted on 11 March 1999 are currently reviewed and new arbitration rules should be enacted before the end of the year 2017. In addition, the duality that exists in the OHADA area in terms of arbitration should be emphasised. In addition to the *ad hoc* arbitration, which entails the application of the Uniform Act of Arbitration as 11 March 1999, the OHADA zone also hosts specific arbitration under the aegis of the CCJA.

[119] Garnett, Gabriel, Waincymer & Epstein *A Practical Guide to International Commercial Arbitration* at 40.

[120] Philippe Fouchard, Emmanuel Gaillard & Berthold Goldman *International Commercial Arbitration* (Kluwer Law International 1999) 633 et seq.

[121] See in particular, to that effect, Article 21 of the ICC's Rules of Arbitration, Article 15 of the OHADA Uniform Act on Arbitration as 11 March 1999, Article 17 of the Arbitration Rules of the Common Court of Justice and Arbitration as 11 March 1999, and Article 35 of the UNCITRAL Arbitration Rules of 2010. To that effect, see Ortscheidt & Seraglini op cit 799; Fouchard, Gaillard & Goldman op cit 785 et seq.

[122] Fouchard, Gaillard and Goldman *International Commercial Arbitration* at 794.

[123] Fouchard, Gaillard and Goldman op cit 801 et seq.

[124] See, in particular, in Algeria, Article 58 of Law No. 13-01 of 20 February 2013 Amending and Supplementing Law No. 05-07 of 28 April 2005 On Hydrocarbons; in Angola, Article 89(2) of Petroleum Law No. 10/04 of 12 November 2004; in Liberia, Article 11(2) of the Petroleum Law, 2002; in Morocco, Article 33 of Law No. 21-90 of 1 April 1992 On the Exploration for and the Exploitation of Hydrocarbon Deposits amended and supplemented by Law No. 27-99 of 15 February 2000; in Sao Tome and Principe, Article 74 of the Framework Law of Petroleum Operations.

[125] See in particular, in Chad, Article 104(2) of Law No. 006-PR-2007 of 2 May 2007 On Hydrocarbons.

(4) The seat of arbitration

20.71 Certain mining and petroleum laws define the seat of arbitration in the clauses related to the settlement of disputes. Among these laws, a distinction must be made as to whether the petroleum and mining laws choose neutrality and establish the seat for arbitration in a state other than that of each of the parties to the arbitration[126] or whether they decide that the arbitration proceeding will take place in the state in question.[127] That said, in most cases, these provisions allow the parties to agree to the seat of arbitration.[128]

20.72 In the absence of legal provisions related to the determination of the seat of arbitration, it is incumbent on the parties to the exploitation agreement to choose the said seat. If they do not make a choice, a distinction must be made between, on the one hand, *ad hoc* arbitration and, on the other, institutional arbitration.[129] In the first scenario -and in the absence of the agreement of the parties- it is generally up to the tribunal to select the seat. In the second scenario, the arbitration rules will usually include terms related to the determination of the seat of arbitration if a choice is not made.[130]

20.73 The mining and petroleum laws' consideration of the arbitration seat should not be perceived as trivial and illustrates the importance thereof in the mind of the legislators. The determination of the seat of arbitration has many consequences.[131]

The seat of arbitration is essential when the parties do not specify the law that applies to the proceedings; in this case, the law of the seat of arbitration will govern the conduct of the arbitration proceedings.[132]

Furthermore, the national court of the seat may intervene to ensure that

---

[126] See in particular, in Namibia, Article 29.3 of the Model Petroleum Agreement of 2007: '[s]uch arbitration, unless the parties otherwise agree, shall take place in London England.'

[127] See, in Botswana, Article 2(4) of the Mines and Minerals Act: '[a]rbitration proceedings shall be conducted in English and shall be held in Gaborone, unless the parties otherwise agree'; in Angola, Article 89(3) of Petroleum Law No. 10/04 of 12 November 2004:

'[o] Tribunal Arbitrai deve funcionar em Angola, aplicar a lei angolana e a arbitragem ser conduzida em língua portuguesa.'

[Free translation into English: '[t]he arbitral tribunal shall have its seat in Angola and apply Angolan law, and arbitration shall be conducted in the Portuguese language'].

[128] See, in the Botswana, Article 2(4) of the Mines and Minerals Act; in the Democratic Republic of Congo, Article 320 of Law No. 007-2002 of 11 July 2002 Enacting the Mining Code:

'l'arbitrage se fait en langue française au lieu convenu par l'État et le titulaire.'

[Free translation into English: 'arbitration will take place in French at the location agreed by the State and the holder.'].

[129] See Ortscheidt & Christophe Seraglini op cit 762.

[130] See Article 18.1 of the ICC's Rules of Arbitration, Article 13 of the Arbitration Rules of the Common Court of Justice and Arbitration of 11 March 1999, and Article 16 of the LCIA Arbitration Rules of 2014.

[131] See Alan Redfern & Martin Hunter *Droit et pratique de l'arbitrage commercial international* 2 ed (LGDJ 1994) 240 et seq.

[132] See Redfern & Hunter op cit 243.

the arbitration proceedings are effective.[133] State courts may therefore be called upon to remedy certain difficulties that may arise throughout the proceedings. Depending on the national laws, the state courts may intervene to help constitute the arbitral tribunal or bring an end to incidents such as death, resignations, unavailability or the challenge or dismissal of the arbitrators. The national court may also order interim and conservatory measures, which, in some cases, may be essential in light of the arbitral tribunal's lack of authority vis-à-vis third parties. The involvement of a national court in these situations is all the more crucial in the framework of *ad hoc* arbitration.

20.74 The determination of the seat of arbitration also has major consequences with regard to state's control over the arbitration award. A petition to have the award set aside may only be brought before the national courts of the seat of the arbitration. Therefore, to ensure that its award is effective, the arbitral tribunal must take the laws of the seat of the arbitration into account to prevent any risk of annulment.[134] It is all the more important to take the legal provisions of the state of the seat of arbitration into consideration as, in most cases,[135] if the courts in the jurisdiction of the seat of the arbitration set aside the award, it may no longer be enforced in a third state. The choice of the seat of arbitration is even more essential as the scope of the control exercised by national courts over the arbitration award may vary from one jurisdiction to another.

20.75 Finally, the choice of the seat of arbitration may have significant consequences when the award is enforced. It is in fact essential to establish the seat of arbitration within a state that has signed the New York Convention on the Recognition and Enforcement of Foreign Arbitral Awards of 1958. The facilitating aspect of the convention could be called into question if the award is rendered in a state that has not signed the convention and is enforced in a country that signed the convention based on reciprocity.[136]

20.76 The choice of the seat of arbitration is far from insignificant. Negotiators and drafters of exploitation agreements do not sufficiently take the legal stakes of this choice into consideration, which frequently leads to major difficulties in particular when precautionary measures must be taken.

(5) Constitution of an arbitral tribunal

20.77 It is essential to provide terms governing the composition of the arbitral tribunal, which can avoid difficulties when the tribunal is established.

---

[133] See Garnett, Gabriel, Waincymer & Epstein op cit 36.

[134] On this issue, see, Ortscheidt & Seraglini op cit 502.

[135] It should be noted that this rule does not apply in certain jurisdictions which enshrine the principle of autonomy of international arbitration in relation to state judicial systems. This is the case in France where the Court of Cassation held that the cancellation of an arbitration award by the courts in the jurisdiction of the seat of the arbitration did not constitute an obstacle to the said award being recognised and enforced aboard (see Civ. 1st, 29 June 2007) 2007 *Rev arb* 507, report of Jean-Pierre Ancel, note Emmanuel Gaillard).

[136] See in particular, Garnett, Gabriel, Waincymer & Epstein op cit 36.

Although the principle that prevails in this matter remains the freedom of choice of the parties in composing the tribunal,[137] certain mining and petroleum laws include terms related to the appointment of the arbitrators, in particular regarding the number of arbitrators or their nationality.

20.78 The choice of the number of arbitrators is important. An arbitral tribunal composed of a single arbitrator will reduce procedural costs and the risk of 'disempowerment' of the co-arbitrators and is easier in terms of availability for arbitration meetings and hearings. Collegial tribunals may guarantee quality, as illustrated by the French adage '*juge unique, juge inique*' ('a single judge, an iniquitous judge'). Although, most often, the parties may freely choose the number of arbitrators, at times, petroleum and mining codes set out the number of arbitrators, whether the tribunal consists of a single arbitrator[138] or three arbitrators—the third one being designated by the two co-arbitrators appointed by the parties.[139]

In order to grant greater independence to the arbitral tribunal, certain mining and petroleum laws or exploitation agreements themselves provide that none of the arbitrators may be of the same nationality as either of the two parties.[140]

In some instances, to facilitate the constitution of the arbitral tribunal, legislatin has gone as far as to include provisions to allow the parties to petition the national court to settle difficulties in constituting the arbitral tribunal.[141]

20.79 Despite certain legal terms related to the composition of the arbitral tribunal, the principle that remains in place in this field is that of the freedom of the parties in constituting the arbitral tribunal, the methods to constitute an arbitral tribunal generally being included in the exploitation agreements, a collegial tribunal is the rule, especially in the extractive sector.

(6) Language of the arbitration

20.80 Exploitation agreements may bring parties together who do not speak the same language. Certain mining and petroleum laws provide for this procedural issue by specifying the language to be used in the arbitration proceedings.[142] It is important to emphasise that all of the mining and

[137] See in particular, Fouchard, Gaillard & Goldman op cit 452; Garnett, Gabriel, Waincymer& Epstein op cit 13.

[138] See in particular, in Botswana, Article 2(2) of the Mines and Mineral Act and Article 2(2) of the Petroleum (Exploration and Production) Act.

[139] See in particular, in Chad, Article 104(3) of Law No. 006-PR-2007 of 2 May 2007 On Hydrocarbons hydrocarbons.

With regard to contracts that provide for an arbitral tribunal composed of three members, see, in particular, Article 39(2) of the Model Petroleum Agreement of the Seychelles of 2013.

[140] See, in particular, in Chad, Article 104(3) of Law No. 006-PR-2007 of 2 May 2007 On Hydrocarbons.

[141] See, in particular, Article 70 of Law No. 96-018 of 4 September 1996 Enacting the Petroleum Code in Madagascar.

[142] With regard to mining and petroleum laws that include provisions on the language of the arbitration, see in particular: in Angola, Article 89 of Petroleum Law No. 10/04 of 12 November 2004:

> '*[o] Tribunal Arbitrai deve funcionar em Angola, aplicar a lei angolana e a arbitragem ser conduzida em língua portuguesa*'.

petroleum laws that set out provisions related to the choice of arbitration language choose the language of the host state.

In the absence of statutory provisions governing the language of arbitration, the parties much reach an agreement. It should be noted that if a choice is not made in the exploitation agreement, the language of the proceedings would be determined by the arbitral tribunal that, unless otherwise indicated, will generally tend to choose the language in which the contract was drafted.[143]

(7) Issues related to the enforcement of the arbitral award

20.81 It seems obvious that the parties to arbitration proceedings, having agreed to the jurisdiction of the arbitral tribunals, subject themselves to the decision of the tribunal in its award.[144] In accordance with this principle, mining and petroleum laws may indicate that the award is binding upon the parties.[145]

20.82 An arbitral award binds the parties, and the party that obtained a favourable award has every interest in ensuring the award is enforced, in particular when the party ruled against does not enforce the award spontaneously.[146] An arbitral award is only subject to forced execution when an *exequatur* decision has been rendered. This requires the intervention of the national court who, unlike the arbitrator, has *imperium* and can order the

---

[Free translation into English: '[t]he arbitral tribunal shall have its seat in Angola and apply Angolan law, and arbitration shall be conducted in the Portuguese language'.]

In Botswana, Article 2(2) of the Mines and Mineral Act: '[a]rbitration proceedings shall be conducted in accordance with the Arbitration Act and conducted in English and shall be held in Gaborone, unless the parties otherwise agree'; in the Democratic Republic of Congo, Article 320 of Law No. 007-2002 of 11 July 2002 Enacting the Mining Code: '*l'arbitrage se fait en langue française au lieu convenu par l'État et le titulaire*'. [Free translation into English: 'arbitration will take place in French at the location agreed by the State and the holder'.]; in Madagascar, Article 71 of Law No. 96-018 of 4 September 1996 Enacting the Petroleum Code: '*[l]a langue de l'arbitrage est le français.*' [Free translation into English: '[t]he language of arbitration is French'].

[143] See, in particular, to that effect, Fouchard, Gaillard & Goldman op cit 678; Garnett, Gabriel, Waincymer, & Epstein op cit 40; also see Article 20 of the ICC's Rules of Arbitration.

[144] See in particular Article 34(6) of the ICC's Arbitration Rules, Article 34(2) of UNCITRAL Arbitration Rules (2010 revised version) and Article 53(1) of the Convention on the Settlement of Disputes Relating to Investments between the State and Nationals of other States.

[145] See in Eritrea, Article 48(5) of Proclamation No. 68/1995 Promoting Development of Mineral Resources and Article 25(2) of Proclamation No. 108/2000, A Revised Proclamation Governing Petroleum Operations; in Ethiopia, Article 76(2) of Proclamation No. 678/2010 Promoting Sustainable Development of Mineral Resources; in Madagascar, Article 72 of Law No. 96-018 of 4 September 1996 Enacting the Petroleum Code.

The binding nature of arbitral awards vis-à-vis the parties is also referred to in the framework of contracts related to the exploitation of natural resources. One example is Article 39(2) of the Seychelles Model Petroleum Agreement of 2013.

[146] 'Enforcement is a positive action taken to compel the losing party to carry out an award that he is unable or unwilling to carry out voluntarily' (Redfern & Hunter op cit 365).

enforcement of the award using the coercive authority vested in it.[147] In some cases, petroleum and mining laws include provisions related to the *exequatur* of arbitral awards.[148] In particular, arbitration decisions are enforceable as of right and bind the parties when they are endorsed by the *exequatur.*[149]

20.83 The *exequatur* of the arbitral award is not always sufficient to ensure that the arbitration decision will be enforced. Arbitration regarding a contract relative to the exploitation of natural resources in fact presupposes the existence of a public entity that may benefit from immunity of jurisdiction and execution.[150] The existence of these types of privileges may represent a barrier to the enforcement of arbitral awards.[151] For this reason and to facilitate enforcement, some laws include terms related to the areas of immunity of the host state. Certain mining and petroleum laws even provide that, in the framework of the enforcement of the arbitral award, the State waive any immunity regarding jurisdiction or enforcement.[152]

---

[147] Redfern & Hunter op cit 365.

[148] See Article 134 of Law No. 2006-17 of 17 October 2006 Enacting the Mining and Mining Taxation Code in the Republic of Benin; Article 320 of Law No. 007-2002 of 11 July 2002 Enacting the Mining Code in the Democratic Republic of Congo; in Madagascar, Article 426 of Decree No. 2006-910 of 19 August 2006 Enacting the Mining Code and Article 72 of Law No. 96-018 of 4 September 1996 Enacting the Petroleum Code.

[149] See, to that effect, Article 134 of Law No. 206-17 of 17 October 2006 Enacting the Mining and Mining Taxation Code of the Republic of Benin; in Madagascar, Article 426 of Decree No. 2006-910 of 19 August 2006 Enacting the Mining Code.

[150] Traditionally, a state's immunity under international law takes two forms: jurisdictional immunity, on the one hand, and immunity from enforcement, on the other. The first allows a state that has jurisdictional immunity not to be subject to the jurisdiction of the courts of another state and the second protects it from any measure to seize its assets or enforce the award (François Knoepfler 'L'immunité d'exécution contre les États—Les états dans le contentieux économique international, III. Le contentieux judiciaire' 2003 *Rev arb* 1017). In addition to national laws, the principle of immunity from execution is also provided in OHADA law in Article 30 of the OHADA Uniform Act as 10 April 1998 On Simplified Recovery Procedures and Measures of Execution:

'*[l]'exécution forcée et les mesures conservatoires ne sont pas applicables aux personnes qui bénéficient d'une immunité d'exécution.*'

[Free translation into English: 'Compulsory execution and protective measures shall not apply to persons enjoying immunity from execution.']

[151] The enforcement of judicial or arbitration decisions has admittedly evolved favourably over the past 30 years towards a reduction in the privilege traditionally granted to States. However, the seizure and acquisition of State assets is often problematic. (François Knoepfler 'L'immunité d'exécution contre les États—Les états dans le contentieux économique international, III. Le contentieux judiciaire' 2003 *Rev arb* 1017).

[152] See, to that effect, Article 320 of Law No. 007-2002 of 11 July 2002 Enacting the Petroleum Code of the Democratic Republic of Congo:

'*[l]es décisions rendues par l'arbitre sont exécutoires et leur exécution peut être demandée devant toute juridiction compétente dans le Territoire National selon la forme prévue par le Code de Procédure Civile Congolais ou dans le pays dont relève le titulaire. En cas d'application des dispositions de l'alinéa précédent, l'État renonce à se prévaloir de toute immunité de juridiction ou d'exécution.*'

[Free translation into English: '[d]ecisions rendered by the arbitrator are enforceable and the execution thereof may be petitioned before any competent court on the National Territory in accordance with the form provided in the Congolese Code of Civil Procedure

20.84 A state's waiver of its immunity of execution may also be contractual.[153] These types of provisions provide significant security to private operators, who often must face obstacles in the enforcement of arbitration awards resulting from the immunities and privileges of states.

As most petroleum and mining laws do not include provisions related to the enforcement of arbitral awards, careful attention must be paid to the drafting of clauses to remedy these difficulties.

---

or in the Country from which the holder originates. If the terms of the preceding paragraph apply, the State waives any jurisdictional immunity and immunity from execution.']

[153] In this regard, a trend in case law that seems to firm up the conditions under which states can waive their immunity from execution should be noted, such as the decision in France of the Court of Cassation, confirming the release from seizures of a state's assets by refusing to recognise the validity of the contractual waiver of the immunity from execution. (See, in particular, to that effect, Laurence Franc-Menget 'Vers un durcissement des conditions de renonciation à l'immunité d'exécution des États étrangers en droit français?' (2013) *Rev arb* 992).

21

INSTRUMENTS TO IMPLEMENT THE TRANSFER OF TECHNOLOGY

G Technological issues

21.1 Mining and petroleum operations are complex projects requiring operators to have considerable technical expertise. The states on the territory on which extractive operations are carried out frequently do not have the technology required to extract their natural resources on their own and resort to private operators. For these states, the purpose is always to subsequently extract the mineral and petroleum resources themselves in a more or less affirmative manner. The host countries and the local State-owned operator must therefore do everything possible to acquire the appropriate technology.

Although multinational mining and petroleum companies endeavour to keep their own technological advantage as long as possible in order to ensure their continued presence in extractive operations, states are making substantial efforts to obtain and master the required technological know-how. The negotiation of exploitation agreements illustrates the way in which private operators and the authorities manage to reconcile their interests.

21.2 For producing countries, technology is a development tool that should not be underestimated. However, its content is variable and frequently ill defined. It can include industrial property rights, accumulated experience or the training of a qualified workforce. Moreover, any actual and sustainable transfer of expertise must be based on a spirit of cooperation and a willingness to work together.

21.3 Although there are certain grounds for criticism, the overall outcome of technology transfer in the mining and petroleum sectors is currently positive. While producing countries acquire the technology required to extract their resources themselves as time progresses, multinational mining and petroleum companies no longer have a monopoly on the technology. It is important to draw attention to the intervention of junior players and small engineering companies used as subcontractors, which appear to represent alternatives to the major petroleum and mining companies.

21.4 The management of the technical issues is expressed in the mining and petroleum codes and in extraction contracts. Numerous provisions encourage the transfer of technology to the benefit of the host country and ensure its effectiveness. These legal and contractual mechanisms exist in order to make the technology transfer effective.

21.5 There are two conditions for ensuring effective transfer of technological expertise: the operator transferring the know-how must possess the appropriate resources in terms of equipment and technology, while the recipient must be in a position to receive it. This statement is not as obvious as it appears and gives rise to numerous difficulties.

The basis for transferring technological know-how lies in the contribution of a material resource and the support of personnel with sufficient technical knowledge.

1 MATERIAL RESOURCES

21.6 Equipment and facilities relating to mining and petroleum operations require major investments and consequently involve considerable financial and economic issues. The producing country, which intends to make use of its natural resource potential itself, must obtain the required material resources. To this end, the mining and petroleum codes and the exploitation agreements generally contain provisions for transferring the ownership of the equipment used in mining and petroleum projects to the host state.

21.7 The transfer of ownership of the equipment to the local State-owned operator raises the question of its location, but also its ownership.

For instance, equipment hired by the private operator is generally subject to special legal and contractual provisions. In principle, rented equipment is not subject to a transfer of ownership to any operator[1] . The scale at which rented equipment is used varies according to the phase of extraction and the location of the projects. Most operators have adopted precise plans in this regard. During the exploration phase, as most of the time they do not have any obligation to establish a company under the laws of the host state, they tend to resort to equipment-leased contracts, thus avoiding the need to deal with import, customs and transport procedures that would delay the project. During the development phase, the decision is taken according to the local tax regime. In general, renting costs are tax-deductible for the year in which they were incurred.

21.8 *A contrario*, if equipment has been purchased, their cost is depreciated over several years. Depreciation is sometimes a very time-consuming and risky solution, mainly due to monetary depreciation.

21.9 Financing of mining and petroleum operations that requires the establishment of third party security and provisions relating to the transfer of assets to the host state must take the existence of guarantees into account[2] ;

[1] In particular, for the Democratic Republic of Congo, see Article 17(3) of the Production Sharing Agreement Between the Democratic Republic of Congo and the Association: Dominion Petroleum Congo & Soco Exploration-Production RDC & La Congolaise des Hydrocarbures on Block V of the Albertine Graben (November 2007).

[2] For Chad, see Article 23(1) of Law No. 2006-PR-2007 of 2 May 2007 On Hydrocarbons:

'[à] l'expiration de la durée de validité du Permis d'Exploitation, ou du recouvrement des investissements par le Contractant relativement aux infrastructures pétrolières, le transfert, notamment de la canalisation de Transport, des ouvrages permettant à la

accordingly, the texts may for example stipulate that the transfer to the host state will only take place once the securities become null and void.[3] A most favourable solution for the host state will be to provide for an automatic transfer of ownership because of the termination of the security relating to the assets. These legal provisions are frequently obscured or overlooked in terms of project financing when the lenders demand collateral on the materials and equipment under the project, which could lead to difficulties.

21.10 Several procedures are applied for transferring the ownership of the equipment used, depending on the type of contract and the location of the project. Two solutions are provided for implementing the transfer of ownership to the local state-owned operator. The first solution distinguishes fixed physical assets from movable physical assets, while the second makes no distinction according to the legal nature of the good. Moreover, the moment of transfer varies considerably depending on the applicable legislation and contractual provisions.

a Transfer conditional on the legal nature of the equipment used

21.11 Procedures for transferring ownership of the equipment to the local State-owned operator can be differentiated according to whether the equipment is an immovable or movable property. This distinction raises difficulties in view of the fact that most frequently there is no provision specifying the criteria for distinguishing between these two categories.[4]

---

poursuite des Opérations Pétrolières, se fait au profit de l'État Tchadien, libre de tout gage, nantissement et de toute hypothèque. Ce transfert de propriété se fait sans charge pour l'État. Lors du transfert, ces ouvrages doivent être opérationnels et en bon état de fonctionnement.'

[Free translation into English: '[o]n expiry of the term of validity of the Extraction Licence or of the recovery of investments by the Contracting Party in relation to petroleum infrastructures, the transmission pipelines and facilities in particular that enable the Petroleum Operations are transferred to the State of Chad free of any pledge, collateral and mortgage. This transfer of ownership takes place free of charge for the State. At the time of the transfer, these facilities must be operational and in good working order.']

[3] In particular, for the Democratic Republic of Congo, see Article 17(2) of the Production Sharing Agreement Between the Democratic Republic of Congo and the Association: Dominion Petroleum Congo & Soco Exploration-Production RDC & La Congolaise des Hydrocarbures on Block V of the Albertine Graben (November 2007).

[4] In uncommon cases, the legal or operating provisions list the movable or fixed assets. For Niger, see Article 17 of Law No. 2007-01 of 31 January 2007 Enacting the Petroleum Code, which provides that:

'*[c]onstituent des immeubles soumis aux dispositions du présent article, les puits, bâtiments, raffineries, machines, équipements, matériels, outillages de sondage, ouvrages utilisés pour l'exploitation des gisements, les infrastructures de stockage, et le transport intérieur des hydrocarbures et tout autre ouvrage, rattachés au sol à perpétuelle demeure. Sont également immeubles par destination, les machines, engins, matériels et outillages non rattachés au sol à perpétuelle demeure, qui sont directement affectés aux opérations pétrolières.*'

[Free translation into English: '[w]ells, buildings, refineries, machinery, equipment, materials, drilling equipment, facilities used for extracting deposits, storage infrastructures and the internal transport of hydrocarbons and any other facilities permanently

i *Equipment owned by a private operator constituting an immovable property*

21.12 It is possible to stipulate the transfer to the State of the immovable property owned by the private operator legally or contractually, generally on expiry of the mining rights.[5] This transfer will in principle take place free of charge, although certain texts stipulate that the transfer implies acquisition of the relevant assets by the State.[6]

This type of clause does not generally make any distinction between real property by nature or by use (a movable asset referred to as immovable as it is attached to an immovable property). With regard to mining and petroleum operations, fixed assets generally include all buildings, outbuildings, wells or galleries.

21.13 Unlike immovable property, movable property used for petroleum and mining operations are intended to be moved in the course of the operations and at the end of the contract period. Their ownership can therefore be transferred to the local State-owned operator.

ii *Equipment owned by a private operator constituting a movable property*

21.14 A transfer of property operated for the profit of the host state can also relate to movable property allocated to mining and petroleum operations.

Some applicable regulations make a distinction between immovable property transferred free of charge and movable property that must be acquired against payment by the host country.

A transfer of movable property to the host state can result from the implementation of a pre-emption right and not a free-transfer of the equipment to the State.

21.15 In the most commonly applied scenario, the transfer is made when the equipment is used permanently on site and its purchase has been included in the operational costs, which is fully justified, as the costs related to the purchase of equipment would have been reimbursed on revenues drawn from sales of the products.

21.16 This permanent use criterion is sometimes replaced by another closely related 'essential use' criterion.

---

fixed to the ground constitute immovable assets subject to the provisions of this Article. Machinery, devices, equipment and tools not permanently fixed to the ground which are directly allocated to petroleum operations are also considered as fixed assets.']

[5] For the Côte d'Ivoire, see Article 44 of Law No. 2014-138 of 24 March 2014 Enacting the Mining Code; in Eritrea, Article 15 of Proclamation No. 108/2000, a Revised Proclamation Governing Petroleum Operations; in Ethiopia, Article 15 of Proclamation No. 295/1986 Regulating Petroleum Operations; in Niger, Article 17 of Law No. 2007-01 of 31 January 2007 Enacting the Petroleum Code.

[6] In particular, for Mali, see Article 97 of Law N 2012-015 of 27 February 2012 Enacting the Mining Code.

b Transfer of the equipment without consideration of the legal nature of the assets

21.17 The equipment may also be transferred to the host state without taking into consideration the legal nature of the asset.[7]

In general, the transfer criterion will no longer be based on the nature of the asset but on whether it is necessary or essential for the mining and petroleum operations to continue.[8]

c Moment of transfer

21.18 Several events may be taken into consideration in order to define the moment of the transfer of ownership of the equipment to the State.

21.19 In most cases, the transfer of ownership of the assets used during the implementation of the works to the host country will take effect at the term of the contract or on expiry of the mining rights.[9] During the term of the exploitation agreement, the private operator will, in principle, retain full ownership rights on equipment used which it has purchased, manufactured or acquired by other means such as through a lease-purchase agreement.

21.20 Certain legal or contractual provisions may also subordinate the transfer of ownership to the depreciation of the assets.[10] Consequently, the operator must have recovered the cost of equipment so that the transfer to the host State can take place.

---

[7] For Burundi, see Article 213 of Order No. 1/138 of 17 July 1976 Enacting the Mining and Petroleum Codes; for Equatorial Guinea, Article 53 of the Mining Act No. 9/2006 of 3 November 2006.

[8] For Algeria, Article 80 of Law No. 05-07 of 28 April 2005 On Hydrocarbons; in Côte d'Ivoire, Article 37, paragraph 2 of Law No. 96-669 of 29 August 1996 Enacting the Petroleum Code; for Morocco, Article 7 of Law No. 21-90 of 1 April 1992 On Prospecting and Extraction of Hydrocarbon Deposits amended and supplemented by Law No. 27-99 of 15 February 2000.

[9] For Algeria, see Article 80 of law No. 05-07 of 28 April 2005 On Hydrocarbons; for Côte d'Ivoire, Article 44 of Law No. 2014-138 of 24 March 2014 Enacting the Mining Code; for Guinea, Article 44 of Law No. L/2014/N°034/AN of 23 December 2014 Enacting the Petroleum Code; for Equatorial Guinea, Article 53 of Mining Act No. 9/2006 of 3 November 2006; for Mali, Article 97 of Law No. 2012-015 of 27 February 2012 Enacting the Mining Code; for Morocco, Article 7 of Law No. 21-90 of 1 April 1992 On Prospecting and Extraction of Hydrocarbon Deposits amended and supplemented by Law No. 27-99 of 15 February 2000; for Niger, Article 17 of Law No. 2007-01 of 31 January 2007 Enacting the Petroleum Code; for the Seychelles, Article 37 of the 2013 Model Petroleum Agreement; for Chad, Article 23 of Law No. 006-PR-2007 of 2 May 2007 On Hydrocarbons.

[10] In particular, for Equatorial Guinea, see Article 53 of the Mining Act No. 9/2006 of 3 November 2006:

> '[w]ith the exception of the goods, equipment and materials mentioned in the previous article all fixed facilities, buildings, machinery, as well as any other goods, in general, owned by the Contractor, which had been used for the Mining Exploitation operations, shall be transferred, in ownership, to the State, at the end of the respective Contracts, in good operating conditions, free of charge, and without any tax assessments or limitations of any kind. The same treatment shall apply to the goods, equipment, materials, fixed facilities, buildings, machinery and any other goods, in general, owned by the Contractor, whose costs had been totally amortized.'

For Guinea, see also Article 44 of Law No. L/ 2014/N°034/AN of 23 December 2014 Enacting the Petroleum Code.

21.21 Less commonly, the transfer of ownership may take place as soon as the equipment enters within the territory of the host State. In this case, the transfer of ownership takes place immediately, even before the equipment has been depreciated or reimbursed.

Such provisions will generally encourage operators to resort to rented equipment. However, a consequence of such provisions is to reduce the risk of nationalisation, as the host State is less tempted to undertake nationalisation as it will already be the owner of the physical equipment. This provision has proved its worth several times in Africa.

21.22 Other provisions make a distinction according to whether the asset was bought in the host country or imported; it becomes the property of the public operator at the moment of purchase if the latter takes place in the host country, or when it enters the host country if it was purchased in another country.

2 HUMAN RESOURCES

21.23 The purchase of materials and equipment in principle does not give rise to difficulties; the market is open to competition and the only restrictions occasionally involve obtaining export licences if the equipment appears in national red lists of high-technology products. On the other hand, it is difficult to be able to choose equipment according to the specific requirements linked to the proper implementation of the project and then be able to operate it. The private operator will therefore make every effort to contribute suitable technical and technological know-how for the project and at the same time provide the support of qualified personnel.

a Flow of appropriate technological expertise

21.24 The flow of appropriate technical expertise has become a determining factor in the negotiation of exploitation agreement and is therefore a priority.

Technical assistance and training of local personnel encourage the flow of an appropriate knowledge. It should be noted that there is considerable vagueness in the perception of what the idea of technical assistance should contain; it is traditionally left to the discretion of the private operator. However, exploitation agreements tend to develop a framework that in principle makes it effective.

b Assistance by qualified personnel

21.25 Assistance by qualified personnel may appear obvious, but in practice, the opposite has turned out to be the case on numerous occasions.

The scale and the economic conditions of this support are generally left entirely to the discretion of the private operator or subcontractors.

21.26 Hiring of expatriate personnel is usually preferred to assist local personnel. The technical qualifications of expatriate personnel are most of the time subject to inspection by the local State operator based on the information it receives. Some contracts contain full lists of personnel, showing the role that each employee will play and his/her main educational qualifications and place of residence.

21.27 In general, in cases of technical problems that were not initially foreseen at the time when the exploitation agreement was drafted, the private operator undertakes to second experts for a limited duration to solve them.

22

Variable effectiveness of technology transfer

22.1 The effectiveness of the obligation for the private operator to transfer technology varies with the circumstances and projects. The content of the technology transfer clauses and a number of external factors can occasionally reduce their effectiveness. Conversely, the effectiveness of the technology transfer obligation can be reinforced, in which case the obligation is similar to a results-based obligation. The legal status of the technology transfer obligation incumbent on the private operator can thus be analysed according to the contracts as a best-effort obligation or a results-based obligation.

1 TECHNOLOGY TRANSFER OBLIGATION WITH REDUCED EFFECTIVENESS

22.2 The technology transfer obligation can be restrictive and doubt can be cast on its effectiveness. Ineffective technology transfer provisions can firstly result from the way the clause is drafted; it may impose limitations on the technology transfer obligation. Apart from the content of the clauses, the effectiveness of the obligation incumbent on the private operator can result from lack of capacity within the host State to receive the transferred technical expertise.

a Limits to transfer due to the content of legal and contractual provisions

22.3 Limits to the transfer may result from setting a restriction due to the occurrence of a specific event or may result from contractual provisions establishing the conditions and content of the transfer.

i *The transfer of technology conditional on the occurrence of a specific event*

22.4 The technology transfer obligation incumbent on the private operator can be subordinated to the occurrence of a particular event. Implementation of the transfer obligation will then depend on the fulfilment of a condition precedent, in particular discovery of the natural resource in commercial quantities. If such a discovery is not made or before it is made, the private operator will only be required to make use of competent local personnel, the exact qualifications of whom are left to its discretion.

The reason for this delay in the transfer of technology is practical. Indeed, personnel will only be required in large numbers as from the development phase, but these personnel will be temporary. On the other hand, when the production begins, the workforce will become permanent. The private operator will therefore not have any real and immediate interest in training local personnel during the exploration, delineation and development phases. This of course could be different for the host State; however, the latter will frequently try to ensure the greatest possible effectiveness by leaving the private operator free to determine its resources.

ii *General provisions applicable to the transfer of technology*

22.5 Frequently the provisions relating to the transfer of technology are drafted much too extensively and leave the private operator with excessive discretionary powers in fulfilling its obligation. The use of terms such as 'to the extent possible' or 'as much as possible'[1] thus relates directly to a best-efforts obligation. Similarly, certain texts merely require the private operator to favour or encourage technology transfer,[2] or even just stipulate an obligation to cooperate in this respect.[3]

In these cases, although the operators may have a will to transfer their technical expertise, this is in no way backed up by clear and explicit legal provisions.

22.6 The reliability of the transferred technological know-how is considerably more complex. In practice, the technical expertise transferred is highly specialised and limited. It should also be noted that frequently technical expertise is merely transferred from a parent company to its subsidiary located in the host State; in this case, it is incorrect to refer to a transfer, to the extent that the development of the technologies as time progresses is a determining factor that must be taken into account.

b Limited effectiveness of the transfer of technology resulting from external factors

22.7 A transfer of technological know-how can be infective due to external factors. This infectiveness can be judged both from the ability of the

---

[1] For example, for Benin, see Article 103 of Law No. 2006-17 of 17 October 2006 Enacting the Mining and Mining Taxation Code; for Guinea, Article 109 of Law No. L/2013/N°053/CNT of 8 April 2013 Amending Law No. 2011/006/CNT of 9 September 2011 Enacting the Mining Code; for Niger, Article 104 of Ordinance No. 93-16 of 2 March 1993 Enacting the Amended Mining Act.

[2] For Chad, see Article 70 of Law No. 001-1995 of 20 June 1995 Enacting the Mining Code: '*[l]es titulaires d'un titre minier ou bénéficiaires d'une autorisation en vertu du Code Minier, ainsi que les entreprises travaillant pour leur compte ou en association avec eux, sont tenus de conduire leurs activités au Tchad en vertu du Code Minier de façon à favoriser les transferts technologiques en faveur des entreprises et de la main d'_uvre tchadiennes.*'

[Free translation into English: '[h]olders of mining rights or beneficiaries of an authorisation under the Mining Code, along with companies working on their behalf or in combination with them, are required to conduct their activities in Chad in accordance with the Mining Code in such a way as to encourage technology transfer to Chadian companies and the Chadian workforce.']

[3] For Namibia, see Article 50 of the 1992 Minerals Prospecting and Mining Act and Article 14 of the 1991 Petroleum (Exploration and Production) Act (amended by Act No. 2 of 1993).

beneficiary of the technology transfer to receive the technological expertise and from the *subsequent* use of the technological expertise transferred.

In order to ensure that technology is effectively transferred, different mechanisms can be used to ensure that the beneficiary of the transfer has the capacity to receive and use the transferred expertise.

i *Capacity of the beneficiary of the transfer to receive the technological expertise*

22.8 In order to be fully effective, a technology transfer to the host state must be accompanied by adequate training of local personnel. This training is essential for expatriates to be replaced by local personnel and will in time provide the State with qualified personnel on its own territory for the implementation of mining and petroleum operations.

22.9 Numerous African countries are aware of the connection between effective technology transfer and adequate training of local personnel and have inserted personnel training provisions into their mining and petroleum legislation. Accordingly, the obligation incumbent on the operator to train the local personnel is generally set out explicitly in the mining and petroleum codes,[4] although certain texts refer back to the provisions of the exploitation agreement.[5] The place allocated to the training of local personnel in the legislation is such that such obligation is occasionally taken into consideration in the process of granting mining rights. Indeed, the mining and petroleum laws can provide that a training programme must accompany the application for mining rights.[6]

22.10 In general, the training obligation incumbent on the private operator involves the elaboration of a training programme.[7] The mining or

[4] However, in some cases, the wording used by the mining and petroleum codes tends to weaken the training obligation at the operator's expense. Accordingly, the training obligation enacted by some of the codes amounts to a best-efforts obligation. This is particularly the case in the Seychelles with Article 18(3) of the 2013 Model Petroleum Agreement: '[t]he Company shall use best efforts to train nationals of Seychelles with respect to and in each Year of its Petroleum Operations, concerning petroleum-related work including technical, administrative, executive and management positions.

[5] For Cameroon, see Article 12 of Law No. 99-013 of 22 December 1999 Enacting the Petroleum Code; for the Côte d'Ivoire, Article 18 of Law No. 96-669 of 29 August 1996 Enacting the Petroleum Code; for Senegal, Articles 17 and 34 of Law No. 98-05 of 8 January 1998 Enacting the Petroleum Code; in Tanzania, Article 10(4) of the 2010 Mining Act.

[6] For Ethiopia, Article 5.2 of Regulations No 182/1994 on Mining Operations; in Ghana, Articles 11 and 50 of the 2006 Minerals and Mining Act.

[7] For Algeria, see Article 15 of the Presidential Decree of 1 March 2003 Enacting the model Mining Contract; for Benin, Article 103 of Law No. 2006-17 of 17 October 2006 Enacting the Mining and Mining Taxation Code and Article 11 of Law No. 2016-18 of 17 October 2006 Enacting the Petroleum Code; for Botswana, Article 12 of the Mines and Minerals Act; for Cameroon, Article 77 of Law No. 99-013 of 22 December 1999 Enacting Petroleum Code; in the Central African Republic, Article 34 of Law No. 09-005 of 29 April 2009 Enacting the Mining Code and Article 60 of Ordinance No 93.007 of 25 May 1993 Enacting the Petroleum Code; for the Republic of Congo, Article 21 of Law No. 24-94 of 23 August 1994 Enacting the Hydrocarbon Code; for the Côte d'Ivoire, Article 134 of Law No. 2014-138 of 24 March 2014 Enacting the Mining Code and Article 53 of Law No. 96-669 of 29 August 1996

petroleum administration enjoys a fairly extensive right to review the drafting of this training programme; accordingly, while certain provisions oblige the operator to submit the training programme to the relevant authority for approval,[8] other texts only stipulate that the administration of the host State must be consulted.[9]

22.11 The content of the training programmes varies; certain mining and petroleum codes can contain provisions in this respect.[10] The texts are also intended to establish the *ratione personae* scope of these training programmes. Although in principle the training programmes are intended for application to local personnel employed by the operator in the course of exploitation projects, the advantages granted under the training programmes can extend to third parties[11] and in particular the personnel of the administration.[12] The relationship between technology transfer and building

Enacting the Petroleum Code; for Djibouti, Article 52 of Law No. 117-AN-05 of 16 October 2005 Enacting the Petroleum Code; for Ghana, Article 50(1) of the 2006 Minerals and Mining Act; for Guinea, Article 54 of Law No. L/2014/N°034/AN of 23 December 2014 Enacting the Petroleum Code and Articles 108 and 109 of Law No. L/2013/N°053/CNT of 8 April 2013 Amending Law No. 2011/006/CNT of 9 September 2011 Enacting the Mining Code; for Lesotho, Article 11(1) of the 2005 Mines and Minerals Act; for Mali, Article 137 of Law No. 2012-015 of 27 February 2012 Enacting the Mining Code; for Mauritania, Article 62 of Law No. 2012/012 Regulating Mining Contracts and Approving a Model Mining Agreement; in Namibia, Article 50 of the 1992 Minerals Prospecting and Mining Act, Article 14 of the 1991 Petroleum (Exploration and Production) Act (amended by Act No 2 of 1993); For Niger, Article 21 of Law No. 2007-01 of 31 January 2007 Enacting the Petroleum Code; In Uganda, Articles 126(1) and 127(1) of the 2013 Petroleum (Exploration, Development and Production) Act; for Senegal, Article 53 of Law No. 98-05 of 8 January 1998 Enacting the Petroleum Code; for the Seychelles, Article 45(2) of the Petroleum Mining Act and Article 18(4) of the 2013 Model Petroleum Agreement; for Sierra Leone, Article 164(4) of the 2009 Mines and Minerals Act; for Swaziland, Article 28 of the 2011 Mines and Minerals Act; in Chad, Article 69 of Law No. 011-PR-1995 of 20 June 1995 Enacting the Mining Code; for Togo, Article 34 of Law No. 96-004 of 26 February 1996 Enacting the Mining Code; for Tunisia, Article 75 of Law No. 2003-30 of 28 April 2003 Enacting the Mining Code and Article 62(2) of Law No. 99-93 of 17 August 1999 Enacting the Hydrocarbon Code; for Zambia, Article 63 of the Petroleum (Exploration and Production) Act, 2008.

[8] For Ethiopia, see Article 28 of Regulations No. 182/1994 On Mining Operations; for Guinea, Article 109 of Law No. L/2013/N°053/CNT of 8 April 2013 Amending Law No. 2011/006/CNT of 9 September 2011 Enacting the Mining Code; in Uganda, Article 126 of the 2013 Petroleum (Exploration, Development and Production) Act.

[9] For South Sudan, see Article 65(5) of the 2012 Petroleum Act; for Zambia, see Article 63(3) of the 2008 Petroleum (Exploration and Production) Act.

[10] With regard to the content of the training programmes, for Guinea, see in particular Article 109 of Law No. L/2013/N°053/CNT of 8 April 2013 Amending Law No. 2011/006/CNT of 9 September 2011 Establishing the Mining Code.

[11] From Morocco, see Article 37 of Law No. 21-90 of 1 April 1992 On Exploration and Extraction of Deposits amended and supplemented by Law No. 27-99 of 15 February 2000, which stipulates that the obligation to contribute to the professional training of national managerial and technical petroleum industry personnel must be at the expense of the operator.

[12] In particular for Mali, see Article 135 of Law N° 2014-138 of 24 March 2014 Enacting the Mining Code:

the capacity of the host state to extract its natural resources on its own then acquires its full meaning.

22.12 Compliance of private operators with the training obligations is ensured by inserting provisions intended to make these obligations effective. Accordingly, mining and petroleum codes, or more frequently exploitation contracts, can include firm targets that must be reached by the operator.[13] Moreover, the texts can establish minimum amounts that must be invested in training of local personnel by the operator or paid to the host state, particularly in the form of a contribution.[14]

22.13 A training fund can be established for the purposes of compiling the amounts intended for training national personnel.[15]

22.14 Certain mining and petroleum codes can also establish incentives to encourage training investments. In particular, this takes place through an option to deduct the amount invested in training programmes from payment of certain taxes.[16]

22.15 Finally, personnel training clauses can be made effective by including sanctions; although such cases are still uncommon, infringement by the operator of its training obligations may lead to the suspension or even the withdrawal of the mining right.[17]

(1) Pre-employment training of local personnel

22.16 It is possible to stipulate that local personnel must be trained before they are employed on mining and petroleum projects, in which case training that can last several weeks will occur before the local personnel is operationally committed.

Occasionally, the private operator organises training sessions on abandoned mines or deposits in order to simulate the works and familiarise trainees with teamwork under real noise and fatigue conditions. Organising pre-employment courses can be found in a number of exploitation contracts; in general, the programme of courses is established in cooperation with all of the operators and with the authorities in the host State. Frequently an

---

'[l]e titulaire du permis d'exploitation est tenu de contribuer au financement du renforcement des capacités des agents de l'Administration minière et à la formation des ingénieurs miniers et géologues ivoiriens. Les modalités de cette contribution sont déterminées par décret.'

[Free translation into English: '[t]he holder of the exploitation permits required to contribute to financing capacity building for agents of the mining Administration and for training mining engineers and geologists from the Côte d'Ivoire. The procedures for making this contribution are determined by decree.']

[13] For Mauritania, see Article 62 of Law No. 2012/012 Regulating Mining Agreements and Approving a Model Mining Agreement.

[14] For the Seychelles, see Article 18(5) of the 2013 Model Petroleum Agreement.

[15] In this connection, for Kenya see Article 11 of the Petroleum Act and, for Sierra Leone, Article 89(1) of the Petroleum Exploration and Production Act of 2011.

[16] For Algeria, see Article 86 of Law No. 05-07 of 28 April 2005 On Hydrocarbons.

[17] In this connection, for Sierra Leone see Article 164(7) of the 2009 Mines and Minerals Act which considers non-fulfilment of training obligations as a 'material breach' that can lead to suspension or revocation of the mineral rights.

amendment to the exploitation contract stipulates the status of the trainees in addition to local regulations.

(2) In-service training of local personnel

22.17 In-service training of local personnel is generally more widespread, particularly because of the lower cost of this method.[18] While local personnel take part in exploitation operations, they can receive additional training on the tasks they are carrying.

Depending on the case in point, this training can begin from the start of the project and can differ during a certain length of time or until the occurrence of an event specified by law or by the exploitation contract.[19]

ii Capacity of the beneficiary of the transfer to make use of the technological expertise

22.18 When the beneficiary of a transfer of technical expertise starts to make use of it by itself, it frequently encounters difficulties that can develop into insoluble issues. Such cases are frequently linked to poorly adapted training.

This finding raises questions about the effectiveness of local personnel training programmes. The effectiveness of the training plans inserted into certain exploitation contracts varies depending on how they are drafted. Besides, the quality and stability of the technology transfer are often not properly monitored.

22.19 Exploitation contracts stipulate one single tangible result: local personnel taking up duties in positions of responsibility, without any real verification of the knowledge they have absorbed.

22.20 In order to improve the way the technological expertise stipulated in the exploitation contract is transmitted, the expatriate personnel are also trained to better understand the human and sociological nature of the local personnel whom they are required to train. Initially this kind of training was seldom applied, but it has become established and uniformly widespread practice. In this connection, a clause inserted in the exploitation contract can stipulate that expatriate personnel must have an appropriate level of fluency in the language spoken in the host State or even occasionally in the area covered by the contract. In addition, management courses can be organised to generate improved in-depth management qualities and a sense of responsibility among the expatriate personnel.

---

[18] A number of texts provide for an in-service personnel training. In this connection, for the Republic of Congo, see Article 101 of Law No. 4-2005 of 11 April 2005 Enacting the Mining Code.

[19] For Mauritania, see Article 62 of Law No. 2012/12 Regulating Mining Agreements and Approving a Model Mining Agreement.

2 TECHNOLOGY TRANSFER OBLIGATION WITH ENHANCED EFFECTIVENESS

22.21 From a historical point of view, a transition from imposed participation to a more intimate and educational relationship corresponds to the beginning of an effective transfer of expertise. Two factors can enhance this effectiveness. The first is in the context of collaboration with the private operator when the latter assumes results-based obligations, while the second only involves the local State operator, or more accurately, what it can do without external aid or assistance after the transfer of expertise.

a Structure of the technology transfer obligation

22.22 As previously seen, prior transfer of basic knowledge is required in order to improve the level of technological expertise of the beneficiary to enable the latter to absorb the basic technological knowledge.

This requires an effort of cooperation irrespective of the natural resource involved. Two situations should be distinguished: in the first case, one of the operators can be less technologically advanced, while in the other, all of the operators have an identical level of technological expertise. The transfer conditions and procedures differ completely between these two cases.

i *Relationship of dependence between operators*

22.23 The private operator holding the technological expertise is aware of the underdeveloped knowledge of the other State-owned operator and undertakes to implement the transfer. Legally, its commitment takes the form of a results-based obligation inserted into an industrial cooperation contract or a technical assistance contract.

(1) General justification of the results-based obligation

22.24 Beyond the obvious advantages that the State can obtain from the transfer of technology, the private operator will also draw benefits from such cooperation. In all cases, it will benefit from additional experience and hence long-term benefits. Moreover, the private operator will not be so visible politically and will therefore not be targeted by potential nationalisation or confiscation measures. This policy can also lead to the signing of new contracts in the host State or in other countries.

22.25 The socio-political impact resulting from the technology transfer is also far from negligible.

(2) Practical applications of the results-based obligation

22.26 Different forms of contractual support can be established in order to ensure fulfilment of the results-based obligation incumbent on the private operator. For instance, reference should be made to implementing technical-economic studies concerning the area covered by the contract, establishing specific training programmes, assistance from experts working for the private operator, etc.

22.27 Technical-economic studies of the area covered by the contract mainly involve interpreting the geological and geophysical data collected, assessing an exploration budget and calculating the economic viability of the project. This involves overall expertise. Obtaining a result in this area will depend on the active participation of the local State operator's personnel. This participation results in joining forces in the work with the private operator, who can provide explanations and advice.

22.28 Compiling special training programmes can also strengthen technology transfer. In order to do so and for them to be fully effective, these programmes must be drafted jointly. They can resume training already in existence or create specific training. The personnel and the local State operator will generally be trained as a matter of priority in the field, either in the host State or in the country of origin of the private operator. In parallel, access to theoretical training is developed in the form of two-way discussions between operators.

22.29 One provision, which is limiting for the private operator and broadly favourable for the host State, involves the insertion of a clause relating to 'the latest global technology improvements'. The local State operator is assured that it will receive the results on an ongoing basis of the latest research of the private operator relating to the technology transferred under the exploitation contract.

22.30 Assistance from the private operator's experts is another feasible contract mechanism. These experts travel and stay in the country in which the exploitation operations take place. Alongside permanently resident experts, others liaise with the head office of the private operator's parent company, undertaking short-duration assignments. The State operator is in principle in charge of providing them with appropriate accommodation and social, administrative and medical assistance.

22.31 In parallel with these mechanisms commonly found in contractual practice in Africa, it is essential to verify that the transfer has genuinely been implemented by the private operator, particularly by seconding personnel for regular visits to the country where the exploitation contract was implemented.

ii *Balanced relationship between operators*

22.32 The issues at stake are different when the operators consider themselves to be at the same technical level and organise joint research or works so that they can provide each other with the benefit of mutual knowledge acquired by the transferred technological know-how.

22.33 This balanced relationship can occasionally be achieved outside any exploitation contract, through a special cooperation agreement. This contractual phenomenon has developed considerably at the beginning of the 1980s.

22.34 We will resume by developing both indications sanctioning this bilateral cooperation, ie organising joint scientific research and joint implementation of exploitation works. Note in advance that this bilateral cooperation is never considered as exclusive.

(1) Organisation of joint scientific research

22.35 In general, operators, or parties in the case of a cooperation contract, must initially pool their respective know-how. If it proves necessary, they then establish a balance by transferring technology and expertise. Subsequently, regular meetings on topics of common interest are generally organised in order to present the development of mutual technical expertise. These updates are supplemented by exchanges of specialised personnel for training purposes, but also with a view to mutual enrichment by comparing reciprocal know-how.

22.36 Joint scientific research forms the basis for joint interventions in third countries for the purposes of implementing exploitation operations.

22.37 Joint work can lead to the discovery of techniques that can be patented. From then on, it is possible to observe that this can occasionally lead to patents being held *in indivis* between the local State operator and the private operator, which gives rise to solid and sustainable links.

(2) Joint implementation of exploitation works

22.38 Two or more companies, which have already benefited from joint experience in training or operations on a mining or petroleum project can bid jointly for exploitation projects within or occasionally even outside their country.

In principle, the economic and technical conditions in which a joint intervention between parties or operators takes place are not established in advance, but are subject to separate contracts, each of them being signed for each operation.

b Spin-offs of the obligation

22.39 The capacity for the beneficiary of the transfer to improve on the technological expertise transferred constitutes a decisive stage in the effectiveness of this transfer. This improvement can take place on a national or international scale. Moreover, technological expertise can again be transferred from its initial beneficiary to others.

i *Improving the technological expertise transferred*

22.40 Henceforward, the main reason for States to resort to petroleum or mining companies to extract their natural resources is technical in nature. In fact, the search for capital is not a genuinely insurmountable obstacle. Moreover, a growing number of specialised publicly-owned companies are showing willingness to take financial risks, though this has not become standard practice. However, the management of a mining or petroleum project cannot be improvised. It involves knowledge acquired in the field through long years of experience. The fact of having worked on projects of a company that has this expertise, even if a substantial transfer of technological expertise has taken place, is far from sufficient. This is mainly due to the specific nature of each project and the current trend towards increasingly difficult operating conditions, as most deposits are now mainly located in relatively inaccessible areas on complex or small-scale geological structures.

In the same way that the distance covered by two drivers of vehicles with the same amount of fuel can vary according to the way they drive; a mining or petroleum project can take more or less time to implement it and it can turn out more expensive or cheaper depending on the company organising the works. Although in one case the differences can be minimal, in another case they will be considerable. With regard to hydrocarbons, poor seismic profile interpretation can lead to errors in setting the drilling objectives, ie can easily lead to the loss of a substantial amount of money and several months of work.

22.41 For countries that wish to extract their natural resources themselves, the essential efforts must now be concentrated on mastering this technological petroleum and mining expertise. The most obvious sign that this target is being achieved is improvement on expertise which has been transferred. The transfer takes place either without external cooperation, or in cooperation with other countries or state bodies.

ii *Consecutive transfer of technological expertise*

22.42 In time, the beneficiary of the transfer can in turn use and transfer the acquired technological expertise to others.

This is why the original exploitation contract occasionally imposes conditions and limits on the use of the transferred technological expertise.

H Development of the host state

22.43 Local development is a major concern of the host states.[20] The latter are fully aware of the growth potential that can result from the exploitation of their natural resources and intend to turn the extractive sector into an economic growth engine and a permanent means of contributing to improving the living conditions of the local populations.[21] As such, the development of the host State assumes considerable importance when exploitation agreements are negotiated.

22.44 The insertion of increasingly numerous provisions relating to local content into exploitation agreements initially originated from the demands of local workers. There have been substantial positive results since the first concession contracts in terms of development of the host State.

However, experience has revealed new challenges arising from the development of a major mining or petroleum sector driving the economy; its growth occasionally tends to exacerbate the disparity in the level of technology between the extractive industry and the rest of the economy and can lead to post-mining problems. It should also be mentioned that local redistribution of the income drawn from extractive industries by the host State has not always been effective.

Lastly, it is in the interest of the State whose industry is largely dependent on natural resources to opt for the diversification of its economy.

22.45 There is now new momentum in the striving to contribute to the development of the host State, mainly driven by civil society, which considers that it does not draw sufficient benefits from the exploitation of natural resources.

For instance, these concerns have been heard in the WAEMU which, at the beginning of 2015, initiated a process of revising the community mining code, in particular for the purposes of strengthening local content and maximising the contribution of the mining sector to the economies of the member states.

22.46 The determination of African states to strengthen the contribution of extractive industries to local development also resulted in legal reforms conducted at the national level.[22]

---

[20] On this topic, see Renouard op cit 96.

[21] See in this respect the Preamble to ECOWAS Directive C/DIR 3/05/09 of 27 May 2009 On the Harmonisation of the Guiding Principles and Policies in the Mining Sector: 'CONSCIOUS that exploitation and transformation of the Mining Product is essential for the socio-economic development of the Member States and that benefits from these activities should be equitably shared among present and future generations'.

[22] As an example, see Law No. 007-2002 of 11 July 2002 Enacting the Mining Code in the Democratic Republic of Congo, which, in its Preamble, states that the collective agreement regime preceding the adoption of the Mining Code was marked by a lack of job creation and a lack of social infrastructure improvement, an absence of integration with other economic sectors and reduced development opportunities for other sectors through a knock-on effect, despite the exorbitant advantages granted to the investors.

The desire to promote the development of the host state is particularly prominent in the latest versions of the mining and petroleum codes, which contain numerous provisions to enhance local development.[23] The establishment of development funds,[24] the signing of special agreements in this field[25] or the implementation of local development plans[26] are now common features of the applicable binding legal framework for mining and petroleum operations. The place granted to development programmes, the drafting of which frequently involves the participation of local communities,[27] can be significant; these programmes are occasionally taken into consideration by the competent authorities when granting mining licences.[28] These mechanisms result from efforts to ensure that income generated by mining and petroleum activities benefits local populations.[29]

22.47 The willingness to involve local populations affected by mining and petroleum operations is increasingly marked.[30] This helps to ensure that

With regard to mining and petroleum codes that establish an obligation on the part of the operators to contribute to the development of the host state, the following provisions has to be mentioned: for Angola, Article 26(2) of Petroleum Act No. 10/04 of 12 November 2004; for Sierra Leone, Article 138 of the 2009 Mines and Minerals Act; for South Sudan, Article 128 of the 2012 Mining Act. At the regional level, Article 16(1) of ECOWAS Directive C/DIR 3/05/09 as 27 May 2009 On the Harmonisation of the Guiding Principles and Policies in the Mining Sector should be mentioned.

[23] Apart from Mining and Petroleum Codes, a number of African states have adopted specific legislation for the purposes of developing local content in their mining and petroleum sectors. For Angola in particular, an example is Order No. 17/09 of 16 June 2009 On Local Content and its Implementation Order, Executive Decree No. 45/10 of 10 May 2010.

[24] The obligation to establish a development fund supported financially by the holders of mining rights and other stakeholders is enshrined in Article 16(7) of ECOWAS Directive C/DIR 3/05/09 as 27 May 2009 On the Harmonisation of the Guiding Principles and Policies in the Mining Sector. With regard to national mining and petroleum codes, as an example, we can mention the following texts: for the Côte d'Ivoire, Article 124 of Law No. 2014-138 of 24 March 2014 Enacting the Mining Code; for Sierra Leone, Article 89 of the 2011 Petroleum Exploration and Production Act; for Burkina Faso, Article 26 of Law No. 036-2015/CNT of 16 June 2015 Enacting the Mining Code of Burkina Faso.

[25] For Nigeria see Articles 116 and 117 of the 2007 Minerals and Mining Act.

[26] For the Côte d'Ivoire see Article 124 of Law No. 2014-138 of 24 March 2014 Enacting the Mining Code; for Mali, See Articles 150 to 153 of Law No. 2012-015 of 27 February 2012 Enacting the Mining Code; for Guinea, see Article 54 of Law No. L/2014/N°034/AN Enacting the Petroleum Code.

[27] In this respect, for Mali see Article 151 of Law No. 2012-015 of 27 February 2012 Enacting the Mining Code.

[28] A particular reference is made to the Democratic Republic of Congo, see Article 69 of Law No. 007-2002 of 11 July 2002 Enacting the Mining Code, which stipulates that the application for an Exploitation Permit must contain a 'plan for the project to contribute to the development of the surrounding communities.'

[29] In this respect, certain texts provide that a percentage of the income generated by the activities of extractive industries must be allocated to the development of local communities (see for Mozambique, Article 20 of Mining Act No. 20-2014 of 18 August 2014 and Articles 19 and 48 of Petroleum Act No. 21-2014 of 18 August 2014).

[30] In particular, for Mozambique see Article 11 of Petroleum Act No. 21-2014 of 18 August 2014 and Article 32 of Mining Act No. 20-2014 of 18 August 2014; for Angola, see Article 196 of Mining Act No. 31/11 of 23 September 2011.

the local populations accept the project and fosters their development. The mining and petroleum projects will be integrated into the local reality and will be a core part of the local landscape.

22.48 Over time, besides the legal provisions favouring the development of the host State, corporate social responsibility regulations have emerged both on a regional[31] and on a national level.[32] Corporate social responsibility, which requires the investor to take account of social, economic and environmental concerns in its operations, enables the full participation of the investors in the development of the host State.[33]

22.49 Although mining and petroleum codes and exploitation agreements allocate an increasingly significant place to the development of the host State, too few provisions take into account the need to encourage the optimum use of the extracted resources within the country. Local processing of these resources indeed constitutes a significant line of development.

---

[31] See Article 11(2) of ECOWAS Directive C/DIR 3/05/09 of 27 May 2009 On the Harmonisation of the Guiding Principles and Policies in the Mining Sector: 'Member States shall ensure that the Corporate Social Responsibility (CSR) and Alternative Livelihoods Programmes (ALP) to be submitted under this article shall be part of the conditions for granting a mining right. [. . .]'].

[32] In the Côte d'Ivoire, see Article 121 of Law No. 2014-138 of 24 March 2014 Enacting the Mining Code; in South Sudan, see Article 128(2) of the 2012 Mining Act: '[t]he Title Holder of a Mineral Title shall implement a Corporate Social Responsibility programme in conformity with Mining Cadastre Office guidelines or in the absence of such guidelines, in accordance with best international practice'; for Gabon, Article 190 of Law No. 011/2014 of 28 August 2014 Regulating the Hydrocarbons Sector in the Republic of Gabon:

'*[l]a responsabilité sociale des entreprises se traduit par l'obligation qui est faite à ces dernières de contribuer aux enjeux de développement durable, à l'amélioration du bien-être des populations locales et à la protection de l'environnement.*'

[Free translation into English: '[c]orporate social responsibility results in an obligation on the part of the companies involved to contribute to sustainable development, improving the well-being of local populations and protect the environment.']

Note that Mozambique has adopted a specific text on corporate social responsibility in the mining industry (Order No. 21/2014 of 16 May 2014).

[33] The legislation in a number of countries provides a definition of the concept of corporate social responsibility. This is particularly the case in Burkina Faso, where Law No. 036-2015/CNT of 16 June 2015 Enacting the Mining Code of Burkina Faso defines this concept in its Article 5 as follows:

'*[l]a responsabilité d'une organisation vis-à-vis des impacts de ses décisions et activités sur la société et sur l'environnement, se traduisant par un comportement transparent et éthique, qui contribue au développement durable, y compris à la santé et au bien-être de la société; prend en compte les attentes des parties prenantes; respecte les lois en vigueur et est en accord avec les normes internationales de comportement, et qui est intégré dans l'ensemble de l'organisation et mis en _uvre dans ses relations.*'

[Free translation into English: '[t]he responsibility of an organisation in terms of the impact of its decisions and activities on society and on the environment, resulting in transparent and ethical behaviour which contributes to sustainable development, including in terms of health and the well-being of society; it takes stakeholder expectations into account; it complies with the laws in force and with international standards of behaviour and is integrated in the organisation as a whole and implemented in its relationships.']

22.50 Supplying the local market, employing national personnel and increasing tax revenues are all reasons which should stimulate host countries to encourage the processing of natural resources within their territory.

23

STIMULATING THE LOCAL ECONOMY

23.1 The local economy benefits from numerous obligations upon the private operator to give preference to national assets, services, companies and workers. Favouring the local content is a significant way to enhance the positive impact of the activities of extractive industries on the local economy. The national economy is also stimulated by injecting capital into the economic systems of the host State.

1 AN OBLIGATION TO GIVE PREFERENCE TO THE LOCAL MARKET

23.2 An obligation to give preference to the local market can be established on two levels, both as part of the exploitation agreement itself and through the signing of service contracts.

a National preference in favour of local equipment and personnel

23.3 In many cases the national preference obligation inextricably involves local goods and the use of local personnel. Whether the preferences relate to the employment of local personnel or locally manufactured goods, the national preference obligation generally falls upon the holders of mining licences or parties to the exploitation agreement. However, the obligation to give preference to local personnel and to purchase local equipment and supplies may in some cases extend to subcontractors[1] or, more generally, to companies working on behalf of the operator.[2]

[1] With respect to subcontractors employed as local personnel, for Cameroon, see Article 77 of Law No. 99-013 of 22 December 1999 Enacting the Petroleum Code; for the Central African Republic, Article 60 of Ordinance No. 93.007 of 25 May 1993 Enacting the Petroleum Code; for the Côte d'Ivoire, Article 134 of Law No. 2014-138 of 24 March 2014 Enacting the Mining Code and Article 53 of Law No. 96-669 of 29 August 1996 Enacting the Petroleum Code; for Guinea, Article 55 of Law No. L/2014/N°034/AN of 23 December 2014 Enacting the Petroleum Code; for Liberia, Article 2.5.9 of the 2002 Petroleum Law and Article 20.2 of the 2000 Minerals and Mining Law; for Mali, Article 137 of Law No. 2012-015 of 27 February 2012 Enacting the Mining Code; for Niger, Article 21 of the Law No. 2007-01 of 31 January 2007 Enacting the Petroleum Code; for Sierra Leone, Article 87 of the 2011 Petroleum Exploration and Production Act; and for South Sudan, Article 65 of the 2012 Petroleum Act. For subcontractors affected by the obligation to use locally manufactured goods, see in particular for Ghana Article 23(12) of the 1984 Petroleum (Exploration and Production) Act; for Sierra Leone, Article 88 of the 2011 Petroleum Exploration and Production Act; and for South Sudan, Article 63 of the 2012 Petroleum Act.

[2] For companies working on behalf of the operator which are concerned by the employment of local personnel, for Djibouti see Article 52 of Law No. 117-AN-05 of 16 October 2005 Enacting the Petroleum Code; for Niger, Article 103 of Ordinance No. 93-16 of 2 March 1993 Enacting the amended Mining Act; for São Tomé and Principe, see Article 57 of the Framework Law on Petroleum Operations; for Senegal, Article 53 of Law No. 98-05 of

The conditions in which this reference obligation applies vary according to whether it relates to local products or local personnel.

i *Purchasing local equipment and supplies*

23.4 Private operators are free in principle to import the goods required for mining and petroleum operations.[3] However, in order to favour national production and hence to stimulate the local economy, legislation and exploitation agreements generally burden the investor with an obligation to give preference to goods manufactured within the host State.[4]

---

8 January 1998 Enacting the Petroleum Code; for Chad, Article 69 of Law No. 011-PR-1995 of 20 June 1995 Enacting the Mining Code.

For companies working on behalf of the operator which are affected by the obligation to use locally manufactured goods, particularly for São Tomé and Principe see Article 58 of the Framework Law on Petroleum Operations.

[3] See Article 105 paragraph 1 of Law No. 2006-17 of 17 October 2006 Enacting the Mining Code and Mining Taxation in the Republic of Benin.

[4] In particular, the following should be mentioned: for Angola, Article 19 of Mining Act No. 31/11 of 23 September 2011 and Article 27 of Petroleum Act No. 10/04 of 12 November 2004; for Botswana, Article 12(1) of the Mines and Minerals Act and Article 10(1) of the Petroleum (Exploration and Production) Act; for the Republic of Congo, Article 101 of Law No. 4-2005 of 11 April 2005 Enacting the Mining Code and Article 22 of Law No. 24-34 of 23 August 1994 Enacting the Hydrocarbons Code; for the Democratic Republic of Congo, Article 273 of Law No. 007-2002 of 11 July 2002 Enacting the Mining Code; for Eritrea, Article 25 of Proclamation No. 68/1995 Promoting the Development of Mineral Resources and Article 12 of Proclamation No. 108/2000, A Revised Proclamation Governing Petroleum Operations; for Ethiopia, Article 34 of Proclamation No. 678/2010 Promoting Sustainable Development of Mineral Resources and Article 12 of Proclamation No. 295/1986 Regulating Petroleum Operations; for Ghana, Article 105 of the 2006 Minerals and Mining Act and Article 23(12) of the 1984 Petroleum (Exploration and Production) Act; for Equatorial Guinea, Article 52 of Mining Act No. 9/2006 of 3 November 2006; for Kenya, Article 9(1) of the Petroleum Act; for Malawi, Article 37(3) of the 1981 Mines and Mineral Act; for Mozambique, Article 22 of Petroleum Act No. 21-2014 of 18 August 2014; for Namibia, Article 50 of the 1992 Minerals Prospecting and Mining Act and Article 14 of the 1991 Petroleum (Exploration and Production) Act (Amended by Act No. 2 of 1993); for Uganda, Article 113(1) of the 2003 Mining Act and Article 125 of the 2013 Petroleum (Exploration, Development and Production) Act; for Rwanda, Article 49 of Law No. 13-2014 of 20 May 2014 On the Exploitation of Mines and Quarries; for São Tomé and Principe, Article 58 of the Framework law on Petroleum Operations; for Sierra Leone, Article 88 of the 2011 Petroleum Exploration and Production Act and Article 163 of the 2009 Mines and Minerals Act; for South Sudan, Article 63 of the 2012 Petroleum Act; for Swaziland, Article 29 of the 2011 Mines and Minerals Act; for Tanzania, Articles 10, 34 and 44 of the 2010 Mining Act; for Togo, Article 34 of Law No. 96-004 of 26 February 1996 Enacting the Mining Code; for Tunisia, Article 75 of Law No. 2003-30 of 28 April 2003 Enacting the Mining Code and Article 62(2) of Law No. 99-93 of 17 August 1999 Enacting the Hydrocarbons Code; for Zambia, Article 63 of the 2008 Petroleum (Exploration and Production) Act; Article 14 of WAEMU Regulation No. 18/2003/CM/UEMOA of 23 December 2003 Enacting the Community Mining Code and Article 11(5) of ECOWAS Directive C/DIR 3/05/09 of 27 May 2009 On the Harmonisation of the Guiding Principles and Policies in the Mining Sector.

With regard to standard contracts containing an obligation to give preference to local goods, the following should be mentioned: for Algeria, Article 14 of the Presidential Decree of 1 March 2003 Enacting the Model Mining Agreementt; for Namibia, Article 23 of the 1998 Model Petroleum Agreement.

23.5 An obligation to purchase locally produced materials and supplies is set out in the vast majority of exploitation agreements. However, this obligation above all results from mining and petroleum codes which aim to strengthen the local content.

The obligation to give preference to local products generally extends over the whole territory of the host State. However, a preference can be given to the immediate periphery of the mining or petroleum site.

Legal and contractual provisions can be of a very general nature with regard to the conditions in which the local preference obligation applies or, on the contrary, can contain precise provisions that leave very little to the discretion of the private operator.

(1) Obligation established in general terms

23.6 The private operator may on its own decide and apply provisions relating to local content according to the criteria stipulated in the exploitation agreement. The local State operator can subsequently request clarification from the private operator on the reasons for its choices. However, once the purchase has been made, this *a posteriori* verification becomes ineffective in practice.

Obligations drafted in these terms are becoming increasingly rare and are being renegotiated for previously signed exploitation agreements.

(2) Obligation with detailed contents

23.7 Other provisions are more complete and stipulate several criteria for the operator to determine under which circumstances it must give preference to local products; among these criteria, the quality, price, performance, technology (particularly in terms of environmental standards), safety, delivery deadlines or product availability should be mentioned.[5] Accordingly, depending on the criteria determined by law or by contract, the private operator

---

[5] For Algeria, see Article 14 of the Presidential Decree of 1 March 2003 Enacting the Model Mining Agreement; for Angola, Article 19 of the Mining Act No. 31/11 of 23 September 2011 and Article 27 of Petroleum Act No. 10/04 of 12 November 2004; for Botswana, Article 12(1) of the Mines and Minerals Act and Article 10(1) of the Petroleum (Exploration and Production) Act; in the Republic of Congo, Article 101 of Law No. 4-2005 of 11 April 2005 Enacting the Mining Code and Article 22 of Law No. 24-94 of 23 August 1994 Enacting the Hydrocarbons Code; for the Democratic Republic of Congo, Article 273 of Law No. 007-2002 of 11 July 2002 Enacting The Mining Code; for Equatorial Guinea, Article 52 of Mining Act No. 9/2006 of 3 November 2006; for Mozambique, Article 41 of Petroleum Act No. 21-2014 of 18 August 2014; for Uganda, Article 113(1) of the 2003 Mining Act; for Rwanda, Article 49 of Law No. 13-2014 of 20 May 2014 On Mine and Quarry Extraction; for São Tomé and Principe, Article 58 of the Framework Law on Petroleum Operations; for South Sudan, Article 63 of the 2012 Petroleum Act; for Tunisia, Article 75 of Law No. 2003-30 of 28 April 2003 Enacting the Mining Code; Article 14 of WAEMU Regulation No. 18/2003/CM/UEMOA of 23 December 2003 Enacting the Community Mining Code and Article 11(5) of ECOWAS Directive C/DIR 3/05/09 as 27 May 2009 On the Harmonisation of the Guiding Principles and Policies in the Mining Sector.

must give preference to national goods. However, the codes can stipulate a margin of preference favouring national goods and requiring the section of local companies if their prices do not exceed those of its foreign competitor within the limits of a certain percentage.[6]

Non-compliance or improper application of these provisions by the private operator can be monitored much more effectively and can incur the contractual liability of the private operator.

23.8 Local taxes applied to imported products play an important role as the host State does not hesitate to grant preferential tax regimes for the benefit of local products.

In practice, in the event of a local market shortage, the private operator may call on local importers.

23.9 The practical results of the obligation to give preference to local products are difficult to evaluate. However, in a large number of situations the construction or manufacture in the host State of the equipment intended for exploitation works does not provide immediate economic justification.

ii *Preference in the employment of local workers*

23.10 Mining and petroleum projects require large numbers of employees with a variety of qualifications. Seasonal fluctuations in the phases of work exert a strong impact on employment. In order to meet manpower needs, the operator may resort to employing expatriate personnel. In this case, the State can ensure that the operator has the possibility of employing the expatriate personnel required for performing the operations.

23.11 For the host State, preferential employment of local workers is a response to the sole concern of ensuring that its economy and its nationals benefit as much as possible from the positive impact resulting from the exploitation of its natural resources.[7] It would be wrong to interpret such provisions as representing an intention to boost or transfer technological expertise. To be more precise, in most situations the insertion of obligations to employ local personnel can only be the consequence of a prior training plan. In the longer term, the aim is to employ qualified local personnel in order to encourage the development of the national economy.

23.12 Numerous mining and petroleum codes contain provisions on the preferential employment of local workers.[8] The workers concerned by this

---

[6] See in particular, in Angola, Article 27 of Petroleum Act No. 10/04 of 12 November 2004, which provides that preference must be given to the use of local resources if the price does not exceed that of foreign competitors by more than 10%. For São Tomé and Principe, See also Article 58(1) of the Framework Law on Petroleum Operations, which applies the same percentage.

[7] Implementation of this objective involves in particular checking the destination of the salaries paid.

[8] The following should be mentioned: for Angola, Article 18 of Mining Act No. 31/11 of 23 September 2011, Article 86 of Petroleum Act No. 0/04 of 12 November 2004 and Article 5 of Executive Order No. 42/10 of 10 May 2010 On the Local Content; for Benin, Article 103 of Law No. 2006-17 of 17 October 2006 Enacting the Mining and Mining Taxation Code and Article 11 of Law No. 2016-18 of 17 October 2006 Enacting the Petroleum Code; for

obligation are generally nationals of the host State. However, a number of texts stipulate that as a matter of priority this preferential treatment must

---

Botswana, Article 10(2) of the Petroleum (Exploration and Production) Act and Article 12(2) of the Mines and Minerals Act; for Cameroon, Articles 12 and 77 of Law No. 99-013 of 22 December 1999 Enacting the Petroleum Code; for the Central African Republic, Article 60 of Ordinance No. 93.007 of 25 May 1993 Enacting the Petroleum Code; for the Republic of Congo, Article 101 of Law No. 4-2005 of 11 April 2005 Enacting the Mining Code and Article 21 of Law No. 24-94 of 23 August 1994 Enacting the Hydrocarbons Code; for the Democratic Republic of Congo, Article 273 of Law No. 007-2002 of 11 July 2002 Enacting the Mining Code; for the Côte d'Ivoire, Articles 18 and 53 of Law No. 96-669 of 29 August 1996 Enacting the Petroleum Code and Article 134 of Law No. 2014-138 of 24 March 2014 Enacting the Mining Code; for Djibouti, Article 52 of Law No. 117-AN-05 of 16 October 2005 Enacting the Petroleum Code; for Eritrea, Article 25 of Proclamation No. 68-1995 Promoting the Development of Mineral Resources and Article 12 of Proclamation No. 108/2000. A Revised Proclamation to Governing Petroleum Operations; for Ghana, Article 105 of the 2006 Minerals and Mining Act; for Guinea, Article 108 of Law No. L/2013/N°053/CNT of 8 April 2013 Amending Law No. 2011/006/CNT of 9 September 2011 Enacting the Mining Code and Article 55 of Law No. L/2014/N°034/AN of 23 December 2014 Enacting the Petroleum Code; for Equatorial Guinea, Article 52 of Mining Act No. 9/2006 of 3 November 2006; for Liberia, Article 2.5.9 of the 2002 Petroleum Law and Article 20(2) of the 2000 Minerals and Mining Law; for Madagascar, Article 7 of Order No. 2006-910 of 19 August 2006 Implementing the Mining Code; for Malawi, Article 25 of the 1981 Mines and Minerals Act and Article 29 of the 1983 Petroleum (Exploration and Production) Act; for Mali, Article 137 of Law No. 2012-015 of 27 February 2012 Enacting the Mining Code; for Mozambique, Article 12 of Petroleum Act No. 21-2014 of 18 August 2014; for Namibia, Article 50 of the 1992 Minerals Prospecting and Mining Act and Article 14 of the 1991 Petroleum (Exploration and Production) Act (amended by Act No. 2 of 1993); for Niger, Article 103 of Ordinance N° 93-16 of 2 March 1993 Enacting the Amended Mining Act and Article 21 of Law No. 2007-01 of 31 January 2007 Enacting the Petroleum Code; for Uganda, Article 113(2) of the 2003 Mining Act and Article 126 of the 2013 Petroleum (Exploration, Development and Production) Act; for Rwanda, Article 48 of Law No. 13-2014 of 20 May 2014 On the Exploitation of Mines and Quarries; for Senegal, Articles 17 and 53 of Law No. 98-05 of 8 January 1998 Enacting the Petroleum Code; for the Seychelles, Article 45(1) of the Petroleum Mining Act; for Sierra Leone, Article 164 of the 2009 Mines and Minerals Act and Article 87 of the 2011 Petroleum Exploration and Production Act; for South Sudan, Article 88 of the 2012 Mining Act and Article 65 of the 2012 Petroleum Act; for Swaziland, Article 28 of the 2011 Mines and Mineral Act; for Tanzania, Article 10(4) of the 2010 Mining Act; for Chad, Article 10 of Law No. 006-PR-2007 of 2 May 2007 On Hydrocarbons and Article 69 of Law No. 011-PR-1995 of 20 June 1995 Enacting the Mining Code; for Togo, Article 37(3) of Law No. 99-003 of 18 February 1999 Enacting the Hydrocarbons Code and Article 34 of Law No. 96-004 of 26 February 1996 Enacting the Mining Code; for Tunisia, Article 75 of Law No. 2003-30 of 28 April 2003 Enacting the Mining Code and Article 62(2) of Law No. 99-93 of 17 August 1999 Enacting the Hydrocarbons Code; for Zambia, Article 63(2) of the 2008 Petroleum (Exploration and Production) Act.

With regard to standard exploration contracts mentioning this preference, see in particular: for Algeria, Article 15 of the Presidential Decree of 1 March 2003 Enacting the Model Mining Agreement; for Mauritania, Article 62 of Law No. 2012/12 Regulating Mining Agreements and Approving the Model Mining Agreement; for the Seychelles, Article 18(1) of the 2013 Model Petroleum Agreement.

With regard to regional texts, the following should be mentioned: Article 16 of WAEMU Regulation No. 18/2003/CM/UEMOA of 23 December 2003 Enacting the Community Mining Code and Article 11(4) of ECOWAS Directive C/DIR 3/05/09 as 27 May 2009 On the Harmonisation of the Guiding Principles and Policies in the Mining Sector.

favour the local communities affected by the mining or petroleum project.[9] As a general rule, preferential employment of local workers is accompanied by a reduction in the expatriate workforce. Accordingly, the objective of replacing expatriate personnel is reminded by legislation in numerous countries.[10]

23.13 The texts applicable to the employment of personnel occasionally contain non-discrimination provisions prohibiting the operator from operating any distinction according to gender, race or religion.[11] Non-discrimination is also ensured by allocating identical salaries and rights between national and expatriate employees in equivalent conditions.[12]

Employment of the local workforce is envisaged in a more or less flexible framework depending on the legal and contractual provisions at stake.

(1) An obligation defined in variable terms

23.14 A preference granted to the local workforce can be established in variable terms; accordingly, the employment of nationals can be left to the discretion of the private operator or can be framed by mandatory provisions.

(a) Employment of the local workforce is left to the discretion of the private operator

23.15 The private operator has a definite interest in employing the local workforce. From a financial point of view, the cost of an expatriate worker can only be higher than the cost of a local worker, for example due to the various expatriate bonuses paid; moreover, frequently the price of qualified personnel is substantially higher in the country of origin of the expatriates than in the host State. Private operators can also benefit from the positive image resulting from the employment of local personnel.

23.16 It should be recognised that the degree of the obligation to give preference to the employment of local personnel varies significantly from one contract to another. Clauses leaving the private operator with room to manoeuvre are necessarily marked with a degree of trust in the judgement capacity of the private operator, while others are less permissive.

The flexibility that can be left to the operator with regard to the employment of local personnel may also result from the legal framework; the legislation in a number of countries stipulates that the operator must give as

---

[9] For instance, for Angola see Article 18 of Mining Act No. 31/11 of 23 September 2011; for Guinea, see Article 108 of Law No. L/2013/N°053/CNT of 8 April 2013 Amending Law No. 2011/006/CNT of 9 September 2011 Enacting the Mining Code.

[10] In this connection, for Algeria see Article 15 of the Presidential Decree of 1 March 2003 Enacting the Model Mining Agreement; for Mali, Article 137 of Law No. 2012-015 of 27 February 2012 Enacting the Mining Code; for Mauritania, Article 62(3) of Law No. 2012/012 Regulating Mining Agreements and Approving a Model Mining Agreement.

[11] For South Sudan, see Article 65(4) of the 2012 Petroleum Act.

[12] For instance, for Angola see Article 86(2) of Petroleum Act No. 10/04 of 12 November 2004 and for São Tomé and Principe, see Article 57(3) of the Framework Law on Petroleum Operations.

much preference as possible to employing the local workforce but leaves a margin of discretion to the private operator.[13]

23.17 Usually this discretion is limited by a number of legal and regulatory criteria. The mining and petroleum codes and, if applicable, the implementation decrees, stipulate the conditions in which private operators must use the local workforce. In this connection, particular mention should be made of the level of experience, qualifications, degrees obtained or the profitability, effectiveness and security of the project.

(b) Employment of the local workforce is subject to binding regulations

23.18 Exploitation agreements or legal and regulatory provisions can set targets to be reached in relation to the employment of local personnel; depending on the case in point, operators can benefit from some flexibility in terms of the timeframe for complying with this obligation.

23.19 The targets to be reached are frequently set out in detail in exploitation agreements and can also appear in national legislation. This is the case of the mining code enacted in Guinea, which sets out in great detail the percentage of nationals who must be employed to the various posts during each operational phase.[14]

23.20 Although the requirement to reach the employment targets is rigid in some cases, a number of exploitation agreements stipulate that the private operator will only be required to reach the employment targets to the extent that they are realistic, depending on the profitability and efficient implementation of the exploitation operations on the relevant projects.

23.21 Verifying the non-availability of local personnel is usually subject to indications specified in the exploitation agreement or in the applicable legislation. This may include action such as advertising jobs in the press[15] or call for tenders in order to check whether or not local personnel are available.[16]

(2) A preference subject to a fluctuating scope

23.22 The scope of the obligation to give preference to the employment of local personnel on the part of the private operator can vary from one state

[13] In this connection, for Ghana see Article 105 of the 2006 Minerals and Mining Act; for Botswana, Article 12 of the Mines and Minerals Act and Article 10 of the Petroleum (Exploration and Production) Act; for Eritrea, Article 12 of Proclamation No. 108/2000, A Revised Proclamation Governing Petroleum Operations; for Uganda, Article 113(2) of the 2003 Mining Act; for Swaziland, Article 28 of the 2011 Mines and Mineral Act; for Zambia, Article 63 of the 2008 Petroleum (Exploration and Production) Act.

[14] For instance, see Article 108 of Law No. L/2013/N°053/CNT of 8 April 2013 Amending Law No. 2011/006/CNT of 9 September 2011 Enacting the Mining Code in the Republic of Guinea.

[15] In this connection, see Article 108 of Law No. L/2013/N°053/CNT of 8 April 2013 Amending Certain Certain Provisions of Law No. 2011/006/CNT of 9 September 2011 Enacting the Mining Code in the Republic of Guinea; for Mozambique, Article 12(4) of Petroleum Act No. 21-2014 of 18 August 2014.

[16] In this respect, see for Chad see Article 53(2) of Law No. 006-PR-2007 of 2 May 2007 On Hydrocarbons.

to another depending on the provisions contained in the mining and petroleum codes. These differences can relate to the positions covered by the local preference or the relevant project phases.

(a) Positions covered by the local preference

23.23 The local preference in favour of nationals applies differentially depending on the jobs involved. A distinction is made between qualified and unskilled positions.

23.24 The preference granted to nationals can be restricted to unskilled jobs. In order to favour the employment of local personnel in jobs not requiring qualifications, the texts occasionally prohibit employing unqualified expatriate personnel. All jobs not requiring any particular qualifications are therefore reserved for nationals of the host State.[17] Such a provision may appear surprising, as private operators have an interest in employing local personnel for this type of employment and qualification.

23.25 The preference given to the local workforce is nevertheless intended to extend to all jobs, whether skilled or not. Numerous texts stipulate that the obligation to favour the employment of local personnel applies for the purposes of allowing access to all jobs corresponding to the desired qualifications,[18] including management positions.[19]

More specific provisions contain details of the management positions that must be occupied by local personnel;[20] in this way they contribute to the purpose of the national preference, ie the development of the host State.

---

[17] In this respect, for Guinea see Article 108 of Law No. L/2013/N°053/CNT of 8 April 2013 Amending Law No. 2011/006/CNT of 9 September 2011 Enacting the Mining Code:
'*[l]e titulaire d'un Titre minier ou d'une Autorisation ainsi que les entreprises travaillant pour son compte sont tenus d'employer exclusivement des Guinéens pour tous les emplois ne nécessitant pas de qualification. La direction du titulaire du Titre minier ou de l'Autorisation pourra réserver certains postes ne nécessitant pas de qualification aux ressortissants de la Communauté locale.*'
[Free translation into English: '[t]he holder of a mining licence or an authorisation and the companies working on its behalf are required to employ only Guineans for all non-qualified jobs. The management of the holder of the mining licence authorisation may reserve certain non-qualified positions to nationals of the local community.']
For Sierra Leone, see also Article 87(2) of the 2011 Petroleum Exploration and Production Act and Article 164 of the 2009 Mines and Minerals Act; for South Sudan, Article 88(2) of the 2012 Mining Act and Article 65(3) of the 2012 Petroleum Act.

[18] In this respect, for Angola see Article 86 of Petroleum Act No. 10/04 of 12 November 2004 and Article 5 of Executive Order No. 45/10 of 10 May 2010 relating to the Local Content; for Niger, Article 103 of Ordinance No. 93-16 of 2 March 1993 Enacting the Amended Mining Act; for São Tomé and Principe, Article 57 of the Framework Law on Petroleum Operations; for the Seychelles, Article 45(1) of the Petroleum Mining Act.

[19] For Guinea, see Article 55 of Law No. L/2014/N°34/AN of 23 December 2014 Enacting the Petroleum Code and Article 108 of Law No. L/2013/N°053/CNT of 8 April 2013 Amending Certain Certain Provisions of the Mining Code.

[20] Accordingly, in Guinea, the Mining Code provides that as from the date when commercial production begins, the assistant general manager must be a Guinean and that at the end of the five-year period following this date, the managing director of the operating company must be a Guinean (Article 108 of Law No. L/2013/N° 053/CNT of 8 April 2013 Amending

(b) Phases of operations affected by the obligation

23.26 The obligation to give preference to local personnel can vary according to the state of progress of the operations and the different project phases.[21]

23.27 *A contrario* certain provisions stipulate that the preferential employment obligation must be applied for nationals irrespective of the phase involved.[22]

23.28 Indeed, numerous texts make no distinction between the different project phases, in which case the obligation to employ local personnel is assumed to apply throughout the duration of the operations.

(3) Practical results of the obligation to give preference to local personnel

23.29 The practical results of the obligation to give preference to the employment of local workers are very difficult to assess. The poor success rate of these provisions in numerous countries is frequently attributed to the fact that the management bodies of companies have become aware of the extremely low level of know-how among the local workforce and hence the importance of the efforts to be made to fill these gaps. The importance of technological expertise and its quick development is also a factor that moderates the limits on the use of expatriate personnel. The adoption of a moderate policy to replace expatriate personnel with local personnel can be explained by a reluctance to risk slowing down operational works.

23.30 In practice, the success of the obligation to give preference to local personnel depends broadly on the establishment of training programmes[23] that include an increasing number of results-based obligations by means of successive and permanent monitoring of the implementation of the transfer of expertise. A verified and monitored transfer of expertise for the benefit of local populations is the only means to ensure that investors will employ a qualified local workforce in the long term.

b National preference in the use of local service providers

23.31 When operators cannot themselves provide or find the equipment and personnel required for the operations, they make use of third parties using subcontractors. These subcontracts, which constitute application contracts, formalise a vertical grouping between the main operators and

---

the Mining Code). With regard to the employment of nationals in management posts for Mozambique, see also Article 12 of the Petroleum Act No. 21-2014 of 18 August 2014.

[21] For Guinea, see Article 108 of Law No. L/2013/N°053/CMT of 8 April 2013 Amending the Mining Code which establishes the minimum quota of Guinean employees by category in the different phases of development of the project.

[22] For Guinea, see Article 55 of Law No. L/2014/N° 034/AN of 23 December 2014 Enacting the Petroleum Code; for Uganda, Article 126 of the 2013 Petroleum (Exploration, Development and Production) Act; for the Seychelles, Article 18 of the 2013 Model Petroleum Agreement; for Sierra Leone, Article 87 of the 2011 Petroleum Exploration and Production Act and Article 164 of the 2009 Mines and Minerals Act; for South Sudan, Article 65(2) of the 2012 Petroleum Act.

[23] With regard to training local workers see para 22.8 above et seq.

accessory operators, the latter appearing as subcontractors to the former. The difficulty for the operator consists in allocating all supplies and services but also ensuring overall coordination.[24]

23.32 In order to favour the development of the local economy, mining and petroleum codes and exploitation agreements frequently prioritise local service providers in mining and petroleum operations. These provisions generally refer to the use of local companies for the provision of services, particularly in relation to construction, supply or procurement contracts.[25]

---

[24] For the concepts of 'operator' and 'contractor' see para 17.12 above et seq.

[25] In this connection, for Algeria, see Article 14 of the Presidential Decree of 1 March 2003 Enacting the Model Mining Agreement; for Angola, Article 19 of Mining Act No. 31/11 of 23 September 2011 and Article 27(1) of Petroleum Act No. 10/04 of 12 November 2004; for Botswana, Article 12(1) of the Mines and Minerals Act and Article 10(1) of the Petroleum (Exploration and Production) Act; for Cameroon, Article 76 of Law No. 99-013 of 22 December 1999 Enacting the Petroleum Code; for the Central African Republic, Article 59 of Ordinance No. 93.007 of 25 May 1993 Enacting the Petroleum Code; for the Republic of Congo, Article 101 of Law No. 4-2005 of 11 April 2005 Enacting the Mining Code and Article 22 of Law No. 24-94 of 23 August 1994 Enacting the Hydrocarbons Code; for the Democratic Republic of Congo, Article 273 of Law No. 007-2002 of 11 July 2002 Enacting the Mining Code; for the Côte d'Ivoire, Article 52 of Law No. 96-669 of 29 August 1996 Enacting the Petroleum Code and Article 133 of Law No. 2014-138 of 24 March 2014 Enacting the Mining Code; for Eritrea, Article 12 of Proclamation No. 108/2000, A Revised Proclamation Governing Petroleum Operations and Article 25 of Proclamation No. 68/1995 Promoting Development of Mineral Resources; for Ethiopia, Article 34 of Proclamation No. 678/2010 Promoting Sustainable Development of Mineral Resources and Article 12 of Proclamation No. 295/1986 Regulating Petroleum Operations; for Gabon, Article 197 of Law N° 05-2000 Enacting the Mining Code; for Ghana, Article 105 of the 2006 Minerals and Mining Act and Article 23(12) of the 1984 Petroleum (Exploration and Production) Act; for Guinea, Article 107 of Law No. L/2013/N°053/CNT of 8 April 2013 Amending Law No. 2011/006/CNT of 9 September 2011 Enacting the Mining Code and Article 54 of Law No. L/2014/N° 034/AN of 23 December 2014 Enacting the Petroleum Code; for Equatorial Guinea, Article 52 of Mining Act No. 9/2006 of 3 November 2006; for Liberia, Article 2.5.8 of the 2002 Petroleum Law; for Mozambique, Article 22(4) of Mining Act No. 20-2014 of 18 August 2014 and Article 41(4) of Petroleum Act No. 21-2014 of 18 August 2014; for Namibia, Article 50 of the 1992 Mineral Prospecting and Mining Act, Article 14 of the 1991 Petroleum (Exploration and Production) Act; for Niger, Article 102 of Ordinance No. 93-16 of 2 March 1993 Enacting the Amended Mining Act and Article 20 of Law No. 2007-01 of 31 January 2007 Enacting the Petroleum Code; for Uganda, Article 113 of the 2003 Mining Act and Article 125 of the 2013 Petroleum (Exploration, Development and Production) Act; for Rwanda, Article 49 of Law No. 13-2014 of 20 May 2014 On the Exploitation of Mines and Quarries; for São Tomé and Principe, Article 58(1) of the Framework Law on Petroleum Operations; for Senegal, Article 53 of Law No. 98-05 of 8 January 1998 Enacting the Petroleum Code; for the Seychelles, Article 45(3) of the Petroleum Mining Act and Article 18(7) of the 2013 Model Petroleum Agreement; for Sierra Leone, Article 88 of the 2011 Petroleum Exploration and Production Act and Article 163 of the 2009 Mines and Minerals Act; for South Sudan, Article 63 of the 2012 Petroleum Act and Article 88(7) of the 2012 Mining Act; for Swaziland, Article 29 of the 2011 Mines and Mineral Act; for Chad, Article 68 of Law No. 011-PR-1995 of 20 June 1995 Enacting the Mining Code; for Togo, Article 37 of Law No. 99-003 of 18 February 1999 Enacting the Hydrocarbons Code and Article 34 of Law No. 96-004 of 26 February 1996 Enacting the Mining Code; for Tunisia, Article 62.2 of Law No. 99-93 of 17 August 1999 Enacting the Hydrocarbons Code and Article 75 of Law No.

23.33 Some more specific provisions stipulate the preferential use of national laboratory services for analysing samples and operational trials.[26] In other cases, national and regional texts may extend the preference given to local companies to public companies.[27]

23.34 In general, the importance allocated to this preference is such that the use of local companies is occasionally taken into consideration in granting mining licences.[28]

23.35 The purpose of these provisions is to promote the use of local companies. In this connection, certain mining and petroleum codes include an obligation for operators to cooperate in creating jobs on a national level[29] and to promote local entrepreneurship,[30] occasionally by establishing a support plan for national companies.[31]

The derivative contract must include a certain number of provisions in line with those contained in the exploitation agreement.

i *Regulating the preferential use of local service providers*

23.36 Mining and petroleum codes generally stipulate a preference in favour of local service providers, while establishing certain conditions. Two variants provide more or less strict control over the choice of service provider by private operators: service contracts can be signed following a joint decision of the operators or can be subject to prior monitoring by the host state.

2003-30 of 28 April 2003 Enacting the Mining Code; for Zambia, Article 63 of the 2008 Petroleum (Exploration and Production) Act.

With regard to the preference given to national services, other national texts make reference to the exploitation agreement: in particular for Tanzania, see Article 10 of the 2010 Mining Act.

The preference given to the local service providers is also recognised on a regional scale by Article 14 of Regulation No. 18/2003/CM/UEMOA of 23 December 2003 Enacting the Community Mining Code and Article 11(5) of ECOWAS Directive C/DIR 3/05/09 of 27 May 2009 On the Harmonisation of the Guiding Principles and Policies in the Mining Sector.

[26] For Mauritania, see Article 65 of Law No. 2012/012 Regulating Mining Agreements and Approving the Model Mining Agreement.

[27] In this connection, see Article 11(5) of ECOWAS Directive C/DIR 3/05/09 as 27 May 2009 On the Harmonisation of the Guiding Principles and Policies in the Mining Sector; for Sierra Leone, Article 163 of the 2009 Mines and Minerals Act; for Ghana, Article 105 of the 2006 Minerals and Mining Act.

[28] For Malawi, see in particular Article 37(3) of the 1981 Mines and Minerals Act and Article 4 of the 1983 Petroleum (Exploration and Production) Act; for Zimbabwe, Article 160 of the Mines and Minerals Act.

[29] See Article 101 of Law No. 4-2005 of 11 April 2005 Enacting the Mining Code in the Republic of Congo.

[30] For Mozambique, see Article 34 of Mining Act No. 20-2014 of 18 August 2014; for Namibia, Article 50 of the 1992 Minerals Prospecting and Mining Act and Article 14 of the 1991 Petroleum (Exploration and Production) Act (amended by Act N° 2 of 1993).

[31] For Guinea see Article 54 of Law No. L/2014/N°034/AN of 23 December 2014 Enacting the Petroleum Code and Article 107 of Law No. 2011/006/CNT of 9 September 2011 Enacting the Mining Code.

(1) Criteria for preferential use of local service providers

23.37 Compliance by operators with the requirement to give preference to local service providers involves introducing detailed criteria to determine in which cases the use of foreign service providers is permissible.

In this way mining and petroleum codes can establish the criteria in reasonable detail. Broadly applied criteria include equivalence in terms of prices, payment conditions, profitability, efficiency, quality, quantity, guarantees, delivery deadlines or security. Accordingly, in equivalent circumstances, private operators must give preference to services provided by local companies.

However, the applicable regulations can establish a margin of preference benefiting local service providers by stipulating that the local company must be selected preferentially in cases where its price does not exceed that of its foreign competitor by more than a certain percentage.[32]

In order to ensure the use of local service providers, certain texts may establish minimum quotas with which the operator must comply when using subcontractors. The mining code in Guinea establishes such an obligation.[33]

(2) Introduction of more or less strict control over the choice of service providers by the operator

23.38 The choice made by operators on the use of a service provider can be more or less restricted depending on whether the service contract is signed following a joint decision of the operators or is subject to monitoring by the other operators.

As previously mentioned, the principle remains that the operator is guaranteed free choice of its subcontractors.[34]

(a) Service contracts signed following a joint decision of the operators

23.39 In general, requests for signing service contracts are made by private operators and have to be justified.

In joint operations, the other operators must be previously consulted and must subsequently receive a copy of the contracts that may have been signed. In general, operators who wish to use a subcontractor must send prior general financial and technical information to the other operators about the companies approached for implementing subcontract works.

Verification of the terms contained in the service contracts essentially concerns the definition and conditions for implementing the subcontracted works.

---

[32] In this connection, for Angola, see Article 27 of Petroleum Act No. 10/04 of 12 November 2004 which stipulates that local service providers must be used preferentially if their price does not exceed that of foreign competitors by more than 10%.

[33] Article 107 of Law No. 2011/006/CNT of 9 September 2011 Enacting the Mining Code mentions the minimum proportion of small and medium-sized enterprises, small and medium-sized industries and companies owned or controlled by Guineans which must intervene at the different stages of the mining project.

[34] See para 17, 2.9 above.

(b) Service contracts signed by one operator without prior monitoring of the other operator

23.40 Outsourcing contracts may be signed without the intervention of the other operators involved in the projects. The extent of this freedom generally depends on their financial scale.

The exploitation agreement stipulates protection to compensate for any use that is hasty or not reasonably justified.

As a general rule, the choice of company will be preceded by a tender procedure, in which the prices, quality and speed of implementation of the services will have been specified. The call for tenders must contain all information on the work to be implemented.

ii *Harmonisation of contractual conditions in the context of outsourcing contracts*

23.41 The implementation of an outsourcing contract, in particular for subcontracted works, can give rise to serious problems and tension between the subcontractor, on the one hand, and the operators, on the other. This situation may have damaging implications for the exploitation agreement itself.

This is why outsourcing contracts and occasionally even mining and petroleum codes generally contain strict provisions with regard to issues such as the confidentiality of information and the subcontractor's liability.

(1) The confidentiality of information related to the project

23.42 The subcontractor is generally under an obligation to maintain confidentiality, not only with regard to information transferred to it for the purposes of carrying out the works, but also with regard to information that it might obtain during the performance.

Any infringement of these obligations incurs the liability of the subcontractor with respect to the operators or even occasionally with respect to the host state.

(2) The contractor's liability

23.43 Numerous problems may appear in the implementation of a service contract related to an exploitation agreement; breaches in its implementation can result in damages for the operators or third parties.

23.44 In principle, in the context of an exploitation agreement, actions on the part of the subcontractor incur the liability of the operator managing the contract which should have performed the subcontracted works; this means that operator's liability can be incurred if the subcontractor is at fault. This is justified by the fact that the operator managing the project retains management and supervisory powers over all operations. In particular, if required it must provide technical assistance to the subcontractor and provide it with information in the light of the information and experience that it has already acquired on the project since the works began; the liability of the subcontractor to the operator managing the project is assessed with regard to these instructions and this support. This principle is enshrined in a number of mining and petroleum codes, which specify that the private operator may use subcontractors under its own liability.[35]

23.45 In view of the principle of the private operator's liability for the actions of its subcontractors, it is important for private operators that the outsourcing contract should reflect the provisions of the exploitation agreement. Accordingly, the outsourcing contract must contain all commitments contained in the exploitation agreement, in particular in terms of obligations, penalties and liability. In this respect reference is made to mirror, back-to-back or pass-through contracts.

23.46 Exploitation agreements may contain special provisions in cases where the subcontractor's application has been imposed by the local authorities or by the local State operator. In this situation, it would be fair for the latter to assume their responsibilities; in particular, it can be stipulated that these subcontractors assume all obligations resulting from the contract at their own expense and agree to fully indemnify the private operators against any prejudice or claim ensuing from their failure to act.

c National preference with regard to the supply of the local market

23.47 Ensuring that the local economy benefits from mining and petroleum projects also requires priority to be given for meeting local needs in terms of natural resources. Giving priority to meeting local economic needs involves the establishment of a preferential obligation to supply the local market.

23.48 Numerous mining and petroleum codes and exploitation agreements contain preferential local market supply provisions.[36] Under the terms

[35] In this connection, for Cameroon, see Article 75 of Law No. 99-013 of 22 December 1999 Enacting the Petroleum Code; for the Central African Republic, Article 58 of Ordinance No. 93.007 of 25 May 1993 Enacting the Petroleum Code; for the Côte d'Ivoire, Article 131 of Law No. 2014-138 of 24 March 2014 Enacting the Mining Code and Article 51 of Law No. 96-669 of 29 August 1996 Enacting the Petroleum Code; for Niger, Article 19 of Law No. 2007-01 of 31 January 2007 Enacting the Petroleum Code; for Togo, Article 37(1) of Law No. 99-003 of 18 February 1999 Enacting the Hydrocarbons Code.

[36] In particular see Article 105, paragraph 4 of Law No. 2016-17 of 17 October 2006 Enacting the Mining and Mining Taxation Code in the Republic of Benin; for Cameroon, Article 79 of Law No. 99-013 of 22 December 1999 Enacting the Petroleum Code; for the Central African Republic, Article 62 of Ordinance No. 93.007 of 25 May 1993 Enacting the Petroleum Code; for the Republic of Congo, Article 23 of Law No. 24-94 of 23 August 1994 Enacting the Hydrocarbons Code; for the Côte d'Ivoire, Article 55 of Law No. 96-669 of 29 August 1996 Enacting the Petroleum Code; for Djibouti, Article 53 of Law No. 117-AN-05 of 16 October 2005 Enacting the Petroleum Code; for Ethiopia, Article 20 of Proclamation No. 295/1986 Regulating Petroleum Operations; for Guinea, Article 139 paragraph 2 of Law No. L/2013/N° 053/CNT of 8 April 2013 Amending Law No. 2011/006/CNT of 9 September 2011 Enacting the Mining Code and Article 41 of Law No. L/2014/N° 034/AN Enacting the Mining Code; for Equatorial Guinea, Article 50 of Law No. 9/2006 of 3 November 2006 Enacting the Mining Code; for Mozambique, Article 11 of Petroleum Act No. 21-2014 of 18 August 2014 and Article 21 of Mining Act No. 20-2014 of 18 August 2014; for Uganda, Articles 121 et seq. of the 2013 Petroleum (Exploration, Development and Production) Act; for Senegal, Article 54 of Law No. 98-05 of 8 January 1998 Enacting the Petroleum Code; for

of this obligation, the operator will have to give priority to meeting the needs of the internal market, in particular by preferentially supplying processing units installed within the country. Once national consumption needs have been met, the operator will be free to dispose of the remainder of the resources as it sees fit.[37]

23.49 Depending on the case in point, the local market can be supplied automatically or on the basis of a decision of the relevant authorities. Automatic supply of the local market can in particular result from a provision obliging the operator to deliver a share of the production intended for local market needs every year, this share being generally set out in the exploitation agreement.[38]

In some cases, the law can stipulate in favour of private operators that the supply obligation can only apply to a maximum percentage of the quantities extracted.

23.50 The price for selling the production can be determined by agreement or by law. Accordingly, the legislation can stipulate that the transfer price must reflect the market price[39] or be determined by mutual agreement between the parties,[40] particularly under the exploitation agreement.[41] However, the reference to the market price is not always applied; accordingly, the legislation in a number of countries can stipulate that the

---

Chad, Article 58 of Law No. 006-PR-2007 of 2 May 2007 On Hydrocarbons; for Togo, Article 37(4) of Law No. 99-003 of 18 February 1999 Enacting the Hydrocarbons Code; for Tunisia, Article 50 of Law No. 99-93 of 17 August 1999 Enacting the Hydrocarbons Code.

With regard to supplying the local market, a number of mining and petroleum codes make reference to the exploitation agreement. This is particularly the case in Botswana, with Article 39 of the Petroleum (Exploration and Production) Act.

[37] For Cameroon, see Article 79 of Law No. 99-013 of 22 December 1999 Enacting the Petroleum Code; for the Côte d'Ivoire, Article 55 of Law No. 96-669 of 29 August 1996 Enacting the Petroleum Code; for Senegal, Article 54 of Law No. 98-05 of 8 January 1998 Enacting the Petroleum Code.

[38] For Gabon, see Article 121 of Law No. 011/2014 of 28 August 2014 Regulating the Hydrocarbons Sector in the Republic of Gabon:

'*[t]out producteur de pétrole brut sur le territoire national est tenu de livrer chaque année sa quotepart de brut déterminée dans le contrat d'hydrocarbures et destiné aux besoins du marché intérieur définis par l'État.*'

[Free translation into English: '[a]ll crude oil producers within the country are required every year to deliver their share of the crude determined in the hydrocarbon contract and intended to meet the needs of the internal market defined by the State.'].

[39] For Djibouti, see Article 53 of Law No. 117-AN-05 of 16 October 2005 Enacting the Petroleum Code; for Senegal, Article 54 of Law No. 98-05 of 8 January 1998 Enacting the Petroleum Code. The Tunisian Hydrocarbon Code, less favourable to the private operator, stipulates that the transfer price is the normal FOB price obtained by the operator for its export sales, reduced by ten per cent (Article 50.1).

[40] For Ethiopia, see Article 20 of Proclamation No. 295/1986 to Regulating Petroleum Operations.

[41] For Chad, see Article 58 of Law No. 006-PR-2007 of 2 May 2007 On Hydrocarbons and for Togo, Article 37(4) of Law No. 99-003 of 18 February 1999 Enacting the Hydrocarbons Code.

transfer price should be calculated according to the price made official by the host state and will match an allowance to the benefit of the latter.[42]

23.51 Apart from the provisions for supplying the local market, a number of texts provide an option for the State to requisition the extracted substances subject to a fair and equitable compensation to the benefit of the private operator.[43]

23.52 The issues related to the obligation to supply the national market are substantial for the local economy in that this mechanism favours the energy independence of the host state.

d Mechanisms favouring effectiveness of the national preference

23.53 In order to ensure compliance with the preference given to the employment of nationals and to the use of local goods and services, a number of states have established mechanisms to ensure the effectiveness of the national preference in mining and petroleum operations.

23.54 Compliance with the preference obligation in particular involves establishing a selection procedure to obtain tenders from local companies. In this respect, a call for tender procedure can be imposed for awarding contracts.[44]

23.55 Encouraging compliance with the local preference obligation also involves the establishment of mechanisms that subordinate the granting of benefits to compliance with the preference requirements. As such, mining and petroleum codes can impose conditions on the application of tax advantages in cases where imported goods could have been supplied locally.[45]

23.56 In order to prevent the operator from circumventing local preference requirements, African states have occasionally established mecha-

---

[42] For Gabon, see Article 121 paragraph 3 of Law No. 011/2014 of 28 August 2014 Regulating the Hydrocarbons Sector in the Republic of Gabon:

> '*[t]out producteur de pétrole brut sur le territoire national est tenu de livrer chaque année sa quotepart de brut déterminée dans le contrat d'hydrocarbures et destiné aux besoins du marché intérieur définis par l'État.*'
>
> [Free translation into English: '[t]he price for transfer by the contractor of the quantity of hydrocarbons intended to meet national industry needs is equal to the price made official by the government of Gabon, matched by an allowance of fifteen per cent (15%) which constitutes a petroleum cost.'].

[43] For Burundi, see Article 193 of Decree-Law No. 1/138 of 17 July 1976 Enacting the Mining and Petroleum Codes:

> '*[d]ans un but d'intérêt général, les substances concessibles extraites peuvent être réquisitionnées par le Gouvernement moyennant une juste et équitable indemnité, pour le ravitaillement national.*'
>
> [Free translation into English: '[f]or the purposes of the public interest, transferable extracted substances may be requisitioned by the Government by means of fair and equitable compensation, for the purposes of provisioning the country.'].

[44] For Mozambique, see Article 22 of the Mining Act No. 20-2014 of 18 August 2014 and Article 41(2) of the Petroleum Act No. 21-2014 of 18 August 2014.

[45] For the Central African Republic, see Article 127 of Law No. 09-005 of 29 April 2009 Enacting the Mining Code.

nisms requiring foreign suppliers of goods and services to work jointly with a local company.[46] The regulations can also stipulate that in the case of imported goods, the private operator will be obliged to act through a local intermediary in order to implement this import.[47]

23.57 States can also introduce monitoring mechanisms, particularly by imposing an obligation on the operator to submit regular reports to the relevant authority to check compliance with the national preference requirements.[48] This monitoring can also involve sending contracts concluded as part of mining and petroleum operations to the relevant authority.

23.58 Ensuring compliance with the preference requirements also involves the establishment of sanctions and redress mechanisms in the event of failure by the private operator to comply with the national preference obligations. A number of mining and petroleum codes provide the option of taking court or administrative action in cases of non-compliance with the national preference requirement.[49] In addition, the applicable regulation can oblige the operator to pay compensation in the event of non-compliance with the national preference requirements.[50] Other, more stringent texts consider that infringing a preference obligation constitutes an offence[51] or may lead to withdrawal of the mining licence.[52] Finally, a number of provisions stipulate that contracts concluded in breach of the provisions establishing a national preference benefiting the local market are null and void.[53]

---

[46] For Mozambique, see Article 22(2) of Mining Act No. 20-2014 of 18 August 2014 and article 41(2) of Petroleum Act No. 21-2014 of 18 August 2014; for Uganda, Article 125(2) of the 2013 Petroleum (Exploration, Development and Production) Act.

[47] For Namibia, see Article 23(1) of the 2007 Model Petroleum Agreement.

[48] With regard to the provision of reports on the employment of local personnel, for Guinea, in particular, see Article 55 of Law No. L/2014/N°034/AN of 23 December 2014 Enacting the Petroleum Code and Article 108 of Law No. L/2013/N°053/CNT of 8 April 2013 Amending Law No. 2011/006/CNT of 9 September 2011 Enacting the Mining Code.

As regards submitting reports on the use of local goods and services, for the Republic of Congo, in particular, see Article 22 of Law No. 24-34 of 23 August 1994 Enacting the Hydrocarbons Code; for Uganda, Article 125(2) of the 2013 Petroleum (Exploration, Development and Production) Act; for Rwanda, Article 49(3) of Law No. 13-2014 of 20 May 2014 On the Exploitation of Mines and Quarries; for Sierra Leone, Article 88(3) of the 2011 Petroleum Exploration and Production Act.

[49] For Angola, see Article 19(2) of Mining Act No. 31/11 of 23 September 2011.

[50] As regards penalties for failing to fulfil the obligations to employ national personnel, for Guinea, in particular, see Article 108 of Law No. L/2013/N°053/CNT of 8 April 2013 Amending Law No. 2011/006/CNT of 9 September 2011 Enacting the Mining Code.

[51] See Article 88(8) of the 2012 Mining Act of South Sudan, which considers that a breach by a private operator of its employment obligations to employ local personnel constitutes an offence.

[52] This is the case for the mining code of Sierra Leone, which stipulates that a violation of employment and personnel training obligations constitutes a breach of a substantial obligation that must be sanctioned by suspending or withdrawing the mining licence (Article 164(7) of the 2009 Mines and Minerals Act).

[53] For Angola, see Article 27(3) of Petroleum Act No. 10/04 of 12 November 2004.

2 CAPITAL INJECTION INTO THE LOCAL ECONOMY

23.59 Capital injected into the local economy can either originate from the results of the exploitation itself or from amounts paid for services and salaries.

a Destination of capital drawn from exploitation

23.60 The destination of the capital drawn from the exploitation of natural resources is an essential issue for the host state. Injection of the products generated by extraction activities into the national economy represents a substantial source of development.

The foreign exchange regime applicable to petroleum and mining operations, although granting numerous guarantees to the private operator, also makes it possible to monitor the reinvestment of the capital generated by extractive industries.

The applicable foreign exchange regulations are determined on a national scale by the mining and petroleum codes, general legislation on foreign exchange regulation, the applicable investment codes and occasionally by specific texts[54] . On a regional level, the application of the relevant supra-national texts[55] should also be noted, the development of which leads to the establishment of increasingly restrictive regulations.

As a general rule, once common tax law has been applied, the profits released will not be freely available to the private operator in all cases, particularly for export. Accordingly, two series of provisions can be identified: provisions granting free movement of capital and those restricting such movements.

i *Provisions granting free movement of capital drawn from exploitation*

23.61 One of the essential concerns of the private operator, as for any investor, is to be able to repatriate the profits generated by mining and petroleum operations. The foreign exchange regime applicable to extractive industries is a decisive factor in the investment decision and assumes considerable importance when exploitation agreements are negotiated.

Mining and petroleum codes take these concerns into account, granting a set of guarantees to the private operator in favour of free movement of the

[54] In Angola, the foreign exchange regime applicable to the mining and petroleum sectors is determined by Notice No. 2/03 of 28 February 2003 issued by the Central Bank of Angola with reference to mining and Law No. 2/12 of 13 January 2012 with regard to petroleum.

[55] Particular reference should be made to the international treaties instituting the Franc zone, the Economic and Monetary Community of Central Africa and the West African Economic and Monetary Union (in particular, for the Republic of Congo, see Article 169 of Law No. 4-2005 of 11 April 2005 Enacting the Mining Code; for the Côte d'Ivoire, Article 172 of Law No. 2014-138 of 24 March 2014 Enacting the Mining Code and Article 81 of Law No. 96-669 of 29 August 1996 Enacting the Petroleum Code; for Niger, Article 105 of Ordinance No. 93-16 of 2 March 1993 Enacting the Amended Mining Act; for Senegal, Article 57 of Law No. 98-05 of 8 January 1998 Enacting the Petroleum Code; for Togo, Article 59 of Law No. 99-003 of 18 February 1999 Enacting the Hydrocarbons Code).

capital drawn from exploitation and particularly a transfer guarantee granted to the private operator.

23.62 Guaranteed transfer of capital covers numerous areas and in particular ensures that the operator can transfer abroad the dividends and products of the invested capital and the product of the liquidation of its assets.[56]

More specifically, the transfer guarantee can also relate to amounts received in compensation for expropriation; in the absence of legal provisions and in the case where payment of this compensation results from the decision of an arbitral tribunal, it is important that such award mentions this free transferability of the amounts received by the private operator.[57]

---

[56] See in particular, for Algeria, Article 175 of Law No. 01-10 of 3 July 2001 Enacting the Mining Act and Article 20(2) of the Presidential Decree of 1 March 2003 Enacting the Model Mining Agreement; for Benin, Article 104 of Law No. 2006-17 of 17 October 2006 Enacting the Mining Code; for Cameroon, Article 110(2) of Law No. 99-013 of 22 December 1999 Enacting the Petroleum Code; for the Central African Republic, Article 80 of Ordinance No. 93.007 of 25 May 1993 Enacting the Petroleum Code and Article 135 of Law No. 09-005 of 29 April 2009 Enacting the Mining Code; for the Republic of Congo, Article 169 of Law No. 4-2005 of 11 April 2005 Enacting the Mining Code; for the Democratic Republic of Congo, Article 15(1) of the Production Sharing Contract between the Democratic Republic of Congo and Dominion Petroleum Congo & Soco Exploration-Production RDC & La Congolaise des Hydrocarbures on Block V of the Albertine Graben (November 2007) and Article 264 of Law No. 007-2002 of 11 July 2002 Enacting the Mining Code; for the Côte d'Ivoire, Article 172 of Law No. 2014-138 of 24 March 2014 Enacting the Mining Code and Article 81 of Law No. 96-669 of 29 August 1996 Enacting the Petroleum Code; for Djibouti, Article 56 of Law No. 117-AN-05 of 16 October 2005 Enacting the Petroleum Code; for Eritrea, Article 39 of Proclamation No. 68/1995 Promoting the Development of Mineral Resources; for Ethiopia, Article 72(2) of Proclamation No. 678/2010 Promoting Sustainable Development of Mineral Resources; for Ghana, Article 30(4) of the 2006 Minerals and Mining Act; for Guinea, Article 185 of Law No. L/2013/N°053/CNT of 8 April 2013 Amending Law No. 2011/006/CNT of 9 September 2011 Enacting the Mining Code and Article 106 of Law No. L/2014/N°34/AN of 23 December 2014 Enacting the Petroleum Code; for Madagascar, Article 158 of Law No. 99/002 of 19 August 1999 Enacting the Mining Code and Article 405 of Decree No. 2006-910 of 19 August 2006 Implementing the Mining Code; for Mali, Article 117 of Law No. 2012-015 of 27 February 2012 Enacting the Mining Code; for Morocco, Article 60 of Law No. 21-90 of 1 April 1992 On Prospecting and Exploitation of Hydrocarbon Deposits amended and supplemented by Law No. 27-99 of 15 February 2000; for Mauritania, Article 99 of Law No. 2010-033 of 20 July 2010 Enacting the Crude Hydrocarbons Code amended by Law No. 2011-044 of 25 November 2011; for Mozambique, Article 67 of Mining Act No. 20-2014 of 18 August 2014; for Niger, Article 105 of Ordinance No. 93-16 of 2 March 1993 Enacting the amended Mining Act and Article 141 of Law No. 2007-01 of 31 January 2007 Enacting the Petroleum Code; for Senegal, Article 57 of Law No. 98-05 of 8 January 1998 Enacting the Petroleum Code; for Chad, Article 63 of Law No. 011-PR-1995 of 20 June 1995 Enacting the Mining Code; for Togo, Article 59(1) of Law No. 99-003 of 18 February 1999 Enacting the Hydrocarbons Code; within the WAEMU, Article 15 of Regulation No. 18/2003/CM/UEMOA of 23 December 2003 Enacting the Community Mining Code; within the ECOWAS, Article 9 of ECOWAS Directive C/DIR 3/05/09 of 27 May 2009 On the Harmonisation of the Guiding Principles and Policies in the Mining Sector.

[57] In this connection, for the Democratic Republic of Congo, see Article 264 of Law No. 007-2002 of 11 July 2002 Enacting the Mining Code; for Madagascar, Article 158 of Law No.

23.63 The private operator is generally granted the option of transferring abroad the amounts required to reimburse debts contracted externally in the context of mining and petroleum operations and in particular for paying approved foreign service providers and suppliers.[58]

23.64 The applicable regulations also ensure that the private operator has a right to open, hold and use bank accounts in local and foreign currencies within the host State and abroad.[59] These accounts allow the private operator to deposit abroad the products of the sale of its share of the production and the funds required for paying suppliers.[60]

99-022 of 19 August 1999 Enacting the Mining Act amended by Law No. 2005-021 of 17 October 2005.

[58] With regard to mining and petroleum codes ensuring that private operators can transfer abroad the funds required to pay subcontractors, suppliers and service providers, the following should be mentioned: for the Central African Republic, Article 135 of Law No. 09-2005 of 29 April 2009 Enacting the Mining Code; for the Democratic Republic of Congo, Article 263 of Law No. 007-2002 of 11 July 2002 Enacting the Mining Code; for Mali, Article 117 of Law No. 2012-015 of 27 February 2012 Enacting the Mining Code.

[59] See Article 104 of Law No. 2016-17 Enacting the Mining and Mining Taxation Code in the Republic of Benin and Article 109 of Law No. 2006-18 of 17 October 2006 Enacting the Petroleum Code in the Republic of Benin; for Cameroon, Article 110(2) of Law No. 99-013 of 22 December 1999 Enacting the Petroleum Code; for the Côte d'Ivoire, Article 172 of Law No. 2014-138 of 24 March 2014 Enacting the Mining Code and Article 81 of Law No. 96-669 of 29 August 1996 Enacting the Petroleum Code; for Djibouti, Article 56 of Law No. 117-AN-05 of 16 October 2005 Enacting the Petroleum Code; for Eritrea, Article 39 of Proclamation No. 68/1995 Promoting the Development of Mineral Resources; for Guinea, Article 106 of Law No. L/2014/N°034/AN of 23 December 2014 Enacting the Petroleum Code; for Liberia, Article 10.24.1 of the 2002 Petroleum Law; for Madagascar, Article 158 of Law No. 99-022 of 19 August 1999 Enacting the Mining Code amended by Law No. 2005-021 of 17 October 2005 and Article 405 of Decree No. 2006-910 of 19 August 2006 Implementing the Mining Code; for Mauritania, Article 99 of Law No. 2010-033 of 20 July 2010 Enacting the Hydrocarbons Code amended by Law N° 2011-044 of 25 November 2011; for Niger, Article 105 of Ordinance No. 93-16 of 2 March 1993 Enacting the amended Mining Act and Article 141 of Law No. 2007-01 of 31 January 2007 Enacting the Petroleum Code; for Senegal, Article 57 of Law No. 98-05 of 8 January 1998 Enacting the Petroleum Code; for Chad, Article 91 of Law No. 006-PR-2007 of 2 May 2007 On Hydrocarbons; for Togo, Article 59 of Law No. 99-003 of 18 February 1999 Enacting the Hydrocarbons Code.

[60] Numerous mining and petroleum codes guarantee that the private operator has the right to pay foreign subcontractors abroad and in foreign currencies for acquiring equipment assets and service provision linked to petroleum operations. For Benin, see Article 109 of Law No. 2006-18 of 17 October 2006 Enacting the Petroleum Code and Article 104 of Law No. 2006-17 of 17 October 2006 Enacting the Mining and Mining Taxation Code in the Republic of Benin; for Cameroon, Article 110(2) of Law No. 99-013 of 22 December 1999 Enacting the Petroleum Code; for the Central African Republic, Article 80 of Ordinance No. 93.007 of 25 May 1993 Enacting the Petroleum Code; for the Republic of Congo, Article 169 of Law No. 4-2005 of 11 April 2005 Enacting the Mining Code; for the Côte d'Ivoire, Article 81 of Law No. 96-669 of 29 August 1996 Enacting the Petroleum Code and Article 172 of Law No. 2014-138 of 24 March 2014 Enacting the Mining Code; for Djibouti, Article 56 of Law No. 117-AN-05 of 16 October 2005 Enacting the Petroleum Code; for Guinea, Article 106 of Law No. L/2014/N°034/AN Enacting the Petroleum Code; for Liberia, Article 10.24.4 of the 2002 Petroleum Law; for Madagascar, Article 158 of Law No. 99-022 of 19 August 1999 Enacting the Mining Code amended by Law No. 2005-021 of 17 October 2005 and Article 405 of Decree No. 2006-910 of 19 August 2006 Implementing the Mining Code; for Mauritania,

The legislation in certain countries also stipulates that the investor can benefit from opening accounts with central banking bodies for special accounts. Accordingly, in the context of the Central Bank of West African States and in predetermined cases, the operator can open an account dedicated to incomes generated by marketing extracted substances and an international payment account for payment of the operator's financial commitments abroad.

23.65 Finally, private operators are generally granted the right to free access to foreign currency at the market rate and to freely convert national into foreign currency.[61] On a regional scale, the guarantee that the national currency and foreign currencies can be freely converted is governed by the international treaties in force, particularly the texts establishing the Franc zone, the Economic and Monetary Community of Central Africa or the West African Economic and Monetary Union.[62]

23.66 The benefits of foreign exchange guarantees granted to private operators frequently extend to third parties to the exploitation agreement. This particularly applies to expatriate personnel, affiliated companies,

---

Article 99 of Law No. 2010-033 of 23 July 2010 Enacting the Hydrocarbons Code amended by Law No. 2011-044 of 25 November 2011; for Niger, Article 141 of Law No. 2007-01 of 31 January 2007 Enacting the Petroleum Code and Article 105 of Ordinance No. 93-16 of 2 March 1993 Enacting the Amended Mining Act; for Senegal, Article 57 of Law No. 98-05 of 8 January 1998 Enacting the Petroleum Code; for Chad, Article 92(1) of Law No. 006-PR-2007 of 2 May 2007 On Hydrocarbons; for Togo, Article 59(1) of Law No. 99-003 of 18 February 1999 Enacting the Hydrocarbons Code.

[61] For Algeria see Article 20(1) of the Presidential Decree of 1 March 2003 Enacting the Model Mining Agreement; For Benin, Article 109 of Law No. 2006-18 of 17 October 2006 Enacting the Petroleum Code of the Republic of Benin; in the Central African Republic, Article 135 of Law No. 09-005 of 29 April 2009 Enacting the Mining Code and Article 80 of Ordinance No. 93.007 of 25 May 1993 Enacting the Petroleum Code; in the Republic of Congo, Article 169 of Law No. 4-2005 of 11 April 2005 Enacting the Mining Code; in the Democratic Republic of Congo, Article 263 of Law No. 007-2002 of 11 July 2002 Enacting the Mining Code; in the Côte d'Ivoire, Article 172 of Law No. 2014-138 of 24 March 2014 Enacting the Mining Code and Article 81 of Law No. 96-669 of 29 August 1996 Enacting the Petroleum Code; for Djibouti, Article 56 of Law No. 117-AN-05 of 16 October 2005 Enacting the Petroleum Code; for Ghana, Article 30(4) of the 2006 Minerals and Mining Act; for Guinea, Article 106 of Law No. L/2014/N°034/AN of 23 December 2014 Enacting the Petroleum Code; for Liberia, Article 10.24.5 of the 2002 Petroleum Law; for Mali, Article 117 of Law No. 2012-015 of 27 February 2012 Enacting the Mining Code; for Mauritania, Article 99 of Law No. 2010-033 of 20 July 2010 Enacting the Hydrocarbons Code, amended by Law No. 2011-044 of 25 November 2011; for Niger, Article 105 of Ordinance No. 93-16 of 2 March 1993 Enacting the Amended Mining Act; for Senegal, Article 57 of Law No. 98-05 of 8 January 1998 Enacting the Petroleum Code; for Chad, Article 91 of Law No. 006-PR-2007 of 2 May 2007 On Hydrocarbons; for Togo, Article 59 of Law No. 99-003 of 18 February 1999 Enacting the Hydrocarbons Code.

[62] For the Republic of Congo, see Article 169 of Law No. 4-2005 of 11 April 2005 Enacting the Mining Code; for the Côte d'Ivoire, Article 172 of Law No. 2014-138 of 24 March 2014 Enacting the Mining Code and Article 81 of Law No. 96-669 of 29 August 1996 Enacting the Petroleum Code; for Niger, Article 105 of Ordinance No. 93-16 of 2 March 1993 Enacting the Amended Mining Act; for Senegal, Article 57 of Law No. 98-05 of 8 January 1998 Enacting the Petroleum Code; for Togo, Article 59 of Law No. 99-003 of 18 February 1999 Enacting the Hydrocarbons Code.

subcontractors and other foreign service providers or suppliers and their foreign employees.[63] However, some mining and petroleum codes do not extend the benefit of the exchange-rate regime to these third parties and limit themselves to making reference to the provisions of the exploitation agreement[64] , where applicable.

ii *Provisions restricting freedom of movement of capital drawn from exploitation*

23.67 More frequently, the host state is highly reluctant in allowing capital drawn from exploitation to leave its territory. It calls for an economic compensation for the exploitation of its natural resource in the form of development of its national wealth. Several solutions have been envisaged to ensure that the state can draw the maximum benefits from the exploitation of its natural resources, in particular the obligation to reinvest part of the profits into the local extractive industry and other national industries or the establishment by certain states of tax relief for local investments.

23.68 With regard to the exchange control regulations, different mechanisms are in place to guarantee that the state has improved control over the funds generated by extractive industries. This is the case for provisions to establish control over the transfers carried out by the private operator to the benefit of its affiliated companies. Notwithstanding the right of free transfer ensured by the host state, mining and petroleum codes can stipulate that intra-group transfers made for the acquisitions of equipment or services rendered must be justified in relation to the prices applied on the market for similar goods and services.[65]

23.69 Foreign investors will not generally be required to repatriate the products drawn from the export of extracted resources to the state and will be able to dispose of them freely; however, a number of texts impose an obligation to maintain in the host state territory or to repatriate the amounts required for mining and petroleum operations to the host state and to pay

[63] For Eritrea, see Article 24(1) of Proclamation No.108/2000, a Revised Proclamation Governing Petroleum Operations; for Ethiopia, Article 24 of Proclamation No. 295/1986 Regulating Petroleum Operations; for Mali, Article 117 of Law No. 2012-015 of 27 February 2012 Enacting the Mining Code; for Chad, Article 63 of Law No. 011-BR-1995 of 20 July 1995 Enacting the Mining Code and Article 92 (2) of Law No. 006-PR-2007 of 2 May 2007 On Hydrocarbons.

[64] For Cameroon, see Article 110(4) of Law No. 99-013 of 22 December 1999 Enacting the Petroleum Code; for the Côte d'Ivoire, Article 81 of Law No. 96-669 of 29 August 1996 Enacting the Petroleum Code; for Guinea, Article 106 of Law No. L/2014/N°34/AN of 23 December 2014 Enacting the Petroleum Code; for Liberia, Article 10.24.7 of the 2002 Petroleum Law; for Niger, Article 142 of Law No. 2007-01 of 31 January 2007 Enacting the Petroleum Code; and for Togo, Article 59(3) of Law No. 99-003 of 18 February 1999 Enacting the Hydrocarbons Code.

[65] See for the Democratic Republic of Congo see Article 265 of Law No. 007-2002 of 11 July 2002 Enacting the Mining Code; for Madagascar, Article 39 of Law No. 2001-031 of 8 October 2002 Enacting the Special Regime for Large-Scale Investments in the Malagasy Mining Sector amended by Law No. 2005-022 of 17 October 2005.

charges, in particular taxes.[66] The applicable regime is generally different for national investors who, in certain cases, do not have an obligation to repatriate the product of exports and extracted resources back to the host state.[67]

23.70 Similarly, exceptions are occasionally made to the principle of free repatriation of amounts drawn from exports, in this way restricting the freedom of movement of capital drawn from exploitation. The legislation in a number of countries stipulates an obligation at the expense of private operators to repatriate receipts obtained from the sale of extracted resources back to the country.[68]

23.71 The restrictions introduced can also apply to the conditions for opening a foreign currency account. A number of mining and petroleum codes subordinate the opening of a foreign currency account to prior authorisation by the competent authority, generally the Minister of Finance.[69]

23.72 Frequently the foreign currency guarantees granted to private operators are coupled with a series of measures to strengthen the control of the state over the funds generated by mining and petroleum activities. This control in particular involves provisions establishing an obligation of information upon the private operator, for example by drafting a statement of imports of convertible currencies and transfers[70] or by regularly sending

---

[66] For Algeria, see Article 55 of Ordinance No. 06-10 of 29 July 2006 Amending and Supplementing Law No. 05-07 of 28 April 2005 On Hydrocarbons; for Benin, Article 109 of Law No. 2006-18 of 17 October 2006 Enacting the Petroleum Code; for Cameroon, Article 110(2) of Law No. 99-013 of 22 December 1999 Enacting the Petroleum Code; for the Central African Republic, Article 80 of Ordinance No. 93.007 of 25 May 1993 Enacting the Petroleum Code; for the Côte d'Ivoire, Article 81 of Law No. 96-669 of 29 August 1996 Enacting the Petroleum Code; for Djibouti, Article 56 of Law No. 117-AN-05 of 16 October 2005 Enacting the Petroleum Code; for Guinea, Article 106 of Law No. L/2014/N°34/AN of 23 December 2014 Enacting the Petroleum Code; for Liberia, Article 10.24.3 of the 2002 Petroleum Law; for Morocco, Articles 55 and 56 of Law No. 21-90 of 1 April 1992 On Exploration and Exploitation of Hydrocarbon Deposits amended and supplemented by Law No. 27-99 of 15 February 2000; for Niger, Article 141 of Law No. 2007-01 of 31 January 2007 Enacting the Petroleum Code; the Sierra Leone, Article 112 of the 2011 Petroleum Exploration and Production Act; for Tunisia, Article 128 of Law No. 99-93 of 17 August 1999 Enacting the Hydrocarbons Code and Article 110 of Law No. 2003-30 of 28 April 2003 Enacting the Mining Code.

[67] For Tunisia, see Article 129 of Law No. 99-93 of 17 August 1999 Enacting the Hydrocarbons Code and Article 111 of Law No. 2003-30 of 28 April 2003 Enacting the Mining Code. In addition, Article 11 of Annex 2 to Regulation No. 09/98/CM/UEMOA of 20 December 1998 On the External Financial Relationships of Member States of the West African Economic and Monetary Union (WAEMU) stipulates that resident economic operators are required to collect and repatriate the entire amounts obtained from the sale of goods abroad and to repatriate them to the paying agent in their country of origin.

[68] For the Côte d'Ivoire, see Article 172 of Law No. 2014-138 of 24 March 2014 Enacting the Mining Code; for Guinea, Article 184 of Law No. L/2013/N°053/CNT of 8 April 2013 Amending Law No. 2011/006/CNT of 9 September 2011 Enacting the Mining Code.

[69] For the Central African Republic, see Article 136 of Law No. 09-005 of 29 April 2009 Enacting the Mining Code; for Ghana, Article 30(2) of the 2006 Minerals and Mining Act.

[70] For Algeria, see Article 55 of Law No. 13-01 of 20 February 2013 Amending and Supplementing Law No. 05-07 of 28 April 2005 On Hydrocarbons.

information on transfers of funds operated between the host state and abroad,[71] by means of an obligation to declare.[72] Monitoring of the established procedures and in particular the obligation to transfer funds through an approved intermediary bank, for example by filling in an exchange document, also ensures improved control by the host state over currency exchange operations.[73]

23.73 In practice, private operators structure their balance sheets so that debts are greater than the equity capital; in this way, very large amounts of money can leave the territory on which exploitation operations are taking place as reimbursement both of the principal and of the interest.

b Payment of salaries

23.74 Payment of salaries to local employees represents a direct flow of capital in the local economy. There are two series of comments on the payment of salaries to local economy. First, a number of provisions contained in exploitation agreements interfere with the freedom to pay salaries and their availability. Secondly, assessment of the impact of salaries on the local economy is a central issue for properly taking certain negotiation and possibly renegotiation factors into account.

i *Strict salary control*

23.75 Salaries can be paid to foreign personnel in local or in foreign currency.[74] In the context of foreign exchange regulations, mining and petroleum codes can grant to the private operator the option of remunerating

[71] For Cameroon, see Article 110(5) of Law No. 99-013 of 22 December 1999 Enacting the Petroleum Code; for Eritrea, Article 24(1) of Proclamation No. 108/2000, a Revised Proclamation Governing Petroleum Operations; for Ethiopia, Article 24(2) of Proclamation No. 295/1986 Regulating Petroleum Operations; for Niger, Article 143 of Law No. 2007-01 of 31 January 2007 Enacting the Petroleum Code; for Chad, Article 93 of Law No. 006-PR-2007 of 2 May 2007 On Hydrocarbons.

[72] For Guinea, see Article 105 of Law No. L/2014/N°034/AN of 23 December 2014 Enacting the Petroleum Code; for Madagascar, Article 35 of Law No. 2001-031 of 8 October 2002 Enacting a Special Regime for Large-Scale Investments in the Malagasy Mining Sector amended by Law No. 2005-022 of 17 October 2005; for Mauritania, Article 101 of Law No. 2010-033 of 20 July 2010 Enacting the Crude Hydrocarbons Code amended by Law No. 2011-044 of 25 November 2011.

[73] In this connection, for the Democratic Republic of Congo, see Article 264 of Law No. 007-2002 of 11 July 2002 Enacting the Mining Code.

[74] For an example of a mining code entitling the private operator to pay salaries and other compensation in foreign currency, see Article 109 of Law No. 2006-18 of 17 October 2006 Enacting the Petroleum Code.

its expatriate personnel directly abroad[75] or transferring amounts abroad as required for paying the expatriate personnel in foreign currency.[76]

Where the salary paid to expatriate personnel is in local currency, in most cases the expatriate personnel will benefit from a guarantee that they can freely convert and transfer all or part of the amount to their country of origin.[77] If applicable, this guarantee can be subordinated to payment by the expatriate personnel of various taxes and contributions in accordance with the applicable.[78] Moreover, a number of mining and petroleum codes go as

[75] For Mauritania, see Article 99 of Law No. 2010-33 of 20 July 2010 Enacting the Crude Hydrocarbons Code amended by Law No. 2011-044 of 25 November 2011; for Guinea, Article 106 of Law No. L/2014/N°034/AN of 23 December 2014 Enacting the Petroleum Code.

[76] In this connection, for Madagascar, see Article 405 of Decree No. 2006-910 of 19 August 2006 Implementing the Mining Code and Article 158 of Law No. 99-022 of 19 August 1999 Enacting the Mining Code amended by Law No. 2005-021 of 17 October 2005.

[77] In this connection, for Benin, see Article 104 of Law No. 2006-17 of 17 October 2006 Enacting the Mining and Mining Taxation Code; for the Central African Republic, Article 137 of Law No. 09-005 of 29 April 2009 Enacting the Mining Code; the Republic of Congo, Article 169 of Law No. 4-2005 of 11 April 2005 Enacting the Mining Code; for the Democratic Republic of Congo, Article 264 of Law No. 007-2002 of 11 July 2002 Enacting the Mining Code; for the Côte d'Ivoire, Article 172 of Law No. 2014-138 of 24 March 2014 Enacting the Mining Code and Article 81 of Law No. 96-669 of 29 August 1996 Enacting the Petroleum Code; for Djibouti, Article 56 of Law No. 117-AN-05 of 16 October 2005 Enacting the Petroleum Code; for Eritrea, Article 39 of Proclamation No. 68/1995 to Promote the Development of Mineral Resources; for Ethiopia, Article 72 of Proclamation No. 678/2010 to promote sustainable development of mineral resources; for Guinea, Article 185 of Law L/2013/N° 053/CNT of 8 April 2013 Amending Law No. 2011/006/CNT of 9 September 2011 Enacting the Mining Code and Article 106 of Law No. L/2014/N°034/AN of 23 December 2014 Enacting the Petroleum Code; for Liberia, Article 10.24.6 of the 2002 Petroleum Law; for Mali, Article 117 of Law No. 2012-015 of 27 February 2012 Enacting the Mining Code; for Mauritania, Article 100 of Law No. 2010-033 of 20 July 2010 Enacting the Crude Hydrocarbons Code amended by Law No. 2011-044 of 25 November 2011; for Niger, Article 142 of Law No. 2007-01 of 31 January 2007 Enacting the Petroleum Code and Article 105 of Ordinance No. 93-16 of 2 March 1993 Enacting the Amended Mining Act; for Senegal, Article 57 of Law No. 98-05 of 8 January 1998 Enacting the Petroleum Code; for Chad, Article 63 of Law No. 011-PR-1995 of 20 June 1995 Enacting the Mining Code; for Togo, Article 59(2) of Law No. 99-003 of 18 February 1999 Enacting the Hydrocarbons Code; see also Article 15 of Regulation No. 18/2003/CM/UEMOA of 23 December 2003 Enacting the WAEMU Community Mining Code.

[78] In this connection, for Benin, see Article 104 of Law No. 2006-17 of 17 October 2006 Enacting the Mining and Mining Taxation Code; in the Central African Republic, Article 137 of Law No. 09-005 of 29 April 2009 Enacting the Mining Code; for Cameroon, Article 110(4) of Law No. 99-013 of 22 December 1999 Enacting the Petroleum Code; for the Republic of Congo, Article 169 of Law No. 4-2005 of 11 April 2005 Enacting the Mining Code; for the Democratic Republic of Congo, Article 264 of Law No. 007-2002 of 11 July 2002 Enacting the Mining Code; for the Côte d'Ivoire, Article 172 of Law No. 2014-138 of 24 March 2014 Enacting the Mining Code and Article 81 of Law No. 96-669 of 29 August 1996 Enacting the Petroleum Code; for Djibouti, Article 56 of Law No. 117-AN-05 of 16 October 2005 Enacting the Petroleum Code; for Guinea, Article 185 of Law No. L/2013/N°053/CNT of 8 April 2013 Amending Law No. 2011/006/CNT of 9 September 2011 Enacting the Mining Code; for Liberia, Article 10.24.6 of the 2002 Petroleum Law; for Mali, Article 117 of Law No. 2012-015 of 27 February 2012 Enacting the Mining Code; for Niger, Article 142 of Law No.

far as extending the benefit of foreign exchange provisions to expatriate employees.[79]

ii *Assessment of the impact of the payment of salaries on the local economy*

23.76 Three main types of remuneration are generally mentioned in exploitation agreements, generally in their appendixes: a daily wage, a monthly salary in local currency and a monthly salary in local currency but indexed to a foreign currency exchange rate. Bonuses and compensation are added to a fixed basic minimum salary either by contract or by regulation with a view in particular to providing protection against illness, unemployment and death. Occasionally, private operators improve the situation of employees by contributing to savings or retirement plans, or even life insurance.

23.77 The nationality of the workers is a source of marked disparities, even if they are nevertheless tending to diminish. Higher salaries paid to foreigners are mainly due to the presence of a majority of expatriate management personnel who benefit from large expatriation and risk bonuses. Although these inequalities persist, states are making significant efforts to prohibit all discrimination and ensure strict compliance with equal rights. Texts applicable to the employment of personnel occasionally contain provisions against discriminatory practices, through which the operator is prohibited from making any distinction on the basis particularly of gender, race or religion.[80] Non-discrimination is also ensured by allocating identical salaries and rights between national and expatriate employees in equivalent conditions.[81]

Foreign expatriate management personnel may be employed as employees of the parent company of the private operator or can be employed directly by its subsidiary in the host state.

2007-01 of 31 January 2007 Enacting the Petroleum Code and Article 105 of Ordinance No. 93-16 of 2 March 1993 Enacting the Amended Mining Act; for Senegal, Article 57 of Law No. 98-05 of 8 January 1998 Enacting the Petroleum Code; for Chad, Article 63 of Law No. 011-PR-1995 of 20 June 1995 Enacting the Mining Code; for Togo, Article 59(2) of Law No. 99-003 of 18 February 1999 Enacting the Hydrocarbons Code; Article 15 of Regulation No. 18/2003/CM/UEMOA of 23 December 2003 Enacting the Community Mining Code.

[79] In this connection, see Article 109 of Law No. 2016-18 of 17 October 2006 Enacting the Petroleum Code. Other mining and petroleum codes refer to contractual provisions on extending benefits to expatriate personnel. In Cameroon this refers to Article 110(4) of Law No. 99-013 of 22 December 1999 Enacting the Petroleum Code; for the Côte d'Ivoire, Article 81 of Law No. 96-669 of 29 August 1996 Enacting the Petroleum Code; for Guinea, Article 106 of Law No. L/2014/N°034/AN of 23 December 2014 Enacting the Petroleum Code; for Liberia, Article 10.24.7 of the 2002 Petroleum Law; for Niger, Article 142 of Law No. 2007-01 of 31 January 2007 Enacting the Petroleum Code; for Togo, Article 59(3) of Law No. 99-003 of 18 February 1999 Enacting the Hydrocarbons Code.

[80] For South Sudan, see Article 65 of the 2012 Petroleum Act.

[81] For São Tomé and Principe, see Article 57(3) of the Framework Law on Petroleum Operations; for Angola, Article 86(2) of Petroleum Act No. 10/04 of 12 November 2004.

Throughout areas where exploitation activities are taking place, states are showing a willingness to diminish or even eliminate these discrepancies. The development of the role of the host states as operational players, particularly with regard to the exploitation of hydrocarbon resources, accentuates this trend.

24

ENRICHING NATIONAL ASSETS OF THE HOST STATE

24.1 Enriching the national assets of the host state has become a major concern of local authorities over several decades until now.[1] It is common to see exploitation agreements with long preambles specifying the importance of integrating the petroleum or mining project in the development of the host state.

24.2 The enrichment of national assets answers to the need to obtain a fair compensation for the exploitation of national wealth by the private operator; the signing of the exploitation agreement and the allocation of mining rights are thus conditional on related economic benefits for the host state.

For private operators in particular, this relates to participation in strengthening the local economic infrastructure and contributing more generally to the development of the host state.

1 REINFORCING THE LOCAL ECONOMY

24.3 The commitments of private operators to strengthen the local economic infrastructure mainly relate to infrastructures linked to extractive activities. However, the cooperation of private operators to the development of the local economic infrastructure frequently extends well beyond the mining and petroleum sectors.

a Reinforcing human and material resources related to extractive industries

24.4 Strengthening the economic infrastructure of local extractive industries requires two complementary actions: one overall activity related to integrating regulations to promote employee protection and the other, more specifically related to the material development of the local economy.

i *The development of rules to promote employee protection*

24.5 The development of modern regulations related to the exploitation of national resources is a priority that must accompany the development of the local extractive industry.

[1] In this regard, see Thomas Wälde 'Transnational Investment in the Natural Resources Industries' 1979 *Law & Pol Int Bus.* 691.

The purpose of these rules and regulations is in particular to ensure the protection of employees working on mining or petroleum projects and, occasionally, their families. For instance, these rules relate in particular to health and safety and social protection.

(1) Safety of employees

24.6 Henceforward, the safety of employees in the course of exploitation of natural resources is subject to particularly stringent regulations. The safety of mining and petroleum operations is granted an increasingly important place; this results particularly in a rapid increase in the number of regulations covering this area. Some legislation refers back to contractual provisions or general health and safety regulations.[2] These regulations can also be supplemented by specific rules applied to health and safety in the extractive sector.[3]

24.7 The importance allocated to employee safety is reflected in particular when granting mining licences. A number of mining and petroleum codes refer to the capacities of the private operator to ensure the safety of employees working on the site as an essential precondition for granting a mining licence.[4] The importance allocated to the safety of exploitation operations is such that failure to comply with the health and safety regulations can lead to withdrawal of the mining licence.[5]

With a view to ensuring employee safety, African legislators have introduced numerous provisions in mining and petroleum codes to ensure compliance with health and safety regulations. Numerous texts refer to the obligation of the holder of a mining licence to implement the works in a way

---

[2] See in particular, Article 53 of Order No. 2008-804 of 31 December 2008 Implementing the Mining and Mining Taxation Code in the Republic of Benin, which refers to the regulations contained in the Labour Law.

[3] See in particular, for Algeria, Law No. 88-07 of 26 January 1998 On Health, Health, Safety and Occupational Medicine; for Kenya, the 2007 Occupational Safety and Health Act; for Uganda, the 2006 Occupational Safety and Health Act. A number of laws are applicable, specifically, to the mining or petroleum sectors; notably for South Africa (ie the 1996 Mine Health and Safety Act).

[4] For examples of mining and petroleum codes that take into account the capacity to ensure compliance with health and safety regulations in the granting of mining licences, for South Africa see sections 17, 23, 43, 75, 80 and 84 of the Mineral and Petroleum Resources Development Act; for Mali, Articles 31 and 39 of Law No. 2012-015 of 27 February 2012 Enacting the Mining Code.

[5] For the Republic of Congo, see Article 91 of Law No. 4-2005 Enacting the Mining Code and Article 61 of Law No. 24-94 of 23 August 1994 Enacting the Hydrocarbons Code; for Ethiopia, Article 44 of Proclamation No. 678/2010 Promoting Sustainable Development of Mineral Resources; for Rwanda, Article 25 of Law No. 13-2014 of 20 May 2014 On the Exploitation of Mines and Quarries.

that fulfils the relevant obligations[6] and even promotes compliance with these regulations.[7]

24.8 A number of mining and petroleum codes go further in specifying that the obligation to comply with these regulations also applies to subcontractors.[8] In a similar vein, it is occasionally incumbent on the operator to ensure that parties working on its behalf comply with the health and safety regulations.[9]

24.9 Ensuring the safety of personnel working on mining and petroleum sites particularly involves the use of suitable methods. Numerous mining and petroleum codes therefore require the operator to implement the works in accordance with best industry practices and taking the measures required to ensure the health and safety of the workers.[10]

---

[6] For Algeria, see Article 152 of Mining Act No. 01-10 of 3 July 2001, Article 8(1) of the Model Mining Agreement adopted by the Presidential Decree of 1 March 2003; for Angola, Article 23 of Petroleum Act No. 10/04 of 12 November 2004; for Benin, Article 126 of Law No. 2006-17 of 17 October 2006 Enacting the Mining and Mining Taxation Code in the Republic of Benin; for Cameroon, Article 78(1) of Law No. 99-013 of 22 December 1999 Enacting the Petroleum Code; for Guinea, Article 145 of Law No. 2011/006/CNT of 9 September 2011 Enacting the Mining Code and Article 56 of Law No. L/2014/N°034/AN of 23 December 2014 Enacting the Petroleum Code; for Madagascar, Article 109 of Law No. 99-022 of 19 August 1999 Enacting the Mining Code amended by Law No. 2005-021 of 17 October 2005; for Mali, Articles 83 and 140 of Law No. 2012-015 of 27 February 2012 Enacting the Mining Code; for Namibia, Article 41(1) of the 1992 Minerals Prospecting and Mining Act and Article 38(1) of the 1991 Petroleum (Exploration and Production) Act; for Niger, Article 22 of Law No. 2007-01 of 31 January 2007 Enacting the Petroleum Code; for the Seychelles, Article 23(2) of the 2013 Model Petroleum Agreement; for Chad, Article 52(1) of Law No. 006-PR-2007 of 2 May 2007 On Hydrocarbons; for Tunisia, Article 71 of Law N° 2003-30 of 28 April 2003 Enacting the Mining Code and Article 59(1) of Law No. 99-93 of 17 August 1999 Enacting the Hydrocarbon Code.

[7] See, in particular, for Angola Article 93 of Mining Act No. 31/11 of 23 September 2011; for Mozambique, Article 15 of Petroleum Act No. 21-1014 of 18 August 2014; for Sierra Leone; Article 142 of the 2009 Mines and Minerals Act.

[8] For Cameroon, see Article 78(1) of Law No. 99-013 of 22 December 1999 Enacting the Petroleum Code; for Guinea, Article 145 of Law No. 2011/006/CNT of 9 September 2011 Enacting the Mining Code; for Mali, Article 140 of Law No. 2012-015 of 27 February 2012 Enacting the Mining Code; for Niger, Article 22 of Law No. 2007-01 of 31 January 2007 Enacting the Petroleum Code; for the Seychelles, Article 23(2) of the 2013 Model Petroleum Agreement.

[9] For the Central African Republic, see Article 61 of Ordinance No. 93.007 of 25 May 1993 Enacting the Petroleum Code; for the Côte d'Ivoire, Article 54 of Law No. 96-669 of 29 August 1996 Enacting the Petroleum Code; for Guinea, Article 56 of Law No. L/2014/N°034/AN of 23 December 2014 Enacting the Petroleum Code.

[10] For Angola, see Article 59 of Mining Act No. 31/11 of 23 September 2011; for Botswana, Article 55(1) of the Petroleum (Exploration and Production) Act; for the the Central African Republic, Article 101 of Law No. 09-005 of 29 April 2009 Enacting the Mining Code; for the Côte d'Ivoire, Article 137 of Law No. 2014-138 of 24 March 2014 Enacting the Mining Code; for Ethiopia, Article 34 of Proclamation No. 678/2010 Promoting Sustainable Development of Mineral Resources; for Ghana, Article 93 of the 2006 Minerals and Mining Act; for Madagascar, Article 33 of Order No. 2004-669 of 29 June 2004 Implementing Law No. 99-010 Governing the Activities of the Upstream Petroleum Sector; for Malawi, Article 48 of the 1983 Petroleum (Exploration and Production) Act; for Niger, Article 121 of Ordinance No. 93-16 of 2 March 1993 Enacting the Amended Mining Act; for Uganda, Article 140 of the

On the site itself, these obligations are visible by means of displays of the rules pertaining to safety and security on site,[11] the establishment of personnel equipment rules[12] or the training of workers on the safety rules.[13]

In addition, the operator's obligation to draft an internal regulation,[14] a programme[15] or a safety plan[16] contributes to improving the safety of personnel on mining and petroleum sites and especially to making this obligation a key factor.

24.10 Beyond preventive measures, the mining and petroleum codes also take account of the need to ensure the safety of workers in the event of an accident. The occurrence of an accident must generally be brought to the attention of the competent authority[17] which, if applicable, triggers the

---

2013 Petroleum (Exploration, Development and Production) Act; for Rwanda, Article 33 of Law No. 13-2014 of 20 May 2014 On the Exploitation of Mines and Quarries; for Sao Tome and Principe, Article 62(1) of the Framework Law on Petroleum Operations; for Sierra Leone, Article 143 of the 2009 Mines and Minerals Act and Article 97 of the 2011 Petroleum Exploration and Production Act; for South Sudan, Article 50 of the 2012 Petroleum Act; for Chad, Article 86 of Law No. 001-PR-1995 of 20 June 1995 Enacting the Mining Code. On a regional level, this regulation is also referred to in Article 11 of Regulation No. 18/2003/CM/UEMOA of 23 December 2003 Enacting the Community Mining Code.

[11] For the Democratic Republic of Congo, see Article 210 of Law No. 007-2002 of 11 July 2002 Enacting the Mining Code.

[12] For Malawi, see Articles 13 and 14 of the Mining (Safety) Regulations.

[13] For Rwanda, see Article 33 of Law No. 13-2014 of 20 May 2014 On the Exploitation of Mines and Quarries; for Sao Tome and Principe, Article 62(2) of Framework Law on Petroleum Operations; for Sierra Leone, Article 142 of the 2009 Mines and Minerals Act; for South Sudan, Article 50(3) of the 2012 Petroleum Act.

[14] In this regard, for Benin, see Article 126 of Law No. 2006-17 of 17 October 2006 Enacting the Mining Code and Mining Taxation in the Republic of Benin; for the Central African Republic, Article 102 of Law No. 09-005 of 29 April 2009 Enacting the Mining Code; for the Côte d'Ivoire, Article 138 of Law No. 2014-138 of 24 March 2014 Enacting the Mining Code; for Madagascar, Article 262 of Order No. 2006-910 of 19 August 2006 Implementing the Mining Code.

[15] In this regard, for the Seychelles see Article 23(1) of the 2013 Model Petroleum Agreement and for South Sudan, Article 52 of the 2012 Petroleum Act.

[16] For Guinea, see Article 57 of Law No. L/2014/N°034/AN of 23 December 2014 Enacting the Petroleum Code.

[17] In this regard, for Cameroon, see Article 78(1) of Law No. 99-013 of 22 December 1999 Enacting the Petroleum Code; for the Central African Republic, Article 103 of Law No. 09-005 of 29 April 2009 Enacting the Mining Code; for the Comoros, Article 40 of Order No. 54-1110 of 13 November 1954 Reforming the Mineral Substances Regime; for the Democratic Republic of Congo, Article 209 of Law No. 007-2002 of 11 July 2002 Enacting the Mining Code; for the Côte d'Ivoire, Article 54 of Law No. 96-669 of 29 August 1996 Enacting the Petroleum Code and Article 139 of Law No. 2014-138 of 24 March 2014 Enacting the Mining Code; for Eritrea, Article 30(3) of Proclamation No. 68/1995 Promoting the Development of Mineral Resources; for Liberia, Article 2.5.10 of the 2002 Petroleum Law; for Madagascar, Article 111 of Law No. 99-022 of 19 August 1999 Enacting the Mining Code amended by Law No. 2005-021 of 17 October 2005; for Niger, Article 22 of Law No. 2007-01 of 31 January 2007 Enacting the Petroleum Code; for Nigeria, Article 84(1) of the 2007 Minerals and Mining Act and Article 140 of the 2011 Minerals and Mining Regulation; for Sao Tome and Principe, Article 62(3) of Framework Law on Petroleum Operations; for Sierra Leone, Article 146 of the 2009 Mines and Minerals Act; for South Sudan, Article 159 of the 2012 Mining Act; for Togo, Article 38 of Law No. 99-003 of 18 February 1999 Enacting

necessary enquiries. In order to remedy any breaches, the applicable texts also stipulate the intervention of the competent authority, which may enact interim measures to restore the safety of employees on sites.[18]

(2) Social protection of employees

24.11 Social benefits can also be addressed under separate contracts of the employment contract itself, in the form of collective agreements. These benefits can extend not only to the families of operators' employees, but also to personnel employed by subcontractors.

24.12 In connection with social protection, the intensive policy of numerous private operators with regard to accommodation for local workers should be noted. This policy, moreover, is taken up by the mining and petroleum codes, which contain provisions on the construction of infrastructures required for accommodating personnel involved in mining and petroleum operations.

ii Reinforcing material resources: the development of local extractive industries

24.13 Industrial integration in the area of natural resources frequently encounters failure. For many years, local processing of extracted resources was not a major concern. There is currently a trend for this phenomenon to be reversed; host states seek to ensure local processing of extracted resources[19] and operators have a major interest in this.

---

the Hydrocarbons Code; for Tunisia, Article 72 of Law No. 2003-30 of 28 April 2003 Enacting the Mining Code and Article 59(4) of Law No. 99-93 of 17 August 1999 Enacting the Hydrocarbons Code; for Chad, Article 93 of Law No. 011-PR-1995 of 20 June 1995 Enacting the Mining Code.

[18] For Algeria, see Article 57 of Law No. 01-10 of 3 July 2001 Enacting the Mining Act; for Burundi, Article 189 of Decree-Law No. 1/138 of 17 July 1976 Enacting the Mining and Petroleum Code; for Cameroon, Article 78(2) of Law N° 99-013 of 22 December 1999 Enacting the Petroleum Code; for the Central African Republic, Article 103 of Law N° 09-005 of 29 April 2009 Enacting the Mining Code; for the Comoros, Article 40 of Order No. 54-1110 of 13 November 1954 Reforming the Mineral Substances Regime; for the Democratic Republic of Congo, Article 208 of Law No. 007-2002 of 11 July 2002 Enacting the Mining Code; for Guinea, Article 146 of Law No. 2011/006/CNT of 9 September 2011 Enacting the Mining Code; for Mali, Article 83 of Law No. 2012-015 of 27 February 2012 Enacting the Mining Code.

[19] In particular, this is set out in Article 21 of Law No. 2012-015 of 27 February 2012 Enacting the Mining Code in Mali: '*[l]es titulaires de titre d'exploitation sont tenus de procéder au traitement, à l'affinage ou à la transformation des produits miniers ou de carrières dans les unités installées au Mali. Cependant et chaque fois que cela s'avère nécessaire, une autorisation pourra leur être accordée par l'Administration chargée des mines pour effectuer ces opérations à l'extérieur du Mali. Les conditions et modalités de réalisation de ces opérations sur un autre site ou à l'extérieur du Mali seront précisées dans la décision autorisant ces opérations.*' [Free translation into English: '[t]he holders of the exploitation licence are required to carry out the treatment, refining or processing of mining or quarrying products in units installed in Mali. However, whenever it proves necessary, authorisation can be granted to them by the mining Administration to carry out these operations outside Mali. The conditions and procedures for implementing these

The development of local extractive industries can take the form of a variety of activities mainly related to the expansion of their material infrastructure, improved knowledge of the natural resource potential of the country and also the financing of scientific research.

(1) Expansion of the local infrastructure

24.14 Proper implementation of exploitation operations in most cases requires the construction of infrastructures, particularly for communication, transport and power generation.

These infrastructure requirements are taken into account not only on a contractual level. Numerous mining and petroleum codes treat issues related to the construction of infrastructures as an obligation at the expense of the investor,[20] even though most applicable legal provisions are limited to granting the right to private operators to build or arrange the building of the infrastructures required for implementing mining and petroleum operations.[21] A distinction can be made between infrastructures built inside the perimeter of the mining licence or the area covered by the exploration

---

operations on another site or outside Mali will be specified in the decision authorising these operations.]'

[20] Very few texts establish an obligation for the investor in this area. In this regard, for the Democratic Republic of Congo, see Article 212 of Law No. 007-2002 of 11 July 2002 Enacting the Mining Code. A number of texts establish a cooperation obligation at the expense of the investor for infrastructure construction and maintenance (eg for Eritrea, see Article 23 of Proclamation No. 68/1995 Promoting the Development of Mineral Resources and for Ethiopia, Article 34 of Proclamation No. 678/2010 Promoting Sustainable Development of Mineral Resources).

[21] In this regard, for South Africa, see section 27(7) of the Mineral and Petroleum Resources Development Act 28 of 2008; for Angola, Article 92 of Mining Act No. 31/11 of 23 September 2011 and Article 28 of Petroleum Act No. 10/04 of 12 November 2004; for Benin, Article 106 of Law No. 2006-17 of 17 October 2006 Enacting the Mining and Mining Taxation Code in the Republic of Benin; for the Democratic Republic of Congo, Article 64 of Law No. 007-2002 of 11 July 2002 Enacting the Mining Code and Article 45 of Decree-Law No. 81-013 of 2 April 1981 Enacting General Legislation On Mines and Hydrocarbons; for the Côte d'Ivoire, Article 59 of Law No. 96-669 of 29 August 1996 Enacting the Petroleum Code; for Djibouti, Article 59 of Law No. 117-AN-05 of 16 October 2005 Enacting the Petroleum Code; for Liberia, Article 9.1.2 of the 2002 Petroleum Law; for Mauritania, Article 54 of Law No. 2010-033 of 20 July 2010 Enacting the Hydrocarbons Code amended by Law No. 2011-044 of 25 November 2011; for Niger, Article 107 of Ordinance No. 93-16 of 2 March 1993 Enacting the Amended Mining Act and Article 13 of Law No. 2007-01 of 31 January 2007 Enacting the Petroleum Code; for Sao Tome and Principe, Article 26 of the Framework Law on Petroleum Operations; for Senegal, Article 60 of Law No. 98-05 of 8 January 1998 Enacting the Petroleum Code; for South Sudan, Article 145 of the 2012 Mining Act; for Chad, Article 72 of Law No. 011-PR-1995 of 20 June 1995 Enacting the Mining Code and Article 47 of Law No. 006-PR-2007 of 2 May 2007 On Hydrocarbons; for Togo, Article 32 of Law No. 96-004 of 26 February 1996 Enacting the Mining Code; for Tunisia, Article 84 of Law No. 99-93 of 17 August 1999 Enacting the Hydrocarbons Code.

A number of mining and petroleum codes mention the right of the holder of a mining licence to occupy the land within the perimeter as required for infrastructure building. In particular for Algeria, see Article 134 of Law No. 01-10 of 3 July 2001 Enacting the Mining Code.

agreement and those built outside it, the construction of the latter generally being subject to prior authorisation from the relevant authority.[22]

24.15 As a general rule, infrastructure building and maintenance is the responsibility of the private operator.[23] In limited circumstances, the building of some infrastructures can be the responsibility of the host state,[24] which in particular will intervene through Public Private Partnerships (PPP).[25] The same can apply for infrastructure maintenance, particularly when the operator does not solely use the infrastructure.[26]

24.16 In parallel, infrastructure requirements are frequently taken into account in the exploitation agreement, which can make the provision of infrastructure mandatory and specifies them in the agreement itself (eg construction of a road infrastructure, construction of a railway often require in the mining industry).

24.17 Reinforcing the local infrastructure does not only refer to transport infrastructures but also to infrastructures necessary to the local processing of the extracted resources.

24.18 This can firstly involve an obligation to build a processing unit, particularly depending on the production thresholds reached. Construction of a processing unit will usually depend on the results of an economic study, assessing in particular the costs for processing the natural resource in the host state with the same in neighbouring countries. Private operators themselves tend to favour the creation of processing units immediately adjacent to the production facilities or at least within the host state.

24.19 The quality of extracted resources is taken into account with regard to the processing of natural resources. If we take the example of crude oil, there are as many quality grades as there are production facilities; accordingly, the proportions of light, medium and heavy fractions will differ according to the quality of this crude oil extracted. It is therefore essential to adapt the refining structure and technology to the quality grades of the crude oil extracted locally.

(2) Financing of scientific or technical research

24.20 The agreement entered into between the host state and the private operator can stipulate that the latter must dedicate an amount to general scientific or technical research, varying according to the level of production.

[22] For Eritrea, see Article 23 of Proclamation No. 68/1995 Promoting Development of Mineral Resources; for South Sudan, Article 146 of the 2012 Mining Act.

[23] In this regard, for Algeria see Article 22(2) of the Presidential Decree of 1 March 2003 Enacting the Model Mining Agreement.

[24] Accordingly, for Burkina Faso, Article 7 of Law N° 036-2015/CNT of 16 June 2015 Enacting the Mining Code of Burkina Faso provides that '*[l]'État a la responsabilité de la mise en place et du renforcement des infrastructures de base* [. . .].' [Free translation into English: '[t]he State is responsible for establishing and strengthening basic infrastructures [. . .].']

[25] For Guinea, see Article 121 of Law No. L/2013/N°053/CNT of 8 April 2013 Amending of Law No. 2011/006/CNT of 9 September 2011 Enacting the Mining Code.

[26] For Algeria, see Article 22(2) of the Presidential Decree of 1 March 2003 Enacting the Model Mining Agreement.

This amount can in particular be allocated to financing or subsidising research conducted in laboratories, design or computing consultancies, experimental stations, pilot workshops, organisations or universities.

24.21 A number of agreements set out lists of possible studies, such as methanol production and export, recovery and export of associated gas, use of nitrogen as fertiliser, petroleum desulphurisation or export of liquefied natural gas, etc. It should be noted that such provisions, originally focusing on hydrocarbons, now also extend to mining resources.

b Economic infrastructure not directly linked to extractive industries

24.22 For a long time, a large proportion of mining and petroleum states have tended to neglect the development of sectors of their economy that are not related to the extractive industry, with the consequence of making countries that produce natural resources increasingly dependent on foreign countries for all logistics surrounding their exploitation.

24.23 The extension by private operators of the economic infrastructure other than that of the extractive industry is currently a major issue for the host state. The industrial agricultural sectors bear witness to the tendency of mining and petroleum projects to exert an indirect impact on other sectors of the national economy.

i *Development of other sectors through the use of the infrastructures*

24.24 Most exploitation agreements open the way to enormous prospects for extending the local economic infrastructure. Exploitation of a natural resource inevitably results in the establishment of a large number of companies, particularly through national preference mechanisms.[27]

However, difficulties may emerge with regard to the use of an infrastructure built under and during the implementation of an exploitation agreement: the host state will strive to ensure that its benefits extend to other activities or other operators who have entered the sector at a later date. The problem may arise with regard to sharing the use of this infrastructure. In order to settle this matter, mining and petroleum codes can contain provisions on the use by third parties of infrastructures built in the course of mining and petroleum operations.[28] As a general rule, the use of these infrastructures by third parties

[27] With regard to the establishment of an obligation to give preference to the local market, see paras 23.2 et seq above.

[28] For Cameroon, see Article 60(1) of Law No. 99-013 of 22 December 1999 Enacting the Petroleum Code; for the Democratic Republic of Congo, Article 213 of Law No. 007-2002 of 11 July 2002 Enacting the Mining Code; for the Côte d'Ivoire, Article 136 of Law No. 2014-138 of 24 March 2014 Enacting the Mining Code; for Eritrea, Article 23(5) of Proclamation No. 68/1995 Promoting Development of Mineral Resources; for Guinea, Article 128 of Law No. 2011/006/CNT of 9 September 2011 Enacting the Mining Code; for Madagascar, Article 114 of Law No. 99-022 of 19 August 1999 Enacting the Mining Code amended by Law No. 2005-021 of 17 October 2005; for Niger, Article 12 of Law No. 2007-01 of 31 January 2007 Enacting the Petroleum Code; for Chad, Article 40 of Law No.

is contingent on the absence of any resulting obstacle for the operator and for the use to be conditional on payment of fair compensation,[29] remuneration[30] or occasionally a licence fee.[31]

24.25 A number of mining and petroleum codes go further and establish a right of transfer of the constructed infrastructures to the benefit of the State on expiry of the exploitation agreement or at the end of the term of validity of the mining licence; the ownership of public utility infrastructures built by the operator will thus be automatically transferred to the host state.[32] In a similar vein, a number of codes and exploitation agreements stipulate that the infrastructure is to become an integral part of the public property.[33]

These different mechanisms extend the benefits of infrastructures built in the course of mining and petroleum projects to other sectors of the economy.

ii *Development of the agricultural sector*

24.26 Agriculture is the sector which has suffered most from the priority given to extractive industries. Nevertheless, a particular benefit for agriculture has been highlighted.

Large international petroleum or mining companies have intended to diversify their services by contributing to the development of local agriculture. International petroleum companies have long contributed to the development of the local agricultural sector on the location of their exploitation sites.

In addition, a number of natural resources, particularly phosphates and hydrocarbons, can make a contribution in the form of fertilisers, pesticides or plastic materials to improve agricultural productivity both in terms of quantity and of quality.

006-PR-2007 of 2 May 2007 On Hydrocarbons; for Togo, Article 45 of Law No. 99-003 of 18 February 1999 Enacting the Hydrocarbons Code.

[29] In particular, for Cameroon see Article 60(1) of Law No. 99-013 of 22 December 1999 Enacting the Petroleum Code; for the Democratic Republic of Congo, Article 213 of Law No. 007-2002 of 11 July 2002 Enacting the Mining Code; for the Côte d'Ivoire, Article 136 of Law No. 2014-138 of 24 March 2014 Enacting the Mining Code; for Madagascar, Article 114 of Law No. 99-022 of 19 August 1999 Enacting the Mining Code amended by Law No. 2005-021 of 17 October 2005; for Togo, Article 45 of Law No. 99-003 of 18 February 1999 Enacting the Hydrocarbons Code.

[30] In particular, for Chad, Article 40 of Law No. 006-PR-2007 of 2 May 2007 On Hydrocarbons.

[31] See, to that effect, for Eritrea, Article 23(6) of Proclamation No. 68/1995 Promoting the Development of Mineral Resources.

[32] In this regard, for the Democratic Republic of Congo, see Article 214 of Law No. 007-2002 of 11 July 2002 Enacting the Mining Code; for Eritrea, Article 23(9) of Proclamation No. 68/1995 Promoting the Development of Mineral Resources; for Guinea, Article 121 of Law No. L/2013/N°053/CNT of 8 April 2013 Amending Law No. 2011/006/CNT of 9 September 2011 Enacting the Mining Code.

[33] For Madagascar, see Article 115 of Law No. 99-022 of 19 August 1999 Enacting the Mining Code amended by Law No. 2005-021 of 17 October 2005.

2 VARIOUS CONTRIBUTIONS TO LOCAL DEVELOPMENT

24.27 In parallel with extending the infrastructure of the host state, there are a variety of contributions to its development which, for private operators, constitute a second series of obligations to contribute to the general development of national assets.

24.28 Historically, private operators have developed policies in favour of local development for the acknowledged purposes of generating a climate of trust and as a defence against nationalisation measures. Contributions by private operators to local development have today been enshrined in legal, regulatory or simply contractual obligations. Besides, the development of corporate social responsibility in the context of mining and petroleum projects contributes to favour the participation of private operators in local development.[34]

General contributions to local development essentially relate to health care and education. They can also relate to other aspects of public life, such as the construction of police buildings, post offices, water pipelines, electric power lines, schools, etc.

a Development of the health care sector

24.29 On the African continent, the life expectancy of local populations varies greatly between one state and another. The contribution of extractive industries to the development of the health care system in the host state is essential. Private operators can indeed improve the health conditions offered locally, either through a legal or a contractual obligation or on the basis of a unilateral decision by the operator itself. In general, mining and petroleum codes do not establish any obligation with regard to the construction of health care facilities; at most, legal provisions mention the possibility to build health care infrastructures in the course of mining and petroleum operations.[35] Improvement of basic social services, particularly the construction of health centres, can also result from a community development plan that the private operator is required to develop under its project.[36]

24.30 The obligations of private operators can also include training local workers in elementary hygiene regulations. Construction of a clinic or a hospital can depend on the number of permanent workers on the site.

24.31 The obligation to develop the health care sector relates to workers employed in the area of operations. In practice, however, the population living in the area of operations can benefit from these infrastructures.

---

[34] With regard to the development of corporate social responsibility in the mining and petroleum sectors, see para 22.48 above.

[35] For Algeria, see Article 22(3) of the Presidential Decree of 1 March 2003 Enacting the Model Mining Agreement; for Eritrea, Article 23(7) of Proclamation No. 68/1995 Promoting the Development of Mineral Resources.

[36] In this regard, for Mali see Article 141 of Order No. 2012-311/P-RM of 21 June 2012 Enacting the Conditions and Procedures for Implementation of the Law Enacting the Mining Code.

b Development of the education sector

24.32 Historically, extractive industries have always attached particular importance to education. Amounts spent by private operators in local education are currently far from negligible.

Education intervention by private operators may or may not be related to a formal contractual obligation.

(1) Legal framework of the private operator's contribution to training

24.33 The contribution of private operators to the construction of an education infrastructure is sometimes mentioned in the texts governing the mining and petroleum sectors. In this regard, a number of texts establish an obligation for the private operator to provide for the construction or improvement of education infrastructures, particularly as part of a development plan.[37] Construction of schools is sometimes included in the framework of a community development plan that must be provided by the operator as a condition for any mining licence application.[38]

(2) Contractual framework of the private operator's contribution to education

24.34 Exploitation agreements frequently impose certain education obligations on private operators.

As from the signing of the exploitation agreement, private operators can be required to make efforts to instruct their local personnel or local nationals in addition to any training obligation in relation to the works performed on site.

24.35 The contribution to education is part of the general obligation for the development of the local production site and its immediate environment. Initially, this involved providing schooling for the children of expatriate personnel, then for the children of all personnel. It is now designed very broadly, because beyond the children of the personnel it encompasses the schooling of a broadly defined area of varying size around the production site.

24.36 These obligations to develop the education sector can also be stipulated outside the exploitation agreement. After signing the exploitation agreement, the private operator informs the local State operator by a separate letter of its intention to provide financial support to the general local education programme. This frequently involves the implementation of previous commitments at the time of negotiating the exploitation agreements.

From a general point of view, the matter of whether or not these contributions to local development and education can be in the form of deductible charges is a major strategic and financial issue at the time of negotiation.

[37] For Mali, see Article 141 of Order No. 2012-311/P-RM of 21 June 2012 Enacting the Conditions and Procedures for Implementing the Mining Code.

[38] For Guinea, see Article 37-II of Law No. L/2013/N°053/CNT of 8 April 2013 Amending Law No. 2011/006/CNT of 9 September 2011 Enacting the Mining Code.

24.37 Petroleum and mining rights in general, and particularly in Africa, is a constantly evolving topic. The laws governing the exploitation of natural resources have to adapt rapidly to economic, political and structural changes. Petroleum and Mining Law constitutes a complex and dynamic subject which is highly specific. At the time of completion of this book, numerous African regulations specifically dedicated to the oil and mining sectors are undergoing change, sometimes on fundamental points and subjects. This extreme volatility of the rules is highly incompatible with the duration and financial issues encountered in extractive projects. It therefore requires skilful professionals who specialise in the sector, both in legal but also technical terms. It also requires practitioners from Africa, as it is true here that good on-the-ground knowledge often makes the difference.

APPENDICES

APPENDIX I

Main regulations applicable to the upstream Oil & Gas sector by country

Nota Bene: *This list does not include related legislation which may apply indirectly to oil operations (environmental law, land law, health and safety law, tax legislation, etc.). It should be noted that the analysis given in this book does not take account of reforms that have been adopted since the end of January 2018.*

Country	Applicable texts	Main reforms and comments
South Africa	• Mineral and Petroleum Resources Development Act, 2002 • Mineral and Petroleum Resources Development Amendment Act, 2008 • Mineral and Petroleum Resources Royalty Act, 2008 • Mineral and Petroleum Resources Royalty (Administration) Act, 2008	Discussions in progress on a reform of the legal framework
Algeria	• Law No. 05-07 of 28 April 2005 On Hydrocarbons • Ordinance No. 06-10 of 29 July 2006 Amending and Supplementing Law No. 05-07 of 28 April 2005 On Hydrocarbons • Law No. 30-01 of 20 February 2013 Amending and Supplementing Law No. 05-07 of 28 April 2005 On Hydrocarbons • Executive Decree No. 06-440 of 2 December 2006 Laying Down the Procedure, Conditions and Methodology for Calculating Tax on Exceptional Profits • Executive Decree No. 07-127 of 5 May 2007 On Delimitation and Classification of the Mining Sector into Zones and the Definition of Prospecting, Exploration and Exploitation Areas • Executive Decree No. 07-130 of 7 May 2007 Laying Down the Procedures for Calculating Amounts of Provisional Monthly Payments Equivalent to Down Payments on Tax on Oil Income • Executive Decree No. 07-183 of 9 June 2007 Laying Down the Selection Procedures for Delimitation of the Areas Subject to Retention Period Requests, Exploitation Areas and Relinquished Exploration Area	Discussions in progress on a reform of the Law On Hydrocarbons of 2005 as amended in 2013

Country	Applicable texts	Main reforms and comments
	• Executive Decree No. 07-184 of 9 June 2007 Laying down the Procedures for Concluding Exploration and Exploitation Agreements and Hydrocarbon Exploitation Agreements Following a Call-for-Tenders • Executive Decree No. 07-185 of 9 June 2007 Laying down the Conditions for Issuing Mining Rights for Exploration Activities and/or Hydrocarbon Exploitation • Executive Decree No. 08-312 of 5 October 2008 Laying Down the Conditions for Approval of Environmental Impact Studies for Activities Falling under the Hydrocarbons Sector • Executive Decree No. 08-314 of 7 October 2008 Amending and Supplementing Executive Decree No. 07-127 of 5 May 2007 On the Delimitation and Classification of the Mining Domain into Areas and the Definition of Prospecting, Exploration and Exploitation Areas • Executive Decree of 30 April 2014 Amending and Supplementing Executive Decree No. 07-130 of 7 May 2007 Laying Down the Procedures for Calculating the Amounts of Provisional Monthly Payments Equivalent to Down Payments on the Tax on Oil Income	
Angola	• Law No. 10/04 of 12 November 2004 On Petroleum Activities • Law No. 11/04 of 12 November 2004 On Petroleum Industry Specific Customs Procedure • Law No. 13/04 of 24 December 2004 On Taxation of Petroleum Activities • Law No. 2/12 of 13 January 2012 On Foreign Exchange Regime Applicable to the Petroleum Companies • Order No. 127/03 of 25 November 2003 On Legal Framework Applicable to the Procurement of Services and Goods by Companies in the Petroleum Industry • Decree No. 48/06 of 1 September 2006 Establishing Rules on Public Tenders for the Procurement of Goods and Services Required for Petroleum Operations • Decree No. 120/08 of 11 December 2008 On Access to Real Estate Property and Land Rights Required for the Performance of Petroleum Activities • Decree No. 1/09 of 27 January 2009 On Petroleum Activities Regulations	

Country	Applicable texts	Main reforms and comments
	• Decree-Law No. 17/09 of 26 June 2009 On Local Content Regulations • Executive Decree No. 45/10 of 10 May 2010 Supplementing Decree-Law No. 17/09 of 26 June 2009 On Local Content Regulations • National Bank of Angola Order No. 20/12 of 25 January 2012 Amending the Foreign Exchange Regime Applicable to Petroleum Companies • Presidential Decree No. 211/15 of 2 December 2015 Establishing Terms and Conditions Applicable to the Oil Exploitation Activities within the Target Development Areas • Presidential Decree No. 109/16 of 26 May 2016 Enacting the Model for Readjustment of the Petroleum Sector's Organization	
Benin	• Law No. 2006-18 of 17 October 2006 Enacting the Petroleum Code in the Republic of Benin	
Botswana	• Petroleum (Exploration and Production) Act (Chapter 67:01)	
Burkina Faso	• No text applicable to the upstream petroleum sector	
Burundi	• Decree-Law No. 1/138 of 17 July 1976 Enacting the Mining and Petroleum Code of the Republic of Burundi • Decree No. 100/130 of 14 December 1982 Establishing Implementing Provisions of Decree-Law No. 1/138 of 17 July 1976 Establishing the Mining and Petroleum Code of the Republic of Burundi	
Cameroon	• Law No. 99-013 of 22 December 1999 Enacting the Petroleum Code • Decree No. 2000-485 of 30 June 2000 Determining the Conditions for Implementing Law No. 99-013 of 22 December 1999 Enacting the Petroleum Code	
Central African Republic	• Ordinance No. 93.007 of 25 May 1993 Enacting the Petroleum Code • Decree Implementing Ordinance No. 93.007 of 25 May 1993 Enacting the Petroleum Code	
Chad	• Law No. 006/PR/2007 of 2 May 2007 Enacting the Hydrocarbons Code • Law No. 002/PR/2014 On the Management of Petroleum Incomes Decree No. 10-796 2010-09-30 PR/PM/MPED Implementing Law No. 006/PR/2007 Enacting the Hydrocarbons Code	

Country	Applicable texts	Main reforms and comments
	• Ordinance No. 001/PR/2010 of 7 January 2010 Amending Law No. 006/PR/2007 Enacting the Hydrocarbons Code and Enacting the Standard Production Sharing Agreement	
Comoro Islands	• Decree No. 13-017/PR of 8 February 2013 Enacting Law No. 12-019/EU of 27 December 2012 Establishing the Petroleum Code	
Congo (Republic of the Congo)	• Law No. 28-2016 of 12 October 2016 Enacting the Petroleum Code	
Democratic Republic of the Congo	• Law No. 15/012 of 1 August 2015 Establishing the General System of Hydrocarbons • Decree No. 16/010 of 19 April 2016 Enacting the Hydrocarbons Regulation	
Côte d'Ivoire	• Law No. 96-669 of 29 August 1996 Enacting the Petroleum Code • Ordinance No. 2012-369 of 18 April 2012 Amending Law No. 96-669 of 29 August 1996 Enacting the Petroleum Code • Decree No. 96-733 of 19 September 1996 On Procedures for Implementing the Petroleum Code	
Djibouti	• Law No. 117-AN-05/5th L of 16 October 2005 Enacting the Petroleum Code	
Egypt	• Mining and Quarries Law No. 86 of 1956 • Law No. 20 of 1987 Organising the Egyptian General Petroleum Company (EGPC) • Law No. 217 of 1980 Organising Natural Gas and Executive Regulations • Law No. 4 of 1988 Regarding Oil Pipelines	
Eritrea	• Proclamation No. 108/2000, A Revised Proclamation to Govern Petroleum Operations • Proclamation No. 109/2000, A Revised Proclamation to Provide for Payment of Income Tax on Petroleum Operations • Legal Notice No. 45/2000, Revised Regulations on Petroleum Operations	
Ethiopia	• Proclamation No. 295/1986, A Proclamation to Regulate Petroleum Operations • Proclamation No. 296/1986, A Proclamation to Provide for Payment of Income Tax on Petroleum Operations • Proclamation No. 226/2000, A Proclamation to Amend the Petroleum Operations Income Tax Proclamation	

Country	Applicable texts	Main reforms and comments
Gabon	• Law No. 011/2014 of 28 August 2014 Enacting the Petroleum Code • Decree No. 0211/PR/MPH of 6 April 2016 On Procedures for Accessing the Petroleum Domain	
Gambia	• Petroleum (Exploration, Development and Production) Act (Chapter 65-02)	
Ghana	• Petroleum (Exploration and Production) Act, 2016	
Guinea	• Law No. L/2014/No.034/AN of 23 December 2014 Enacting the Petroleum Code	
Guinea-Bissau	• Law No. 4/2014 of 15 April 2014 Regulating Liquid and Gaseous Hydrocarbon Prospecting, Exploration, Production and Transport Activities and their Supervision	
Equatorial Guinea	• Hydrocarbons Law No. 8/2006 of 3 November 2006 of the Republic of Equatorial Guinea • Petroleum Regulation No. 4/2013 of 20 June 2013 • Model Petroleum Production Sharing Contract	
Mauritius	• Petroleum Act, 1970 • Petroleum (Licence and Lease) Regulations	
Kenya	• Petroleum (Exploration and Production) Act (Chapter 308) • Petroleum (Exploration and Production) Regulations, 1984 • Petroleum (Exploration and Production) (Training Fund) Regulations, 2006 • Petroleum Exploration, Development and Production (Local Content) Regulations, 2014	
Lesotho	• No text applicable to the upstream petroleum sector	
Liberia	• Petroleum Law, 2002 • National Petroleum Policy, 2012 • Act to Amend and Restate the New Petroleum Law of Liberia 2002 Thereby Establishing the National Oil Company of Liberia, 2014	
Libya	• Law No. 25 of 155 Enacting the Petroleum Law (as amended)	
Madagascar	• Law No. 96-018 of 4 September 1996 Enacting the Petroleum Law • Law No. 2004-003 of 23 June 2004 On the Liberalization of the Petroleum Sector	Reform in progress
Malawi	• Petroleum (Exploration and Production) Act (Cap: 61:02)	

Country	Applicable texts	Main reforms and comments
	• Liquid Fuels and Gas (Production and Supply) Act (Cap 50:03)	Reform of the petroleum sector is currently under discussion
Mali	• Law No. 2015-035 of 16 July 2015 On the Organization of the Exploration, Exploitation and Transport of Hydrocarbons • Decree No. 2016-0272/P-RM of 29 April 2016 Implementing Law No. 2015-035 of 16 July 2015 On the Organization of the Exploration, Exploitation and Transport of Hydrocarbons	
Morocco	• Law No. 21-90 of 1 April 1992 Relating to the Exploration and Exploitation of Hydrocarbon Deposits Amended and Supplemented by Law No. 27-99 of 15 February 2000 • Decree No. 2-93-986 of 3 November 1993 Implementing Law No. 21-90 Relating to the Exploration and Exploitation of Hydrocarbon Deposits • Decree No. 2-99-210 of 16 March 2000 amending and Supplementing Decree No. 2-93-786 of 3 November 1993 Implementing Law No. 21-90 Relating to the Exploration and Exploitation of Hydrocarbon Deposits	
Mauritania	• Law No. 2010-033 of 20 July 2010 Enacting the Petroleum Code • Law No. 2011-044 of 25 November 2011 Amending and Supplementing Certain Provisions of Law No. 2010-33 of 20 July 2010 • Law No. 2015/0016 Amending and Supplementing Certain Provisions of Law No. 2010-33 of 20 July 2010 • Decree no. 2011-229 of 2 October 2011 Establishing Exemption from the Call-for-tender Procedure • Decree No. 2011-230 of 2 October 2011 Establishing the Procedures for Implementing Articles 7, 8, 12, 18 and 29 of the Petroleum Code • Decree No. 2011-240 of 27 October 2011 Relating to the Annual Contribution Intended for Training and Improvement, for the Monitoring of Oil Operations and the Promotion of the Oil Sector • Decree No. 286-2011 of 15 November 2011 Enacting the Standard Exploitation Production Agreement	
Mozambique	• Law No. 21/2014 of 18 August 2014 Enacting the New Petroleum Law • Law No. 27/2014 of 23 September 2014 Enacting the New Petroleum Tax Law	

Country	Applicable texts	Main reforms and comments
	• Decree No. 34/2015 of 31 December 2015 Enacting the Petroleum Operations Regulations • Decree No. 56/2010 of 22 November 2010 Enacting the Environmental Regulations for Petroleum related Activities • Decree No. 63/2011 of 7 December 2011 Enacting the Regulations on the Hiring of Expats for Oil and Mining Sectors • Ministerial Order No. 272/2009 of 30 December Enacting the Petroleum Operations and Infrastructures Licensing Regulations • Ministerial Order No. 31/2014 of19 March 2014 Enacting the Licensing of Technicians for Petroleum Activities Regulations	
Namibia	• Petroleum (Exploration and Production) Act, 1991 • Petroleum (Taxation) Act, 1991 • Petroleum (Exploration and Production) Amendment Act, 1993 • Petroleum Laws Amendment Act, 1998 • Model Petroleum Agreement, 2007	
Niger	• Law No. 2007-01 of 31 January 2007 Enacting the Petroleum Code • Law No. 2014-06 of 16 April 2014 Amending and Supplementing Law No. 2007-01 of 31 January 2007 Enacting the Petroleum Code • Decree No. 2007-082/PRN/MME of 28 March 2007 relating to the procedures for Implementing Law No. 2007-01 of 31 January 2007 Enacting the Petroleum Code.	A new petroleum code has been adopted by the Parliament in August 2017.
Nigeria	• Petroleum Act • Petroleum Profits Tax Act • Niger Delta Development Commission Act • Nigerian Oil and Gas Industry Content Development Act • Associated Gas Re-injection Act • Deep Offshore and Inland Basin Production Sharing Contracts Act • Petroleum (Drilling and Production) Regulations • Petroleum (Drilling and Production) (Amendment) Regulations • Mineral Oils (Safety) Regulations • Marginal Fields Operations (Fiscal Regime) Regulations	Reform of the petroleum sector is currently under discussion
Uganda	• Petroleum (Exploration, Development and Production) Act, 2013	

Country	Applicable texts	Main reforms and comments
	• Petroleum (Exploration, Development and Production) (National Content) Regulations, 2016 • National oil and gas policy for Uganda, 2008	
Rwanda	• Law No. 13/2016 of 2 May 2016 Governing Petroleum Exploration and Production Activities	
São Tomé and Principe	• Law No. 16/2009 of 4 November 2009 Enacting the Framework Law on Oil Operations • Law No. 8/2004 of 29 December 204 Enacting the Framework Law on Oil Incomes • Law No. 57/2009 of 30 December 2009 Organizing the ZEE into areas and blocks • Law No. 15/2009 of 4 November 2009 On Petroleum Taxation (amended) • Law No. 11/2006 of 29 December 2006 Organizing the Petroleum Taxation Commission • Treaty between the Federal Republic of Nigeria and The Democratic Republic of São Tome and Principe Creating a Joint Development Zone of 26 June 2004 • Decree-Law No. 5/2004 of 30 June 2004 Establishing the National Petroleum Agency	
Senegal	• Law No. 98-05 of 8 January 1998 Enacting the Petroleum Code	Discussions in progress on a potential reform of the legal framework
Seychelles	• Petroleum Mining Act, 1976 (Chapter 167) • Petroleum (Taxation) Act, 2008 • Petroleum Taxation (Amendment) Act, 2013 • Seychelles Model Petroleum Agreement, 2013	
Sierra Leone	• Petroleum Exploration and Production Act, 2011	
Somalia	• Petroleum Law, 2008	
Sudan	• Petroleum Resources Act, 1972 • Petroleum Resources Regulations, 1973 • Petroleum Resources (Amendment) Act, 1975 • Petroleum Wealth Act, 1998	
South Sudan	• Petroleum Act, 2012 • Petroleum Health, Safety and Environmental Management System and Plans Regulations, 2015	
Swaziland	• No text applicable to the upstream petroleum sector	
Tanzania	• Petroleum Act, 2015 • Oil & Gas Revenue Management Act, 2015 • Petroleum (Local Content) Regulations, 2017	

Country	Applicable texts	Main reforms and comments
Togo	• Law No. 99-003 of 18 February 1999 Enacting the Petroleum Code	
Tunisia	• Law No. 99-93 of 17 August 1999 Enacting the Petroleum Code • Law No. 2002-23 of 14 February 2002 Supplementing Law No. 99-93 of 17 August 1999 Enacting the Petroleum Code • Law No. 2004-61 of 27 July 2004 Supplementing Law No. 99-93 of 17 August 1999 Enacting the Petroleum Code • Law No. 2008-15 of 18 February 2008 Supplementing Law No. 99-93 of 17 August 1999 Enacting the Petroleum Code • Decree No. 2000-713 of 5 April 2000 Determining the Composition and Operation of the Hydrocarbon Consultative Committee • Decree No. 2000-946 of 2 May 2000 Determining the Geographic Coordinates and Reference Point Numbers of Peaks of Elementary Areas Relating to Hydrocarbon Rights • Decree No. 2000-1322 of 13 June 2000 Determining the Procedures for Calculating and Applying Report 'R' relating to the Determination of Rates of the Proportional Duty on the Production and Tax on Profits • Order of the Minister of Industry of 15 February 2001 Determining the Procedures for Submitting and Reviewing Applications for Hydrocarbon Rights	
Zambia	• Petroleum (Exploration and Production) Act, 2008 • Petroleum (Exploration and Production) (General) Regulations, 2011 • Petroleum (Exploration and Production) (National Petroleum Company) Regulations, 2011	
Zimbabwe	• Mines and Minerals Act (Chapter 21:05) • Indigenisation and Economic Empowerment Act (Chapter 14:33)	

Appendix II

Main regulations applicable to the upstream mining sector by country

Nota Bene: *This list does not include related legislation which may apply indirectly to mining operations (environmental law, land law, health and safety law, tax legislation, etc). It should be noted that the analysis given in this book does not take account of reforms that have been adopted since the end of January 2018.*

Country	Applicable texts	Main reforms and remarks
South Africa	• Mineral and Petroleum Resources Development Act, 2002 (as amended) • Mining Titles Registration Act, 2003 • Mining Titles Registration Regulations (Government Notice of 14 May 2004) • Mineral Technology Act, 1989 • Mine Health and Safety Act, 1996 • Mineral and Petroleum Resources Royalty Act, 2008. • Mineral and Petroleum Resources Royalty (Administration) Act, 2008 • Regulations pertaining to the Financial Provision for the Rehabilitation and Closure and Post Closure of Prospecting, Exploration, Mining or Production Operations, 2014 (GN 940 of 2014) • Codes of Good Practice for the Minerals Industry (GN 446 of 2009)	Discussions in progress on a reform of the legal framework
Algeria	• Law No. 01-10 of 3 July 2001 Enacting the Mining Law • Ordinance No. 07-02 of 1 March 2007 Amending and Supplementing Law No. 01-10 of 3 July 2001 Enacting the Mining Law • Law No. 14-05 of 24 February 2014 Amending the Mining Law • Executive Decree No. 02-65 of 6 February 2002 Defining the Conditions and Procedures for Granting Mining Rights • Executive Decree No. 02-66 of 6 February 2002 Determining the Conditions for Tendering of Mining Rights • Executive Decree No. 02-472 of 24 December 2002 Determining the Criteria for Reductions Applicable to the Extraction Fee • Presidential Decree of 1 March 2003 On the Model Mining Convention	

Country	Applicable texts	Main reforms and remarks
	• Executive Decree No. 03-199 of 3 May 2003 Establishing the Model of Specifications [*Cahier des charges*] for Mining Activities • Executive Decree No. 06-257 of 30 July 2006 Amending And Supplementing Executive Decree No. 03-256 of 22 July 2003 Determining the List of Specific Equipment Exempted from VAT and/or Customs Duties, Taxes and Fees • Order of 26 July 2003 Relating to the Method for Determining the Quantity of Mineral Substances Extracted • Circular of 22 March 2016 Establishing the Conditions and Modalities for Implementing the Preference in Favor of Landowners • Notice of April 2016 issued by the Mining Administration and Relating to the Surface Fee	
Angola	• Law No. 31/11 of 23 September 2011 Enacting the Mining Law • Order No. 2/03 of 28 February 2003 of the National Bank of Angola Establishing the Foreign Exchange Regime for Holders of Mineral Rights • Order No. 255/14 of 28 January 2014 of the Ministry of Geology and Mines On Monitoring of Posting of Bonds and Payments of Surface Fee and Royalties under the Mining Law • Presidential Decree No. 2/14 of 2 January 2014 On the Market Regulation Agency for Gold • Presidential Decree No. 36/12 of 5 March 2012 On the Internal Organization and Functioning of the Mineral Negotiations Committee • Presidential Decree 158/16 of 10 August 2016 Approving the Mineral Administrative Infringements Regime • Presidential Decree 231/16 of 8 December 2016 Which Classifies Rare Metals and Rare Earth Elements as Strategic Minerals	
Benin	• Law No. 2006-17 of 17 October 2006 Enacting the Mining Code and Mining Taxation in the Republic of Benin • Decree No. 2008-804 of 31 December 2008 Implementing the Mining Code and Mining Taxation in the Republic of Benin • Ministerial Order No. 016 of 20 December 2011 Related to the Granting of Quarries Opening and Exploitation Authorizations	

Country	Applicable texts	Main reforms and remarks
	• Ministerial Order No. 87 of 5 August 2014 Determining the Minimum and Maximum Perimeter of an Exploration Permit and Fixing the Mining Cadastral Unit • Ministerial Order No. 108/MERPMEDER/DC/SGM/CYJ/CTRPM/DGM/SA of 13 November 2014 Establishing the Fixed Duties Relating to the Granting, Renewal and Transfer of Mining Titles in the Republic of Benin • Decree No. 2002-580 of 31 December 2002 Regulating Artisanal Mining in the Republic of Benin	
Botswana	• Mines and Minerals Act (Chapter 66:01) • Mineral Rights in Tribal Territories Act (Chapter 66:02) • Unwrought Precious Metal Act (Chapter 20:03) • Mines, Quarries, Works and Machinery Act (Chapter 44:02) • Precious and Semi-Precious Stones Act (Chapter 66:03) • Explosive Act (Chapter 24:02)	
Burkina Faso	• Law No. 036-2015/CNT of 16 June 2015 Establishing the Mining Code of Burkina Faso • Decree No. 2015-885/PRES-TRANS Implementing the Mining Code • Decree No. 2017-0023 of 23 January 2017 Fixing Taxes and Mining Royalties • Decree No. 2017-0024 of 23 January 2017 On the Organization, the Functioning and the Collection Methods of the Mining Fund for Local Development • Decree No. 2017-0034 of 26 January 2017 On the Organization, the Functioning and the Collection Methods of the Financing Fund Promoting Geological and Mining Exploration and Supporting Earth Sciences Trainings • Decree No. 2017-0035 of 26 January 2017 Establishing a Standard Model of Mining Convention • Decree No. 2017-0036 of 26 January 2017 On the Management of Mining Permits and Authorizations • Decree No. 2017-0047 of 1 February 2017 On the Organization, the Functioning and the Collection Methods of the Fund for the Rehabilitation and Security of Artisanal Mining Sites and the Fight Against the Use of Prohibited Chemical Products	

Country	Applicable texts	Main reforms and remarks
	• Decree No. 2017-0068 of 15 January 2017 on the Organization, the Functioning and the Collection Methods of the Fund for the Rehabilitation of Mining Sites and Mines Closure	
Burundi	• Law No. 1/21 of 15 October 2013 Establishing the Mining Code • Ministerial Ordinance No. 760/540/770/1757 of 26 December 2013 Determining the Annual Contribution for the Rehabilitation of Sites for the Extraction of Minerals by Artisanal Mining Methods, Quarries as well as Purchase and Sale Agencies of Minerals Extracted by Artisanal Mining Methods • Ministerial Ordinance No. 760/540/1758/2013 of 26 December 2013 Establishing the Administrative and Fiscal Conditions for Granting and Renewal of Prospecting Permits, Exploration Licences, Exploitation Licences for Mines and Quarries, Artisanal Exploitation Licences and Approvals for Agencies for the Purchase and Sale of Mineral Substances Extracted in Burundi	
Cameroon	• Law No. 2016/017 of 14 December 2016 Enacting the Mining Code	
Cape Verde	• Law-Decree No. 3/2015 of 6 January 2015 Enacting the New Legal Regime for the Exploitation of Mineral Resources	
Central African Republic	• Law No. 09-005 of 29 April 2009 Establishing the Mining Code • Decree No. 09-126 of 30 April 2009 Determining the Conditions for Enforcement of Law No. 09-005 of 29 April 2009 Establishing the Mining Code	
Comoros	• Decree-Law No. 54-1110 of 30 November 1954 Reforming the Mineral Substance Scheme • Decree No. 58-9 of 2 January 1958 Implementing Decree-law No. 54-1110 of 30 November 1954 Amended and Supplemented Relating to the Mineral Substances Scheme, the Nationality Conditions to be Fulfilled by Natural Persons or Companies and their Managers to Carry Out Mining Activity	
Congo (Republic of the Congo)	• Law No. 4-2005 of 11 April 2005 Enacting the Mining Code • Law No. 24-2010 of 30 December 2010 Determining the Rates and Rules for the Collection of Duties on Mining Rights	

Country	Applicable texts	Main reforms and remarks
	• Decree No. 2007-274 of 21 May 2007 Determining the Conditions for Prospection, Exploration and Exploitation of Mineral Substances and those for Carrying out Administrative Monitoring • Decree No. 2007-293 of 31 May 2007 Determining the Technical Rules for Exploiting Quarries of Geomaterials	
Democratic Republic of the Congo	• Law No. 007/2002 of 11 July 2002 Enacting the Mining Code • Law No. 18/001 of 9 March 2018 Amending and Supplementing Law No. 007/2002 of 11 July 2002 Enacting the Mining Code • Decree No. 038/2003 of 26 March 2003 Enacting the Mining Regulation • Decree No. 011/26 of 20 May 2011 On the Obligation to Publish All Contracts Related to Natural Resources • Interministerial Order No. 3154/CAB.MIN/MINES/01/2007 and No. 031/CAB/MIN/FINANCES/2007 of 9 August 2007 Establishing the Rates of Duties, Taxes and Fees to be Collected at the Initiative of the Minister of Mines • Interministerial Order No. 0536/CAB.MIN/MINES/01/2008 and No. 276/CAB/MIN/FINANCES/2007 of 2 December 2008 Amending and Supplementing Interministerial Order No. 3154/CAB.MIN/MINES/01/2007 and No. 031/CAB/MIN/FINANCES/2007 of 9 August 2007 Establishing the Rates of Duties, Taxes and Fees to be Collected at the Initiative of the Minister of Mines	
Côte d'Ivoire	• Law No. 2014-138 of 24 March 2014 Enacting the Mining Code • Decree No. 214-397 of 25 June 2014 Implementing the Mining Code • Ordinance No. 2014-148 of 26 March 2014 On Surface and Proportional Taxes relating to Mining Activities • Decree No. 2014-632 of 22 October 2014 Determining Fixed duties on Mining Activities • Ministerial Order No. 002/MIM/CAB of 11 January 2016 Relating to Procedures for Assignment and Renewal of Mining Licences and Authorizations	
Djibouti	• Law No. 138/AN/16/7th L of 23 July 2016 Enacting the Mining Code	

Country	Applicable texts	Main reforms and remarks
Egypt	• Law No. 198 of December 2014 Enacting the Mining Law	
Eritrea	• Proclamation No. 68/1995, A Proclamation to Promote the Development of Mineral Resources • Proclamation No. 69/1995, A Proclamation to Provide for Payment of Tax on Income from Mining Operations • Proclamation no. 165/2011, A Proclamation to Amend the Mining Law Proclamation No. 68/1995 • Legal Notice No. 19/1995, Regulations on Mining Operations	
Ethiopia	• Proclamation No. 678/2010, A Proclamation to Promote Sustainable Development of Mineral Resources • Proclamation No. 816/2013, A Proclamation to Amend the Mining Operations Proclamation • Proclamation No. 651/2009, A Proclamation to Promote and Regulate Transactions of Precious Minerals • Mining Income Tax Proclamation No. 53/1993, A Proclamation to provide for Payment of Tax on Income from Mining Operations • Proclamation No. 23/1996, Mining Income Tax (Amendment) Proclamation • Proclamation No. 802/2013, Mining Income Tax (Amendment) Proclamation	
Gabon	• Law No. 17/2014 of 30 January 2015 Enacting the Mining Law	
Gambia	• Mines and Quarries Act (Chapter 64-01) • Minerals Rules (Chapter 121)	
Ghana	• Minerals and Mining Act, 2006 (Act 703) • Minerals and Mining (Amendment) Act, 2010 (Act 794) • Minerals and Mining (Amendment) Act, 2015 (Act 900) • Minerals and Mining (General) Regulations, 2012 (L.I 2173) • Minerals and Mining (Support Services) Regulations, 2012 (L.I 2174) • Minerals and Mining (Compensation and Resettlement) Regulations, 2012 (L.I 2175) • Minerals and Mining (Licensing) Regulations, 2012 (L.I 2176) • Minerals and Mining (Explosives) Regulations, 2012 (L.I 2177)	

Country	Applicable texts	Main reforms and remarks
	• Minerals and Mining (Health, Safety and Technical) Regulations, 2012 (L.I 2182)	
Guinea	• Law No. L/2011/006/CNT of 9 September 2011 Enacting the Mining Code • Law No. L/2011/005/CNT of 11 August 2011 On the Constitution and Management of Mining Assets • Law No. L/2013/53/CNT of 8 April 2013 Amending Law No. L/2011/006/CNT of 9 September 2011 Enacting the Mining Code • Decree No. D/2011/218/PRG/SGG Establishing a Mining Assets Company • Decree No. D/2014/015/PRG/SGG Enacting a Standard Mining Agreement • Decree No. D/2015/016/PRG/SGG Amending Decree No. D/2011/218/PRG/SGG Establishing a Mining Assets Company and Repealing Decree No. D/2012/093/PRG/SGG Establishing Interim Measures for Managing the Guinean Mining Assets Company	Enactment of regulations implementing the mining code expected
Guinea-Bissau	• Law No. 3/2014 of 29 April 2014 Enacting the New Mining Code	
Ecuatorial Guinea	• Law No. 9/2006 of 3 November 2006 Enacting the Mining Code • Ministerial Order No. 1/2014 of 26 September 2014 Enacting the New Local Content Regulation	
Mauritius	• Minerals Act, 1966	
Kenya	• Mining Act, 2016 • Mining (Reporting of Mining and Mineral Related Activities) Regulations, 2017 • Mining (National Mining Corporation) Regulations, 2017 • Mining (Strategic Minerals) Regulations, 2017 • Mining (Mine Support Services) Regulations, 2017 • Mining (Community Development Agreement) Regulations, 2017 • Mining (Award of Mineral Rights by Tender) Regulations, 2017	
Lesotho	• Mines and Minerals Act, 2005	Reform announced by the Government in 2016
Liberia	• Mineral and Mining Law, 2000	
Libya	• Law No. 2 of 1971 Enacting the Mining Law	

Country	Applicable texts	Main reforms and remarks
Madagascar	• Law No. 99-022 of 19 August 1999 Establishing the Mining Code Amended by Law No. 2005-021 of 17 October 2005 • Law No. 2001-031 of 8 October 2002 Establishing the Special Scheme for Large Mining Investments in Madagascar • Decree No. 2006-910 of 19 August 2006 Implementing the Mining Code • Decree No. 2003-784 of 8 January 2003 Determining the Conditions for Enforcement of Law No. 2001-031 Establishing the Special Scheme for Large Mining Investments in Madagascar • Decree No. 98-394 of 28 May 1998 Enacting the National Mining Policy • Ministerial Order No. 2007-21985 of 20 December 2007 Establishing the Modalities for Recovery of the Royalties and Mining Rebates • Ministerial Order No. 19560-2004 On Suspension of the Granting of Mining Permits and Forest Permits in Reserved Areas • Ministerial Order No. 12032-2000 on the Regulation of the Mining Sector in Terms of Environmental Protection	Reform in progress
Malawi	• Mines and Minerals Act, 1981 • Mines and Minerals Policy, 2013	Discussions in progress on a reform of the legal framework
Mali	• Law No. 2012-015 of 27 February 2012 Enacting the Mining Code • Decree No. 2012-311/P-RM of 21 June 2012 Implementing the Mining Code • Decree No. 2013-690/P-RM of 28 August 2013 Amending Decree No. 2012-311/P-RM of 21 June 2012 Implementing the Mining Code	Discussions in progress on a reform of the legal framework
Morocco	• Law No. 33-13 Relating to Mines Enacted by Dahir No. 1-15-76 of 1 July 2015 • Decree No. 2-15-390 of 20 April 2016 Implementing Law No. 33-30 Relating to Mines and Establishing the Procedure for Granting Mining Rights • Law No. 27-13 of 9 June 2015 Relating to the Exploitation of Quarries Enacted by Dahir No. 1-15-66	
Mauritania	• Law No. 2008-011 of 27 April 2008 Enacting the Mining Code • Law No. 2009-026 of 7 April 2009 Amending Law No. 2008-01 of 27 April 2008 Enacting the Mining Code	Discussions in progress on a reform of the legal framework

Country	Applicable texts	Main reforms and remarks
	• Law No. 2012-014 Repealing and Replacing Certain Provisions of Law No. 2008-011 of 27 April 2008 Enacting the Mining Code • Law No. 2012/012 Regulating Mining Agreements and Enacting the Standard Mining Agreement • Law No. 2014-008 of 27 April 2014 Amending Law No. 2008-01 of 27 April 2008 Enacting the Mining Code • Decree No. 2008-159 of 4 November 2008 On Mining and Quarrying Rights • Decree No. 015.2009/PM/MIM Amending and Supplementing Certain Provisions of Decree No. 2008-159 of 4 November 2008 On Mining and Quarrying Rights • Decree No. 2009-131 of 20 April 2009 On Mining Police	
Mozambique	• Law No. 6/2011 of 11 January 2011 On Rules and Requirements of Licensing, Production, Storage, Distribution and Transport of Explosive Substances • Law No. 20/2014 of 18 August 2014 Enacting the New Mining Law • Law No. 28/2014 of 23 September 2014 Enacting the New Mining Tax Law • Decree No. 31/2015 of 31 December 2015 Implementing Law No. 20/2014 • Decree No. 26/2004 of 20 August 2004 Enacting the Environmental Regulations for Mining Activities • Decree No. 45/2004 of 29 September 2004 Enacting the Environmental Impact Evaluation Process Regulations • Decree No. 61/2006 of 26 December 2006 On Technical Health and Safety Regulations • Decree No. 62/2006 of 26 December 2006 Enacting the Mining Law Regulations • Decree No. 42/2008 of 4 November 2008 Amending Decree No. 45/2004 of 29 September 2004 • Decree No. 20/2011 of 1 June 2011 On Mineral Products Distribution Regulations • Decree No. 40/2013 of 21 August 2013 On Explosive Substances Regulations • Decree No. 35/2014 of 1 August 2014 Amending of Decree No. 40/2013 of 21 August 2013	

Country	Applicable texts	Main reforms and remarks
	• Decree No. 25/2015 of 20 November 2015 On Sale of Diamonds, Precious Metals and Gemstones Regulations, Repealing Decree No. 20/2011 of 1 June 2011. • Order No. 21/2014 of 16 May 2014 On Social Responsibility of Extractive Industry of Mineral Resources • Ministerial Order No. 189/06 of 14 December 2006 On Rules on Environmental Management of Mining Activities • Ministerial Order No. 92/2007 of 11 July 2007 On Rules and Requirements for Approval of Technicians Capable of Preparing Prospecting and Research Reports for Mining Projects	
Namibia	• Minerals (Prospecting and Mining) Act, 1992 • Minerals (Prospecting and Mining) Amendment Act, 2008 • Diamond Taxation Proclamation Repeal Act, 1995 • Diamond Act, 1999	
Niger	• Ordinance No. 93-16 of 2 March 1993 Enacting the Mining Law, as amended by Ordinance No. 99-48 of 5 November 1999 and Law No. 2006-26 of 9 August 2006 • Law No. 2014-08 of 16 April 2014 Amending and Supplementing Ordinance No. 93-16 of 2 March 1993 Enacting the Mining Law modified by Law No. 2006-26 of 9 August 2006 • Ordinance No. 2017-03 of 30 June 2017 Amending and Supplementing Ordinance No. 93-16 of 2 March 1993 Enacting the Mining Law • Law No. 2008-30 of 3 July 2008 Granting Incentive for Investments in Major Mining Projects • Decree No. 2006-265/PRN/MM/E of 18 August 2006 Implementing the Mining Law • Order No. 76/MME/E/DM of 12 September 1995 Enacting the Conditions for Liquidation and Recovery of the Tax on Artisanal Mining • Order No. 53/MME/MF of 1 August 2000 On the Annual Land Royalties • Order No. 70/MME/TM of 5 August 2004 On the Code of Conduct on Artisanal Mining Exploitation Sites (EMA) Monitored and Inspected by the Administration • Order No. 41/MMP/TM of 2 May 2007 On the Transfer of Mining Rights	

Country	Applicable texts	Main reforms and remarks
Nigeria	• Minerals and Mining Act, 2007 • Minerals and Mining Regulations, 2011 • Mines and Quarries (Controls of Buildings, etc.) Act, 2004 • National Minerals and Metals Policy, 2008	On 31 August 2016, the Government approved a new roadmap for the growth and development of the mining sector. One of the key elements is the creation of an independent mining industry regulation body.
Uganda	• Mining Act, 2003 • Mining Act (Commencement) Instrument, 2004 • Mining Regulations, 2004	
Rwanda	• Law No. 13-2014 of 20 May 2014 On the Exploitation of Mines and Quarries • Ministerial Order No. 003/MINIRENA/2015 of 24 April 2015 On the Procedures for Applying for, Granting and Using Mining and Quarrying Permits	
São Tomé and Principe	• No regulation applicable to the mining sector	
Senegal	• Law No. 2016-32 of 8 November 2016 Enacting the Mining Code • Decree No. 2017-459 of 20 March 2017 Implementing the Mining Code	
Seychelles	• Minerals Act (Chapter 129)	
Sierra Leone	• Mines and Minerals Act, 2009 • Mines and Minerals Operational Regulation, 2013 • Environment Protection (Mines and Minerals) Regulations, 2013	Discussions in progress on a reform of the legal framework
Somalia	• Mining Code, 1984	
Sudan	• Mineral Resources and Mining Development Act, 2007 • The Mineral Wealth and Mining Resources Development Act, 2015 • Mines and Quarries Act, 1972 • Mining and Quarries Regulation, 1973	
South Sudan	• Mining Act, 2012 • Mining (Mineral Title) Regulations 2015	
Swaziland	• Mines and Minerals Act, 2011 • Explosive and Diamonds Act, 2012	
Tanzania	• Natural Wealth and Resources (Permanent Sovereignty) Act, 2017 • Natural Wealth and Resources (Review and Re-Negotiation of Unconscionable Terms) Act, 2017	

Country	Applicable texts	Main reforms and remarks
	• Written Laws (Miscellaneous Amendments), Act 2017 • Mining Act, 2010 • Mining Regulations (Amendment), 2012	
Chad	• Law No. 011-PR-1995 of 20 June 1995 Enacting the Mining Code • Decree No. 821/PR/MMEP/95 of 20 October 1995 Implementing the Mining Code and Mining Taxation	Discussions in progress on a reform of the legal framework
Togo	• Law No. 96-004 of 26 February 1996 Enacting the Mining Code • Law No. 2003-012 of 4 October 2003 Amending and Supplementing Law No. 96-004 of 26 February 1996 Enacting the Mining Code • Law No. 2011-008 of 5 May 2011 On the Contribution of Mining Company to Regional and Local Content	Discussions in progress on a reform of the legal framework
Tunisia	• Law No. 2003-30 of 28 April 2003 Enacting the Mining Code	
Zambia	• Mines and Minerals Development Act, 2015 • Mines and Minerals Development (Amendment) Act, 2016 • Mines Acquisition (Special Provisions) Act • Mines Acquisition (Special Provisions) (No. 2) Act	
Zimbabwe	• Mines and Minerals Act (Chapter 21:05) • Indigenisation and Economic Empowerment Act (Chapter 14:33)	

Appendix III

Constitutions	
South Africa	Constitution of the Republic of South Africa, adopted in October 1996
Algeria	Constitution of the Republic of Algeria, adopted in November 1996
Angola	Constitution of the Republic of Angola, adopted in January 2010
Benin	Constitution of the Republic of Benin, adopted in December 1990
Botswana	Constitution of Botswana, adopted in March 1966
Burkina Faso	Constitution of Burkina Faso, adopted in June 1991
Burundi	Constitution of the Republic of Burundi, adopted in March 2005
Cameroon	Constitution of Cameroon, adopted in June 1972
Cape Verde	Constitution of the Republic of Cape Verde, adopted in May 2010
Central Africa Republic	Constitution of the Central African Republic, adopted in December 2004. The National Transition Council adopted a draft outline Constitution in 2015.
Comoro Islands	Constitution of the Comoro Islands, adopted in May 2001
Congo	Constitution of the Republic of the Congo, adopted in January 2002
Democratic Republic of the Congo	Constitution of the Democratic Republic of the Congo, adopted in December 2006
Côte d'Ivoire	Constitution of the Côte d'Ivoire, adopted in July 2000
Djibouti	Constitution of the Republic of Djibouti, adopted in September 1992
Egypt	Constitution of the Arab Republic of Egypt, adopted in January 2014
Eritrea	Constitution of Eritrea, adopted in May 1997
Ethiopia	Constitution of the Federal Democratic Republic of Ethiopia, adopted in December 1994
Gabon	Constitution of the Republic of Gabon, adopted in March 1991
Gambia	Constitution of the Republic of Gambia, adopted in August 1996
Ghana	Constitution of the Republic of Ghana, adopted in May 1992
Guinea	Constitution of the Republic of Guinea, adopted in May 2010
Guinea-Bissau	Constitution of the Republic of Guinea-Bissau, adopted in May 1996
Equatorial Guinea	Constitution of the Republic of Equatorial Guinea, adopted in January 1991
Mauritius	Constitution of Mauritius, adopted in March 1968
Kenya	Constitution of Kenya, adopted in May 2010
Lesotho	Constitution of Lesotho, adopted in April 1993
Liberia	Constitution of the Republic of Liberia, adopted in July 1986
Libya	Constitutional declaration of 3 August 2011
Madagascar	Constitution of the Fourth Republic of Madagascar, adopted in December 2010
Malawi	Constitution of the Republic of Malawi, adopted in May 1994
Mali	Constitution of the Republic of Mali, adopted in January 1992
Morocco	Constitution of the Kingdom of Morocco, adopted in July 2011
Mauritania	Constitution of the Islamic Republic of Mauritania, adopted in July 1991 and amended in June 2006

Constitutions	
Mozambique	Constitution of the Republic of Mozambique, adopted in November 2004
Namibia	Constitution of the Republic of Namibia, adopted in February 1990
Niger	Constitution of the Republic of Niger, adopted in October 2010
Nigeria	Constitution of the Republic of Nigeria, adopted in May 1999
Uganda	Constitution of the Republic of Uganda, adopted in October 1995
Rwanda	Constitution of the Republic of Rwanda, adopted in May 2003
São Tomé and Principe	Constitution of the Republic of São Tomé and Principe, adopted in March 1990
Senegal	Constitution of the Republic of Senegal, adopted in January 2001
Seychelles	Constitution of the Republic of Seychelles, adopted in June 1993
Sierra Leone	Constitution of Sierra Leone, adopted in August 1991
Somalia	Temporary constitution of the Federal Republic of Somalia, adopted in August 2012
Sudan	National interim constitution of the Republic of the Sudan, adopted in July 2005
South Sudan	Constitution of the Republic of South Sudan, adopted in July 2011
Swaziland	Constitution of the Kingdom of Swaziland, adopted in July 2005
Tanzania	Constitution of the Republic of Tanzania, adopted in April 1977
Chad	Constitution of the Republic of Chad, adopted in March 1996
Togo	Constitution of the Republic of Togo, adopted in September 1992
Tunisia	Constitution of the Tunisian Republic, adopted in January 2014
Zambia	Constitution of Zambia, adopted in August 1991
Zimbabwe	Constitution of Zimbabwe, adopted in December 2013

Appendix IV

REGIONAL TEXTS

WAEMU

WAEMU Treaty of 10 January 1995

Regulation No. 04/96/CM/UEMOA of 20 December 1996 Establishing the Joint Accounting Reference Material within the WAEMU, as amended by Regulation No. 05/2013/CM/UEMOA of 28 June 2013 and Regulation No. 005/2014/COM/UEMOA of 31 May 2014

Regulation No. 18/2003/CM/UEMOA of 23 December 2003 Enacting the Community Mining Code

Regulation No. 08/CM/UEMOA of 26 September 2008 Adopting the Rules for the Avoidance of Double Taxation within the UEMOA and the Rules for Assistance in Taxation

Directive No. 02/98/CM/UEMOA of 22 December 1998 On the Harmonisation of the Laws of the Member States On the Value Added Tax, amended by Directive no. 02/2009/CM/UEMOA of 27 March 2009

Directive No. 07/2001/CM/UEMOA of 26 November 2001 Establishing the Harmonized Regime for Advance Payments of Profit Tax within WAEMU

Directive No. 01/2008/CM/UEMOA of 28 March 2008 On the Harmonization of the Modalities for the Determination of the Taxable Result for Legal Entities within WAEMU

ECOWAS

Revised Treaty of the ECOWAS dated 24 July 1993

Directive C/DIR 3/05/09 of 27 May 2009 On the Harmonisation of Guiding Principles and Policies in the Mining Sector

Additional Act A/SA.3/01/10 Amending Article 9 of the ECOWAS Treaty, as amended by Additional Protocol A/SP.1/06/06

Additional Act A/SA.18/02/12 On the Adoption of ECOWAS's Mineral Resource Development Policy (PDRMC)

CEMAC

Act No. 5/66 UDEAC-49 of 13 December 1966 Enacting the CEMAC Taw Convention on Mutual Administrative Assistance in Tax Matters

Appendix 2 to Directive No. 02/01/UEAC-050-CM-06 Rewing Act No. 3/72-153-UDEAC of 22 December 1972 Instituting Corporate Tax

Act No. 10/88-UDEAC-257 of 7 December 1988 On Harmonisation of Registration Duties

Directive No. 1/99/CEMAC-028-CM-03 of 17 December 1999 Harmonizing the Legislations of the Member States on Value Added Tax (VAT) and Excise Duty

Regulation No. 17/99/CEMAC-20-CM-03 of 17 December 1999 Enacting the Investment Charter

OHADA

OHADA Uniform Act of 10 April 1998 Establishing the Organisation of Simplified Recovery and Enforcement Procedures
OHADA Uniform Act of 11 March 1999 Relating to Arbitration
OHADA Uniform Act of 24 March 2000 The Organization and Harmonization of Business Accounting
OHADA Uniform Act of 15 December 2010 On Securities
OHADA Uniform Act of 30 January 2014 Relating to Commercial Company Law and the Economic Interest Group
OHADA Uniform Act of 26 January 2016 Relating to Accounting and Financial Information

African Union

African Union Convention on the prevention and fight against corruption of 2003

Organisation of African Unity

African Charter of Human and Peoples' Rights of 1981

APPENDIX V

INTERNATIONAL TEXTS

Resolutions

Resolutions of the General Assembly of the United Nations

Resolution 421 (V) of the Assembly General of 4 December 1950—Draft International Covenant on Human Rights and Measures of Implementation: Future Work of the Commission on Human Rights

Resolution 523 (VI) of the Assembly General of 12 January 1952—Integrated Economic Development and Commercial Agreements

Resolution 626 (VII) of the Assembly General of 21 December 1952—Right to Exploit Freely Natural Wealth and Resources

Resolution 1314 (XIII) of 12 December 1958—Recommendations concerning International Respect for the Rights of Peoples and Nations to Self-Determination

Resolution 1803 (XVII) of the Assembly General of 14 December 1962—Permanent Sovereignty over Natural Resources

Resolution 2200 A (XXI) of 19 December 1966—International Covenant on Economic, Social and Cultural Rights, International Covenant on Civil and Political Rights and Optional Protocol to the International Covenant on Civil and Political Rights

Resolution 3281 (XXIX) of 12 December 1974—Charter of Economic Rights and Duties of States

OPEC Resolutions

OPEC Resolution XVI.90 of 24 June 1968—Declaratory Statement of Petroleum Policy in Member Countries

International conventions

New York Convention on the Recognition and Enforcement of Foreign Arbitral Awards of 1958

Washington Convention on the Settlement of Investment Disputes between States and Nationals of Other States of 1965

International Convention on Civil Liability for Oil Pollution Damage of 1969

Vienna Convention on the Succession of States in respect of Treaties of 1978

United Nations Convention on the Law of the Sea of 1982

Seoul Convention Establishing the Multilateral Investment Guarantee Agency of 1985

APPENDIX VI

Case Law

Index of arbitral awards

Arbitral awards given under the aegis of ICSID

AGIP SpA v Republic of the Congo (ARB/77/1) (Award of 30 November 1970) 1982 *Rev crit DIP* 92

Benvenuti and Bonfant v Congo (ARB/77/2) Award of 8 August 1980) 21 *ILM* 1982, at 740

Fedax v Venezuela (AR /96/3) (Decision on jurisdiction of 11 July 1997) 1999 *JDI* at 276 obs E Gaillard

Antoine Goetz et al v Burundi (ARB/95/3) (Award of 10 February 1999) (2000) 15 *ICSID Rev* 457

Robert Azinian, Kenneth Davitian & Ellen Baca c. Mexique (ARB(AF)/97/2) (Award of 1 November 1999) Available from: http://www.italaw.com/sites/default/files/case-documents/ita0057.pdf

Alex Genin et al v Estonie (ARB/99/2) (Award of 25 June 2001) (2002) 17 *ICSID Rev* 395.

Salini Costruttori SpA and Italstradee SpA v Maroc (ARB/00/4) (Decision of 23 July 2001 on jurisdiction) 2002 *JDI* 196 obs E GAILLARD

Noble Ventures Inc v Roumanie (ARB/01/11) Award of 12 October 2005 Available from: http://www.italaw.com/sites/default/files/case-documents/ita0565.pdf

CMS Gas Transmission Company v The Republic of Argentina (ARB/01/8) Award of 12 May 2005 Available from: http://www.italaw.com/sites/default/files/case-documents/ita0184.pdf

Enron Corporation and Ponderosa Assets, LP v The Republic of Argentina (ARB/01/3) Award of 22 May 2007 Available from: http://www.italaw.com/sites/default/files/case-documents/ita0293.pdf

Parkerings-Companiet AS v Lituanie (ARB/05/8) Award of 11 September 2007 Available from: http://www.italaw.com/sites/default/files/case-documents/ita0619.pdf

Duke Energy International Peru Investments v The Republic of Peru (ARB/03/28) Award of 18 August 2008 Available from: http://www.italaw.com/documents/DukeEnergyPeruAward_000.pdf

El Paso Energy International v The Republic of Argentina (ARB/03/15) Award of 31 October 2011 Available from: http://www.italaw.com/sites/default/files/case-documents/ita0270.pdf

Occidental Petroleum Corporation and Occidental Exploration and Production Company v Republic of Ecuador (ARB/06/11) Award of 8 October 2012 Available from: http://www.italaw.com/sites/default/files/case-documents/italaw1094.pdf

Arbitral awards given under the aegis of the ICC

Award made in ICC case no. 1526 (1968), *Mojzez Lubelski v Burundi* 1974 *JDI* 915 note Y Derains

Award made in ICC case no. 1512/24 (1971), YB Com Arb 1975 128

Award made in ICC case no. 3093/3100 (1979) 1981 *JDI* 951 obs Y Derains

Award made in ICC case no. 3493 (1983), *Spp (Middle East) and Southern Pacific Properties Ltd v Arab Republic of Egypt and Egyptian General Company for Tourism and Hotels* 22 ILM 752 (1983)

Award made in ICC case no. 4230 (1985). Cited by I Vaugon, 'Expertise et médiation dans le contexte du contentieux extractif' [Evaluation and mediation within the context of the extractive dispute], in *Le contentieux extractif*, ICC 2015 203

Award made in ICC case no. 4761 (1987) 1987 *JDI* at 1012 obs S Jarvin

Award made in ICC case no. 5953 (1989), Compendium of ICC Arbitration Awards 1986-1990, comm. Y Derains

Ad hoc arbitration awards

Power and Traction Finance Company Ltd v Greece, award of 18 March 1930 (Award not published. Extract from Sébastien Manciaux 'Changement de législation fiscale et arbitrage international' (2001) *Rev arb* at 311)

Petroleum Development Limited v the Sheikh of Abu Dhabi (Award of 18 August 1951) 195118 *ILR* 144

Jordan Investment Ltd. v Sojutzneftexport (Award made on 19 June 1958 by the External Trade Arbitration Committee of the USSR) 1960 *Rev arb* at 92

Saudi Arabia v Arabian American Oil Co. (ARAMCO) (Award of 23 August 1958) 1963 *Rev crit DIP* 272

Sapphire International Petroleum Limited v National Iranian Oil Company (Award of 15 March 1963) 1964 *Int'l & Comp LQ* 1011

British Petroleum v Government of the Arab Republic of Libya (Award of 10 October 1973) 1980 *Rev arb* 117

Texaco Overseas Petroleum Co and California Asiatic Oil Co v Government of the Arab Republic of Libya (Award of 19 January 1977) 1977 *JDI* 350

Libyan American Oil Company (LIAMCO) v Government of the Arab Republic of Libya (Award of 19 January 1977) 1977 *JDI* 350

Petrola *Hellas SA v Greece* (Award of 22 April 1978) (1986) 11 YB Com Arb 105

Elf Aquitaine Iran v National Iranian Oil Co (Preliminary award of 14 January 1982) 1984 *Rev arb* 401

American Independent Oil Co. (AMINOIL) v Kuwaiti Government (Award of 24 March 1982) 1982 *JDI* 869

Award made in case no. 109/1980, *All Union Foreign Trade Association Sojuzneftexport v Joc Oil* Ltd, of 9 July 1987, YB Com Arb 1993 92

Ad hoc award made in the case of *Company B (having its registered office in State A) v État A* (1989) 7 (4) *ASA Bulletin* 382

Ad hoc arbitration award (UNCITRAL)

Himpurna California Energy Ltd v PT (Persero) Perusahaan Listruik Negara (Award of 4 May 1999) 2000 YB Com Arb 13

Methanex Corporation v United States (Final award on jurisdiction and the substance of 3 August 2005) Available from: http://www.italaw.com/sites/default/files/case-documents/ita0529.pdf

Decisions by the Permanent Court of International Justice (PCIJ) and the International Court of Justice (ICJ)

Case of Serbian borrowings (Order of 12 July 1929 Series A No 20) Available from: http://www.icj-cij.org/pcij/serie_A/A_20/62_Emprunts_Serbes_Arret.pdf

Case of the Chorzów Plant (Germany v Poland) (Order of 13 September 1928 Series A No 17) Available from: http://www.icj-cij.org/pcij/serie_A/A_17/54_Usine_de_Chorzow_Fond_Arret.pdf

Case of Anglo-Iranian Oil Co (United Kingdom v Iran) (Order of 22 July 1952) 1952 Compendium 93 Available from: http://www.icj-cij.org/docket/files/16/1997.pdf

Decisions by the African Commission of Human and Peoples' Rights

Communication no. 155-96 13 October 2001: *Social and Economic Rights Action Center (SERAC) and Center for Economic and Social Rights (CESR) v Nigeria* Available from: http://www.achpr.org/files/sessions/30th/comunications/155.96/achpr30_155_96_fra.pdf

Decisions by the French jurisdictions

Court of Appeal of Paris (Additional chamber 1), 12 July 1984, *Arab Republic of Egypt v Southern Pacific Properties Ltd, Southern Pacific Properties (Middle East)* 1986 *Rev arb* 75 note by Ph Leboulanger

French Supreme Court, 1st Civ. Div, 29 June 2007, *Rev arb* 2007 at 507, report by J-P Ancel, note by E Gaillard

Bibliography

Books

Mohamed Said Ait Chaalal *Pays producteurs de pétrole et compagnies internationales* (Imprimerie La Concorde 1977)

Jean Alisse *L'obligation de renseignement dans les contrats* (Doctorate Thesis Paris II Panthéon-Assas 1975)

Serge Bandoki *Le droit minier et pétrolier en Afrique* (Édilivre-APARIS 2008)

Robert Brasseur *Législation et fiscalité internationales des hydrocarbures—Exploration et Production* (Éditions Technip 1975)

Nadine Bret-Rouzaut & Jean-Pierre Favennec *Recherche et production du pétrole et du gaz: réserves, coûts, contrats* 2nd ed (Éditions Technip 2011)

Jean-Pierre Boivin & Claude Lavaire *Carrières et Granulats* (Le Moniteur 2008)

Dominique Carreau & Patrick Julliard *Droit international économique* 4th ed (Dalloz 2010)

Henry Cattan *The Law of Oil Concessions in the Middle East and North Africa* (Oceana Publications, 1967)

Jean Combacau & Serge Sur *Droit International Public* 8th ed (Montchrestien 2008)

Gérard Cornu *Vocabulaire juridique* (Presses Universitaires de France 1987)

Patrick Daillier, Mathias Forteau & Alain Pellet *Droit International Public* 8th ed (LGDJ 2009)

Albert Geouffre de la Pradelle & Nicolas Politis *Recueil des arbitrages internationaux* vol II (Éditions internationales 1957)

Jean-Marie Deleuze *Le contrat de processus de transfert de technologie, know-how* (Masson 1982)

Jean Devaux-Charbonnel *Principes et applications du droit minier des hydrocarbures* (Éditions Technip 1975)

Mustafa Erkan *International Energy Investment Law Stability through Contractual Clauses* (Kluwer Law International 2011)

Arghyrios A Fatouros *Government Guarantees to Foreign Investors* (Columbia University Press 1962)

Alain Feneon *Droit des sociétés en Afrique: Ohada* (LGDJ 2015)

John D Finnerty *Project Financing: Asset-Based Financial Engineering* 2nd ed (John Wiley & Sons Inc 2007)

Phillipe Fouchard, Emmanuel Gaillard & Berthold Goldman *International Commercial Arbitration* (Kluwer Law International 1999)

Gérard Fouilloux *La nationalisation et le Droit international public* (LGDJ 1962)

Zhiguo Gao *International Petroleum Contracts: Current Trends and New Directions* (Graham & Trotman/Martinus Nijhoff 1994)

Richard Garnett, Henry Gabriel, Jeff Waincymer & Judd Epstein *A Practical Guide to International Commercial Arbitration* (Oceana Publications 2000)

Kamal Hossain *Law and Policy in Petroleum Development—Changing Relations Between Transnationals and Governments* (Nichols Publishing Company 1979)

Rudolf Isay *Le droit minier des principaux États civilisés* (Éditions Marcel Giard 1930)

Philippe Janin & Barthélémy Mercadal *Les contrats de coopération inter-entreprise* (Éditions Juridiques Lefebvre 1974)

Khalil Khalilian *Investissements privés étrangers et souveraineté économique* (Doctoral Thesis Paris II Panthéon-Assas 1978)

Lin Ken Wong *The Malayan Tin Industry to 1914* (University of Arizona Press 1965)

Zouhair A Kronfol *Protection of Foreign Investment—A Study in International Law* (Leiden 1972)

George Lenczowski *Oil and State in the Middle East* (Cornell University Press 1960)

Sébastien Manciaux *Investissements étrangers et arbitrage entre États et ressortissants d'autres États* (Litec 2004)

Jean-Pierre Mattout *Droit bancaire international* 4th ed (Revue Bancaire 2009)

Louis Millot & François-Paul Blanc *Introduction à l'étude du droit musulman* (Dalloz 2001)

John Bassett Moore *History and Digest of International Arbitration* (Government Printing Office 1898)

Arnaud de Nanteuil *Droit International de l'Investissement*, Éditions A. Pedone, Paris, 2014

Alan Redfern & Martin Hunter, *Droit et pratique de l'arbitrage commercial international* 2nd ed (LGDJ 1994)

Seymour J Rubin *Private Foreign Investment Legal and Economic Realities* (Johns Hopkins Press Baltimore 1956)

Jérôme Ortscheidt & Christophe Seraglini *Droit de l'arbitrage interne et international* (Montchrestie 2013)

Fiona Scott & Claus-Peter Martens *International Project Finance* (Transnational Publishers 2000)

Simon G Siksek *The Legal Framework for Oil Concessions in the Arab World* (Middle East Research and Publishing Center 1960)

Yves Simon & Samir Mannaï *Techniques Financières Internationales* 7th ed (Economia 2002)

Richard L Sklar *Corporate Power in an African State: The Political Impact of Multinational Mining Companies in Zambia* (University of California Press 1975)

David N Smith & Louis T Wells *Negotiating Third World Mineral Agreements* (Ballinger Publications 1975)

Alexandre Stärker *Des concessions aux concentrations pétrolières dans les pays du Moyen-Orient* (Doctoral Thesis: Université de Lausann 1974)

Frédéric Teulon *La nouvelle économie mondiale* (Presses Universitaires de France 1993)

Arlète Tonye *Pratique juridique des financements structurés en Afriqu* (L'Harmattan 2010)
Shavarsh Toriguian *Legal Aspects of Oil Concessions in the Middle East* (Hamaskaine Press 1972)
Joe Verhoeven *Droit International Public* (Larcier 2000)
Dow Votaw *The Six-Legged Dog, Mattei and ENI—A Study of Power* (University of California Press 1964)

Collective Works

Études offertes à Berthold Goldman (Litec 1982)
Etudes à la mémoire de Fernand-Charles Jeantet (LexisNexis Litec 2010)
La concurrence des organisations régionales en Afrique, under the direction of Matthieu FAU-NOUGARET (L'Harmattan 2012)
Le contentieux extractif, under the direction of Achille Ngwanza & Gilles Lhuilier (ICC 2015)
Les Grands articles de doctrine de l'OHADA parus au Recueil Penant de 2000 à 2013 Alain Feneon (Juris Africa 2013)
Les marchés internationaux des matières premières, under the direction of Claude Mouton & Philippe Chalmin (Economica 1982)
Liber Amicorum Bernardo Cremades Miguel Angel Fernández-Ballesteros David ARIAS (eds), (La Ley 2010)
Liber Amicorum Raymond Ranjeva, L'Afrique et le droit international: variations ou l'organisation internationale (Pédone 2013)
Mélanges en l'honneur de Paul Lagarde (Dalloz 2005)
Mélanges offerts à Charles Rousseau (Pédone 1974)
Mélanges offerts à Jacques Maury (Dalloz 1960)
Transfert de technologie et développement, Conference held at the University of Dijon published in (Librairies Techniques 1977)
Natural Resource Investment and Africa's Development, F N Botchway (ed) (Edward Elgar Publishing 2011)
Private investors abroad—problems and solutions in international business in 1967, under the direction of V Shook Cameron (Southwestern Legal Foundation, Matthew Bender 1967)
Proceedings of the second international symposium on uranium supply and demand, conference held in Londong under the direction of the Uranium Institute from 22 to 24 June 1977 (Mining Journal Books Ltd 1977) 177
The transnational law of international commercial transactions, Studies in Transnational Economic Law, Vol II, under the direction of Norbert Horn, Clive M Schmitthoff & Richard M Buxbaum (Kluwer 1982)

Contributions to Collective Works

Henri Battifol 'Subjectivisme et objectivisme dans le droit international privé des contrats' in Jacques Maury (ed) *Mélanges offerts* (Dalloz, Paris, 1960) at 39

Jean-Paul Beraudo 'Faut-il avoir peur du contrat sans loi?' in Paul Lagarde (ed) *Mélanges en l'honneur* (Dalloz 2005) 93

Georges R Delaume 'Le financement international des transferts de technologie et le développement: le rôle de la Banque mondiale' in *Transfert de technologie et développement* (Librairies Techniques 1977) 267

Hans-Michael Donndorf 'Co-operative Ventures with Developing Countries—An Assessment of Recent Experience' in *Proceedings of the second international symposium on uranium supply and demand* (Mining Journal Books Ltd 1977) 177

Henry Cattan 'Present Trends in Middle Eastern Oil Concessions and Agreements' in *Private investors abroad—problems and solutions in international business in 1969* (Matthew Bender 1969)140

Isidore Costade 'Publication des contrats et confidentialité de l'industrie extractive' in *Le contentieux extractif* (ICC 2015) 159

Gilles Darmois 'Contingentements de l'OPEP et équilibre économique des contrats', in *Le contentieux extractif* (ICC 2015) 74

Hamid G Gharavi & Marie-Lame Bizeau 'L'expropriation indirecte dans les conflits entre contractants extractifs' in *Le contentieux extractif* (ICC 2015) 167

Elie Kleiman 'Les nationalisations' in *Le contentieux extractif* (ICC 2015) 84

Mamadou Konate 'Le cadre juridique de la renégociation' in *Le contentieux extractif* (ICC 2015) 41

Florent Lager 'Les Dispute Board dans le secteur extractif' in *Le contentieux extractif* (ICC 2015) 213

Thierry Lauriol 'Financement de projets miniers dans les États parties à l'OHADA: l'apport des titres miniers en garantie' in *Les Grands articles de doctrine de l'OHADA parus au Recueil Penant de 2000 à 2013* (Juris Africa 2013) 218

Thierry Lauriol & Sophie Da Cunha 'La règlementation minière en Afrique de l'Ouest: une multiplicité des normes' in *Études à la mémoire de Fernand-Charles Jeantet* (LexisNexis Litec 2010) 261

Thierry L'Auriol & Émilie Raynaud 'La rupture unilatérale des contrats d'exploitation' in *Le contentieux extractif* (ICC 2015) 38

Honoré Le Leuch 'Analyse économique et renégociation des contrats extractifs' in *Le contentieux extractif* (ICC 2015) 30

Valère Ndior 'Société civile et bailleurs de fonds internationaux: Un même combat ?' in *Le contentieux extractif* (ICC 2015) 114

Michael D. Nolan & Frédéric G. Caivano 'Limits of Consent—Arbitration without Privity and Beyond' in *Liber Amicorum Bernardo Cremades* (La Ley 2010) 873

Evaristus Oshionebo 'Fiscal regimes for natural resource extraction: implications for Africa's development' in *Natural Resource Investment and Africa's Development* F N Botchway (ed) (Edward Elgar Publishing 2011) 200

Alan C. Page 'Transnational Mining Contracts' in *The transnational law of international commercial transactions* Studies in Transnational Economic Law Vol 2 (Kluwer Deventer 1982). 224

Hery Ranjeva 'L'obligation de réhabilitation d'un site d'extraction minière à la charge de l'investisseur' *Liber Amicorum Raymond Ranjeva, L'Afrique et le droit international: variations ou l'organisation internationale* (Pédone 2013) 549

Cécile Renouard 'De la contribution de l'industrie extractive au développement socio-économique des populations riveraines' in *Le contentieux extractif* (ICC 2015) 96

Jean Robert 'Contrats et arbitrage international dans le domaine des matières premières' in *Les marchés internationaux des matières premières* (Economica1982) 73

Julien Rochette 'Activité pétrolière et gazières en offshore et protection de l'environnement' in *Le contentieux extractif* (ICC 2015) 125

Clive Schmitthoff 'Nature and Evolution of the Transnational Law of Commercial Transactions', in *The transnational law of international commercial transactions*, Studies in Transnational Economic Law, Vol 2 (Kluwer Deventer 1982) 19

Philippe Sebille-Lopez 'Origines et règlements politiques des conflits dans le secteur des hydrocarbures' in *Le contentieux extractif* (ICC 2015) 181

Eduardo Silva-Romero & Audrey Caminades 'Actualité des clauses de stabilisation', in *Le contentieux extractif* (ICC 2015) 51

Isabelle Vaugon 'Expertise et médiation dans le contexte du contentieux extractif', in Le contentieux extractif (ICC 2015) 200

Prosper Weil 'Principes généraux du droit et contrats d'État' in *Études offertes à Berthold Goldman*, (Litec, Paris 1982) 387

Prosper Weil 'Les clauses de stabilisation ou d'intangibilité insérées dans les accords de développement économique' in *Mélanges offerts à Charles Rousseau* (Pédone 1974) 301

Legal articles

Abdullah Al Faruque 'Typologies, Efficacy and Political Economy of Stabilisation Clauses: A Critical Appraisal' (2007) 5 (4) *OGEL* 4

Abdullah Al Faruque 'Validity and Efficacy of Stabilisation Clauses' (2006) 23 (4) *Journal of International Arbitration* 317

Abdullah Al Faruque 'Adaptation of Petroleum Contract by Third Party: An Overview' (2006) 3 (4) *TDM* 3

Sa'ad Allam 'The Legal Aspects of the Governmental Participation in Petroleum Concessions' (1973) 1 *Arab Petroleum Directory* 176

Pascal Ancel 'L'efficacité procédurale des clauses de conciliation ou de médiation' 2003 *D* 1386

James N D Anderson 'The Moslem Ruler and Contractual Obligations' (1958) 33 *NYUL Rev* 917

Eduardo Jiménez de Arechaga 'International Law in the Past Third of a Century' (1978) 159 *RCADI* at 1

Samuel Asante, 'Stability of Contractual Relations in the Transnational Investment Process' (1979) 28 *Int'l & Comp LQ* 401

Henri Batiffol 'La sentence ARAMCO et le Droit international privé' 1964 *Rev crit DIP* 647

Suzanne Bastid 'Le Droit international public dans la sentence arbitrale de l'ARAMCO' 1961 *AFDI* 300

Walid Ben Hamida 'L'incidence des intérêts moratoires sur l'exécution des sentences arbitrales dans les pays arabes' 2012 *Rev arb* 539

Klaus Peter Berger 'Renegotiation and Adaptation of International Investment Contracts the Role of Contract Drafters and Arbitrators' (2004) 1 (4) *TDM*

Dominique Berlin 'Les Contrats d'États et la protection des investissements internationaux' 1987 *DPCI* 187

Piero Bernardini 'Stabilization and adaptation in oil and gas investments' (2008) 1 (1) *Journal of World Energy Law & Business* 98

Christopher Boog 'How to Deal with Multi-tiered Dispute Resolution Clauses—Note—6 June 2007—Swiss Federal Supreme Court'2008 *ASA Bulletin* 103

Jean-Albert Boon 'Problèmes juridiques soulevés par la conclusion de consortiums internationaux'1964 *Ann Fac Dt Liège* 373

Maurice Bourquin 'Arbitration and Economic Development Agreement' 1960 *Bus Lawyer* 860

John Perry Bullington 'The Land and Petroleum Laws of Mexico' (1928) 22 *AJIL* 50

Geneviève Burdeau 'Droit international et contrats d'États. La sentence Aminoil/Koweït du 24 mars 1982' (1982) *AFDI* 454

Gérard Cohen-Jonathan 'L'arbitrage Texaco-Calasiatic c. Gouvernement Libyen' 1977 *AFDI* 452

Alfredo de Jesus 'The prodigious story of the lex petroleum and the rhinoceros philosophical aspects of the transnational legal order of the petroleum society' (2012) 1 *TPLI Series on Transnational Petroleum Law* 1

Kenneth S Carlston 'International Role of Concession Agreements' (1957) 52 *NWULR* 618

Claude Champaud 'Les méthodes de groupement des sociétés' (1967) 20 *RTD Com*. 1004

Alain Charlet, Bertrand Laporte & Grégoire Rota-Graziosi 'La fiscalité minière en Afrique de l'Ouest et du Centre' (2013) 48 *Revue de droit fiscal* 11

Jean Charpentier 'De la non-discrimination dans les investissements' 1963 *AFDI* 35

Allan Cree 'Status of Petroleum Exploration in Indonesia' (1972) 10 *Exp Eco Pet Ind* 123

Michel Cremieux 'Le financement des infrastructures: le rôle des investisseurs financiers' (1999) 51 (51) *Revue d'économie financière* 217

Nicolas David 'Les clauses de stabilité dans les contrats pétroliers. Question d'un praticien' 1986 *JDI* 79

Michel de Juglart 'L'obligation de renseignement dans les contrats' 1945 *RTD Civ* 1

Georges R Delaume 'Des stipulations de droit applicable dans les accords de prêt et de développement économique et de leur rôle' 1968 *RBDI* 336

Georges R Delaume 'Excuse for Non-Performance and Force Majeure in Economic Development Agreements' 1971 *Columbia Journal of Transnational Law* 242

Deloitte 'La réforme du régime juridique des succursales des sociétés étrangères en droit des sociétés commerciales OHADA' (2014) 2 *Revue Juridique et fiscale d'Afrique Francophone* 8

Deloitte 'La reconnaissance et l'efficacité des clauses de stabilisation dans les contrats miniers' March 2015 *Revue juridique et fiscale d'Afrique francophone* 2

Robert Delson 'Nationalization of the Suez Canal Company: Issues of Public and Private International Law' 1957 *Columbia Law Review* 9

Jean Devaux-Charbonnel 'L'accord pétrolier franco-iranien conclu le 27 août 1966 entre la Société Nationale Iranienne des Pétroles (SNIP) et le Groupe de l'Entreprise de Recherches et d'Activités Pétrolières (ERAP)' 1966 *AFDI* 798

Ahmed Sadek El-Kosheri 'Le régime juridique créé par les accords de participation dans le domaine pétrolier' (1975) 147 *RCADI* 221

Malak El-Shishini 'Les contrats quasi-internationaux/Base juridique et droit applicable' 1968 *Rev Egypt DI* 81

Georges FISCHER 'La souveraineté sur les ressources naturelles' 1962 *AFDI* at 516

Peter Fischer 'Historic Aspects of International Concession Agreement' 1972 *Grotian Soc Pap* at 222

Laurence Franc-Menget 'Vers un durcissement des conditions de renonciation à l'immunité d'exécution des États étrangers en droit français ? note sous Cass. civ. 1re, 28 mars 2013' 2013 *Rev arb* 992

Dana H Freyer 'Practical Considerations in Drafting Dispute Resolution Provisions in International Commercial Contracts—A US Perspective' (1998) 15 (4) *Journal of International Arbitration* 7

Marcel Fontaine 'Les clauses de force majeure dans les contrats internationaux' 1979 *DPCI* 469

Philippe Fouchard 'L'adaptation des contrats à la conjoncture économique' 1979 *Rev arb* 67

Philippe Fouchard 'L'arbitrage Elf Aquitaine Iran c. National Iranian Oil Company, une nouvelle contribution au droit international de l'arbitrage' (1984) *Rev arb* 333

Efrain Friedmann 'Energy Activities of the World Bank' 1978 *NRF* 91

Morley Friedman 'The Validity and Efficacy of Choice of Law Clauses in International Contracts' (1969) 5 *LLJ* 71

Léna Gannage 'Le contrat sans loi en droit international privé' (2007) 11.3 *Electronic Journal of Comparative Law* Available from: http://www.ejcl.org/113/article113-10.pdf

Andrea Giardina 'Clauses de stabilisation et clauses d'arbitrage: vers l'assouplissement de leur effet obligatoire?—Les états dans le contentieux économique international, I. Le contentieux arbitral' 2003 *Rev arb* 647

Werner Goldschmidt 'Transactions between States and Public Firms and Foreign Private Firms' (1972) 136 *RCADI* 203

John J Gorman 'Commercial Hardship and the Discharge of Contractual Obligations under American and British Law' (1980) 13 *Vand J Transnat'l L* 107

John Gotanda 'Renegotiation and Adaptation Clauses in Investment Contracts, Revisited' (2003) 36 *Vand J Transnat'l L* 1461

Jean Honorat 'Rôle effectif et rôle concevable des quasi-contrats en droit actuel' 1969 *RTD Civ* 653

Philippe Kahn 'Force majeure et contrats internationaux de longue durée' 1975 *JDI* 467

Philippe Kahn 'Contrats d'État et nationalisations: les apports de la sentence du 24 mars 1982' 1982 *JDI* 844

Philippe Kahn 'Souveraineté de l'État et règlement du litige. Régime juridique du contrat d'État' 1985 *Rev arb* 641

François Knoepfler 'L'immunité d'exécution contre les États—Les états dans le contentieux économique international, III. Le contentieux judiciaire' 2003 *Rev arb* 1017

Lazare Kopelmanas 'Le rôle de l'expertise dans l'arbitrage commercial international ' 1979 *Rev arb* 205

Stefan Kröll 'The Renegotiation and Adaptation of Investment Contracts' (2004) 1 (3) *TDM* Available from: www.transnational-dispute-management.com

Xavier Lafarge 'L'efficacité des clauses de conciliation ou de médiation' 2000 *Rev arb* 377

Lars O Lagerman 'Choice of Forum Clauses in International Contracts: What Is Unjust and Unreasonable?' 1978 *Int Lawyer* 779

Jean-Flavien Lalive 'Un grand arbitrage pétrolier entre un gouvernement et deux société privées étrangères' 1977 *JDI* 319

Jean-Flavien Lalive 'Contrats entre États ou entreprises étatiques et personnes privées—Développements récents' (1983) 181 *RCADI* 9

Ole Lando 'Renegotiation and Revision of International Contracts' 1980 *GYIL* 37

Thierry Lauriol 'L'OHADA, le temps pour un acte uniforme minier' (July-September 2015) 892 *Penant* 281

Charles Leben 'Les modes de coopération entre pays en développement et entreprises multinationales dans le secteur de la production des matières premières minérales' 1980 *JDI* 539

Charles Leben 'L'évolution de la notion de contrat d'État' 2003 *Rev arb* 629

Charles Leben 'La théorie du contrat d'état et l'évolution du droit international des investissements' (2003) 302 *RCADI* 200

Philippe Leboulanger 'Le régime juridique des investissements pétroliers en Égypte' 1972 *Rev Egypt Dr Int* 112

Pierre Legoux 'L'évolution des idées dans les législations minières des pays neufs' 1975 *Rev JPIC* 75

Pierre Legoux 'Legal Aspects of Mining Development in Developing Countries' 1981 *NRF* 167

Sophie Lemaire 'La mystérieuse "umbrella clause" (Interrogations sur l'impact de la clause de respect des engagements sur l'arbitrage en matière d'investissements)' 2009 *Rev arb* 479

Henry Lesguillons 'Frustration, Force Majeure, Imprévision, Wegfall der Geschaftsgrundlage' 1979 *DPCI* 507

Andrew Loewinger 'Multilateral Funding for Mineral Exploration in the Third World' (1980) 14 (6) *Journal of World Trade* 469

Jean-Marc Loncle & Damien Philibert-Pollez 'Les clauses de stabilisation dans les contrats d'investissement' (2009) 3 *RDAI* 267

Abdelkrim Mehani 'Les accords algéro-français du 29 juillet 1965 sur les hydrocarbures' 1966 *R Alg SJPE* 267

Sébastien Manciaux 'Changement de législation fiscale et arbitrage international' 2001 *Rev arb* 311

Abdul F Munir Maniruzzaman 'Choice of Law in International Contracts—Some Fundamental Conflict of Laws Issues' (1999) 16 (4) *Journal of International Arbitration* 141

Francis A Mann 'The Proper Law of Contracts Concluded by International Persons' (1959) 35 *BYIL* 34

Pierre Mayer 'La neutralisation du pouvoir normatif de l'État en matière de contrats d'État' 1986 *JDI* 5

Joseph Russel Milton 'International Consortia: Definition, Purpose and the Consortium Agreement' (1979–1980) 3 (2) *FILJ* 121

M V A Naidu 'On the Choice of Law in Contracts: The Doctrine of Autonomy of Parties' (1973) 13 (4) *IJIL* 594

Roger Nevry 'La garantie autonome à première demande: droit OHADA, droit français'(2007) 3 *Revue de droit des affaires internationales* 313

Suzy H Nikiema 'Indemnisation de l'Expropriation', International Institute for Sustainable Development, 2013, Best Practices Series Available from: http://www.iisd.org/pdf/2013/best_practice_compensation_expropriation_fr.pdf

Yves Nouvel 'Les mesures équivalant à une expropriation dans la pratique récente des tribunaux arbitraux'2002 *RGDIP* 79

Yves Nouvel 'L'indemnisation d'une expropriation indirecte' (2003) 5 (3) *International Law Forum* 198

Ucheora Onwuamaegbu 'Resolution of oil and gas disputes at ICSID' (2004) 21 (1) *News From ICSID* Summer 2004 Available from: https://icsid.worldbank.org/apps/ICSIDWEB/resources/Documents/News%20from%20ICSID%20Vol%2021%20No%201.pdf

Huseyin Pazarci 'La responsabilité internationale des États à l'occasion des contrats conclus entre États et personnes privées étrangères' 1975 *RGDIP* 354

Stéphane Piedelièvre 'Garantie à première demande' 2008 *Rep com Dalloz* (updated: September 2011) Available from: https://www.dalloz.fr/documentation/Document?id=DZ/ENCY/COMR

Stéphane Piedelievre 'Les nouvelles règles de la CCI relatives aux garanties à première demande' 2010 *RTD Com* 513

Rouhollah K Ramazani 'Choice of Law Problems and International Oil Contracts: A Case Study' (1962) 11 *Int'l & Comp LQ* 503

Patrick Rambaud 'Un arbitrage pétrolier: la sentence LIAMCO' 1980 *AFDI* 274

Willis L M Reese 'Power of Parties to Choose Law Governing their Contract' 1960 *Proc ASIL* 49

Julien Rochette 'Forage en eux troubles—Pour une régulation internationale des activités pétrolières et gazières offshore' 2014 *Institute for Sustainable Development and International Relations, Policy Brief* (14 February 2014) Available from: http://www.iddri.org/Publications/Forage-en-eaux-troubles-Pour-une-regulation-internationale-des-activites-petrolieres-et-gazieres-offshore?fr

Julien Rochette, Matthieu Wemaëre, Lucien Chabason & Sarah Callet 'En finir avec le beau pétrole—Pour une meilleure régulation des activités pétrolières et gazières offshore' *Institute for Sustainable Development and International Relations, Policy Brief* (14 February 2014) Available from: http://www.iddri.org/Publications/Collections/Analyses/Study0114_JR%20et%20al_offshore_FR.pdf

Jeswald W Salacuse 'Renegotiating International Projects Agreements' (2000) 24 (4) *Fordham International Law Journal* 1319

John J Schanz, 'The United Nations' Endeavour to Standardize Mineral Resource Classification' 1980 *NRF* 307

Joanna Schmidt 'Négociation et conclusion de contrats' 1982 *D* 229

Hartley William Shawcross 'Le problème des investissements à l'étranger en droit international' (1961) 102 *RCADI* 365

Ibrahim Shihata 'L'agence multilatérale de garantie des investissements' (1987) 33 *AFDI* 601

Muthucumaraswamy Sornarajah 'The Myth of International Contract Law' (1981) 15 *JWTL* 187

Brigitte Stern 'Trois arbitrage un même problème, trois solutions: les nationales pétrolières libyennes devant l'arbitrage international' 1980 *Rev arb* 3

Thierry Tanoh 'Le financement de projet en Afrique subsaharienne' (2006) 235 *Banque & Stratégie* 14

Catharine Titi 'Les clauses de stabilisation dans les contrats d'investissement: une entrave au pouvoir normatif de l'État d'accueil' (2014) 2 *JDI* 6

John Vafai 'Production Control in the Petroleum Industry: A Critical Analysis' 1971 *Santa Clara Lawyer* 189 Available from: http://digitalcommons.law.scu.edu/lawreview/vol11/iss2/1

Pierre Van Ommeslaghe 'Les clauses de force majeure et d'imprévision (hardship) dans les contrats internationaux' 1980 *RDIDC* 7

Thomas Wälde 'Negotiating for Dispute Settlement in Transnational Mineral Contracts: Current Practice, Trends, and an Evaluation from the Host Country's Perspective'(1911) 7 *Denver JILP* 33

Thomas Wälde 'Transnational Investment in the Natural Resources Industries' 1979 *Law & Pol Int Bus* 691

E H Wall 'The Iranian Italian Oil Agreement of 1957'(1958) 7 *Int'l & Comp LQ* 736

Prosper Weil 'Problèmes relatifs aux contrats passés entre un État et un particulier'(1969) 128 *RCADI* 95

Wilhelm Wengler 'Immunité législative des contrats multinationaux' 1971 *Rev crit DIP* 637

Hasan S Zakariya 'The Petroleum Lending Programme of the World Bank' (1983) 17 (6) *Journal of World Trade* 471

Press articles

'Gold, Diamonds Hold Steady but Copper Suffers' *African Business* February 2015 at 35

'Nigeria's black gold' *African Mining* January–February 2015 at 28

'Mali: gold and good people' *African Mining* January–February 2015 at 33

'Sénégal—Lazareff Le Bars propose des quotas d'exploitation' *Africa Mining Intelligence* 9 June 2015 at 2

'Les class actions menacent les compagnies minières!' *Africa Mining Intelligence* 7 April 2015 at 1

'Burkina Faso: Kaboré arrêté pour plus de transparence?' *Africa Mining Intelligence* (343) 21 April 2015 at 2

'Congo-K Fiscalité: la FEC cite la Zambie en contre-exemple', *African Mining Intelligence* (345) 26 May 2015 at 1

'Guinée: Le Comité de revue clos en mars' *Africa Mining Intelligence* (363) 1 March 2016 at 2

'Maroc: San Leon effectue des tests positifs sur l'huile de schiste de Timahdit, selon le procédé Enefit', *Agence Ecofin* (online) 26 January 2015 Available from: http://www.agenceecofin.com/gaz-de-schiste/2601-26076-maroc-san-leon-effectue-des-tests-positifs-sur-l-huile-de-schiste-de-timahdit-selon-le-procede-enefit

Antoine Izambar 'Chute du prix du pétrole: pourquoi l'OPEP peut exploser?' *Challenges* (online) 20 January 2016 Available from: http://www.challenges.fr/challenges-soir/20160119.CHA4090/chute-du-prix-du-petrole-pourquoi-l-opep-peut-exploser.html

Stéphane Ballong 'Quand le cours du pétrole a entraîné le naira dans sa chute' *Jeune Afrique* (online) 1 December 2014 Available from: http://economie.jeuneafrique.com/entreprises/entreprises/energie/23635-quand-le-cours-du-petrole-a-entraine-le-naira-dans-sa-chute.html

Bloomberg 'Imminent Boom?' *Mining Weekly* (vol 20 no 2) 24-30 January 2014 at 18

Jean-Louis Borloo (statement taken by Christophe Châtelot) 'Il faut un plan Marshall pour électrifier l'Afrique' *Le Monde* (online) 3 March 2015 Available from: http://www.lemonde.fr/afrique/article/2015/03/03/il-faut-un-plan-marshall-pour-electrifier-l-afrique_4586493_3212.html#D5uu9B4dfIfI2iJ8.99)

Meredith Campion & Tim Scales 'New dawn for Africa in 2015' *Australia's Playdirt* (vol 1 Issue 223) December 2014—January 2015 at 61

Terence Creamer 'Risk Factor' *Mining Weekly* (vol 21 no 2) 23–29 January 2015 at 26

Tiemoko Diallo & Bate Felix 'Mali eyes $9.54 billion rail projects' *Australia's Playdirt* (vol 1 Issue 223) December 2014—January 2015 at 46 Available from: http://www.lesechos.fr/finance-marches/marches-financiers/021847361615-petrole-projet-daccord-sur-un-gel-des-niveaux-de-production-a-doha-1214379.php

Max Falque 'Gaz de Schiste: le problème de fond est la propriété des droits sur le sous-sol', Institut Turgot Available from: http://blog.turgot.org/index.php?post/Falque-Gaz-de-schiste-2

Grégory Gautier (statement taken by Alain Faujas) 'Les terres rares sont le pétrole du xxie siècle', *Le Monde* (online) 16 June 2014 Available from: http://www.lemonde.fr/economie/article/2014/06/16/c-est-le-moment-d-investir-dans-les-terres-rares_4438932_3234.html

'Le Mali fait le ménage dans son secteur minier' *Jeune Afrique* (online) 1 August 2014 Available from: http://www.jeuneafrique.com/7832/economie/le-mali-fait-le-m-nage-dans-son-secteur-minier/

'Mauritanie: Glencore repousse le démarrage du projet minier d'Askaf' *Jeune Afrique* (online) 28 October 2014 Available from: http://economie.jeuneafrique.com/entreprises/entreprises/mines/23383-mauritanie-glencore-repousse-le-demarrage-du-projet-minier-daskaf.html

'Chute des cours du pétrole: le Gabon envisage de réajuster son budget 2015' *Jeune Afrique* (online) 9 January 2015 Available from: http://economie.jeuneafrique.com/entreprises/entreprises/energie/23874-chute-des-cours-du-petrole-le-gabon-envisage-de-reajuster-son-budget-2015.html

'Tullow Oil réduit à nouveau ses dépenses d'exploration' *Jeune Afrique* (online) 15 January 2015. Available from: http://economie.jeuneafrique-.com/entreprises/entreprises/energie/23906-tullow-oil-reduit-a-nouveau-ses-depenses-dexploration.html

'Pour le président algérien, l'exploitation du gaz de schiste n'est pas encore à l'ordre du jour' *Jeune Afrique* (online) 28 January 2015 Available from: http://economie.jeuneafrique.com/entreprises/entreprises/energie/23989-pour-le-president-algerien-lexploitation-du-gaz-de-schiste-qnest-pas-encore-a-lordre-du-jourq.html

Christophe Le Bec 'En Guinée, le doute plane à nouveau sur le mont Simandou' *Jeune Afrique* (online) 20 October 2014 Available from: http://economie.jeuneafrique.com/entreprises/entreprises/mines/23279-en-guinee-le-doute-plane-a-nouveau-sur-le-mont-simandou.html

Christophe Le Bec 'Chute du cours du pétrole: qui perd, qui gagne en Afrique?' *Jeune Afrique* (online) 18 November 2014 Available from http://economie.jeuneafrique.com/entreprises/entreprises/energie/23542-chute-du-cours-du-petrole-qui-perd-qui-gagne-en-afrique-.html

Jim Lennon 'Africa—coping with the new realities' *Metal Bulletin Magazine, African Mining Special Issue* February 2015 at 62

'Regional Overview: East Africa' *Magic* 2014 at 18

'Africa: A future Center of Gravity For Global Mining' *Magic* 2015 at 6

'Regional Overview: East and Central Africa' *Magic* 2015 at 82

'International View' *Metal Bulletin Magazine* African Mining Special Issue February 2015 at 26

'West African iron ore reassessed' *Metal Bulletin Magazine* African Mining Special Issue February 2015 at 43

'Taking Safety more seriously in Africa' *Mining Review Africa* (Issue 1) January 2015 at 72

Jasmine NG 'Goldman Calls End to Iron Age After 'Dramatic' Drop in Ore Price', *BloomberdBusiness* (online), 10 September 2014. Available from: http://www.bloomberg.com/news/articles/2014-09-10/goldman-calls-end-to-iron-age-after-dramatic-drop-in-ore-price

Izak Odendaal 'Outlook on oil', *Mining Prospectus* (Issue 20) February 2015 at 25

Ina Opperman 'Commodities forceast 2015' *Mining* December 2014 at 58

Dominic Piper 'African exploration down and more support is needed' *Australia's Playdirt* (vol 1 Issue 223) December 2014—January 2015 at 92

Amir Rabah 'L'Algérie autorise l'exploitation du gaz de schiste', *Jeune Afrique* (online) 23 May 2014 Available from: http://economie.jeuneafrique.com/entreprises/entreprises/energie/22175-lalgerie-autorise-lexploitation-du-gaz-de-schiste.html

Chloé Rondeleux 'Face à la chute des cours du pétrole, l'Algérie change de ton' *Jeune Afrique* (online), 17 December 2014 Available from: http://economie.jeuneafrique.com/entreprises/entreprises/energie/23751-devant-la-chute-des-cours-du-petrole-l-algerie-change-de-ton.html

Ilan Solomons 'Private equity's role in financing mine projects to be addressed' *Mining Weekly* (vol 21 no 2) 23—29 January 2015 at 42

Ilan Solomons 'Countries seek to ensure increased mineral benefits' *Mining Weekly* (vol 21 no 2) 23—29 January 2015 at 60

Reports

African Development Bank, Organisation for Economic Co-operation and Development, United Nations Development Programme, Economic Commission for Africa *African Economic Outlook 2013: Structural Transformation and Natural Resources* (2013) Available from: http://www.africaneconomicoutlook.org/fileadmin/uploads/aeo/2013/PDF/AEO_2013_THEMATIC_EDITIONS_WEB.pdf

African Development Bank, Organisation for Economic Co-operation and Development, United Nations Development Programme, Economic Commission for Africa *African Economic Outlook 2017: Entrepreneurship and Industrialisation* (2017) Available from: https://www.afdb.org/fileadmin/uploads/afdb/Documents/Publications/AEO_2017_Report_Full_English.pdf

African Development Bank *Le gaz de schiste et ses implications pour l'Afrique* 2013 Available from: http://www.afdb.org/fileadmin/uploads/afdb/Documents/Project-and-Operations/Le%20Gaz%20de%20Schiste%20et%20ses%20implications%20pour%20l'Afrique%20et%20la%20Banque%20africaine%20de%20d%C3%A9veloppement.pdf

World Bank *Le potentiel transformateur de l'industrie minière. Une opportunité pour l'électrification de l'Afrique subsaharienne* (2015) Available from: https://openknowledge.worldbank.org/bitstream/handle/10986/21402/9781464804861.pdf?sequence=4

World Bank *Paying Taxes 2015* (2015) Available from: http://www.pwc.com/gx/en/paying-taxes/pdf/pwc-paying-taxes-2015-low-resolution.pdf

BP *BP Statistical Review of World Energy* (June 2016) Available from: http://www.bp.com/content/dam/bp/pdf/Energy-economics/statistical-review-2014/BP-statistical-review-of-world-energy-2014-full-report.pdf

Comité interministérielle pour la Coopération International et le Développement (CICID—French Interministerial Committee for International Cooperation and Development) *Ressources minérales et développement en Afrique. Document d'orientation stratégique* (May 2005) Available from: http://www.diplomatie.gouv.fr/fr/IMG/pdf/618_Resources_minerales_DOS.pdf

Commissariat Général au Développement Durable (French General Commission on Sustainable Development) *Chiffres clés de l'énergie, Édition 2014* (February 2015) Available from: http://www.developpement-durable.gouv.fr/IMG/pdf/Rep_-_Chiffres_cles_energie_2014.pdf.

Economic Commission for Africa *Les ressources minérales et le développement de l'Afrique—Rapport du Groupe d'études international sur les régimes miniers de l'Afrique* (2011) Economic Commission for Africa and African Union at 50 Available from: http://ti.au.int/en/sites/default/files/minirals%20and%20africas%20development%20FRE.pdf

Ernst & Young *Funding challenges in the oil and gas sector* (2014) Available from: http://www.ey.com/Publication/vwLUAssets/EY-Funding-challenges-in-the-oil-and-gas-sector/$FILE/EY-Funding-challenges-in-the-oil-and-gas-sector.pdf

Ernst & Young *EY's attractiveness survey Africa 2015 Making choices* (2015) Available from: http://www.ey.com/Publication/vwLUAssets/EY-africa-attractiveness-survey-2015-making-choices/$FILE/EY-africa-attractiveness-survey-2015-making-choices.pdf

MIGA *Rapport annuel 2013* Available from: http://www.miga.org/documents/MIGA_AR13_French.pdf

OECD *CleanGovBiz: Integrity in practice* (2014) Available from: http://www.oecd.org/cleangovbiz/49693613.pdf

OECD *Rapport de l'OCDE sur la corruption transnationale, Une analyse de l'infraction de corruption d'agents publics étrangers* (2014) Available from: http://www.keepeek.com/Digital-Asset-Management/oecd/governance/rapport-de-l-ocde-sur-la-corruption-transnationale_9789264226623-fr#page1

OPEC *Annual Statistical Bulletin 2014* (2014) Available from: http://www.opec.org/opec_web/static_files_project/media/downloads/publications/ASB2014.pdf

INDEX

H

I

S

www.ingramcontent.com/pod-product-compliance
Lightning Source LLC
LaVergne TN
LVHW060013210926
848481LV00013B/3697
* 9 7 8 1 4 8 5 1 2 7 0 7 9 *